TUNNELLING ASIA'2000

International Conference

TUNNELLING ASIA'2000

26–29 September, New Delhi, India

Need for Accelerated Underground Construction — Issues and Challenges

Editors

S.P. KAUSHISH
T. RAMAMURTHY

Organised by

Ō **Adhering Committee of**
International Tunnelling Association (India)

and

 **Central Board of Irrigation and Power**
Malcha Marg, Chanakyapuri, New Delhi 110 021, India

A.A. BALKEMA/ROTTERDAM/BROOKFIELD/2000

SPONSORS

International Tunnelling Association
Gammon India Limited
Skanska International Civil Engineering AB
LARSEN & TOUBRO LIMITED
ECC — CONSTRUCTION DIVISION

CO-SPONSORS

Nathpa Jhakri Power Corporation Limited
Tehri Hydro Development Corporation Limited
Delhi Metro Rail Corporation Limited
Bharat Heavy Electricals Limited
Impregilo SpA
Statkraft SF
Directorate General Border Roads
RITES Limited
Committee of the International Society for Rock
    Mechanics (India)
Indian Committee on Large Dams

CO-ORGANIZERS

National Hydroelectric Power Corporation Limited
National Thermal Power Corporation Limited
Patel Engineering Limited
Jaiprakash Industries Limited
Hindustan Construction Company Ltd.
ALSTOM Power India Limited
Samsung Corporation
ICICI

SUPPORTERS

Bhakra Beas Management Board
Electrowatt Engineering

ISBN 90 5809 228 3

A.A. Balkema, P.O. Box 1675, 3000 BR Rotterdam, Netherlands
Fax:+31.10.4135947; e-mail: balkema@balkema.nl
Internet site: http://www.balkema.nl

Distributed in USA and Canada by
A.A. Balkema Publishers, 2252 Ridge Road, Brookfield, Vermont 05036, USA
Fax: 802.276.3837; e-mail: Info@ashgate.com

FOREWORD

Subsurface space has become a fruitful domain for many authorities, in charge of infrastructure projects, town developments and industrial installations, while minimizing the effect on the existing environment. Looking ahead, tunnelling is assured of an expanding future as the need to avoid urban congestion forces us to put more utilities/facilities underground. The worldwide tunnelling activity continues to increase to develop and upgrade the infrastructure and facilities needed to meet the ever-growing demands from irrigation, hydropower generation, drinking water, industrial needs, highway and railway tunnels, etc., at the same time preserving urbanised areas, natural resources and the environment.

What in the past was treasure hunting or search for shelter has become in modern life an opportunity to relieve the congestion of the cities, to promote communication, to accommodate specific activities, to improve working conditions or even to provide an environment for housing. The huge development of the mining industry stimulated the imagination of engineers and encouraged them to extend and adapt mining methods for construction of communication tunnels.

Efficient mass transit systems have become indispensable to the healthy economic activity of modern cities. The major cities of the advanced nations have constructed various mass transit systems, including railway systems, from an early stage, and these have added to the economic development of those cities.

In India, considerable activities in the field of tunnelling are in progress, for the execution of water resources projects for irrigation, hydropower generation, building of roads in mountainous areas, subsurface excavations for underground railway and for mining purposes. Several high earth-rockfill dams have been constructed and are under construction in the country. Tunnels have been constructed in India under unfavourable geological conditions in many of the hydropower projects. Subsurface excavation for the construction of the metro in the city of Calcutta has been successfully completed. Such works are being planned in other metropolitan cities as well. Delhi Metro work will be in full gear by 2002.

Tunnelling has been for a long time an empirical rather than an exact engineering science. The main reason for this has been the nature of the host environment in which tunnels and underground structures have been and still are excavated - the often unpredictable geology and ground conditions. It is only recently, mainly over the last four decades, that tunnelling has progressed from the stage of art or craftsmanship to the stage of engineering science. This has been primarily due to the application of rock and soil mechanics principles to tunnel behaviour and due to development of highly advanced tunnel boring machines.

The present conference is being organized with a view to providing current state of knowledge and experience gained in the recent large-scale tunnel and other underground projects to benefit the future projects.

More than 60 papers contributed by experts actively involved in tunnelling are published in the proceedings, contain their valuable experiences, use of best of the technologies, the resulting problems and their solutions adopted for their mitigation and performance on ground. We express our sincere gratitude to all the authors for their contributions.

The Central Board of Irrigation and Power (CBIP) expresses sincere thanks to the International Tunnelling Association (ITA), co-organisers and other national and international organizations for sponsoring/cosponsoring the conference, as well as assisting us in many ways. We are grateful to the Chairmen and Members of the Organizing and Technical Committees for their guidance and advice.

It is hoped that ample input of information and experience, both from national and international scene, through the presentation of technical papers, keynote lectures and discussions will be available, which may enable some conclusions to be drawn to recommend for concrete action for short-term as well as for long-term perspective.

S.P. KAUSHISH
Secretary
Central Board of Irrigation and Power

New Delhi
September 2000

CONTENTS

METRO TUNNELLING

TUNNELLING IN COMPLEX GEOLOGICAL SETTINGS

SHOTCRETING AND MICRO TUNNELLING

SPECIAL TUNNELLING TECHNIQUES

INNOVATIVE CONSTRUCTION METHODOLOGIES

SPECIAL ASPECTS

AUTHOR INDEX

ROCK MASS CLASSIFICATION, TESTING AND MEASUREMENT

NUMERICAL APPROACHES ALONGWITH A CASE STUDY

B. PRABHAKAR

Scientist

Central Mining Research Institute Regional Centre, Nagpur, India

SYNOPSIS

Two approaches to numerical modelling of rock masses can be identified, both recognizing geological structures as being discontinuous due to joints, faults and bedding planes. A continuum approach treats the rock mass as a continuum intersected by a number of discontinuities, while a discontinuum approach views the rock mass as an assemblage of independent blocks or particles. Sound rock mechanics practice for mining and other ground excavation requires effective techniques for predicting rock mass response to mining and excavation activities. A particular need is for methods which allow parametric studies to be undertaken quickly and efficiently, so that a number of operationally feasible mining options can be evaluated for their geomechanical soundness. Alternatively, parameter studies may be used to identify and explore geomechanical appropriate mining strategies and layouts. The paper presents in brief different approaches of numerical modelling such as Finite Element Method, Boundary Element Method, Finite Difference Method, and Displacement Discontinuity Method and their limitations also. The paper also presents a typical case study which is studied and instrumented by author and it gives a relevant and new idea for further such type of hazardous project.

1.0 INTRODUCTION

To design underground openings such as tunnels, pillars, panels, stops, caverns, rock stops and others, is one of the vital challenges at present. Several factors affect the stability of underground openings such as properties and stresses in the rock mass, design geometry, rate of extraction and others. Any rational procedure for the design of openings, should include field, laboratory and theoretical studies of rock behaviour. One of such design requirements of underground openings, involves stability analysis in which the stress distribution and the displacement of various points around the opening should be evaluated.

Closed-form solutions are available for regular geometry shapes of the openings such as circular, rectangular and elliptical based on the theory of elasticity for idealised rock conditions. Classical closed-form analysis, however, is based on two main assumptions namely the rock behaves as a continuum and other discontinuum. Numerical modelling, however, is essential for the rock behaves as discontinuum. This is advantageous for complex geometry and three dimensional numerical modelling.

The Finite Element Method (FEM) and the Boundary Element Method (BEM), however, are two well estabilised numerical approaches used for analysis of underground openings. Both the approaches have their own limitations. FEM is advantageous in analysing geotechnical behaviour where the geometry and boundary conditions change due to excavation or support system.

The BEM is advantageous in analysing the behaviour in the infinite domain of the grounds. The utilization of both FEM and BEM can be used by coupling both namely FEBEM. Finite Difference Method is not so popular for static elastic analysis of rock mass vis-a-vis FEM and BEM. This method, however, is simple to programme and use also. Displacement Discontinuity Method (DEM) is mostly useful for slit like opening. The paper also presents a case study of Maneri Bhali Stage - II tunnel by applying one of the numerical methods. At the Dhanarigad site, up-stream face of tunnel has exhibited extremely difficult tunnelling conditions. The closures were very high and rock loads appeared to be enormous. The supports got twisted very badly and the excavated portion near the face collapsed. Some collapsed portion had been cleared and fresh supports had been installed. These fresh supports have also experienced closures similar to the closures occurred soon after initial excavation. Instrumentation was done to design optimum support and smooth tunnelling.

2.0 CONTINUUM MODEL

Continuum models use the two classic theories of elasticity and plasticity to calculate the stresses and displacements induced in the initially stressed rock following excavation. They assume that displacements are continuous everywhere within the rock mass and necessarily involve idealization and simplification of its geometrical and mechanical properties. For simple excavation shapes and boundary conditions, closed- form and pseudo closed-form solutions may be obtained to problems in elasticity and more rarely, plasticity . For more complex excavation shapes and boundary conditions, numerical methods notable the finite element method (FEM) and boundary element method (BEM) must be used.

Selection of a particular numerical method, for modelling of mine excavations, is very important. Accuracy of prediction regarding stability or instability very much depends on the choice of the method. Here, a brief description of the methods and their limitations are enumerated.

2.1 Finite Element Method

In this method, the whole domain of the problem is discretised in small elements (Fig. 1). Stiffness of each element is calculated using the material property of rockmass which is contained in the element. Stiffness of each element is assembled to the "global stiffness matrix" [K]. The following equation of equilibrium is solved for unknown displacements {U} of the element nodes :

$$[K] \{U\} = \{F\}$$

Where, {F} is the nodal force matrix, strains and stresses in each element are calculated using the nodal displacements. The above equation may be linear or non-linear depending on its constitutive behaviour of the rockmass *i.e.*, linearly elastic, elasto-plastic or plastic.

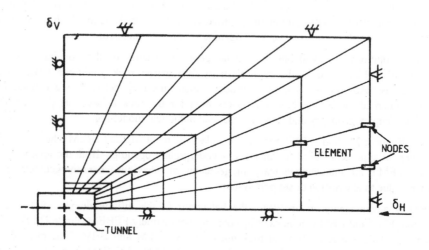

Fig. 1 : 2-D Finite element discretisation of a tunnel

Different materials may be used in the elements. Different types of rockmasses may be easily simulated, for examples, dolomite and shear zones in a mine. Non-linear load deformation behaviour can be easily incorporated. Thus, progressive failure of mine pillars, for example may be simulated. Filling of excavation can also be simulated.

Input data preparation is time consuming and arduous. Some softwares called "preprocessors" are available for preparing the input data interactively on computer terminals. If the domain of the problem is big and complex then this method takes a lot of computer time. Hence, this becomes costlier in terms of computer time and manpower as problem becomes larger and geometrically complex.

2.2 Boundary Element Method

The BEM differs from FEM in the sense that here only the boundary of the problem domain is discretised (Fig. 2). For example, only the surface of the mine opening is to be discretised in small elements. When the opening is at a depth of approximately 10 times its maximum dimension, it does not influence the surface. Otherwise, the surface is also to be discretised. BEM uses the known fundamental solution of stresses and displacements due to a point source acting on an infinite body. The formulation results into a system of simultaneous equations. The equations are solved for unknown displacements and stresses at the boundary. Then, these quantities are calculated at the field points *i.e.*, over the whole domain.

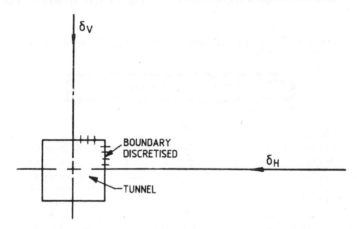

Fig. 2 : Boundary element discretisation of a tunnel

Input data preparation job is reduced to a minimum. Suitable for large and three dimensional problems where elastic solution serves the purpose.

This method is ideal for a linearly elastic isotropic and homogeneous material for which fundamental solutions are well known. It is not suitable for inelastic or anisotropy material. Therefore, it is not used for studying progressive failure behaviour.

Simulation of different types of rock is difficult. Filling of excavation can be simulated with difficulty.

2.3 Finite Difference Method

This method is not so popular for static elastic analysis of rockmass as FEM and BEM are. In this method, like FEM, the whole domain is discretised into small regular elements (Figs. 3 and 4). Over each element, the differential equation of equilibrium is approximated while in FEM the solution is approximated. This also results into a system of simultaneous equations which are generally solved by iteration methods.

This method is simple to programme and use. This method could be easily handled at site also to know the immediate behaviour of openings.

This method is very much useful to solve the problem of anisotropy, non-homogeneous and non-linear material properties of rock masses for elastic and plastic also. The importance of utilization of this method among other numerical approaches, as shown in Table 1 by Gnileen, 1989.

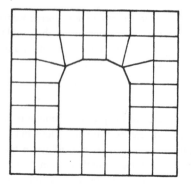 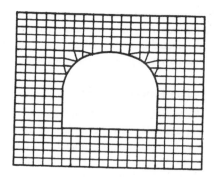

Fig. 3 : Tunnel created with simple INI commands   **Fig. 4 : Arck created with Gen commands**

Table 1 : Numerical methods for tunnel engg. (Gnileen, 1989)

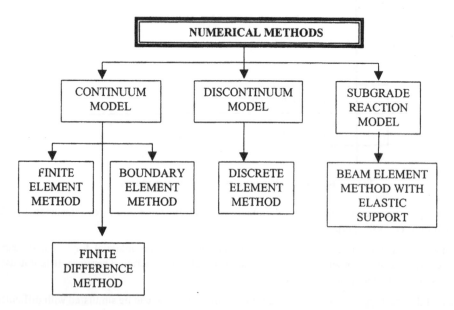

The formulation results into a conditionally stable problem. It means that the convergence of the solution at different stages of iteration to a true solution depends on the size of the elements and the size of the load steps. Moreover, the convergence also depends on the equation solution method.

2.4 Displacement Discontinuity Method

This is suitable for slit like opening, for example, a goaf in coal seam (Fig. 5). The body outside the seam is taken to be elastic, isotropic and homogeneous like BEM. Here, the fundamental solution, like BEM is used for the elastic body. Seam may have linear or non-linear behaviour. Filling of goaf may be simulated with the greatest ease. In DDM, discretisation of only the seam of orebody is needed.

3.0 DISCONTINUUM MODELS

Discontinuum models treat the rockmass as an assembly of discrete blocks, thereby recognising essentially the discontinuous nature of rockmasses and the fact that displacements may be not continuous within them in particular, around the periphery of an excavation (Fig. 6).

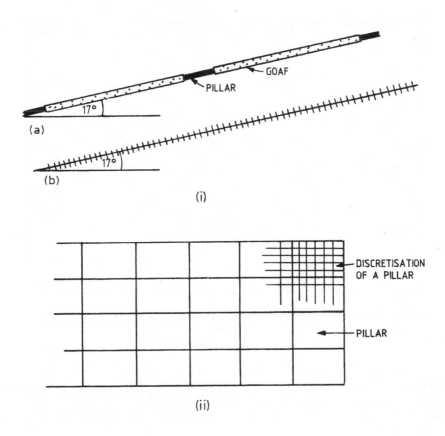

Fig. 5 : Displacement discontinuity discretisation of a 2-D & 3-D problems in a coal seam

Numerical models which give solution of discontinuous nature or rockmass are called "Distinct Element Method" (DEM). In the numerical methods for the continuum model, the discretisation of boundary or domain is only conceptual. In DEM, we consider that the whole domain is discretised in blocks by the presence of natural joints. In this method, rotation of the blocks and sliding of one block over other can be simulated. Thus, the real mechanical behaviour of jointed rockmasses may be studied using DEM. In FEM and FDM methods, the effect of joints may be accounted implicitly by modifying the material properties which does not reveal the realistic failure behaviour of the blocky rock under low stresses, for example, near the surface or in the open cast mines.

In the DEM, the realistic failure behaviour of the blocky rock under low pressure could not be revealed.

5

4.0   A CASE STUDY OF MANERI BHALI STAGE-II PROJECT

The Maneri-Bhali Hydroelectric Scheme Stage-II, having an installed capacity of 304 MW, envisages the construction of a 16 kms long tunnel of 6.0m diameter standard horse shoe section. The tunnel is being driven from four faces, two at both ends and two from an Intermediate Adit at Dhanarigad. The geological formations include quartzite, metavolcanics, limestone, dolomite and epidiorites. The rocks have a large number of open joints, faults and longitunal and cross shear zones. The geological section of the tunnel is shown in Fig. 7.

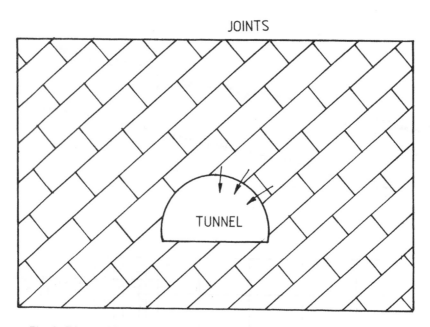

Fig. 6 : Discrete element blocks around a tunnel defined by two sets of joints

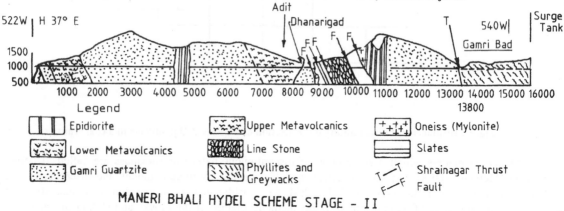

Fig. 7 : Diagrametic geological section along the tunnel alignment

The Dhanarigad Adit meets the Head Race Tunnel at a distance of 7969m from the inlet end. Metabasic rock formation continued upto a distance of 795m on the upstream face of Intermediate Adit. The rock is folded in this reach and its behaviour changes according to the folding and joint pattern and occurrence of shear zones and clay seams. The tunnelling upto 762m did not pose major problem except for a short reach near Ch. 753m where the rock was poor with horizontal joints charged with nominal head of water. Fore

6

poling had to be resorted to for advance of the heading . However, when the tunnel had been excavated upto 772m , there was sudden in rush of water from the right side near the invert at Ch. 776m. This resulted in collapse of right side of tunnel section. After the entrapped water was released, the space created by rock fall was packed backfill concrete behind the steel supports and tunnelling proceeded.

Inspite of all the precautions of careful drilling and blasting, a 12 to 15 m deep cavity was formed at Ch. 786 m. This location is close to the joint of metabasic and quartzite rock formations. The metabasic rocks are folded and the joints dip 60^0 to 70^0 towards the excavated section. The supports of 150 x 150 mm H joint had already been erected at a spacing of 0.5 m to 0.9 m and the blocking concrete was being done when the cavity was formed. Fore poles of R.S. joints extending upto Ch. 793 m were installed but before the mucking operations could be completed, another rock fall with heavy flow of water occurred on 13th December, 1984. The failed muck filled the tunnel to a considerable distance which extended upto 40 m behind the collapsed face. After channellising the flow of incoming water, the removal of muck was accomplished upto Ch. 775m. However, upheaval of tunnel invert to an extent of 1m was observed and the supports collapsed in a length of about 32 m (Ch. 761 to 793 m). Grouting was then carried out by 6 m deep holes at a pressure of 20 - 25 kg/cm^2 . Supports of ISHB 150 at a spacing of 45 cm c/c with a bottom strut have now been erected and blocking concrete done upto inner flange of the supports. The excavated bore has been enlarged to accommodate a 60 m thick lining. Finite difference method has been used by assuming some rock parameters here.

4.1 Used Rock Properties

The rock properties have been used in calculation of stresses and displacements, and recommended also for elastic model of tunnels as mentioned below :

Young's Modulus, $E = (2.6)x10^5$ kg/cm^2

Unit weight of rock, $\gamma = 2.7$ gm/cc

Possion's ratio, $\mu = 0.25$

Shear stress (calculated), $Sh = 10.4e9$

Bulk modulus (calculated), $Bu = 17.33e9$

For plastic (squeezing) model, the following rock properties have been used in estimation of stresses and displacements for tunnels, and these properties have been recommended also.

Cohesion, $c = 1.5$ MPa

Angle of friction, $\phi = 30^0$

Tensile strength, $\sigma_t = 1$ MPa

In these two models, vertical and horizontal stresses have been calculated by applying the formulae :

Vertical stress $= \gamma H$

and

Horizontal stress $= (0.4 + 300/H) \gamma H$

in this equation $K = (0.4+300/H)$ has been put.

5.0 NUMERICAL MODELS

Title
Maneri stage II tunnel
grid 20, 38
mod elas
gen 0,0 0,16 16,16 16,0   r = 1.1, 1.1   i = 1,9 j = 1,9

7

```
gen 0,16 0,254 16,254 16,16          r = 1.1, 1.07        i = 1,9   j = 9,39
gen 16,0 16,16 45,16 45,0            r = 1.001, 1.1       i = 9,21 j =1,9
gen 16,16 16,254 45,254 45,16        r = 1,001, 1.07      i = 9,21 j = 9,39
gen arc 0,0 3.5,0 90
wind −5,20 −5,20
mark i = 1,3 j =1
mark j = 1,3 i =1
set gravity 9.81
fix y j=1
fix x y i=21
fix x i=1


prop dens = 2700 Sh=10 , 8e9 bu 17.33e9
ini sxx. −10.8e6 var 0,2.743e6
ini syy -6.86e6 var 0,6.86e6


his nstep 5
his xdis i=8 j=8
his ydis i =8 j=8
his unbal i=1 j=1
mod null region (1,1)

Title
Title
Maneri stage II tunnel
grid 20, 38
mod elas
gen 0,0 0,16 16,16 16,0             r = 1.1, 1.1         i = 1,9   j = 1,9
gen 0,16 0,254 16,254 16,16         r = 1.1, 1.07        i = 1,9   j = 9,39
gen 16,0 16,16 45,16 45,0           r = 1.001, 1.1       i = 9,21 j = 1,9
gen 16,16 16,254 45,254 45,16       r = 1.001, 1.07      i = 9,21 j = 9,39
gen arc 0,0 3.5,0 90
wind −5,20 −5,20
mark i = 1,3 j = 1
mark j = 1,3 i = 1

set gravity 9.81
fix y j=1
fix x y i=21
fix x i=1

prop dens = 2700 Sh=10.8e9 Bu 17.33e9
prop coh=1.5e6 ten=1e6 fri=30

ini sxx. −10.8e6 var 0,2.743e6
ini syy -6.86e6 var 0,6.86e6

his nstep 5
his xdis i = 8 j = 8
his ydis i = 8 j = 8
his unbal i=1 j = 1
mod null region (1,1)
```

For preparing numerical models of tunnels for squeezing and non-squeezing grounds these rock properties have been used and details of programme have been mentioned. Various steps, output and calculated results of stress and displacement of Maneri Stage-II tunnel have been described in the end of the paper. One may use these rock properties while preparing numerical models for underground openings by using Finite Different Method in Flac programme and results are shown in Figs. 8 and 9 on the basis of numerical model.

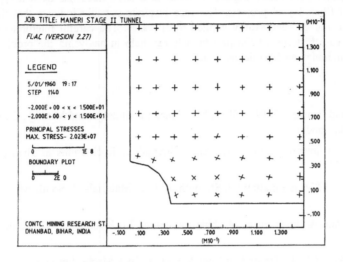

Fig. 8 : Graphical presentation of principal stresses

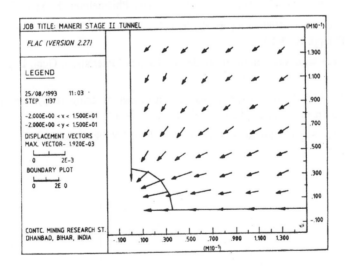

Fig. 9 : Graphical presentation of displacement

6.0 CONCLUSIONS

It is utmost important to remember that numerical model output, obtained from computer, very much depend on the parameters given to the model. If input parameters are wrong or their approximation is not realistic, then results are not reliable at all. Therefore, utmost care was taken in evaluating input data before evaluation of output from computer. Further, confidence on the results had been put depending upon confidence on the input data. In design, the numerical model was used as an important tool which reasonably predicts rockmass behaviour.

Tunnelling through young formations of Himalayan rocks having mixed lithology, varying tectonic behaviour and trapped water reservoir at considerable head may lead to occasional hazards that have to be tackled by the field engineers on the merit of the case. Formation of cavities due to flowing of large quantities of crushed rock together with heavy ingress of water at a large head has been a typical phenomena of tunnelling at Maneri Bhali Hydel projects. The case histories strongly indicate that the general rules for counter-measures can not be successfully applied to all cases. The careful interpretation of the observed phenomena and the case histories of the past tunnels can help in analysing the problem and to adopt the counter- measures required in each particular case.

REFERENCES

Brady B.H.G. and Brown E.T 1985. Rock Mechanics for Underground Mining , George Allen & Unwin (publishers) Ltd., London.

Barton N.R. and Choubey V. 1977. The Shear Strength of Rock Joints in Theory and practice, Rock Mech, 10, pp.1-54.

Bieniawski Z.T. 1974. Estimating the Strength of Rock Materials, J. South Afr. Inst. Min. Metal, 74, pp. 312-320.

Franklin J.A. 1985. A Direct Shear Machine for Testing Rock Joints, ASTM Geotech. Testing J., 8(1), pp. 25-29.

Franklin J.A. 1971. Triaxial Strength of Rock Materials, Rock Mech. (Springer), 3, pp. 86-98.

Goodman R.E. 1970. The Deformability of Joints in Determination of the In-situ Modulus of Deformation of Rock, STP 477 (Am. Soc. Test. Materials, Philadelphia, Pa.) pp. 174-196.

Hoek E. and Brown E.T. 1980. Emperical Strength Criterion for Rock Masses, J. Geot. Engg. Div., ASCE, 106 (GT9), pp. 1013-1035.

Johnston I.W. 1985. The Strength of Intact Geomechanical Materials, J. Geot. Engg. Div., ASCE, 3(6), pp. 730-749.

Laubscher D.H. 1984. Design Aspects and Effectiveness of Support System in Different Mining Conditions, Trans. Instn. Mining & Metal. Vol. 93, pp. 127-140.

Sheorey P.R. 1985. Support Pressures Estimation in Failed Rock Conditions, Engg. Geology, Vol. 22, pp. 127-140.

PREDICTION OF ENGINEERING BEHAVIOUR OF JOINTED BLOCK MASS

T. RAMAMURTHY   **K. SESHAGIRI RAO**

Indian Institute of Technology
New Delhi, India

MAHENDRA SINGH

M.N.R. Engineering College
Allahabad, India

SYNOPSIS

Strength and modulus of jointed rock mass is important for the design of slopes, foundations, underground openings and anchoring systems. The intact rock material may be significantly strong, but the failure of the mass may be governed either by sliding along the joints, shearing and splitting of intact material or rotation of block elements forming the rock mass. The engineering response of the jointed mass is thus governed by both the intact rock and the properties of the joints. The best way to assess the strength and modulus of the rock mass is to conduct the field tests. These tests are however very expensive and time consuming and some tie not feasible also. In the present article a methodology has been presented where the properties of the jointed block mass have been correlated to that of the intact rock through mapping of the joints in the field. The methodology is based on test results of large number of experiments conducted on jointed block mass under uniaxial stress state.

1.0 INTRODUCTION

Most of the civil engineering works located on or in rock mass experience confining pressure that is very low as compared to the strength of the intact rock. For such cases, the rockmass may be treated to be under uniaxial stress for preliminary feasibility studies for comparison of alternate sites. Strength and deformational characteristics of rock mass are required for designing the structures. It is observed that these engineering properties are very complex due to different modes of failure, mainly governed by joint configuration and extent of confinement. At relatively higher confining pressure, the modes are easy to predict and therefore a better prediction of engineering properties is possible whereas under low confining or under uniaxial stress condition the modes are complex and large variation in strength and deformability may result. The present paper discusses a methodology for predicting strength and modulus of deformation of regularly jointed block mass based on few simple tests in the laboratory and mapping of joints in the field.

2.0 BACKGROUND

Extensive tests have been conducted on jointed model materials and natural rocks of varying strength and joint configuration by various researches at IIT Delhi. Yaji (1984) studied the effect of roughness and inclination of joints on the response of jointed cylindrical specimens under uniaxial and confined state. The studies were carried out on plaster of Paris, sandstone and granite. Arora (1987) also did extensive testing of jointed cylindrical specimens with variable frequency and orientation of joints under uniaxial and triaxial stress conditions. He used plaster of Pairs, Jamrani sandstone and Agra sandstone. Roy (1993) also used cylindrical specimens and adopted two sets of joints that were close or filled with two types of gouge materials. Singh (1997) extended the work to specimens of jointed block mass, containing at least about 260 elemental blocks. The mass consisted of three orthogonal sets of joints. The joint set III was always kept vertical to make the problem two dimensional (Fig. 1). The joint set 1 was continuous and at an inclination 'θ' with the horizontal, whereas the joint set II was stepped at variable stepping 's'. Various combinations of orientation 'θ', stepping 's' (Fig. 1) and geometry of the block forming the specimen (Fig. 2) were adopted to achieve variety of modes of failure commonly occurring in nature. The tests were conducted under uniaxial stress condition and the deformations in all the three directions were measured with increase in load.

3.0 FAILURE MODES

It was observed by Singh (1997) that under unconfined state, a regularly jointed mass may have any one of the following, as dominating mode initiating the failure of the mass :

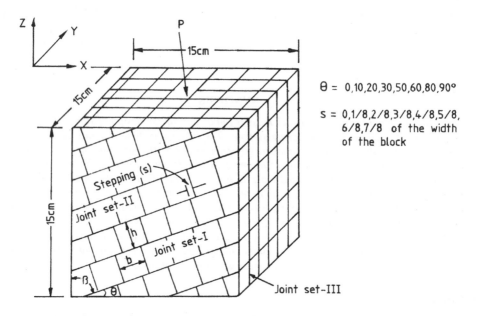

θ = 0,10,20,30,50,60,80,90°

s = 0,1/8,2/8,3/8,4/8,5/8, 6/8,7/8 of the width of the block

Fig. 1: Configurations of type — A specimens

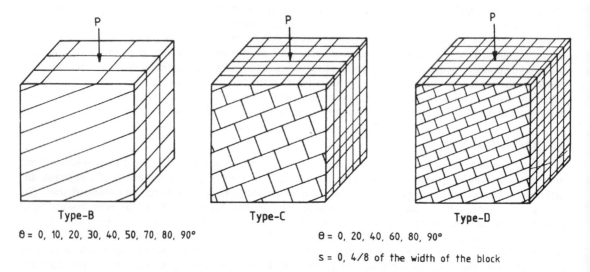

Type-B
θ = 0, 10, 20, 30, 40, 50, 70, 80, 90°

Type-C

Type-D

θ = 0, 20, 40, 60, 80, 90°

s = 0, 4/8 of the width of the block

Fig. 2 : Configurations of type — B, C and D specimens

3.1 Splitting

In this case the material fails due to tensile stresses developed inside the elemental blocks. Large number of minute cracks parallel to the loading direction were found with no sign of shearing.

3.2 Shearing

There are one or more than one shearing planes that are inclined with respect to loading direction. The fracture plane may or may not pass through pre-existing joints. Fractured material is formed due to shearing.

3.3 Rotation

The failure is initiated by rotation of blocks. Generally the translation of the specimen as a whole also takes place. The blocks remain intact unless the deformations become large.

3.4 Sliding

The material fails due to sliding along the continuous joints. The mode is associated with large deformations, stick-slip phenomenon and poorly defined peak in stress-strain curve.

4.0 DETERMINATION OF STRENGTH AND MODULUS OF DEFORMATION

Based on extensive testing of natural and artificial intact and jointed rocks, the concept of joint factor was proposed by Ramamurthy and co-workers (Arora, 1987; Ramamurthy and Arora, 1994). The joint factors (J_1) was defined as a function of frequency of joints, orientation of joints and shearing strength along the joint. The engineering properties of the jointed rock was linked to that of intact rock through the joint factor. Singh (1997) extended the concept to the regularly jointed block mass sufficiently large to have minimum scale effects. Another parameter which is very important for a block mass, $i.e.$, failure mode was also incorporated by Singh (1997) while extending the concept to the mass. Based on the tests conducted on the specimens of jointed block mass, a method wad evolved to compute the strength and tangent modulus of the mass under uniaxial stress condition. In the following, the step by step procedure is given for estimating the strength (σ_{cj}) and tangent modulus (E_j) of the jointed block mass based on failure mode.

4.1 Step 1 : Assess the Mode of Failure

The problem is treated in two dimension taking two joint sets at a time, one of which is continuous. In case there are more than two sets of joints considerably affecting the behaviour of the mass, different combinations can be worked out and most conservative results could be adopted. From mapping of joints the orientation of continuous joints will be available. Extent of interlocking of the mass $i.e.$, nil, low, intermediate, high or very high may be assigned to the mass based on visual inspection or one's personal experience of the mass. Corresponding to the orientation of joints and the extent of interlocking, the failure mode may be assessed from Table 1.

TABLE 1
Failure mode of jointed block mass

90	Shearing							
80	*Rotation*							
70								
60								
50				*Sliding*				
40								
30							*Shearing*	
20								
10								
0				*Splitting*				
θ^0 ↑	0	1/8	2/8	3/8	4/8	5/8	6/8	7/8
s →	Nil	Low		Medium		High		V. High
	Extent of Interlocking							

13

4.2   Step 2 : Compute Joint Factor

Draw a section of the mass to the scale with assumed stepping as per the extent of interlocking. Draw potential failure surfaces are as shown in Fig. 3 and compute the Joint Factor as given below.

$$J_f = \frac{J_n}{n.r} \qquad \qquad ...(1)$$

Where, J_n is the number of joints/m in the direction of loading; n is joint inclination parameter and depends on inclination β, of the joint with the direction of loading (Table 2), and 'r' is the joint strength parameter equal to tan ϕ_j; where ϕ_j is friction angle along the critical joints. The friction angle ϕ_j is obtained by testing two blocks in direct shear test under very low normal stress without damaging the angularities. The joint inclination parameter n is assigned from Table 2, and J_n is computed as given below :

TABLE 2
Joint inclination parameter n (Ramamurthy, 1993)

Orientation of Joint β^0	0	10	20	30	40	50	60	70	80	90
Inclination parameter n	0.82	0.46	0.11	0.05	0.07	0.30	0.46	0.64	.82	.95

(i)   For specimens failing due to splitting, potential failure surfaces are considered as shown in Fig. 3a. A vertical line is drawn on the section, and it is counted as to how many surfaces belonging to joint sets Arora (1987) and Baoshu, Huoyao and Hanmin (1986) are intersected by this vertical line. By trial, the position of the vertical line giving maximum intersection points for both the surfaces Arora (1987) and Baoshu, Huoyao and Hanmin (1986) is found. The maximum value is considered critical and used to compute J_n in Eqn. 1.

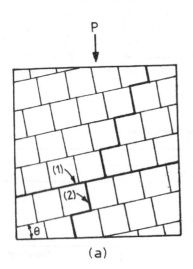 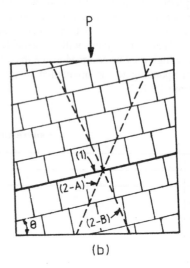

Fig. 3 : Potential failure surfaces for computing J_f

(ii)   For the specimens failing due to shear the method is similar to splitting if the continuous joints are horizontal, vertical or slightly inclined with horizontal (θ≈10°). For other orientations two alternate shearing planes are considered as shown in Fig. 3b. The shearing plane having its orientation β (with vertical) near to 45°ϕ/2 is considered critical.

(iii)   For rotational mode of failure the method is similar to that for shearing if the joints are stepped. If both the joint sets are continuous, the J_f is computed along both of them and maximum is chosen as critical.

(iv)   For sliding mode of failure, only the continuous joints are considered for computing J_f. If both the sets are continuous J_f is computed for both of them and higher value is considered critical.

4.4   Step 3 : Compute the Strength and Tangent Modulus of the Mass

Based on the test results from the present study and also from the literature (Brown, 1970; Brown and Trollope 1970; Ladanyi and Archambault, 1972; Einstein and Hirschfeld, 1973; Yaji 1984; Baoshu et al., 1986; Arora 1987; Roy, 1993; and Yang and Huang, 1995), following equations are suggested to compute the strength and tangent modulus of the jointed block mass (Figs. 4 and 5):

$$\sigma_{cj}/\sigma_{cj} = \exp(a \cdot J_f) \qquad \qquad \ldots (2)$$

$$E_{jl} E_j = \exp(b \cdot J_f) \qquad \qquad \ldots (3)$$

Where 'a' and 'b' are empirical constants and can be assigned from Table 3 according to the mode of failure; σ_{cj} and σ_{ci} are the strength values of jointed mass and intact rock respectively and E_j, E_i are the tangent moduli of jointed mass and intact rock respectively.

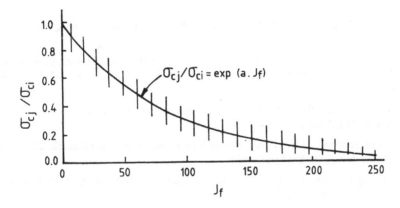

Fig. 4 : Variation of strength with J_f

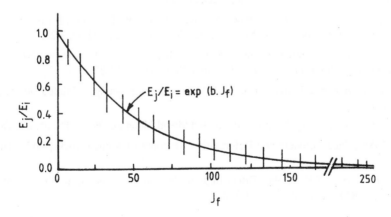

Fig. 5 : Variation of tangent modulus with J_f

TABLE 3
Values of empirical constants 'a' and 'b'

Mode of failure	Value of 'a'	Value of 'b'
Splitting	-0.0123	-0.020
Shearing	-0.0122	-0.020
Rotation	-0.0250	-0.040
Sliding	-0.0400	-0.060

5.0 CONCLUSIONS

The bearing capacity of ground and modulus deformation are the two most important parameters required at preliminary stage for feasibility of the site in jointed rock masses. The best prediction on these aspects can only be available from actual field testing, which is very costly and time consuming. For preliminary studies an approach which is based on simple laboratory tests and mapping of joints would be the best solution. The methodology developed under the present study can assess the required properties of the mass under unconfined state upto desired accuracy. For prediction of the values under confined state the strength criteria suggested by Ramamurthy and Arora (1994) can be used.

REFERENCES

Arora, V.K. (1987), Strength and Deformational Behaviour of Jointed Rocks, Ph.D. Thesis, IIT New Delhi, India.

Baoshu, G., Huoyao, X. and Hanmin, W. (1986), An Experimental Study on the Strength of Jointed Rock Mass, Proc. of Int. Symp. on Engineering in Complex Rock Formations, 3-7 Nov., Beijing, China, pp. 190-198.

Brown, E.T. (1970), Strength of Models of Rock with Intermittent Joints, Jl. of Soil Mech. and Found. Div., Proc. ASCE, 96 (SM6), 1935-1949.

Brown, E.T. and Trollope, D.H. (1970), Strength of a Model of Jointed Rock, Jl. of Soil Mech. and Found. Div., Proc. ASCE, 96(SM2), 685-704.

Einstein, H.H. and Hirschfeld, R.C. (1973), Model Studies on Mechanics of Jointed Rock., Jl. of Soil Mech. and Found. Div., Proc. ASCE, 90, 229-248.

Ladanyi, B. and Archambault, G. (1972), Evluation of Shear Strength of a Jointed Rock Mass., Proc. 24th Intl. Geological Congress, Montreal, Section 13D, 249-270.

Ramamurthy, T. and Arora, V.K. (1994), Strength Prediction for Jointed Rocks in Confined and Unconfined states. Int. Jl. Rock Mech. Min. Sci. and Geomech. Abstr., 31(1), 9-22.

Ramamurthy, T. (1993), Strength and Modulus Response of Anisotropic Rocks. Chapter 13, Comprehensive Rock Engineering, Vol. I, Pergamon Press, U.K. 313-329.

Roy N. (1993), Engineering Behaviour of Rock Masses Through Study of Jointed Models, Ph.D. Thesis, IIT Delhi, India.

Singh, M. (1997), Engineering Behaviour of Jointed Model Materials, Ph.D. Thesis, IIT, Delhi, India.

Yaji, R.K. (1984), Shear Strength and Deformation Response of Jointed Rocks, Ph.D. Thesis IIT Delhi, India.

Yang, Z.Y. and Huang, T.H. (1995), Effect of Joint Sets on the Anisotropic Strength of Rock Masses, Proc. 8th Cong. ISRM, Japan, Vol. 1, 367-370.

ASSESSMENT OF IN-SITU STRESSES AND DEFORMABILITY OF ROCK MASS IN WATER SUPPLY TUNNELS IN MUMBAI, MAHARASHTRA

B.K. SAHA
Senior Research Officer

A.K. GHOSH
Senior Research Officer

J.M. SHIRKE
Chief Research Officer

I. AZARAIAH
Joint Director

Central Water and Power Research Station, Pune, India

SYNOPSIS

Investigations carried out for assessment of in-situ stresses and deformability of rock mass in three tunnels of Mumbai Metropolitan water supply project are described in this paper. The tunnels passed through multiple rock formations such as volcanic breccia, tuff, amygdular basalt, massive basalt. These formations are often intercalated by inter-trappean shale. Excavated at an average depth of 60 m from the ground level, these tunnels are meant for feeding additional 455 million litre per day of fresh water to major service Reservoirs in the city located at Malabar Hill and Bhandarwada Hill. Adequate tunnel lining is required to be designed to avoid seepage intrusion and to provide stability. Information on in-situ stresses and deformability of rock mass is sought after for designing optimum thickness of tunnel lining. From the flat jack tests at several locations selected to represent the different rock mass zones, it was found that the rock mass in different zones were of soft variety and in-situ stresses and deformability of rock mass varied significantly. The vertical component of in-situ stresses varied from 0.9 Mpa to 4.19 Mpa in tuff breccia while the horizontal component varied from 1.8Mpa to 11.02 Mpa in tuff zone itself. The values for all rock types can be generalised between 1.6 to 2.8 MPa with the average value being around 2.0 MPa. The Deformation modulus of the rock mass varied from minimum value of 4.20 Gpa in tuff breccia to a maximum value of 37.70 Gpa in massive basalt zones. The above results provide crucial information for designing suitable concrete lining in the tunnels.

1.0 INTRODUCTION

Metropolitan Mumbai is located on the coast of Arabrian Sea at longitude of 18°/72° 56'N and at latitude 51' E and one of the most densely populated cities in the world. The requirement of water for domestic and industrial use is increasing by leaps and bounds. The need to augment the existing water supply position is a long felt requirement. However, laying additional pipe lines for this purpose is practically impossible proposition since such pipe line has to pass through densely populated areas. Mumbai Municipal Corporation has appropriately worked out a scheme to augment the water supply through series of underground tunnels (Fig. 1). Conventional drilling and blasting method if adopted would cause vibrations and result in damaging standing buildings over the tunnels. So all the three tunnels have to be drilled by Tunnel Boring Machine (TBM). 3.87 km long Mahalaxmi to Malabar Hill Tunnel, the 4 km long tunnel between King's Circle to Sewree and 5 km long tunnel from Ruparel College to Dr. E. Moses Road are executed using TBM (Fig. 1). Under this scheme it is proposed to augment the existing water supply position by additional 455 million liters of water per day. The tunnels traversed through several rock mass formations such as volcanic breccia, tuff, amygdular basalt, masive basalt intercalated by intra-trappean shale. Considerable seepage was observed at several locations through tuff breccia zones and inter-trappean shale layers. Overbreaks and roof collapses apparently due to seepage conditions were also experienced in the tunnels. In order to design suitable concrete lining to provide stability and water tight conditions to the tunnels, assessment of in-situ stresses and deformability of rock mass was sought after. Flat Jack tests were conducted at several locations covering all the formations of the rock mass and in-situ stresses and deformability were assessed. Engineering properties of rock materials were also evaluated by conducting laboratory tests on rock cores. The investigations carried out and findings there off are presented in this paper.

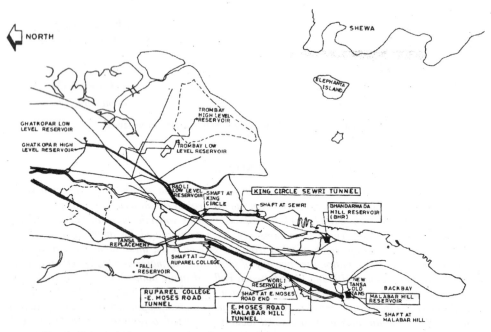

Fig. 1 : Integrated water supply schemes including TBM drilled tunnels in Mumbai

2.0 GEOLOGICAL FORMATIONS

From the bore hole logging it was seen that the rock is overlaid by about 6 to 10 m overburden. Groundwater level is struck within 2 to 3 m from the ground surface. The predominant rocks met in exploratory bore holes revealed presence of Cretaceo-Eocene volcanic basalt and breccia. The pyroclastic material such as tuff, volcanic breccia, volcanic ash and structures like pillow lava etc. was encountered at many locations with considerable thickness indicating intrusion at latter stage. Inter-trappean shale was also encountered as a intermittent layer within the trap flows. Variants of rock types encountered in the tunnels are depicted Figs. 2 to 5.

3.0 WATER SUPPLY TUNNELS

3.1 Mahalaxmi to Malabar Hill Tunnel

The 3870 m long tunnel of 3.5 m diameter passes through formations of amygdular basalt and tuff breccia of soft variety with sporadic patches of compact basalt and weathered amygdular basalt (Table 1). Half the portion of tunnel passes below the Malabar Hill and Cumbela Hill and other half passes below the Wellington Golf Course and Mahalaxmi Race Course. From the logging of the bore holes it is understood that amygdular basalt and soft volcanic breccia as a weathered rock is available above the crown level of the tunnel. The overburden consisting of soil and weathered to fresh rock mass varies from 45-80 m.

3.2 King's Circle and Sewree Tunnel

The 4 km long and 3.6 m diameter tunnel between King Circle to Sewree passes through zones of volcanic breccia and tuff breccia (Table 2). Two shafts and eight bore holes were available to examine rock type and sub-surface conditions. The average depth of the tunnel was about 60 m from ground level. Below the thin pile of overburden rock exposed in 6 bore holes is massive basalt. The strata at shaft and bore hole reveals presence of volcanic breccia throughout the excavated depth of about 60 m. Thin layers of tuff and shale is encountered in three bore holes.

Fig. 2 : Calcite filled joints showing leakage–Amygdular Basalt

Fig. 3 : Angular fragments of basalt in zeolitic and glassy groundmass volcanic breccia

3.3 Ruparel College to Dr. E. Moses Road

The 5 km long and 3.6 m diameter tunnel passed through various rock formations consisting of massive basalt, amygdular basalt, tuff, volcanic breccia, tuff breccia and inter-trappean shale (Table 3). Structure like pillow lava is conspicuously visible in this tunnel face. Joints are not persistent, devoid of definite pattern except in shale and massive basalt where they are parallel to bedding planes. Two shafts and six bore holes were available for examination. Rock overburden varies from 50-60 m all along the length of the excavated tunnel.

Fig. 4 : Shale layer with seepage through bedding planes

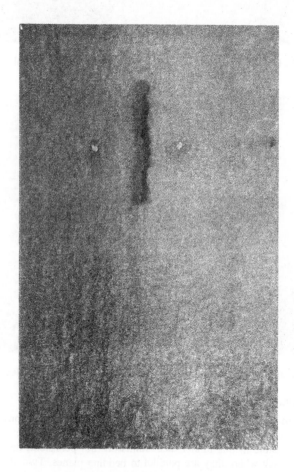

Fig. 5 : View of slot and embedded reference pins for flat jack in Amygdular Basalt

3.4 Water Seepage

Seepage of water in all the three tunnels during excavation was of the order of 3000 to 6000 lit/min. Heavy local seepage of about 500-1000 lit/min. was also experienced in some stretches. Cement grouting was carried out to minimize seepage flow during excavation. In Malabar Hill tunnel, Chemical grouting was resorted to in some stretches where cement grouting was not effective. In the King's circle tunnel also, leakage encountered was about 4500 lit./min. and was tackled by cement grouting. However, in Ruparel college tunnel, the leakage varied in range of 4000-5000 lit/min. Cement grouting with additives like bentonite, sodium silicate etc. were not effective. Therefore, chemical grouting with polyurethane based chemical was carried out at locations of heavy seepage. Although the grouting was effective locally, overall inflow of seepage water did not significantly decrease indicating pervious nature of tunnelled rock mass. Though a few overcuts were seen, the general conditions of all the three tunnels were satisfactory. Cut face of rock is smooth and has preserved the internal fabric of rock such as joint planes, pillow structures and bedding planes etc.

4.0 FIELD INVESTIGATION

Excavation of underground openings disturb the pre-existing in-situ stress field in the rock mass and new state of stress is induced around the boundaries of the openings. In order to design a suitable support system, prior knowledge of in-situ stresses and deformability of the rock mass is essential.

Field investigations for all the tunnels were carried out after the tunnels were excavated by the Flat Jack Method (IS 13946C Part 4), 1994, ISRM 1987) which is ideally suited when openings are available and facilitates evaluation of deformability modulus (E_m) and in-situ stresses. The method is comparatively inexpensive and permits a better statistical assessment of the properties based on large number of test locations.

The method consists of fixing two reference pins of good quality stainless steel into rock surface at a known distance of about 25 cm between them. A thin slot of size 30 cm x 30 cm x 4 cm is cut across the centre of the pins by drilling a series of overlapping holes into the rock surface. This process relieves rock surface of the stresses that are originally existed across the slot. Because of the stress relief, the sides of the slot converge. After the slot is made, the distance between the two pins are measured again to find out the amount of convergence. The flat jack is then embedded tightly into the slot by means of cement mortar. After the mortar in the slot has set and hardened, the jack is pressurized using hydraulic pump to neutralize the convergence. The stress at neutralisation, commonly designated as cancellation pressure, is corrected for the influence of the slot size, reference pin position and stress acting parallel to the major axis of the slot to obtain stress acting normal to the slot. One flat jack, therefore yields stress in one direction. it is, however to be noted that the measured stresses are induced on the boundary of the opening after the opening has been made.

By knowing the stress concentration factors at the tested locations, it is possible to evaluate the in-situ stresses that existed at the site before the opening was made. The E_M of the rock mass can be evaluated using the stress-displacement envelope obtained during the loading/unloading cycles of pressurization of the slot.

$$E_M = \frac{Pc_0}{w}[(1-\mu)(a_0\frac{y}{c_0}) + \frac{1+\mu}{a_0})]$$

where, P= Stress applied

c_0= Half length of flat jack

w= Half displacement of the reference pins during flat jack test

y= Half initial distance between two reference pins

μ=Poisson's ratio of the rock mass

$$a_0= (1-\frac{y^2}{c_0^2})^{1/2}$$

The measured induced stress (P_θ) tangential to the boundary of the tunnel and the induced stresses (P_H) in the direction parallel to the axis of the tunnel, the in-situ vertical (σ_v) and horizontal stress (σ_H) and the Static modulus of deformation (E_M) values from the three sites are given in Table 1, 2 and 3.

21

TABLE 1
Mahalaxmi to Malabar hill tunnel flat jack test results

Tunnel length	Chainage (m)	Rock formation	Test locations at chainage (m)	Induced Stresses P_θ P_H (MPa)		In-Situ Stresses σ_V σ_H (MPa)		Modulus of deformation (GPa)
	0-500	Amygdular basalt	270	1.60	2.20	1.60	2.10	8.0
	500-1980	Tuff breccia	550	2.10	2.10	1.70	2.90	8.0
			1130	1.60	3.80	3.20	5.80	6.0
			1440	1.10	1.30	0.90	1.80	9.0
3870	1980-3120	Amygdular basalt	2440	2.90	2.10	2.00	3.00	8.0
	3120-3450	Volcanic breccia	*	*	*	*	*	*
	3450-3870	Amygdular basalt	3540	3.30	2.10	1.90	2.50	8.0

TABLE 2
King's circle to Sewree tunnel flat jack test results

Tunnel length (m) Chainage	Chainage (m)	Rock formation	Test locations at chainage (m)	Induced stresses P_θ P_H (MPa)		In-Situ stresses σ_V σ_H (MPa)		Modulus of deformation E_M (GPa)
	0-1098	Volcanic Breccia	556	1.70	6.60	4.00	10.20	10.20
	1098-1722	Tuff	1099	1.50	3.50	2.30	5.40	15.00
4000			1722	1.80	7.00	4.20	11.00	12.50
	1722-4000	Volcanic Breccia	2527	1.70	5.90	3.60	9.30	14.30
			3434	1.50	3.60	2.30	5.70	12.00

5.0 LABORATORY STUDIES

Laboratory studies on rock core samples were carried out to evaluate density (γ), static modulus of elasticity (E_L) and unconfined compressive strength (σ_c). Summarised results from the three sites are given in (Table 4). Values of E_M are also included for comparison with E_L.

The in-situ vertical stresses found at these three project sites nearly agree with the overburden stresses. The average ratio of in-situ horizontal stress to vertical stress was found to be 2.0.

Tunnel length (m) chainage	Chainage (m)	Rock formation	Test locations at chainage (m)	Induced stresses P_θ (MPa)	P_H	In-Situ stresses σ_V (MPa)	σ_H	Modulus of deformation E_M (GPa)
	0-400	Massive basalt	*	*	*	*	*	*
	400-640	Shale	635	0.90	2.50	1.20	3.20	15.00
	640-1100	Amygdular basalt (pillow lava)	910	1.30	1.90	1.10	2.30	8.90
5000	1100-1160	Shale	1150	2.70	3.60	2.20	4.10	10.00
	1160-3200	Massive basalt	2830	1.90	3.10	2.00	4.20	37.70
			3110	2.60	4.30	2.60	5.40	33.30
	3200-3450	Shale	3425	1.80	3.10	2.20	4.20	9.80
	3450-4000	Massive Basalt	3765	2.30	3.10	2.10	4.10	35.20
	4000-5000	Tuff breccia	4692	0.90	3.10	1.70	4.40	4.20

Note : In Ruparel college tunnel, shale layers of 2 to 4 m are encountered as intermittent layers in basalt and breccia.

TABLE 4
Laboratory test results

Project	Rock type	Modulus of deforma-tion E_M GPa	Modulus of elasticity E_L GPa	Unconfined Compressive Strength σ_c MPa	Density Υ kN/m^3	Water absorption w (%)	$\dfrac{E_M}{E_L}$	Remarks
Mahalaxmi to Malabar Hill	Amygdular basalt	8.00	11.80 (4.10-18.8)	15.00	21.90 (20.0-23.7)	6.90 (2.50-9.50)	0.68	
Tunnel	Tuff breccia	7.70 (6.0-9.0)	12.60 (4.10-45.20)	12.00 (9.0-19.0)	22.90 (19.7-26.8)	8.80 (2.40-14.40)	0.61	Values in the parenthesis indicate range.
Kings Circle to Sewree hill tunnel	Volcanic breccia	12.20 (10.20-14.30)	15.10 (2.10-28.20)	32.00 (11-104)	23.00 (20-26)	5.90 (1.8-10.7)	0.81	
	Tuff breccia	13.80 (12.50-15.0)	16.80	28.00 (12-44)	25.00	2.90 (2.1-3.6)	0.82	
Ruparel college to Dr. E. Moses road	Shale	12.50 (9.80-15.0)	21.0	30.00 (22.0-38.0)	24.00 (23.5-24.5)	2.60 (2.05-3.10)	0.60	
	Amygdular basalt (pillow lava)	8.90	*	*	19.0	7.10 (6.80-7.30)	*	
	Massive basalt	35.50 (33.3-37.7)	73.50	176.00 (158.0-194.0)	29.00	0.25 (0.15-0.35)	0.48	
	Amygdular basalt	26.90 (14.80-37.7)	*	*	22.25 (22.0-23.0)	3.80 (2.00-5.60)	*	
	Tuff breccia	4.20	*	*	*	*	*	

6.0 CONCLUSIONS

In case of Malabar hill tunnel and King circle tunnel major rock types encountered are tuff, tuff breccia and volcanic breccia while at Ruparel college tunnel major rock types are massive and amygdular variety of basalt.

The flat jack method used in these studies is a versatile method recommended by ISRM as one of the methods for evaluating two dimensional in-situ stress field.

The deformation modulus of the rock mass of harder variety basalt was found to be 31.0 Gpa while for the softer variety the values found were of the order of 4 Gpa to 15 Gpa and these values are less than the values of the elastic modulus of rock core samples tested in the laboratory.

The measured induced stress around the boundary of the tunnels was found to be compressive and of moderate intensity having values ranging from 0.9 MPa to 3.3 MPa.

The average in-situ vertical stress and horizontal stress were found to be 2.3 MPa and 4.8 MPa respectively.

Detailed investigations to examine the state of stress in the rock mass around the underground structures are essential since these parameters provide vital input for designing optimum thickness of lining of the tunnels.

ACKNOWLEDGEMENTS

The authors are grateful to Mr. R. Jeyaseelan, Director, CWPRS for his encouragement. The authors are also grateful to Mr. B.M. Rame Gowda, Ex. Joint Director for his invaluable contribution towards investigations. The assistance of Mr. V.Ramakrishna, Assistant Research Officer, Mr. S.M. Kengale, Mr. H.R. Bhujbal, Mr. J.M. Deodhar, Laboratory Assistants during the field investigations and Ms. S.S. Apte for typing assistance are acknowledged with thanks.

REFERENCES

Rame Gowda B.M., Datta R., Ghosh A.K., Mokhashi S.L. (1989). Study of In-situ Stresses and Deformability of Rock Mass at Kuttiyadi Tunnel, Kerala., Proc. 55th CBIP R & D Session, Srinagar, India.

IS 13946 (Part 4)-1994, Determination of Rock Stress; Code of Practice: Part 4, Using Flat Jack Technique.

ISRM (1987), Suggested Methods for Rock Stress Determination. Int. J. Rock Mech and Min. Sec. Vol. 24, pp 53-74.

Saha B.K., Ghosh A.K., Ramakrishna V., Govindan S. and Shirke J.M. "Assessment of Deformability and In-situ Stresses in Deccan Trap Rock Masses", International Conference on Rock Engineering Techniques for Site Characterisation, 6-8 December, 1999, Banglore.

Khatkhate V.R. and Ramachandran B., "Tunnelling in the Metropolitan City of Mumbai Water Supply Projects" Tunnelling Asia '97, 20-24 January, 1997, New Delhi, India.

DEFORMABILITY OF ROCK MASS BY DIFFERENT METHODS INSIDE UNDERGROUND DESILTING CHAMBER

RAJBAL SINGH
Chief Research Officer

A.K. DHAWAN
Joint Director

Central Soil and Materials Research Station, Olof Palme Marg, Hauz Khas, New Delhi, India

SYNOPSIS

In all 19 deformability tests which includes 3 plate loading tests (PLT), 15 Goodman jack tests (GJT) and 1 plate jacking test (PJT) using borehole extensometers were conducted for the evaluation of deformability of rock mass (modulus of deformation, E_d and modulus of elasticity, E_e) in the drift for the proposed underground desilting chamber of Tala Hydroelectric Project, Bhutan. The tests were conducted using five cycles of loading and unloading with a maximum applied stress level of 5.0 MPa in PLT and PJT and 7.5 MPa in GJT. The modulus value increases in general and ratio of E_e/E_d decreases with the increase in stress level in all the methods. The decrease in moduli ratio shows the closing of joints at high stress level. The moduli values in horizontal direction perpendicular to drift axis are slightly higher than moduli values in vertical direction as well as in horizontal direction parallel to drift axis. The variation in modulus values in all the three direction by GJT shows that the rock mass is anisotropic. The modulus of deformation in vertical direction evaluated from Goodman Jack tests is slightly higher than that evaluated from plate loading tests. However, there is significant difference in the modulus of deformation by plate jacking test. The modulus values by PJT are 2 to 3 times higher than those obtained by GJT and PLT.

1.0 INTRODUCTION

The development of hydropower potential of river Wangchu has been under consideration in various stages since 1961. The Royal Government of Bhutan plans to harness the hydropower potential of about 2500 MW in a phased manner on river Wangchu. Commencing hydropower generation by installing mini hydel project in 1968, a significant stride in this direction has been made by commissioning of Chukha Hydroelectric Project of 336 MW (4x84 MW) in 1985. With the memorandum of understanding between Royal Government of Bhutan (RGB) and the Government of India, the investigation work and detailed project reports (DPRs) were completed for 4 major projects in Bhutan (Tala 1020 MW, Wangchu 900 MW, Bunakha 180 MW and Sankosh 1525 MW). Two medium projects (Kurichu 45 MW and Basochu 46 MW) and several mini hydel projects are under construction. The Tala Hydroelectric Project Authority (THPA) has taken up the construction of Tala Hydroelectric Project, Bhutan.

The modulus of deformation of rock mass is an important engineering parameter required for the stability analysis and design of rock structures. Different equipment and techniques are used to arrive at the design modulus value. Since any error in the estimation of modulus values results in multiplication of its effects in the analysis, it is necessary to know about the reliability of testing equipment and procedures.

In the above background, a number of tests using different equipments/methods were conducted by the Central Soil and Materials Research Station (CSMRS), New Delhi in the same rock mass for the determination of modulus of deformation. The following tests/methods were used :

- Plate jacking test (PJT) with displacement measurement by borehole extensometers,
- Plate loading test (PLT) with surface displacement measurement, and
- Goodman jack test (GJT) inside drill holes.

This paper deals with the evaluation of deformability characteristics of rock mass by Plate Load Test (PLT), Goodman Jack Test (GJT) and Plate Jacking Test (PJT) inside desilting chamber at Tala Hydroelectric Project, Bhutan.

2.0 TALA HYDROELECTRIC PROJECT

The Tala Hydroelectric Project envisages the construction of a 92 m high concrete gravity dam across the Wangchu river in Western Bhutan near Honka (3 km downstream of existing Chukha Tail Race Tunnel). The project proposes to generate 1020 MW (6 x 170 MW) of power. The proposed powerhouse is underground with the size of the main cavern 206 m x 19 m x 44.5 m. The underground powerhouse is located near Tala village. The other main features of the project are 3 desilting chambers each of size 250 m x 13.92 m x 13.5 m and a 22.64 km long, 6.8 m dia modified horse-shoe shaped concrete lined head race tunnel (HRT).

3.0 GEOLOGY OF DESILTING CHAMBER

The geology of the project has been described in detail by the resident geologist and can be referred from the project documents THPA (1998), Chaudhary (1999). The proposed desilting arrangement comprises three chambers of 13.92 m (w) x 13.5 m (h) x 250 m (l) and is aimed to remove particles larger than 0.2 mm size. The desilting arrangement is proposed underground in the right bank hills with steep cliff and escarpments. The hill mass rises continuously from the river Wangchu at El. 1291m to more than 1000 m with 40^0 to 70^0 slopes. The proposed desilting chamber complex area has a lateral cover of 100 m to 160 m and a vertical cover of 80 m to more than 200 m. Three desilting chambers along with rock pillars in between would involve an area of about 100 m width and 350 m length.

The rocks exposed in the area are predominantly muscovite-biotite gneiss with minor bands of biotite schist, quartzite and calc silicate rocks. The area encompassing the desilting chamber has one synformal and one antiformal structure. The general foliation trend is N 10^0 W - S 10^0 E with 36^0 N 80^0 E dips. The alignment of desilting chamber cavity has been tentatively selected in N 15^0 W - S 15^0 E direction taking into account the major structural features, topographical control and hydraulics. This orientation is expected to maintain the overall stability of the hill mass and provide favourable position requiring minimum rock support problems on the roof and walls. The tunnel quality 'Q' is good to very good for gneiss, augen gneiss and quartzite except in sheared and jointed rock. The Q values for schists and calc silicate rock are generally fair to good. The shear zone with more than 0.50 m thickness has poor to very poor Q values.

The portal of the testing drift is located on the right abutment of the dam axis at El 1338 m. It has been excavated for a length of 45 m in S 20^0 W direction and 11 m in a crosscut at RD 12 m in S 70^0 E direction. The drift lies essentially in biotite gneiss with quartz bounding. The general trend of foliation is in N 10^0 W - S 10^0 E' 36^0 N 80^0 E direction. This drift has further been excavated up to a length of 193 m to reach inside the desilting chamber. The deformability tests were conducted in the end of the drift. The rock mass in this reach of the drift is quartz mica gneiss with quartz bounding. The rock is fresh moderately jointed and thinly foliated. A 15 cm thick shear plane filled with clay gauze and quartz bounding starting from RD 155 m is extended up to the end of the drift.

4.0 DEFORMABILITY TEST

It was proposed to determine the deformability of rock mass by plate load tests and Goodman jack test in the drift. Three numbers of plate load tests, 15 Goodman jack tests and 1 Plate jacking tests using borehole extensometers were conducted for the determination of deformability of rock mass at different RD's in 5 cycles.

4.1 Plate Loading Test

4.1.1 *Test Procedure*

At the selected test site, the rock surface in the bottom and top of the drift was smoothened by chiselling to obtain parallel faces, about 5 cm more than the diameter of the test plate. The test plate of 60 cm diameter was used. A 2 cm thick and 60 cm dia concrete pad was constructed in the bottom for test plate and about 5 cm thick and 60 cm dia pad was constructed at the top to take the reaction for loading. Both the pads were kept parallel to each other.

The testing equipment was installed with 2.5 cm thick and 60 cm dia plates in the bottom and then 45 cm and 30 cm dia plates were placed. Then the hydraulic jack of 200 t capacity was placed. The gap between top pad and jack was filled by aluminium alloy pipes with a M.S. plate of 2.5 cm thick and 30 cm dia on the top. The remaining small gap was filled by applying seating load or by moving out plunger. The displacement measuring unit was installed by using four extensometers with an accuracy of 0.01 mm. The assembly of the plate loading test is shown in Fig. 1.

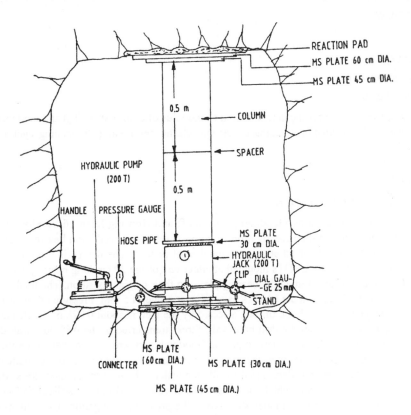

Fig. 1 : Plate loading test assembly inside a testing drift

4.1.2 *Calculation*

The modulus of deformation for the loading cycle by considering total deformation of a particular cycle and modulus of elasticity by considering elastic deformation or deformation of unloading was calculated by using the following equations :

$$E = \frac{Pm\,(1 - v^2)}{W\sqrt{A}} \qquad \qquad \ldots (1)$$

where,

E = Modulus of deformation/Elasticity in kg/cm^2
P = Applied load in kg
v = Poisson's ratio

m   =   Constant depending upon the shape of plate (m = 0.95 for square plate and m = 0.96 for circular plate)

W   =   Deformation corresponding to load in cm

A   =   Area of plate in cm^2

For biotite schist rock mass, the Poisson's ratio is assumed as 0.25 and diameter of the circular plate is 60 cm. Hence, the Eq. 1 may be reduced to :

$$E = 0.016926 \ P/W \qquad \qquad \ldots (2)$$

Or in terms of applied stress, the above equation can be written as :

$$E = 47.857 \ \sigma/W \qquad \qquad \ldots (3)$$

where,

P is load in kg

W is deformation in cm

σ is stress in kg/cm^2

The Eqn. 3 can be utilised for the determination of modulus of deformation (E_d) and modulus of elasticity (E_e) based on the total deformation (loading cycle) and elastic deformation (unloading cycle) of particular cycle respectively.

4.2   Goodman Jack Tests

4.2.1 *Goodman Jack and Test Procedure*

The drill hole jack designed by Goodman, Harlomoff and Horning licensed by Slope Indicator Co., Seattle, USA, has been used in the investigation. It consists of two curved rigid bearing plates of angular width 90 degrees, which can be forced apart by a number of pistons. The device is used inside an NX size (76 mm diameter) drill hole. Two LVDTs mounted at either end of the 20 cm long bearing plates measure the displacement. Two return pistons retract the bearing plates to their original position. The total piston travel of the equipment is about 12.5 mm and the LVDTs have a linear range of 5 mm. Pressure of the order of 70 MPa can be applied by the jacks. The volume of rock affected by the jack is about 0.028 m^3 (1 cft.) and extends to about 114 mm into rock away from the drill hole wall. Stress transferred to the drill hole wall depends upon the particular model used for the tests. Two types of Goodman jack are used to determine the modulus of deformation of rock mass depending upon the type of rock mass. For hard rocks the Goodman jack model 52101 is used in which the stress transferred to the drill hole walls is 93 % of the applied stress whereas for the soft rocks the model 52102 is used in which the stress transferred to the drill hole walls is 55 % of the applied stress. The Goodman jack of soft rock model was used in the present investigation. The tests were conducted in two mutually perpendicular directions inside each drill holes.

4.2.2 *Calculation*

The modulus of deformation is calculated using the following relationship given by Goodman et al. (1968) :

$$E = 0.86 \frac{\Delta P}{\dfrac{\Delta D}{D}} K(\nu, \beta) \qquad \qquad \ldots (4)$$

where

E	=	the modulus of deformation/elasticity (kg/cm^2).
ΔP	=	Pressure increment (kg/cm^2).
ΔD	=	Diametral displacement increment (cm).
D	=	Diameter of drill hole (cm).
K (ν, β)	=	Constant depending upon Poisson's ratio (ν) and angle of loaded arc (β).

The percentage of applied stress transferred to the drill hole walls is 55% (as per manufacturer). The provision of 55% was included in the applied stress. The modulus of deformation for Poisson's ratio of 0.25 and loaded arc of 45 degrees has been calculated using the following relationships :

$$E = 0.86 \text{ x } 7.6 \text{ x } 1.254 \frac{\Delta P}{\Delta D} \qquad \qquad ... (5)$$

$$= 7.6224 \frac{\Delta P}{\Delta D}$$

where,

0.86	=	3-D effect
0.55	=	Hydraulic efficiency, 55% for soft rock jack (not included in Eqn. 5 as the stress applied was 55% higher)
7.6	=	Diameter of NX size drill hole
1.254	=	K (ν, ß) for ν = 0.25 and ß = 45° using Goodman's chart

4.3 Plate Jacking Test

The schematic diagram of the plate jacking set up is shown in Fig. 2 along with a typical illustration of installation of anchors and extensometers inside the drill hole. It comprises hand pumps, hydraulic jacks, flat jacks, multiple point borehole extensometers with anchors and the measuring system with displacement transducers and a 12 channel digital readout unit with an accuracy of 0.001 mm. The capacity of the system is 7 MPa uniaxial pressure.

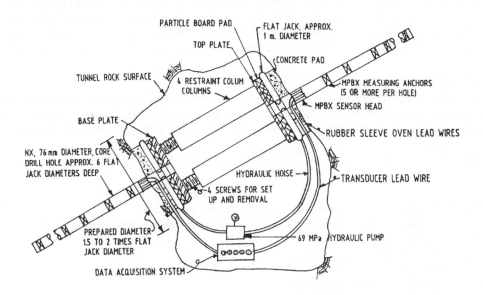

Fig. 2 : Schematic diagram of the plate jacking test set up

4.3.1 Site Preparation

The plate jacking tests were conducted by applying load in the direction normal to drill holes. The rock surface of the drift at the test locations were carefully prepared by removing all loose rock material by chiselling within a diameter of 150 cm around the drill holes. The loading surfaces were kept concentric. Nx size (76 mm diameter) instrumentation drill holes of about 6 m depth were drilled at the prepared surfaces. Both the drill holes were aligned carefully so that they are normal to surface and are in line with each other.

Concrete pads using rich mix are cast around the drill holes to ensure smooth transfer of load from the flat jacks to the rock mass. The pads were allowed to cure for about seven days to obtain sufficient strength prior to commencement of the test.

4.3.2 *Equipment Installation*

The extensometers with the help of anchors were installed at suitable locations inside the drill holes. The location of anchors were decided after careful examination and logging of drill hole cores. Care was taken so that the anchors were not placed on joints. The last anchor in the drill hole is kept about 20 - 40 cm below the rock surface just to avoid blasting effects in the drift. The deepest anchor was located at a depth of 480 cm (about 6 times the diameter of the flat jack) from the rock surface in order to provide a fixed point to which the movement of all the extensometers can be referred. In all six to seven anchors were installed in each instrumentation drill hole, which accommodated five to six extensometers in each drill hole. The gap between the flat jack assembly and base and the top plates was filled up by special particle board made of wooden chips and resin, fabricated to accommodate the flat jack configuration on one side and the base plate on the other side.

4.3.3 *Test Procedure*

After all the components were installed, the system was checked for the actual test. The loading was applied through the flat jack system by manually operated hydraulic pump. It was tried to maintain the rate of loading as 0.4 MPa/min and the load was applied in cycles of 1, 2, 3, 4 and 5 MPa of loading and unloading the pressure every time to zero. However, the modulus values have been calculated for the cycles of 2, 3, 4 and 5 MPa. The first cycle was not considered for evaluation of deformability as the closing of joints due to blasting and some settlement of loading assembly takes place in loading and unloading. The load was maintained for 5 minutes at the stage of initial loading, incremental loading and maximum loading, while the intermediate load increments were maintained for one minute. The tests were conducted according to the suggested method by ISRM (1979). However, time dependent deformability of rock mass was not determined.

4.3.4 *Calculation*

Deformation measurement for the various load cycles are utilised to compute deformation modulus according to appropriate formula. The modulus of deformation has been calculated for each cycle of loading and unloading. The equation utilised for this purpose is given below by utilising the following formula :

$$ W_z = \frac{2P(1-v^2)}{E}\left[(a^2+z^2)^{\frac{1}{2}}-z\right] - \frac{Pz(1+v)}{E}\left[z(a^2+z^2)^{-\frac{1}{2}}-1\right] \qquad ...(6) $$

where

W_z = Displacement in the direction of applied pressure (cm)
z = Distance from the loaded surface to the point where displacement is measured (cm)
P = Applied pressure (in MPa)
A = Outer radius of flat jack (cm)
v = Poisson's ratio
E = Modulus of rock mass (in MPa)

After substituting the appropriate values of a, z and v, the Eqn. 6 can be written as :

$$ W_z = \frac{P}{E}(K_z) \qquad ...(7) $$

The modulus of deformation (E_d) can be determined by the following formula :

$$ E_d = P\left[\frac{K_{z1}-K_{z2}}{W_{z1}-W_{z2}}\right] \qquad ...(8) $$

Where, K_{z1} and K_{z2} are constants at depth z_1 and z_2, respectively. Similarly, W_{z1} and W_{z2} are deformations measured between depths z_1 and z_2. The Eqn. 8 can be utilised for the determination of modulus of deformation (E_d) and modulus of elasticity (E_e) based on the total deformation (loading cycle) and elastic deformation (unloading cycle) of particular cycle respectively.

5.0 RESULTS AND DISCUSSION

In all 19 deformability tests which includes 3 plate loading tests (PLT), 15 Goodman jack tests (GJT) and 1 plate jacking test (PJT) using borehole extensometers were conducted in drift for the evaluation of deformability of rock mass (modulus of deformation, E_d and modulus of elasticity, E_e in desilting chamber). The drift is located on the right abutment of dam at EL 1338 m. The drift has been excavated up to a length of 193 m. The tests were conducted using five cycles of loading and unloading with a maximum applied stress level of 5.0 MPa in PLT and PJT and 7.5 MPa in GJT. The modulus of deformation of rock mass was calculated by taking total deformation of loading cycle at a particular applied stress level. The modulus of elasticity of rock mass was calculated by considering deformation of unloading cycle at a particular applied stress level.

5.1 Plate Loading Test

A total of 3 plate loading tests were conducted inside drift at the RDs 156.50 m, 181.65 m and 184.65 m. The typical stress versus deformation curve has been shown in Fig. 3. The average values of moduli of deformation and moduli of elasticity have been presented in Table 1 along with variation at the applied stress level of 2, 3, 4 and 5 MPa. The modulus value increases and ratio of E_e / E_d decreases with the increase in stress level. The average value of modulus of deformation shows increase from 1.86 GPa to 2.28 GPa with an average value of 2.09 GPa at 5 MPa stress level. The modulus of elasticity varies from 2.10 GPa to 3.63 GPa with an average value of 2.67 GPa. The ratio of moduli of elasticity and deformation (E_e / E_d) is 1.70 at an applied stress level of 2 MPa and decreases to 1.28 at an applied stress level of 5 MPa.

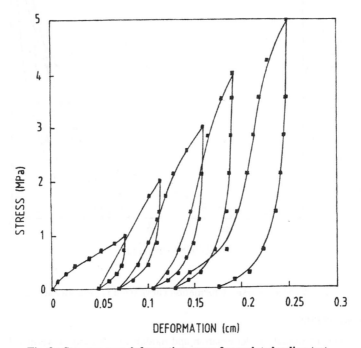

Fig. 3 : Stress versus deformation curve from plate loading test

31

TABLE 1

Average value of modulus of deformation and elasticity by plate load test

Stress MPa	Modulus of deformation, E_d (GPa)			Modulus of elasticity, E_e (GPa)			Ratio E_e/E_d Average
	Minimum	Maximum	Average	Minimum	Maximum	Average	
2	0.95	1.45	1.13	1.37	2.28	1.92	1.70
3	1.19	1.58	1.38	1.61	2.61	2.07	1.50
4	1.41	2.20	1.78	1.95	3.04	2.46	1.39
5	1.86	2.28	2.09	2.10	3.63	2.67	1.28

5.2 Goodman Jack Test

A total of 15 Goodman jack tests (GJT) were conducted inside 3 Nx size drill holes up to a depth of 6 m each (one horizontal drillhole and 2 vertical drill holes in upward and downward direction in drift at RD 188.40 m). The typical stress versus deformation curve has been shown in Fig. 4. The Goodman jack tests were conducted in three directions, *i.e.*, vertical direction and horizontal direction parallel and perpendicular to the axis of drift. This is the main advantage with GJT to conduct the tests in the desired direction to know the anisotropy of rock mass.

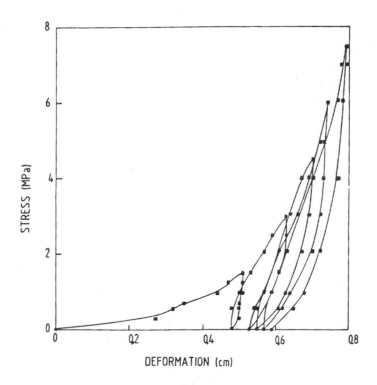

Fig. 4 : Stress versus deformation curve from Goodman Jack test

5.2.1 *Vertical Deformability*

The average values of moduli of deformation and moduli of elasticity in vertical loading direction, *i.e.*, GJT1 to GJT4 have been presented in Table 2 at the applied stress level of 3, 4.5, 6 and 7.5 MPa. The modulus values increases and ratio of E_e / E_d decreases with the increase in stress level. The average value of modulus of deformation shows increase from 1.98 GPa to 4.98 GPa with an average value of 3.60 GPa at 7.5 MPa stress level. The modulus of elasticity varies from 2.08 GPa to 5.42 GPa with an average value of 3.99 GPa. The ratio of moduli of elasticity and deformation (E_e / E_d) is 1.26 at an applied stress level of 3.0 MPa and decreases to 1.11 at an applied stress level of 7.5 MPa.

TABLE 2
Average moduli values in vertical direction by Goodman Jack test

Stress (MPa)	Modulus of deformation, E_d (GPa)			Modulus of elasticity, E_e (GPa)			Ratio E_e / E_d Average
	Minimum	Maximum	Average	Minimum	Maximum	Average	
3.0	1.25	2.62	2.22	1.47	3.83	2.80	1.26
4.5	1.53	3.56	2.02	1.70	3.74	3.06	1.52
6.0	1.78	3.83	3.25	1.92	5.25	3.54	1.09
7.5	1.98	4.98	3.60	2.08	5.42	3.99	1.11

5.2.2 *Horizontal Deformability Parallel To Drift*

The average values of moduli of deformation and moduli of elasticity in horizontal loading direction parallel to drift axis (*i.e.*, GJT 5 to GJT 11) have been presented in Table 3 at the applied stress level of 3, 4.5, 6 and 7.5 MPa. The modulus values increases and ratio of E_e / E_d decreases with the increase in stress level. The average value of modulus of deformation shows increase from 2.83 GPa to 7.33 GPa with an average value of 3.74 GPa at 7.5 MPa stress level. The modulus of elasticity varies from 2.04 GPa to 8.31 GPa with an average value of 4.37 GPa. The ratio of moduli of elasticity and deformation (E_e / E_d) is 1.29 at an applied stress level of 3.0 MPa and decreases to 1.17 at an applied stress level of 7.5 MPa. The moduli values in horizontal direction parallel to drift are slightly higher than moduli values in vertical direction.

TABLE 3
Average moduli values in horizontal direction (parallel to drift) by Goodman Jack test

Stress (MPa)	Modulus of deformation, E_d (GPa)			Modulus of elasticity E_e (GPa)			Ratio E_e / E_d Average
	Minimum	Maximum	Average	Minimum	Maximum	Average	
3.0	1.61	2.93	2.31	1.85	4.98	2.98	1.29
4.5	2.41	5.75	3.36	2.41	6.80	3.78	1.13
6.0	2.49	7.12	3.75	2.77	7.12	4.02	1.07
7.5	2.83	7.33	3.74	2.04	8.31	4.37	1.17

5.2.3 *Horizontal Deformability Perpendicular to Drift*

The average values of moduli of deformation and moduli of elasticity in horizontal loading direction parallel to drift axis, *i.e.*, GJT 12 to GJT 15 have been presented in Table 4 at the applied stress level of 3, 4.5, 6 and 7.5 MPa. The modulus values increases and ratio of E_e / E_d decreases with the increase in stress level. The average value of modulus of deformation shows increase from 3.12 GPa to 4.98 GPa with an average value of 3.82 GPa at 7.5 MPa stress level. The modulus of elasticity varies from 3.12 GPa to 4.79 GPa with an average value of 3.82 GPa. The ratio of moduli of elasticity and deformation (E_e / E_d) is 1.21 at an applied stress level of 3.0 MPa and decreases to 1.00 at an applied stress level of 7.5 MPa. The rock mass has behaved perfectly elastic in this case. The moduli values in horizontal direction perpendicular to drift axis are slightly higher than moduli values in vertical direction as well as in horizontal direction parallel to drift axis.

TABLE 4
Average moduli values in horizontal direction (perpendicular to drift) by Goodman Jack test

Stress (MPa)	Modulus of deformation, E_d (GPa)			Modulus of elasticity, E_e (GPa)			Ratio E_e / E_d
	Minimum	Maximum	Average	Minimum	Maximum	Average	Average
3.0	1.92	2.77	2.17	1.85	3.56	2.63	1.21
4.5	2.41	3.94	3.02	2.41	4.40	3.16	1.05
6.0	2.49	4.15	3.21	2.77	4.47	3.43	1.07
7.5	3.12	4.98	3.82	3.12	4.79	3.82	1.00

5.2.4 *Moduli Values by GJT in Desilting Chamber Drift*

Based on 15 Goodman jack tests (GJT) conducted inside 3 Nx size drillholes in drift at RD 188.40 m, the average values of moduli of deformation and moduli of elasticity, *i.e.*, GJT 11 to GJT 13 have been presented in Table 5 at the applied stress level of 3, 4.5, 6 and 7.5 MPa. The modulus values increases and ratio of E_e / E_d decreases with the increase in stress level.

TABLE 5
Average moduli values in desilting chamber by Goodman Jack test

Stress (MPa)	Modulus of deformation, E_d (GPa)			Modulus of elasticity, E_e (GPa)			Ratio E_e / E_d Average
	Minimum	Maximum	Average	Minimum	Maximum	Average	
3.0	1.00	3.83	2.33	1.19	5.54	3.14	1.35
4.5	1.47	5.75	3.16	1.53	6.80	3.48	1.10
6.0	1.78	7.12	3.60	1.78	7.12	3.92	1.09
7.5	1.98	7.33	4.06	2.04	8.31	4.24	1.04

The average value of modulus of deformation shows increase from 1.98 GPa to 7.33 GPa with an average value of 4.06 GPa at 7.5 MPa stress level. The modulus of elasticity varies from 2.04 GPa to 8.31 GPa with an average value of 4.24 GPa. The ratio of moduli of elasticity and deformation (E_e / E_d) is 1.35 at an applied stress level of 3.0 MPa and decreases to 1.04 at an applied stress level of 7.5 MPa.

5.3 Plate Jacking Test

Due to time constraint and difficulty in drilling, it was possible only to conduct one plate jacking test (PJT) with borehole extensometers inside drift at the RDs 188.40 m. The moduli of deformation and elasticity have been shown in Table 6 for the applied stress in vertical direction. The stress versus deformation curves has been shown in Fig. 5 for all the five cycles. The variation of deformation with depth from different extensometers is shown in Fig. 6.

The extensometers were installed in both vertical downward and upward drill holes and the loading was applied in both the direction by flat jacks on the top and bottom of the loading assembly. However, the deformation could not be measured properly due to failure of top pad. The loading could also not be transferred properly in the upward direction due to lot of steel reinforcement used for the construction of top pad because of time constraint. Hence, the moduli values have been calculated only for the downward vertical drill hole.

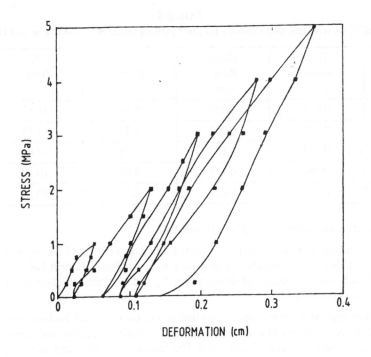

Fig. 5 : Stress versus deformation curve from plate jacking test

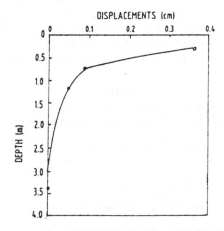

Fig. 6 : The variation of displacements with depth

The values of moduli of deformation and moduli of elasticity have been presented in Table 6 at the applied stress level of 1, 2, 3, 4, and 5 MPa. The modulus of deformation increases/decreases (in general it increases, but the variation in this particular test may be due to the failure of upper concrete pad), modulus of elasticity decreases and ratio of E_e / E_d decreases with the increase in stress level. The magnitude of modulus of deformation is 8.70 GPa and the modulus of elasticity is 10.14 GPa at a stress level of 5 MPa. The ratio of moduli of elasticity and deformation (E_e / E_d) is 1.76 at an applied stress level of 1 MPa and decreases to 1.12 at an applied stress level of 5 MPa.

TABLE 6
Moduli of deformation and elasticity for plate jacking test at R.L. 1342.70 m and 188.40 m

Stress (MPa)	Total deformation (cm)	Elastic rebound (cm)	E_d (GPa)	E_e (GPa)	Ratio E_e / E_d
1	0.051	0.029	8.63	15.17	1.76
2	0.107	0.067	8.22	13.13	1.59
3	0.135	0.110	9.78	12.00	1.23
4	0.194	0.173	9.07	10.17	1.12
5	0.253	0.217	8.70	10.14	1.12

5.4 Comparison of Deformability by GJT, PLT and PJT

In all 19 deformability tests which includes 3 plate loading tests (PLT), 15 Goodman Jack tests (GJT) and 1 plate jacking test (PJT) using borehole extensometers were conducted in drift DR-2 for the evaluation of deformability of rock mass (modulus of deformation, E_d and modulus of elasticity, E_e in the proposed underground desilting chamber).

The stress was applied in only vertical direction in the case of plate loading and plate jacking tests. However, the stress by Goodman Jack tests was applied in vertical direction and horizontal direction, *i.e.*, parallel and perpendicular to the axis of drift DR-2. For comparing all the three methods, at least three plate jacking tests should have been conducted, as this is a large size test as compared to PLT and GJT.

A comparison of moduli values in vertical direction from PLT, GJT and PJT has been shown in Table 7. The variation in moduli values with direction, *i.e.*, anisotropy of rock mass by GJT has also been shown in Table 7. The moduli values in horizontal direction perpendicular to drift axis are slightly higher than moduli values in vertical direction as well as in horizontal direction perpendicular to drift axis. The modulus values increases in general and ratio of E_e / E_d decreases with the increase in stress level in all the methods. The decrease in moduli ratio shows the closing of joints at high stress level.

TABLE 7
Anisotropy of rock mass and comparison of deformability by PLT and PJT at applied stress level 5 MPa and by GJT at stress level 6 MPa

Loading direction w.r.t. drift DR-2	Modulus of deformation, E_d (GPa)			Modulus of elasticity, E_e (GPa)		
	PLT	GJT	PJT	PLT	GJT	PJT
Perpendicular	-	3.82	-	-	3.82	-
Parallel	-	3.74	-	—	4.37	-
Vertical	2.09	3.60	8.70	2.67	4.00	10.14

The modulus of deformation in vertical direction evaluated from Goodman Jack tests (3.25 GPa at 6 MPa stress) is slightly higher than that evaluated from plate loading tests (2.09 GPa at 5 MPa stress). The moduli values by plate loading test are low due to the deformation measurement at the surface of the drift which is disturbed due to blasting. However, there is significant difference in the modulus of deformation (8.70 GPa) in the case of plate jacking test. The modulus values by PJT are 2 to 3 times higher than those obtained by GJT and PLT. This is due the fact that the loaded area is large in PJT and displacements are measured inside drill hole in the comparatively undisturbed rock mass. The low modulus value by GJT is due to the fact that loaded area in GJT is much smaller than PJT as also concluded by Singh *et al.* (1994) and Sharma *et al.* (1989). Heuze and Amadei (1985) have suggested by trial and error method for improving the moduli values obtained by borehole jack method. They tried to increase the value of constant K factor (Eqn. 4) which was also discussed by Singh *et al.* (1994). Beiniawski (1989) tried to compare the rock deformability by GJT with Petite Seismique and·flat jack methods.

6.0 CONCLUSIONS

On the basis of 19 deformability tests conducted in drift for the evaluation of deformability of rock mass (modulus of deformation, E_d and modulus of elasticity, E_e) in desilting chamber by 3 plate loading tests (PLT), 15 Goodman Jack tests (GJT) and 1 plate jacking test (PJT) using borehole extensometers, the following conclusions are drawn :

- The modulus values increases in general and ratio of E_e / E_d decreases with the increase in stress level in all the methods. The decrease in moduli ratio shows the closing of joints at high stress level.
- The moduli values in horizontal direction perpendicular to drift axis are slightly higher than moduli values in vertical direction as well as in horizontal direction parallel to drift axis. The variation in modulus values in all the three direction by GJT shows that the rock mass is of anisotropic. It can be concluded that the rock mass is slightly anisotropic.
- The modulus of deformation in vertical direction evaluated from Goodman Jack tests (3.25 GPa at 6 MPa stress) is slightly higher than that evaluated from plate loading tests (2.13 GPa at 5 MPa stress). However, there is significant difference in the modulus of deformation (8.70 GPa) by plate jacking test. The modulus values by PJT are 2 to 3 times higher than those obtained by GJT and PLT.

ACKNOWLEDGEMENT

The authors are thankful to all the team members of CSMRS, WAPCOS and THPA. The authors are also thankful to Director, CSMRS for permitting to publish this paper.

REFERENCES

Bieniawski, Z.T. (1989), "A Comparison of Rock Deformability Measurement by Petite Seismic, the Goodman Jack and Flat Jacks" Rapid Excavation and Tunnelling Conference, Atlanta.

Chaudhary, A.K. (1999), "Brief Geology of Dsilting Chamber".

Goodman, R.E. and Tran, V.K. (1968), "The Measurement of Deformability in Boreholes", Department of Geological Engineering, University of California.

Heuze, F.E. and Amadei, B. (1985), "NX Borehole Jack : A Lesson in Trial and Errors", Int. J. Rock Mech. Min. Sci. and Geomech. Abstracts, Vol. 22, No. 2.

ISRM, (1979) "Suggested Methods for Determining In-situ Deformability of Rock", Int. J. Rock Mech. Min. Sci. and Geomech. Abstracts, Vol. 16, No. 3, pp. 195-214.

Sharma, V.M., Singh, R.B. and Chaudhary, R.K. (1989), "Comparison of Different Techniques and Interpretation of the Deformation Modulus of Rock Masses", Indian Geotechnical Conference (IGC-89), Vishakhapatnam.

Singh, R.B, Hari Dev, Dhawan, A.K. and Sharma, V.M. (1994) "Deformability of Rock Mass by Plate Jacking and Goodman Jack tests", Indian Geotechnical Conference-94, Warangal, pp. 385-388.

THPA (1998), "Bidding Document for Contract Package Number C1 Construction of Civil Works for Dam, Intake, Desilting Arrangement and Head Race Tunnel from Thiyomachu, Tala Hydroelectric Project (1020 MW)".

OBSERVATIONAL APPROACH FOR STABILITY OF TUNNELS

A. SWARUP   R.K. GOEL   V.V.R. PRASAD

Central Mining Research Institute, Regional Centre, Roorkee, India

SYNOPSIS

To understand the behaviour of rock mass in diverse fields of geo-mining and geo-engineering conditions, is one of the most difficult problems of rock engineering in spite of rapid strides made in the recent years in the field of geomechanics. A great deal of experience is now available and scientific studies have been carried out progressively to enrich the state of art. Various techniques have been developed to predict the behaviour of rock mass for designing underground structures. The paper presents assessment of the rock mass behaviour around underground openings using a simple observational approach which can be effective in predicting the tunnel stability. The approach is based on estimation of strains from closure data obtained by monitoring of tunnels and evaluating them by "hazard warning levels". Case histories of tunnels excavated in Himalayas are analysed to assess the applicability of the technique for tunnels in Himalayan conditions. From the results obtained it can be assessed that the approach can be used for predicting stability of tunnel immediately after taking measurements.

1.0 INTRODUCTION

In the construction of tunnels and underground openings in complex rock formations, rock behaviour and its influence on tunnels driving plays a vital role in determining the state of stability. The excavation of tunnel through rocks disturbs the existing state of stresses. New stresses are developed which causes the rock to move inwards. The inward movement or closure may vary, depending upon the rock quality and the induced stresses. This movement, when contained by installation of supports, results in development of rock loads.

It is not always possible to predict the closure and the rock loads. The prediction process depends upon the knowledge of accurate geology and physical properties of the rock masses. However the accuracy of these parameters varies considerably, primarily due to the uncertainties involved, which are largely due to the discontinuous and variable nature of rock mass. The rock mass contains bedding planes, joints, faults and other structural fractures which render them discontinuous and often control their engineering behaviour.

In order to overcome this difficulty, field observations of deformation behaviour are carried out during the excavation for assessing the stability of the tunnels. There are many types of instruments available for in-situ monitoring of tunnels. The most commonly used are the closure measurements using tape extensometers and the borehole extensometers. These are reliable and simple to use in the field.

2.0 TUNNEL INSTRUMENTATION

Normally in a tunnel, only a few types of rocks are encountered and there may be a few faults. It is possible to monitor the behaviour of different rocks and fault zones with the help of suitable instruments. As tunnelling progresses, some closure bolts may be installed to know the radial displacements at the tunnel periphery (Fig. 1). The closures should be measured using tape extensometers. Similarly borehole extensometers can also be installed in addition to closure bolts in order to estimate the extent of rock mass displacement inside the rock mass.

The instruments should be installed soon after the excavation of the tunnel section. The installation programme should coincide with that of support placement. The observations of the instruments should be taken regularly. The frequency of observation should be high in the beginning, but lessened progressively with the passage of time.

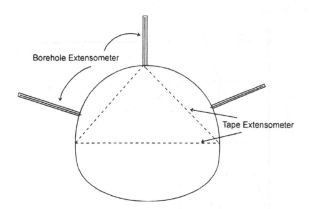

Fig. 1 : Displacement measurements using tape extensometers and borehole extensometers

3.0 OBSERVATIONAL APPROACH

Initial support design in normally done on the basis of the data collected during the investigation. When constructing a tunnel, field measurements are usually carried out, not only for monitoring the stability of surrounding rocks, but also for the adequacy of support structures such as shotcrete, rock bolts and steel ribs. Both the original design for the support structures and construction procedures are then evaluated considering the results of the field measurements. They are then modified if necessary. In order to use this procedure successfully in the construction of tunnels, the interpretation of field measurement results is extremely important, and its results must be analysed properly without delay for assessing the support measures and construction procedures.

For interpreting the field measurement results properly, Sakurai (1997) has proposed a method based on the concept of strain. In this method, the stability of tunnels is assessed by comparing the strain occurring in rocks around tunnel with the allowable value of strain. The strain distribution in tunnel is obtained directly from the results of displacement measurements, such as extensometer measurements.

4.0 HAZARD WARNING LEVEL

Sakurai (1997) proposed a "hazard warning level" which can assess the stability of structure immediately after taking measurements by simply comparing the measured values to the hazard warning level. This hazard warning level is based on the concept of critical strain proposed by Sakurai (1981), which can successfully be used for assessing the results of displacement measurements in tunnels. When the measured values remain smaller than the hazard warning level, the stability of the structure is confirmed. If the measured values are predicted to become greater, then the engineers must take some action to stabilise the structures and to modify the original design.

4.1 Critical Strain

The critical strain for intact rocks defined as the ratio of uniaxial compressive strength to modulus of elasticity, and it may be used as an allowable value of strain, which can successfully be used for assessing the results of displacement measurements in tunnels, such as crown settlement, convergence and extensometer measurements.

$$\varepsilon_0 = \frac{\sigma_c}{E} \qquad \qquad \ldots (1)$$

where σ_c is uniaxial compressive strength and E is Young's modulus. It is noted that this critical strain ε_0 generally differs from strain at failure ε_f (Fig. 2) and is smaller than the strain at failure. In case of in-situ rock mass the Eqn. 1 may be written as :

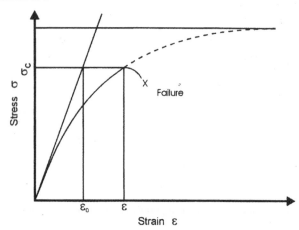

Fig. 2 : Stress-strain relationship and definition of critical strain (ε_0)

$$\varepsilon_{0R} = \frac{\sigma_{cR}}{E_R} = \frac{m\sigma_c}{nE} = \left(\frac{m}{n}\right)\varepsilon_0 \qquad \qquad \ldots (2)$$

where σ_{cR} is uniaxial compressive strength of rock mass and E_R is the modulus of deformation of rock mass. The constants m and n, respectively are non-dimensional reduction factors for the strength and deformability of rock masses with joints compatible with the laboratory specimens and may range between 0 to 1.0. The ratio m/n varies between 1.2~3.0 based on the finding of Sakurai (1983). If m = n, then Eqn. 2 becomes :

$$\varepsilon_{0R} = \frac{\sigma_c}{E} = \varepsilon_0 \qquad \qquad \ldots (3)$$

This means that the critical strain of in-situ rock masses is the same as that of intact rock with no joints. That is the critical strain of in-situ rock masses is almost the same order of magnitude as that of intact rocks, though both uniaxial strength and Young's modolus of intact rocks largely differ from those of in-situ rock masses. This is because the effects of joints are cancelled out by taking the ratio of the two, although the uniaxial strength and Young's modulus are both greatly influenced by existing joints. Therefore in engineering practice it may be possible to use the value of critical strain of intact rocks as a hazard warning level for monitoring the stability of tunnels.

In order to verify the applicability of the hazard warning level for assessing the stability of tunnels, some displacement measurements were carried out and the stains occurring around the tunnels as a result of excavation are calculated from measured displacements by Eqn. (4).

$$\varepsilon_\theta = \frac{u_c}{a} \qquad \qquad \ldots (4)$$

where

u_c is the measured value of crown settlement; and
a is the tunnel radius.

The strains calculated by Eqn. (4) are plottee in relation to the uniaxial strength of soils and rocks as shown in Fig. 4. The dotted lines indicates the upper and lower bounds for the critical strain obtained from laboratory tests (Sakurai 1981). The numbers given beside the data indicate the sort of difficulties encountered during the excavation of tunnels, while the data with no numbers are those tunnels excavated with no serious problems. The types of difficulties are classified as follows :

(1) difficulties in maintaining tunnel face;
(2) failure or cracking in shotcrete;
(3) buckling of steel ribs;
(4) breakage of rock bolts;
(5) fall-in of roof;
(6) swelling at invert; and
(7) miscellaneous.

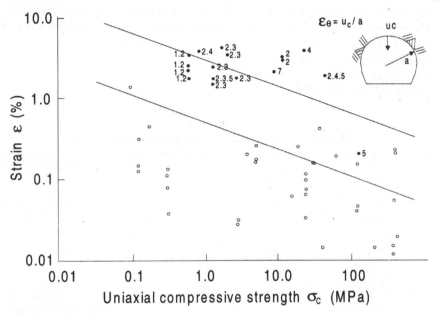

Fig. 3 : Relationship between the measured strain (obtained from crown settlement) and hazard warning levels (after Sakurai, 1997)

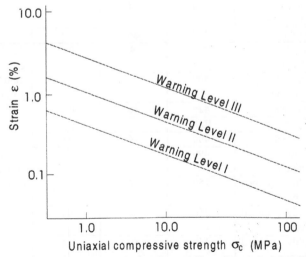

Fig. 4 : Hazard warning levels for assessing the stability of tunnels (after Sakurai, 1997)

It can be observed from Fig. 3 that those tunnels whose observed stain were smaller than the lower bound of the critical strain were stable or had no excavation problems. Whereas, when strain approach the upper bound then tunnelling problems occur. Based on the above analysis Sakurai (1997) classified the hazard warning level into three stages in relation to the degree of stability as shown in Fig. 4.

Fig. 4 shows three levels of warning zone with warning level I indicates generally no tunnelling problem whereas warning level III leads to various tunnelling problems as listed above. A few Indian tunnelling cases have been analysed in the light of this approach and discussed below.

5.0 CASE STUDY

The closure data of the following projects located in the Himalayas were taken up for analysis in order to verify the accuracy of this hazard warning method. The critical strains were estimated (Table 1) from the measured displacements and their relationship with uniaxial compressive strength of rock was plotted.

1. Giri Hydel Project
2. Maneri Bhali Hydel Stage-I Scheme
3. Maneri Bhali Hydel Stage-II Scheme
4. Salal Hydroelectric Project

TABLE 1
Details of critical strains of various projects in Himalayas

Project	Rock Type	Rock Cover m	UCS, MPa	Critical Strain, (%)	Major Tunnelling Problems
Giri Project	Phyllites	440	12	41.3	High squeezing pressures; plastic flow of rocks; floor heaving; Support deformations and failure.
	Phyllites	300	12	29.8	
	Phyllites	300	12	24.1	
	Slates	360	16.5	5.9	
	Siltstone	200	16.5	14.9	
	Siltstone	200	16.5	8.0	
Maneri Bhali-I	Sheared Metabasics	450	11	6.3	Water in-rush, cavity formation and support failure.
	Crushed Quartzities	750	15	4.2	
	Foliated Metabasics	250	33	0.3	
	Quartzites	250	26	0.8	
Maneri Bhali-II	Sheard metabasics	200	9	9.0	Sudden ingress of water, running ground, cavity formation and mild to medium squeezing ground.
	Sheard Metabasics	400	9	2.2	
	Metabasics	425	16	0.3	
	Metabasics	1000	16	0.7	
	Greywacke	400	24	0.6	
Salal Project	Blocky Dolomite	490	100	0.6	No frequent tunnelling problems encountered except a major collapse with water in-rush.
	Highly jointed dolomite	250	80	1.0	
	Highly jointed dolomite	250	80	1.1	
	Highly jointed dolomite	440	80	0.4	

It can be observed from the Fig. 5 that the critical strains are generally above the Warning Level-I and with a few above the Warning Level-III. Thus, it can interpreted that the tunnels analysed are not stable and some corrective measures are required to stabilise them. The tunnels whose critical strains are lying above Warning Level-II have already had severe tunnelling problems and failure has occurred (Table 1), thus validating that the hazard warning can predict the stability of tunnels. The tunnels lying between levels I and III indicate that some measures are necessary to stabilise the tunnel and the tunnel may not have failed and those lying below the level I are safe and stable thus not requiring any safety measures. The numbers 1.3.5.7 and Fig. 5 indicate tunnelling difficulties as classified above.

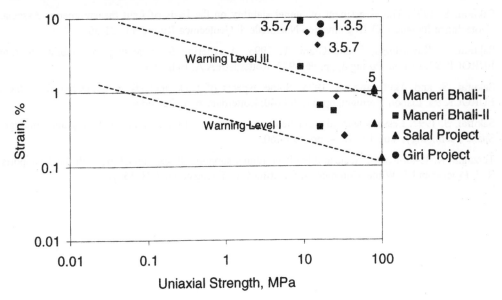

Fig. 5 : Relationship between measured strain (from extensometers) and hazard warning levels

6.0 CONCLUSIONS

With the results of the above study it can be concluded that the hazard warning level method of stability analysis is a very promising means of achieving the rational design of tunnels. According to the proposed technique, the stability of openings can be assessed directly from measured displacements. The hazard warning levels provide an important tool for interpreting the field measurements and they must be determined prior to construction, so that the measured values can be assessed immediately after taking them.

REFERENCES

Bieniawski, Z.T. 1989. Engineering rock mass classification. John Wiley and Sons. Inc.

Dube, A.K. and Singh, B. 1969-1977. Periodic progress reports of geomechanical studies in Giri Hydel Project. Unpublished report of Central Mining Research Institute, Dhanbad.

Dube, A.K., Jethwa, J.L. and Singh, B. 1983. Analysis of instrumentation data of Maneri-Uttarkashi Tunnel, U.P. Unpublished report of Central Mining Research Institute, Dhanbad.

Dube, A.K. et. al. 1989. Assessment of rock mass behaviour in the head race tunnel of Maneri Bhali Project Stage-II, Distt. Uttarkashi, Garhwal, India. Unpublished report of Central Mining Research Institute, Dhanbad.

Dube, A.K.; Goel, R.K.; Swarup, A. and Kumar, P. 1995. Geotechnical instrumentation for monitoring the rock mass behaviour in tail race tunnel-II, Salal Hydroelectric Project, J&K. Unpublished report of Central Mining Research Institute, Dhanbad.

Mahatab, M.A. and Grasso, P. 1992. Geomechanics principles in the design of tunnels and caverns in rocks, Elsevier Science.

Sakurai, S. 1981. Direct strain evaluation technique in construction of underground openings. Proc. 22nd US Symp. Rock. Mech., Cambridge, Massachusetts, M.I.T., pp. 278-282.

Sakurai, S. 1983. Displacement measurements associated with the design of underground openings. Proc. Int, Symp. Field Measurements in Geomechanics, Zurich, Switzerland, pp. 2:1163-1178. Rotterdam:A.A. Balkenia.

Sakurai, S. 1993. The assessment of tunnel stability on the basis of field measurements. Associazone Geotechnica Italiana - XVIII Convegno Nazionale di Geotecnica-Rimini, pp. 21-30.

Sakurai, S. Kawashima, I. and Otani, T. 1995. A criterion for assessing the stability of tunnels, EUROCK'93, Lisbon, Portugal, pp. 969-973. Rotterdam:A.A. Balkema.

Sakurai, S. and Akutagawa, S. 1995. Some aspects of back analysis in geotechnical engineering. EUROCK'93, Lisbon, Portugal, pp. 1133-1140. Rotterdam:A.A. Balkema.

Sakurai, S. 1997. Lessons learned from field measurements in tunnelling. Tunnelling and Underground Space Technology, Vol. 12, No. 4, pp. 453-460.

Terzaghi, K. 1946. Rock defects and loads on tunnel supports. Rock Tunnelling with Steel Supports, ed. R.V. Proctor and T. White, Commercial Shearing Co., Youngstown, OH, 15-99.

ROCK MASS ANALYSIS

FINITE ELEMENT ANALYSIS OF UNDERGROUND CAVERNS OF NATHPA JHAKRI HYDEL PROJECT

M.N. BAGDE

Scientist

Central Mining Research Institute
Regional Center, Shankar Nagar, Nagpur, India

SYNOPSIS

In the present study, analysis of powerhouse caverns was carried out by FEM using rock properties obtained from laboratory testing and data available from field investigations with aim to perform rock support interaction analysis for effective support design. The displacements and stresses were obtained from the analysis. The resultant stress condition in the rock mass surrounding the cavern was obtained by adding the in situ stress to induced stress due to excavation. The principal stresses and factors of safety were then calculated. To check failure in the rock mass, Hoek and Brown failure criterion was used. Displacements, principal stresses and factor of safety were computed for in situ stress ratio of 0.56 and for two different loading conditions. Variations in stresses and displacements values for machine hall obtained considering the proposed loading conditions was found to be negligible, which showed the interaction between the two caverns of an insignificant consideration. The differences in the values of principal stresses and displacements along the boundary and different sections of the machine hall are too small to take into account. Gauss points along the boundaries of the two caverns either had the factor of safety one or less than that showed tension failure meant the requirement of support system for the caverns.

1.0 INTRODUCTION

In developing and undeveloped countries, the great need for effective utilization of water resources had resulted into construction of large dams, underground powerhouses, tunnels and other associated structures during the past few decades. Many multipurpose river valley projects are being planned, designed and constructed for irrigation, power generation and water supply. A river valley project consisted of dams, underground structures and canals. The underground structures are the tunnels for conveyance of water and powerhouse caverns for installation of machinaries for power generation. The knowledge of stresses and displacements in the surrounding rock due to excavation for the existing geological conditions is essential for design of underground structures. Nathpa - Jhakri hydroelectric project located in the middle reaches of Satluj valley (Kinnaur district, Himachal Pradesh, India) is the largest power tunnel project in Asia. It envisaged construction of a 60.5 m high concrete gravity dam on river Satluj at Nathpa, a 27.78 km long and 10.15 m diameter head race tunnel on the left bank of river Satluj and an underground powerhouse at Jhakri with an installed capacity of 1500 MW. The machine hall was 20 m wide, 49 m high and 211 m long, while transformer hall was 18 m wide, 29 m high and 198 m long.

Finite Element Method (FEM) is the most popular present day technique of numerical solutions. This methods simulates the very complex situations *e.g.* non-homogeneous media, non-linear and material behaviour, initial stress conditions, spatial variations in material properties, arbitrary geo-metrics and discontinuities, etc. For this purpose, an already developed computer programme in Fortran was used for the analysis of underground openings under plane strain conditions. The failure criteria proposed by Hoek and

Brown (1980) was adopted in the finite element programme. In the present study, analysis of powerhouse caverns *i.e.*, machine hall and transformer hall were carried out by FEM using rock properties obtained from laboratory testing and data available from field investigations with aimed to perform rock support interaction analysis for effective support design.

2.0 FINITE ELEMENT FORMULATION AND COMPUTER PROGRAM

FEM is being increasingly used to solve many geotechnical engineering problems due to its versatility and the availability of high-speed digital computers. It had been described in detailed by a number of authors Desai and Abel (1972), Anderson et al., (1966), Barla (1970), etc. It is used for foundation analysis and design of underground caverns, rock slopes, embankments, etc.

2.1 Finite Element Method

In FEM, the body is divided into equivalent systems of smaller bodies. Starting with small body, formulation is obtained and by combination of these small bodies, formulation for the entire body is obtained. The process essentially is going form a part to whole. Three different approaches are available to solve any problem when finite element method is employed. These are the stiffness, force and mixed method. Out of these, stiffness method was adopted in the present study. In these approach displacements are primary unknowns while strains and stresses are secondary unknowns.

2.2 Basic Steps in FEM

The various steps adopted for the analysis of a problem using FEM were as to follows.

The FEM consisted of eight basic steps (Desai, 1970).

1. Discretisation of the structure,
2. Selection of approximation function,
3. Definition of strain-displacement and stress-strain relationships,
4. Derivation of element equations,
5. Assembly of element equations and introduction of boundary conditions,
6. Computation of primary (displacements) unknowns,
7. Computation of secondary unknowns (strains and stresses), and
8. Interpretation of results.

2.3 Computer Programme Details

The computer programme in FEM2D available with following features was used for analyzing the underground cavern of Nathpa-Jhakri hydel project using linearly elastic material model.

2.3.1 *Main Programme*

The main segment controls the order and sequence in which subroutines are called to perform specific tasks. The global stiffness matrix was modified by incorporating boundary conditions. This section also prints the displacements and stresses. In total, there were 17 subroutines in the programme.

1. Subroutine INPUT : It reads most of the input data and echoes it. Some of the important control data such as maximum number of elements. Maximum numbers of nodes, number of materials and number of different materials properties are read. It reads nodal connectivities, nodal coordinates, nodal boundary conditions and material properties.

2. Subroutine GAUSSP : This gives the Gaussian point coordinates and the respective weights used for numerical integration.

3. Subroutine SHAP2D : This subroutine computes the shape functions and their local derivatives for the 8-noded elements.

4. Subroutine DERV2D : The global coordinates, Jacobian matrix, determinant of Jacobian matrix and global derivatives of shape functions are calculated at the Gaussian points of the elements in this subroutine.

5. Subroutine INSTR : This subroutine computes *in situ* stresses. It has provision for uniform and gravity stresses.

6. Subroutine MESHCHG : The data for mesh and material changes are read in this subroutine. The numbers of elements and/or nodes, which are to be added/deleted, are read.

7. Subroutine PRESUR : The equivalent nodal load vector due to surface pressure is computed. This section reads the number of sides with surface pressure, type of pressure and magnitude of pressure at nodes.

8. Subroutine EXCAVT : This computes equivalent nodal loads due to excavation of the elements. This subroutine calls subroutine MESCHG and PLOAD.

9. Subroutine BMAT2D : It generates strain-displacement transformation matrix (B).

10. Subroutine DMAT2D : The elasticity matrix (D) is generated in this subroutine.

11. Subroutine DBMAT2D : The matrix [D] and [B] are multiplied in this subroutine.

12. Subroutine STIF : This calculates the element stiffness matrix.

13. Subroutine ASEMBL : This assembles element stiffness matrices to obtain global stiffness matrix.

14. Subroutine SOLVE : This forward elimination of stiffness matrix is carried out using Gauss-elimination approach.

15. Subroutine RESOLV : The displacements are computed for nodal load vector by using back substitutions.

16. Subroutine STRESS : It computes the strains and stresses at the Gauss points.

17. Subroutine PRINC : It computes principal stresses and their direction at Gauss points.

3.0 GEOLOGY

3.1 Regional Geology

The rock type present in the project area comprised variety of metamorphic rocks like augen gneiss, granite gneiss, biotite schist, quartz mica schist, chlorite muscotive schist with amphibolites (basic intrusives) and pegmatite (acid intrusives). Quartz veins are also present in abundance. These rocks belonged to Jeori-Wangtu Gneissic complex of Pre-Cambrian Age (1000 million years) and form the basement of Rampur group of rocks lying at higher stratigraphic position. The rocks of the project area are openly folded. The axial regions of the folds are fractured. A number of anticlines and synclines had been identified in the project area.

3.2 Geology at Power House Site

At the underground powerhouse site at Jhakri, the rocks expected are essentially quartz mica schist and biotite schist with thin bands of sericite schist. The general strike of schistosity was almost across the flow of river with dips at steep angles towards upstream *i.e.*, 20°-60° in N20° W to N10° E direction. There were many minor folds and drag folds, which plunge towards $N30^0$ W direction by about $30°$. The most prominent sets of joints recorded in the powerhouse area are the followings (Ashraf *et.al.*, 1991).

Joint, strike	$N10^0$ E - $S10^0$ E, dip 75^0 $N80^0$ E
Joint, strike	$N70^0$ W - $S70^0$ E, dip 72^0 $S20^0$ W
Joint, strike	E - W, dip 35^0 N
Joint, strike	$N25^0$ W - $N25^0$ E, dip 42^0 $N65^0$ E
Joint, strike	$N44^0$ E - $S44^0$ W, dip 52^0 $N46^0$ W
Joint, strike	$N55^0$ E - $S55^0$ E, dip 50^0 - 70^0 in $N30^0$ E
Joint, strike	$N70^0$ W - $S70^0$ E, dip 72^0 in $S20^0$ W

3.3 Rock Mass Classification

On the basis of the detailed engineering geological observations made in the exploratory drift, rock mass was classified by NGI's Q system proposed by Barton *et al.*, (Ashraf *et al.*, 1991). the Barton's approach by virtue of a variety of factors connected with the stability of underground powerhouses appeared to be considerably useful in connection with the proposed powerhouse at Jhakri. The rock mass description at the powerhouse drift and rating for each at the rock units are summarized in Table 1.

TABLE 1
Rock quality designation (RQD) and tunnelling quality index (Q) of the rock mass in Jhakri power house area (Ashraf *et al.*, 1991)

	Tunnelling Media	RQD	Q
1.	Quartz mica schist - moderately jointed to massive	80-100	9.77-25
2.	Quartz mica schist - closely jointed to thinly foliated	55-70	1.08-5.5
3.	Biotite schist - thinly foliated	0-21	0.24-0.75
4.	Sericite schist - completely sheared and crushed rock	0-5	0.01

3.4 Material Properties used for Analysis

The rock mass at the location of powerhouse caverns comprised mostly quartz mica schist. Intact rock samples were tested in laboratory and modulus of elasticity and Poisson's ratio were 8300 MPa and 0.24 respectively. The rock mass at the site was classified using Q-system. The Q varied between 1.08 and 5.50 (Table 1). These Q values were converted at Rock Mass Rating (RMR) using the following expression (Bieniawski, 1976).

RMR = 9 in Q + 44

Two values of RMR calculated were 45 and 60. From the values of RMR, the deformation modulus of rock mass was computed as,

E = 2 RMR - 100 (RMR > 50) (Bieniawski,1978)

$E = 10^{(RMR-10/40)}$ (RMR < 50) (Serafim and Pereira, 1983)

Substituting, RMR = 45 and 60, one obtained E_m = 7.5 and 20 GPa respectively. An average value of 12.5 GPa was adopted in the analysis.

The material properties used in the analysis were as follows :

Young's modulus	= 12,500 MPa
Poisson's ratio	= 0.25
Unit weight	= 0.02697 MN/m^3

In situ stress ratio, K_0 = 0.56

Overburden depth   = 300 m (above the crown of proposed machine hall cavern)

Empirical constants of Hoek and Brown criterion for rock mass m = 0.014 and s = 0.00049

4.0 ANALYSIS AND RESULTS

Two powerhouse caverns to be excavated at Jhakri are machine hall and transformer halls. The rock at the location of powerhouse caverns was quartz mica schist and rock cover above the crown of the proposed machine hall cavern was 300 m. Finite element analysis of the two caverns had been carried out and a result obtained from analysis were discussed. The variation of stresses and displacements along the two boundaries and along different sections in the surrounding rock were also presented.

4.1 Machine Hall and Transformer Hall

Fig. 1 showed the location and sections of two caverns. Both the caverns are D shaped. Dimensions of the caverns were as follows :

Machine hall :   Length = 211 m, width = 20 m and height = 49 m.

Transformer hall :   Length = 198 m, width = 18 m and height = 29 m.

The excavation programme was to excavate first the machine hall and then the transformer hall.

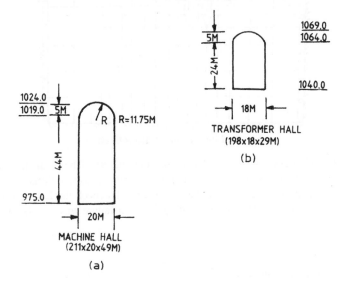

Fig. 1 : Boundaries of the two powerhouse caverns

4.2 Loading Conditions

The excavation of the cavern was simulated by applying equal and opposite forces as calculated from the *in situ* stress condition on the face of the excavation boundary. The loading conditions adopted were designated as follows :

  (i)   Case I : Excavation of the machine hall cavern,

 (ii)   Case II : Excavation of the transformer hall after the excavation of machine hall.

4.3   Analysis

The discretisation scheme used for the analysis with eight-noded iso-parametric elements is showed in Fig. 2. In all 620 elements and 1943 nodes were used. Two sampling points in each direction were adopted for Gauss - Legendre quadrature in calculating the stiffness matrix and load vector of the elements. Plain strain condition was assumed to prevail. Analysis were carried out with simulation of excavation for both the cases. In simulation of excavation, equivalent nodal load due to *in situ* stresses was calculated along the excavation boundary and applied in opposite direction to had stress free boundary.

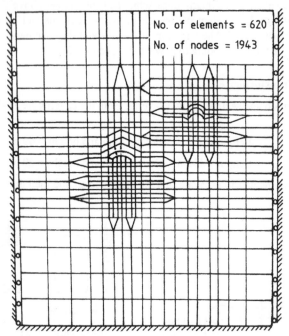

Fig. 2 : Finite element discretisation

The displacements and stresses were calculated from the analysis. The resultant stress condition in the rock mass surrounding the cavern was obtained by adding the *in situ* stress to induced stress due to excavation. The principal stresses and factors of safety were then calculated. To check the failure in the rock mass, Hoek and Brown failure criterion was used. The factor of safety for shear failure was calculated as,

Factor of safety = σ_{1f} / σ_1

where,

$\sigma_{1f} = \sigma_3 + (m\sigma_c\sigma_3 + s\sigma_c^2)^{1/2}$.

σ_1 and σ_3 = major and minor principal stresses.

σ_c = compressive strength of the rock sample.

m and s = empirical constants of Hoek and Brown criterion for rock mass.

A check for tension failure was also made in the rock mass.

5.0   RESULTS

Fig. 3 showed the eight sections AA' to HH' across the cavern boundaries, along which the dispalcements and principal stresses had been computed and plotted.

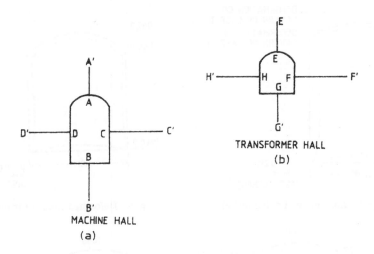

Fig. 3 : Sections across the cavern boundaries

5.1 Deformed Shapes of Caverns

Figs. 4 and 5 showed the deformed shapes of the machine hall and the transformer hall respectively. For machine hall cavern, both the loading conditions (Case I and Case II) were considered and for transformer hall, only Case II was considered. The horizontal and vertical displacement values in the caverns at typical nodes (Fig. 6) were given in Table 2. In machine hall cavern, the maximum horizontal displacements of left and right walls were 1.54 cm and 1.65 cm respectively for Case I. For Case II, the right wall deformation reduced to 1.57 cm. Vertical crown displacement of 1.25 cm for Case I increased to 1.4 cm for Case II. Displacements in left wall and in invert portion remained the same for both the cases. The maximum horizontal displacement of right wall of transformer hall was 0.82 cm at node 1272. Vertical displacement of its crown was 1.2 cm.

TABLE 2
Displacement of the caverns in cms

	Machine hall cavern				Transformer hall cavern		
Node No.	Case I		Case II		Node No.	Case III	
	X	Y	X	Y		X	Y
486	-0.06	1.46	-0.034	1.48	1166	-0.23	-0.32
482	0.32	0.48	0.34	0.48	1163	-0.11	-0.043
690	1.5	-0.12	1.56	-0.16	1269	-0.51	-0.48
927	0.67	-0.65	0.72	-0.7	1455	-0.21	-0.87
931	-0.047	-1.25	-0.38	-1.3	1458	-0.06	-1.2
935	-0.78	-0.64	-0.67	-0.64	1461	-0.38	-0.76
698	-1.65	-0.11	-1.57	-0.065	1272	-0.82	-0.3
490	-0.44	0.49	-0.4	0.53	1169	0.38	0.17

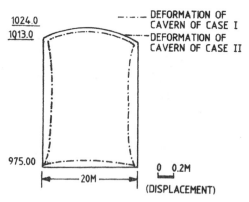

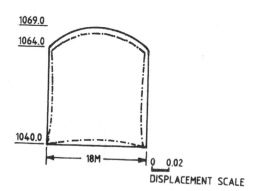

Fig. 4 : Deformation shape of machine hall

Fig. 5 : Deformed shape of transformer hall

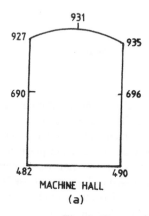

MACHINE HALL
(a)

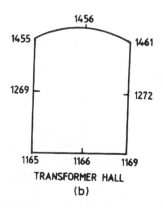

TRANSFORMER HALL
(b)

Fig. 6 : Caverns boundaries showing typical nodes

5.2 Displacement along Different Sections

For the machine hall horizontal displacements found to be reduced substantially away from the cavern, along sections BB', CC', DD' and along AA', and it showed decreasing-increasing trend. The vertical displacement also found to be reduced away from the cavern along AA' and BB' sections and along CC' and DD' sections. The displacement showed increasing-decreasing-increasing trend, for both the cases. For the transformer hall, the horizontal dispalcements decreased along sections FF' and HH' but it showed increasing-decreasing trend along GG' and EE' sections. The vertical displacement reduced along EE' and GG' away from the cavern and showed increasing-decreasing trend along section FF'. Along section HH', vertical displacement showed decreasing-increasing-decreasing trend away from the cavern.

5.3 Principal Stresses

Figures 7 and 8 showed the variation of principal stresses along the boundaries of the machine hall and transformer hall caverns respectively. In machine hall, maximum major and minor principal stress occurred at Gauss points near the corners of the invert, the values of which are 24.5 and 14.5 MPa respectively. The minor principal stresses in the two walls of the opening were too small to be plotted. In transformer hall also, the maximum major and minor principal stresses occurred at Gauss points near the corners of the invert are 18.5 and 3.5 MPa respectively. In the machine hall for all sections, except CC', the values of major and minor principal stresses remained same for both the cases (Case I and caseII) and along CC', the principal stresses showed a small change after the excavation of transformer hall (Case II). Hence both the cases were considered for section CC'. The variation of major and minor principal stresses showed similar trend along

54

sections FF' and HH'. For both the sections, the maximum value of major principal stress was 10 MPa, which occurred at 6 m away from the cavern. Minor principal stress increased rapidly up to nearly 45 m away from the cavern, where its value was 4.0 MPa for section FF' and 4.7 MPa for section HH'. Maximum major principal stress value of 5.0 MPa occurred at nearly 20 m away from cavern.

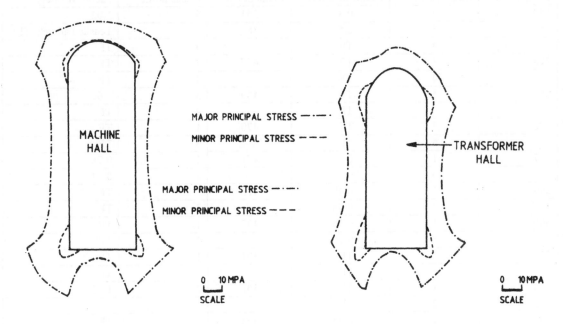

Fig. 7 : Principal stress along the boundary of machine hall

Fig. 8 : Principal stresses along boundary of transformer hall

5.4 Factor of Safety

Table 3 showed values of factor of safety at typical Gauss points along the boundaries of the caverns (Fig. 9). For machine hall, at each Gauss point, the factor of safety was either less than one or it showed tensile failure. Some results are observed in transformer hall, barring few Gauss points, which showed factor of safety slightly greater than one.

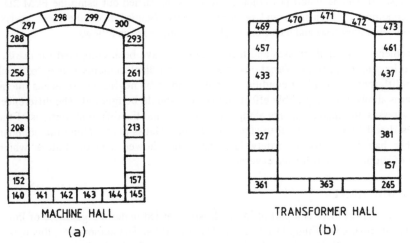

MACHINE HALL
(a)

TRANSFORMER HALL
(b)

Fig. 9 : Cavern boundaries showing typical Gauss points

TABLE 3
Factor of safety at typical Gauss points along the cavern boundaries

Machine hall			Transformer hall		
Element No.	Gauss point	Factor of safety	Element No.	Gauss point	Factor of safety
140	4	0.8	361	4	1.02
152	3	0.55	387	3	Tensile Failure (*TF)
	4	TF		4	TF
208	3	TF	457	3	TF
	4	TF		4	TF
256	3	TF	469	3	1.66
	4	TF		4	TF
288	3	TF	470	1	TF
	4	TF		3	TF
297	1	0.77	471	1	0.78
	3	0.60		3	0.74
299	1	0.48	472	1	TF
	3	0.48		3	TF
300	1	0.56	437	1	TF
	3	0.70		2	TF
261	1	TF	391	1	TF
	2	TF		2	TF
213	1	TF	362	2	1.12
	2	TF			
157	1	0.55	363	2	TF
	2	TF		4	TF
145	2	0.78			
143	2	TF			
	4	TF			

* TF - Tensile Failure

6.0 CONCLUSIONS

The analysis of Nathpa-Jhakri powerhouse caverns were carried out using the FEM 2D program under plain strain conditions. Conclusions and recommendations obtained from the FEM analysis using intact properties from laboratory tests and data from field investigations are as follows.

Displacements, principal stresses and factor of safety have been computed for *in situ* stress ratio of 0.56 and two loading conditions. Variations in stresses and displacements values for machine hall were obtained for different two loading conditions were found to be negligible. It means interaction between the two caverns showed the negligible effect and since it should be ignored. The differences in the values of principal stresses and displacements along the boundary and the different sections of the machine hall for the Cases II and I were too small to take into account. Gauss points along the boundaries of the two caverns either had the factor of safety less than one showed tension failure which showed the requirement of support system for the caverns.

ACKNOWLEDGEMENT

The study presented there is a part of M. Tech work carried under the guidance of Prof. A. Vardarajan and Prof. K.G. Sharma, Civil Engg. Deptt., IIT in Delhi in 1994. The author takes this opportunity to thank them for their supervision and guidance. The author is also sincerely thankful to the head of the institution for his kind permission to publish the paper. The views expressed herein are of the author and not necessarily of the institution he belongs to.

REFERENCES

Anderson, H.W. and Dodd, J.S. (1966), Finite element method applied to rock mechanics, in Proc. Ist Congress, ISRM, Vol. II, 317-322.

Ashraf, Z. and Chowdhary, A.K. (1991), Comprehensive geotechnical report on the studies carried out for Nathpa-Jhakri Hydel Project, Shimla and Kinnuar district, Himachal Pradesh, Report. GSI, Part-I of II.

Barla, G. (1970), Stress distribution around underground opening in engineering applications, in Proc. 2nd Congress of ISRM, Vol. IV.

Bieniawski, Z.T. (1974), Geomechanics classification of rock masses and its application in tunneling, in Proc. 3rd Int. Congress of ISRM, 27-32.

Bieniawski, Z.T. (1984), Rock mechanics design in mining and tunnelling, Balkema publications.

Bieniawski, Z.T. (1989), Engineering rock mass classification, Wiley, New York.

Desai, C.S. and Abel, J.F. (1972), Introduction to the finite element method, Van Nostrand Reinhold Publication, 477p.

Hoek, E. and Brown, E.T. (1980), Underground excavations in rock, Institution of Mining and Metallurgy, London.

EQUIVALENT MATERIAL MODELLING TO PREDICT THE STABILITY OF UNDERGROUND OPENINGS

A.H. GHAZVINIAN
Head, Rock Mechanics Group

Deptt. of Mining Engineering
Tarbiat Modaress University, Tehran, Iran

K.K. GUPTA
Associate Professor

T. RAMAMURTHY
Emeritus Professor

Deptt. of Civil Engineering
Indian Institute of Technology, New Delhi, India

SYNOPSIS

Estimation of the state of stability is the prime objective in the designing of underground openings. Equivalent material modelling technique has the special advantage of providing a physical feel of the problem. This technique has been used for studying the deformational behaviour and roof support pressure of underground openings. Several equivalent material models were constructed and tested to represent depths of 71, 142 and 284 m from the ground surface, a sedimentary formation in India.

These models were instrumented by LVDT's to measure the deformation around rectangular openings. Roof support pressures were measured with the help of miniature hydraulic jacks. Immediate roof failure, subsequent recurrent failures and total roof collapse were observed by widening the width and height of openings for various overburdens.

Experimental results showed that deformation is dependent on the width of opening; height of opening does not effect significantly. The height of overburden above the opening has significant influence on the development of roof pressure for deformation condition. From the analysis of the results, stable, recurrent failure and total collapse zones have been defined based upon roof deformation, width of opening, compressive/bending strength of the formation and in-situ stress. It is possible to predict in advance, the limits for the type of collapse of rock mass in the roof of an underground opening. Equations in terms of non-dimensional parameters have been developed for the prediction of the roof deformation, the pressure distribution on the roof span, the average roof pressure and the variation of roof pressure with roof deformation in the stratified formation. The findings and the suggested zones of stability are helpful in understanding the ground response and in monitoring the underground excavation.

1.0 INTRODUCTION

For the design of underground openings, estimation of the roof deformations and the roof pressures are of the prime importance. There are various approaches for their evaluations, viz., analytical methods, numerical methods, empirical methods and equivalent materials modelling technique. Amongst these methods equivalent material modelling is more advantageous to get a physical feel of both the qualitative and quantitative understanding of the problem. Therefore, this approach has been preferred in the present work.

Amongst the various parameters controlling the stability of underground openings, so far the effect of the size of the opening under various heights of overburden above the opening has not been attempted. Further, the roof pressure distribution and also the effect of size of opening on the roof pressure for various heights of overburden above the opening has also not been studied to enable designing of an optimum support system.

Keeping the above facts in view, a systematic study was planned; (i) to develop an understanding of the effect of width and height of rectangular openings on the stability in terms of deformations of its surrounding rock mass in stratified formation under various overburdens, (ii) to study the effect of the width of rectangular openings on the intensity and distribution of roof pressures development under various overburdens.

2.0 REVIEW OF LITERATURE

The evaluation of displacements around an underground opening and estimation of pressures are important steps in the design of the opening and predicting its stability. The magnitudes of the redistributed stresses after the excavation of the opening are governed by a variety of factors such as shape and size of the opening, geological conditions, mechanical properties of the overburden, stress state, the methods of excavation and the period during which the opening is left unsupported. There are various approaches available for the analysis of rock mass behaviour around the opening in the literature. These approaches could be classified as : analytical or closed form solutions, numerical methods, empirical methods and equivalent material modelling.

2.1 Analytical Method

An analytical approach requires the properties of the material, the state of stresses involved and simplification of the mathematics and physics of the problem by making of the regular assumptions. These methods can further be grouped as; (a) elastic solutions, and (b) elasto-plastic solutions. Considering elastic solutions, closed form analysis for different tunnel shapes, e.g., circular, elliptical, rectangular, and rectangular with rounded corners, in an infinite mass have been discussed by Savin (1961), Obert and Duvall (1967), Poulos and Davis (1973), Jaeger and Cook (1976) and Goodman (1980). In the case of lined circular tunnels, the problem obtaining stresses and displacements have been considered by Burns and Richard (1964) and Hoeg (1968). The results of Burns and Richard have been summarized by Krizek and Key (1971) in the form of dimensionless plots of deflections, thrust, moment and shear forces.

The problem of stress distribution and displacements around a circular tunnel has been solved by Pender (1980). For elasto-plastic cases, closed form solutions are difficult to obtain for different shapes and are confined to circular openings only. The early theories, adopted the Mohr-Coulomb yield criterion for rock mass. Solution for stresses around circular tunnels are presented by Obert and Duvall (1967) assuming plane strain condition. Bray (1967) used Coulomb-Navier criterion, and Goodman (1980) adopted Mohr-Coulomb criterion to evaluate radial and tangential stresses and radial displacement around circular openings. Brown et al. (1983) used the non-linear Hoek and Brown (1980) criterion of yielding. They gave a closed form solution for the elastic-brittle-plastic ground and suggested a step wise numerical solution for elastic-strain softening plastic ground. They divided the post peak strains into two regions and related them by experimental parameters or by the associated flow rule.

Sharma (1985) suggested modification in the Brown et al. (1983) approach by incorporating integration within the thin annular rings around the tunnels. A parametric study of various factors, viz., peak strength, residual strength, modulus of elasticity, rate of strain softening and dilation characteristics of rock mass affecting ground convergence, radius of broken zone and stresses distribution was carried out in order to establish the relative importance of these factors in the design of tunnel. It is seen that the modified procedure is more efficient as the iteration cycle converges faster and the results are closer to those obtained from closed form solutions.

2.2 Numerical Methods

Numerical methods are subdivided into : finite difference method, finite element method (FEM), finite element method with infinite elements, boundary element method and coupled finite element and boundary element methods.

FEM is more commonly used amongst the available methods. In this method, many complex features could be simulated in the analysis, e.g., shape, geological conditions, support system and interaction of opening with support system and also other factors which affect the displacements and stresses around the openings. Even with the rapid advancement of this method, still, some limitations are dominant, *e.g.*, (a) it is very difficult to establish a realistic constitutive (stress-strain) law which means it is not possible to simulate behaviours of all joints or discontinuities and approximations are involved in representing the material behaviour, (b) evaluation of stresses at the corner of the opening is very difficult and does not yield good results, and (c) actual failure and its mode could not be predicted.

2.3 Empirical Methods

Rock mass classification systems developed by Barton, Lien and Lunde (1974), Bieniawski (1974), Deere (1964), Lauffer (1958), Terzaghi (1946), etc., relate the experience encountered at previous projects to the conditions prevailing at a proposed site.

3.0 EQUIVALENT MATERIAL MODELLING

In equivalent material modelling, the proto system or the original ground formation is simulated in the laboratory. The original rock formation are replaced by artificial materials. This approach has been used in many countries namely Australia, East and West Germany, Hungary, India, Poland, UK, USA, USSR and at other places.

In India extensive work has been done in this field at Central Mining Research Station (Dhanbad), since 1964. This research centre is having wide range of findings for mining thick coal seams (Singh and Singh 1976, 1977, 1978, 1980), mining under waterlogged areas (Singh and Singh 1973), caving characteristics of roof rock mass under different conditions (Singh and Singh 1979), and others.

Many materials have been developed and used by various investigators. These are plaster, paraffin, vaseline, portland cement, gypsum, lime, etc. sand, mica, chalk, talc, clay, etc., have been used as filler (Singh, 1981).

To evolve an appropriate design approach, the initial step is to study the prototype, isolate its constituents parameters and then select parameters which are relevant to the problem.

4.0 EXPERIMENTAL WORK

The experimental work comprises developing suitable equivalent material, design of equivalent formation and thereafter, constructing and testing of several models for the study of roof deformation behaviour and roof pressure development.

To perform the studies, the following equipment and instruments were designed and fabricated; Ghazvinian (1990).

(i)   apparatus to determine the strength of weak equivalent material specimens by line load test and bending strength test (tensile and bending strength of as low as 0.37 kg/cm^2, and least count of 0.01 kg/cm^2),

(ii)   miniature hydraulic jacks along with oil-water interface chamber,

(iii)   surcharge pressure loading system,

(iv)   frame (size : 280 × 150 × 20 cm) and model construction accessories,

(v)   split moulds of 2 × 2 × 5 and 2 × 2 × 12 cm for preparation of prismatic specimens.

To select the appropriate equivalent materials, as many as 36 mixes of plaster of paris, sand and mica powder were investigated.

Around 700 prismatic specimens prepared from these mixes were tested for compressive strength, tensile strength using indirect methods. The results of these tests provided a wide range of equivalent materials.

Equivalent formation was designed for simulating sedimentary rock mass. Geometrical dimensions of the proto formations were scaled down to 1 : 50. The same scale was considered in using Bukingham – pi dimensional relationships to evaluate the equivalent strengths.

A preliminary model of $200 \times 20 \times 20$ cm size was tested for beam test and cantilever tests to estimate a weakening coefficient. This weakening coefficient was used to estimate the mass strength from the specimen strengths.

Hence equivalent strength of each formation of the mass was calculated as follows :

$$s_m = \alpha_L \cdot K \cdot \gamma_m / \gamma_p \cdot s_p \qquad \qquad \dots (1)$$

where

s_m =   the model material strength,

α_L =   geometrical scale = 1 : 50,

K =   the weakening coefficient = 0.5 (obtained from preliminary model test),

γ_m =   the density of model material,

γ_p =   the density of proto material, and

s_p =   the proto material strength.

Suffix m and p are used for denoting the model and the proto system, respectively.

By finding a suitable matching of the equivalent materials for the desired strengths of the formations with due consideration of close proximity of compressive strength and tensile strength, the equivalent formations were designed. Only some portion of the overlying stratas (27 m) were simulated in these models. To bring the proposed opening under the influence of desired height of overburden the required surcharge pressures were applied on the top of the formation as described later.

A total number of eight similar models of 280 cm long, 20 cm wide and 92 cm high were constructed and tested for the two distinct set of experiments to measure deformations or pressures.

1. First set was for roof deformational behaviour, 5 models were tested for heights of overburden above the opening corresponding to proto overburden (H_p) of 71, 142 and 284 m. The surcharge pressure was applied by means of a long rubber bag encased in a stiff aluminium channel casing. The bag was filled with water under the desirable pressure and connected to a self compensating mercury pot system to maintain constant pressure. In this phase of experiments, the effect of width and height of a rectangular opening on roof deformations were studied for various heights of overburden. The roof deformations were studied for various heights of overburden. The roof deformations were measured during the various stages of excavation of the opening by LVDTs placed in rows/columns (Fig. 1) above the openings and recorded by a programmable data logger. The width of the openings were increased at the rate of 4 cm per 35 minutes cycle (*i.e.* corresponding to 2 m per 4 hours cycle in proto) and heightenings were carried out at the rate of 2 cm per 35 minutes cycle. Immediate roof failure, subsequent recurrent failures and total roof mass collapse were observed for increasing widths of opening and different heights of overburden. In one of the models the effect of height of opening was also studied.

2. Second set of experiments were for roof pressure investigations for which 3 models were tested for the heights of overburden above the opening corresponding to 71, 142 and 284 m. In each model a rectangular opening was made with the fixed height of the opening of 10 cm and the width was increased. With the increase of the width of the opening, the roof span was supported for nil roof deformation (within the experimental limitations), by specially designed and fabricated miniature hydraulic jacks (Ghazvinian, 1990). The pressures exerted from these jacks to the roof of the opening were considered as a measure of the roof pressure at various locations for various increasing widths.

5.0 RESULTS AND ANALYSIS

5.1 Roof Deformational Behaviour

1. The width of the opening affects the stability significantly in terms of roof deformation, *i.e.*, deformation increases with the increase of the width of opening, (Fig. 2). The height of the opening, (at least for unsupported height to width ratios of the opening from 0.125 to 2.15), does not affect the roof deformation (Fig. 3). It was also observed that the height of overburden above the opening has significant influence on the magnitude of roof deformation. The roof deformation increases with the increase of height of overburden above the opening for the same width (Fig. 4).

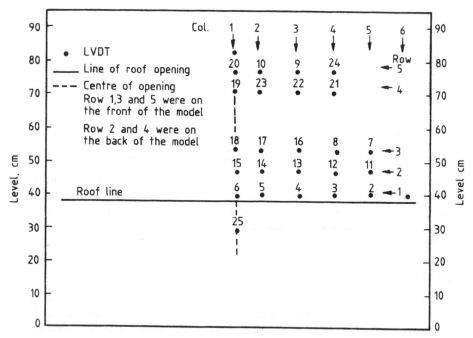

Fig. 1 : Typical configuration of deformation measuring points (LVDTs)

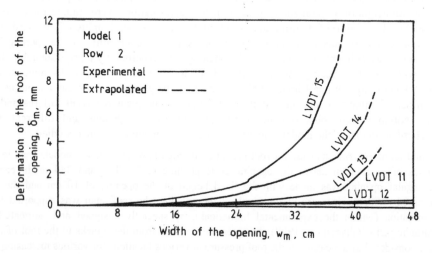

Fig. 2 : Variation of roof deformation, δ_m, at measuring points with the width of the opening w_m

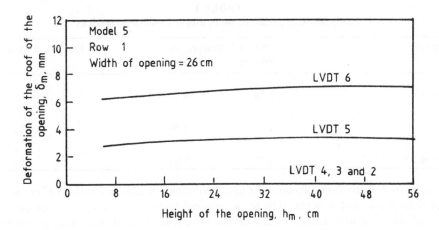

Fig. 3 : Variation of roof deformation, δ_m, at measuring points with the height of opening

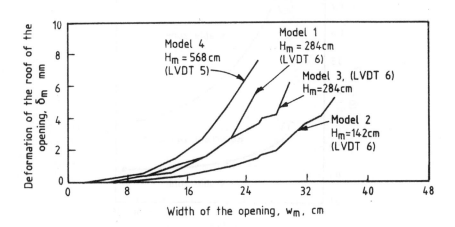

Fig. 4 : Comparison of deformation recorded by LVDTs positioned 2 cm above the roof at middle of the opening

2. The width of the opening, w_p, the corresponding deformation, δ_p, and the in-situ stress, P_{0p} (subscript p denotes proto conditions), are related through the Eqns. 2, 3 and 4 given below :

$$\delta_p = A_p . e\, (B_p . w_p) \qquad\qquad \ldots (2)$$

where

$$A_p = 0.19 \times 10^{-2} + 0.96 \times 10^{-4}\, P_{0p} \qquad\qquad \ldots (3)$$

and

$$B_p = 0.23 + 0.12 \times 10^{-2}\, P_{0p} \qquad\qquad \ldots (4)$$

The roof deformations were predicted for the case records of three tunnelling projects available in the literature, and predicted deformation (using Eqns. 2 to 4), compared well with the observed deformations (Table 1).

TABLE 1
Comparison of predicted and observed roof deformations

| Name of tunnel | Roof deformation, cm | | Percentage variation |
	Predicted	Observed	
H. Dwight, D. Eisenhower	1.58	1.3	21.5
Tunnel Giri Hydel	4.67	7.56	38.2
Tunnel Chhibro-Khodri	2.43	3.96	38.6

3. Stable, recurrent fall and total collapse zones have been clearly demarcated in terms of ratios $(\delta/w)^{1/2}$ (δ is the roof deformation, and w is the width of the opening) and $(\sigma_c/P_0)^{1/2}$ (Fig. 5), to enable to predict in advance the nature of expected failure of rock mass overlying the roof of an underground opening. The critical values of δ/w could be obtained from the model test and utilized in the field during excavation to control the stability of the roof mass. A similar zoning has also been proposed on the basis of the bending strength of the immediate roof material. This zoning will be more useful for stratified formations.

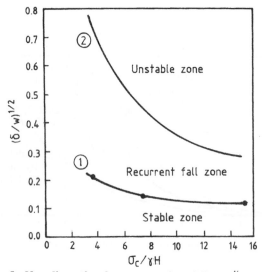

Fig. 5 : Non-dimensional representation of $(\delta / w)^{1/2}$ with $\sigma_c / \gamma H$

5.2 Roof Pressure Development Studies

1. The width of the opening and the height of overburden above it have significant influence on the development of roof pressure (p) for nil roof deformation conditions. The maximum roof pressure is developed at the mid span and it gradually decreases towards the side walls of the openings, resulting in a near bell shaped roof pressure distribution (Figs. 6 and 7).

2. The pressure distribution over the roof span could be predicted by using a general polynomial equation of fourth order as :

$$y = a_0 - a_1x + a_2x^2 - a_3x^3 + a_4x^4 \qquad \qquad ... (5)$$

where,

y = Non-dimensional parameter, $p/p_0 \cdot w_{max} / w$ (w_{max} is the final width of the opening. In tunnelling case where the openings are excavated to a single size then $w_{max} = w$, hence $w_{max} / w = 1$).

x = Non-dimensional parameter, D (w/2), where D is the distance from the middle of the opening, and a_0, a_1, a_2, a_3, a_4 = Coefficients of fourth degree polynomial.

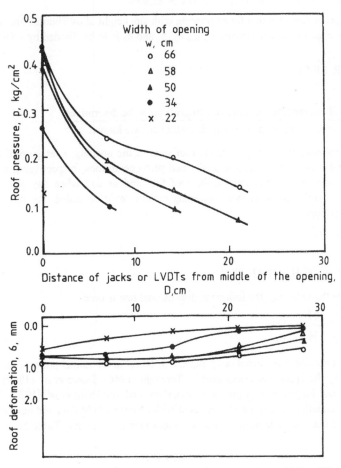

Fig. 6 : Variation of roof pressure and roof deformation from the middle of the opening

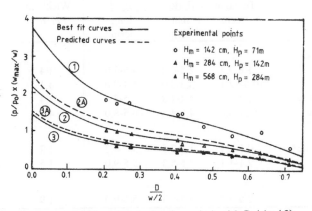

Fig. 7 : Variation of (p / p_o) x (w_{max} / w) with D / (w / 2)

65

The coefficients a_0, a_1, a_2, a_3, a_4 of a reference polynomial are calculated for $H_p = 71$ m. This reference polynomial in the present investigation was developed as;

$$y = 3.723 - 15.17x + 40.43\,x^2 - 52.07x^3 + 22.89x^4 \qquad \ldots (6)$$

To find out the predicting equation for an opening with any height of overburden above the opening (at least up to 300 m, tested), each of the coefficients a_0, a_1, a_2, a_3 and a_4 is to be divided by a factor, F, as per Eqn. 7, given below :

$$F = 0.5R + 0.53 \qquad \ldots (7)$$

where

$$R = \frac{\text{Height of overburden for which coefficients are to be determined}}{\text{Height of overburden for which coefficients are known}}$$

3. To have a control over the most part of the roof span of the opening, average roof pressure, p_{av} (further, subscripts m and p are used to denote model and proto conditions, respectively), may be required to be assessed in advance in order to control the roof fall, and as and when any sign of greater distress is observed in the middle region further supporting system could be added. For this purpose the following equations are developed.

$$p_{mav} = Z \cdot w_m \qquad \ldots (8)$$

where,

$$Z = 2.35 \times 10^{-6}\,H_m + 2.33 \times 10^{-3} \qquad \ldots (9)$$

Now, to evaluate the value p_{pav} the following dimensional law is used :

$$\frac{p_{mav}}{p_{pav}} = \frac{\gamma_m \cdot H_m}{\gamma_p \cdot H_p} = \frac{p_{om}}{p_{op}} \qquad \ldots (10)$$

The average roof pressures predicted using Eqns. 8 to 10 for seven tunnel project cases are within the range with those proposed by the approaches suggested by Terzaghi (1946), Deere et al. (1969), Barton et al. (1974, 1975) for short-term and long-term roof pressure estimation, and Wickham et al. (1972). The predictions made through the proposed equations are in good agreement with observed field data and have shown better matching as compared to the predictions made from the classifications mentioned above (Table 2).

TABLE 2
Comparison of predicted and observed roof pressures on tunnels

Name of tunnel Rock mass description	Predicted roof pressure in kg/cm²			Wickham et. al	Proposed emperical equation	Field observed roof pressure kg/cm²
	Terzaghi Deere	Barton short term	et. al long term			
Chhibro-Khodari Tunnel						
crushed red shale+	1.8-3.4	2.5-4.2	4.6-7.0	1.4	3.3	3.1
crushed red shale*	10.3-22.1	3.3-5.0	5.4-9.0	4.1	9.5	10.8
soft and plastic black clay+	1.5-3.3	4.4-4.8	6.5-8.0	1.4	3.2	3.2
soft and plastic black clay*	5.2-10.0	4.0-4.8	6.5-8.0	4.1	9.6	11.5
Giri Hydel Tunnel						
very blocky and seamy slate*	0.7-2.3	1.8-2.5	3.5-4.2	1.2	4.8	2.0
crushed phyllites*	2.1-4.1	3.5-5.3	5.5-9.0	1.5	3.7	2.0
Loktak Hydel Tunnel						
crushed shale*	2.9-5.4	3.5-5.3	5.6-9.0	1.9	5.4	5.4

* Highly squeezing   + Moderately squeezing

The amount of roof deformation that could occur for an unsupported opening under certain conditions and the intensity of the support pressure (for the same opening) for nil roof deformation could be estimated using the Eqns. 11 to 14.

For $\qquad$ p_{av} / p_0

$$\delta/w = M_1\, e^{(N_1 \cdot P_{av} / P_0)} \qquad\qquad \dots (11)$$

where

$$M_1 = \text{a constant} = 0.187 \times 10^{-2}, \text{ and}$$

$$N_1 = -1.235 + 0.063 H_p \; (H_p \text{ in metres}) \qquad\qquad \dots (12)$$

For $\qquad$ p_{max} / p_0

$$\delta / w = M_2\, e^{(N_2 \cdot P_{max} / P_0)} \qquad\qquad \dots (13)$$

where

$$M_2 = \text{a constant} = 0.155 \times 10^{-2}, \text{ and}$$

$$N_2 = -0.47 + 0.23 H_p \; (H_p \text{ in metres}) \qquad\qquad \dots (14)$$

The above equations are developed by correlating the non-dimensional ratio of δ/w (where opening is unsupported) with p_{av} / p_0 and p_{max} / p_0 (when opening is supported) separately, (p_{max} = the roof pressure at the mid span). These equations could also be used to estimate the optimum support requirement. The roof span could be allowed to undergo some known or desired deformation, or the support be designed for the unreleased roof pressure.

6.0 CONCLUSIONS

Important facts regarding roof mass behaviour were deduced from the experiments. Stable, recurrent failures and total collapse zones have been clearly demarcated to enable to predict, in advance, the nature of expected failure of rock mass in the roof of an underground opening. Several equations in terms of non-dimensional parameters have been developed for the prediction of the roof deformation, the roof pressure distribution on the roof span, the average roof pressure and the variation of roof pressure with deformation in the stratified formation. For establishing the precise stability conditions of a specific case, the coefficients of these equations could be obtained by testing 2, or preferably 3, models simulating the given formation.

Comparison is made between the predicted and the actual values of the roof deformation and the roof pressure from the available case histories, and found to be in good agreement. This shows that the approaches developed could be used for predicting the stability conditions in complicated geological settings in advance.

The findings are helpful in understanding the ground response above an excavated underground opening. The relationships developed and the suggested zones of stability are of significant practical utility to the practicing rock engineers.

REFERENCES

Barton, N., Lien, R. and Lunde, J. (1974). Engineering Classification of Rock Masses for the Design of Tunnel Support. Rock Mechanics, Vol. 6, No. 4, pp. 189-236.

Barton, N., Lien, R. and Lunde, J. (1975). Estimation of Support Requirements for Underground Excavations. Proc. 16th Symposium on Rock Mechanics, University of Minnesota, Minneapolis, U.S.A., pp. 163-177.

Bieniawski, Z.T. (1974). Geomechanics Classification of Rock Masses and its Applications in Tunnelling. Proc. 3rd Int. Cong. Rock Mech., Vol. 1, pp. 27-32.

Bray, J.W., (1967). A Study of Jointed and Fractured Rock Theory of Limiting Equilibrium. J. Rock Mech. and Engg. Geol., Vol. 5, pp. 197-216.

Brown, E.T., Bray, J.W., Ladanyi, B. and Hoek, E. (1983). Ground Response Curves for Rock Tunnels. J. Geotech. Engineering Division, ASCE, Vol. 109, No. 1, pp. 15-39.

Burns, J.Q. and Richard, R.M. (1964). Attenuation of Stresses for Buried Cylinders. Proc. Symp. On Soil-Structure Interaction, Univ. of Arizona, Tempe, pp. 278-392.

Deere, D.U. (1964). Technical Description of Rock Cores for Engineering Purposes. Rock Mechanics and Engineering Geology, Vol. 1, No. 1, pp. 17-22.

Deere, D.U., Peck, R.B., Monsees, J.W. and Schmidt, B. (1969). Design of Tunnel Liner and Support System. University of Illinois, Prepared for the Office of High Speed Ground Transportation, U.S. Department of Transport, Contract No. 3-0152.

Ghazvinian, A., (1990). Prediction of Stability of Underground Openings by Equivalent Material Modelling. Ph.D. Thesis, IIT Delhi, New Delhi, India, pp. 229.

Goodman, R.E. (1980). Introduction to Rock Mechanics. John Wiley and Sons, New York.

Hoeg, K. (1968). Stresses Against Underground Structural Cylinders. J. of Soil Mech. and Found. Engg. Div., Proc. ASCE 94, SM4, pp. 833-858.

Hoek, E. and Brown, E.T. (1980). Underground Excavations in Rock. Institution of Mining and Metall., London.

Jaeger, J.C. and Cook, N.G.W. (1976). Fundamental of Rock Mechanics. Chapman and Hall, London.

Krizek, R.J. and Kay, J.N. (1971). Material Properties affecting Soil Structure Interaction of Underground Openings. Soil Structure Interaction Symposium, Highway Research Board No. 413, pp. 833-858.

Lauffer, H., (1958). Gebirgsklassifiziering fur den stollenbau. Geologie and Bauwesen, Vol. 24, No. 1, pp. 46-51.

Obert, T., Duvall, W.I. (1967). Rock Mechanics and the Design of Structures in Rock. John Wiley and Sons, Inc., New York.

Pender, M.J. (1980). Elastic Solutions for a Deep Circular Tunnel. Geotechnique, Vol. 30, No. 2, pp. 216-222.

Poulos, H.G. and Davis, E.H. (1973). Elastic Solutions for Soil and Rock Mechanics. John Wiley and Sons, Inc., New York.

Savin, G.N. (1961). Stress Concentrations around Holes. Translated from Russian by E. Gros, p. 430, Pergamon, Oxford.

Sharma, V.M. (1985). Prediction of Closure and Rock Loads for Tunnels in Squeezing Grounds. Ph.D. Thesis, Deptt. of Civil Engineering, Indian Institute of Technology, New Delhi, India.

Singh. T.N. (1981). Laboratory Model Simulation of Mine Strata Deformation. M.S. Thesis, University of Newcastle Upon Tyne, New Castle Upon Tyne, England.

Singh, T.N. and Singh, B. (1973). Feasibility of Mining 7m Thick XI/XII Seam under High Flood Level of Damodar River at Sudamih Mine. C.M.R.S. Report, Dhanbad, India.

Singh, T.N. and Singh, B. (1976). Ground Movement due to Mining of a Thick Seam. Third International Symposium on Mine Surveys, Leoben, Austria.

Singh, T.N. and Singh, B. (1977). Model Studies of Strata Movement around a Longwall Face. J. Mines, Met. Fuels, Feb. pp. 39-44.

Singh, T.N. and Singh, B. (1978). Extraction of Thick Coal Seam, Field and Model Investigation. Eleventh Common Wealth Mining and Metallurgy Congress, Hongkong.

Singh T.N. and Singh B. (1979). Ground Movement Associated with Thick Steep Seam Mining. Fourth International Symp. For Mine Surveys, Aachen, West Germany.

Singh, T.N. and Singh, R.D. (1980). The Feasibility of Horizontal and Inclined Slicing of a Thick Coal Seam in Conjunction with Hydraulic Stowing. Symposium on Cut and Fill Mining, Lulea, Sweden.

Terzaghi, K. (1946). Rock Defects and Loads on Tunnel Support. Rock Tunnelling with Steel Supports, eds. R.V. Proctor and T. White, Commercial Shearing Co., Youngstown, Ohio, pp. 15-99.

Wickham, G.E., Tiedemann, H.R. and Skinner, E.H. (1972). Determination Based on Geologic Predictions. Proc. Rapid Excavations and Tunnelling Conference, AIME, New York, pp. 43-64.

EFFECT OF DEPTH ON SUPPORT PRESSURES AND CLOSURES IN TUNNELS

R.K. GOEL
Central Mining Research Institute
Regional Centre, CBRI
Roorkee, India

J.L. JETHWA
Central Mining Research Institute
Regional Centre, Shankar Nagar
Nagpur, India

SYNOPSIS

Tunnel depth or the overburden is an important parameter for planning and designing the tunnels. The quantitative effect of tunnel depth on support pressures and closures has been presented in the paper. For the study, the empirical correlations developed by Goel (1994) have been used.

The study shows that the depth has considerable effect on support pressures and closures in squeezing ground conditions. Under non-squeezing ground conditions, on the other hand, less or practically there is no effect of overburden on support pressures. It has also been highlighted that the depth effect on support pressure is more in comparison to tunnel closure and it increases with decrease in rock mass quality.

1.0 INTRODUCTION

It is known that the *in situ* stresses are influenced by the depth below the ground surface. It is also learned from the theory that the support pressure and the closure for tunnels are influenced by the *in situ* stresses. Therefore, it is recognized that the depth of tunnel or the overburden is an important parameter while planning and designing the tunnels.

Most of the earlier empirical approaches for predicting support pressures do not consider the tunnel depth. Barton *et al.* (1974) have considered tunnel depth in terms of parameter SRF. Experiences in Indian tunnels have shown that SRF did not take care the stress conditions adequately specially under squeezing ground conditions. Goel (1994) has developed empirical correlations for predicting support pressures and closures for tunnels. In addition to other parameters, tunnel depth has also been used in these correlations. Effect of tunnel depth on support pressure and closure, therefore, has been studied by using these correlations and presented in the paper.

For better appraisal, initially the correlations for estimating support pressures and closures are presented in brief in the following paragraphs.

2.0 PREDICTION OF SUPPORT PRESSURE

For developing the correlations, the support pressures were actually measured in the tunnels at twenty-five test sections covering both the non-squeezing and the squeezing ground conditions and tunnel width from 2 m to 14 m. Complete data files on all 25 tunnel sections are presented in Goel *et al.* (1995). Out of these, 15 tunnel sections were from the non-squeezing ground conditions. Since the rock mass response varies with the ground condition, as evident from the ground reaction curve concept, it was considered necessary to develop different correlations for changed ground conditions.

Considering tunnel depth, tunnel radius, tunnel closure (to take into account the support stiffness in case of squeezing conditions), and rock mass number N, defined as stress-free Barton's Q (SRF=1), following set of correlations has been developed for predicting ultimate support pressure.

For non-squeezing conditions

$$p_{el} = (0.12 \cdot H^{0.1} \cdot a^{0.1} / N^{0.33}) - 0.038 \text{ MPa} \qquad \dots (1)$$

For squeezing conditions

$$p_{sq} = (f/30) [10^{(H^{0.6} \cdot a^{0.1}/50 \cdot N^{0.33})}] \text{ MPa} \qquad \dots (2)$$

where,

p_{el}	=	ultimate support pressure in non-squeezing ground in MPa,
p_{sq}	=	ultimate support pressure in squeezing ground in MPa,
f	=	correction factor for tunnel closure (see Table 1),
H	=	tunnel depth in metres,
a	=	tunnel radius in metres,
N	=	rock mass number given as
		$(RQD \cdot J_r \cdot J_w) / (J_n \cdot J_a)$

RQD, J_r, J_w, J_n, and J_a are the parameters defined by Barton *et al.* (1974) for estimating rock mass quality Q.

Detailed procedure of developing Eqn. 2 is discussed by Goel *et al.* (1995). Estimated support pressures from Eqns. 1 and 2 for tunnel sections under non-squeezing and squeezing ground conditions have been compared with the measured values and a correlation coefficient of 0.95 and 0.97 respectively is obtained.

TABLE 1
Correction factor f for tunnel closure in Eqn. 2 (Goel *et al.*, 1995)

Sl. No.	Degree of squeezing	Normalized tunnel closure, %	Correction factor, f
1.	Very mild squeezing $(270 N^{0.33} \cdot B^{-0.1} < H < 360 N^{0.33} \cdot B^{-0.1})$	1 - 2	1.5
2.	Mild squeezing $(360 N^{0.33} \cdot B^{-0.1} < H < 450 N^{0.33} \cdot B^{-0.1})$	2 - 3	1.2
3.	Mild to moderate squeezing $(450 N^{0.33} \cdot B^{-0.1} < H < 540 N^{0.33} \cdot B^{-0.1})$	3 - 4	1.0
4.	Moderate squeezing $(540 N^{0.33} \cdot B^{-0.1} < H < 630 N^{0.33} \cdot B^{-0.1})$	4 - 5	0.8
5.	High squeezing $(630 N^{0.33} \cdot B^{-0.1} < H < 800 N^{0.33} \cdot B^{-0.1})$	5 - 7	1.1
6.	Very high squeezing $(800 N^{0.33} \cdot B^{-0.1} < H)$	>7	1.7

3.0 PREDICTION OF TUNNEL CLOSURE

In addition to the influence on the excavation size, the tunnel closure also effect the design of the support systems specially under squeezing ground conditions where high tunnel closures are not uncommon. For example - Giri - Bata Tunnel of India where closure as high as 20 per cent of tunnel size has been measured.

At sixty locations the tunnel closures were measured using the tape extensometer with an accuracy of up to $\pm$ 0.1 mm and least count 0.01 mm. Out of these, at 23 tunnel sections squeezing ground conditions were encountered. The complete data file including the other parameters is presented in Goel (1994). The other parameters selected for developing the correlations of estimating tunnel closures are - rock mass number N, tunnel depth, tunnel radius and stiffness of the support system.

Using the values of all the four selected parameters and the measured tunnel closures from 37 tunnel sections under non-squeezing and 23 tunnel sections under squeezing ground conditions, a regression analysis was performed and the following set of empirical correlations were developed for estimating the radial tunnel closure :

For non-squeezing ground condition

$$u_{el} \; = \; (a \cdot H^{0.61}) / (28 \; N^{0.4} \; K^{0.35}) \; \text{cm} \qquad\qquad \text{... (3)}$$

For squeezing ground condition

$$u_{sq} \; = \; (1 / 10.5) \; (a^{1.12} \; H^{0.8}) / (N^{0.27} \; K^{0.62}) \; \text{cm} \qquad\qquad \text{... (4)}$$

where,

u_{el} = estimated radial tunnel closure in cm under non-squeezing ground condition,

u_{sq} = estimated radial tunnel closure in cm under squeezing ground condition,

K = effective support stiffness in MPa,

H, a and N have already been defined in Eqns. 1 and 2.

The estimated tunnel closures from Eqns. 3 and 4 have been compared with the measured values. The comparison shows a good agreement with correlation coefficients of 0.91 and 0.94 for non-squeezing and squeezing ground conditions respectively.

4.0 EFFECT OF TUNNEL DEPTH

Firstly, the effect of tunnel depth on support pressures and then on tunnel closures is discussed. The discussion is presented based on the ground conditions.

During the parametric study, to ensure that either non-squeezing or squeezing ground condition is maintained, the correlations suggested by Goel et al. (1995a) and given in Table 1 have been used.

4.1 Depth Effect on Support Pressures

4.1.1 Non-Squeezing Ground Condition

Eqn. 1 shows that the effect of tunnel depth on support pressure in non-squeezing ground conditions is moderate. e.g. for tunnel radius as 6 m and support stiffness as 10 MPa the rise in support pressure on increasing tunnel depth from 100 m to 700 m in good rock masses represented by rock mass number N as 40 will be approximately 50 per cent (0.029 MPa to 0.044 MPa). This rise under non-squeezing condition may be considered moderate since the actual support pressure values are small.

4.1.2 Squeezing Ground Condition

Eqn. 2 provides ultimate tunnel support pressure in squeezing ground conditions. This equation shows that the tunnel depth has a significant effect on the support pressures. In fact, the effect of tunnel depth on support pressure increases as the rock mass deteriorates. For the study, Fig. 1 has been drawn using Eqn. 2 for three types of rock masses called as moderately good, fair and poor represented by N values of 10, 1 and 0.5 respectively. It can be seen in Fig. 1b that in case of fair rock masses (N = 1), the support pressure increases 4 times (400 per cent) when the tunnel depth is increased from 300 m to 700 m, whereas for poor rock masses (N = 0.5) as shown in Fig. 1c, the support pressure increases by 4.5 times (450 per cent) with the same increase of tunnel depth (300 to 700 m). This effect of tunnel depth is for 1.5m tunnel radius. The effect of depth for tunnels with a radius of 6m is higher as can be seen in Figs. 1a, 1b and 1c. This is perhaps because the confinement decreases and the degree of freedom for the movement of rock blocks increases in larger tunnels.

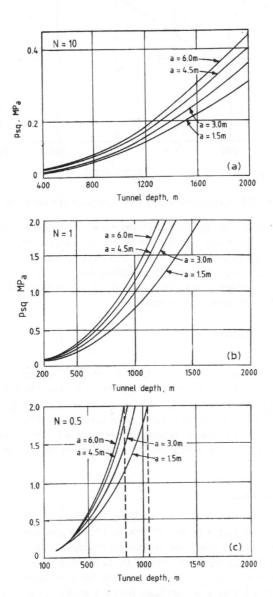

Fig. 1 : **Variation of predicted support pressure p_{sq} with tunnel depth for (a) N = 10, (b) N = 1 and (c) N = 0.5 (after Goel, 1994)**

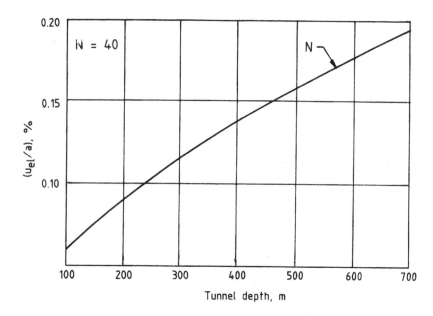

Fig. 2 : Variation of normalised tunnel closure under non-squeezing ground condition (u_{el}) with tunnel depth

For fair and poor rock masses in Figs. 1b and 1c, the maximum value of support pressure on Y - axis has been taken as 2 MPa, a value for which supports can be designed without much of difficulties. It is inferred that the support pressure would be greater than 2 MPa and the supporting costs would be very high in poor rock masses when the depth exceeds 800 m for a 12 m wide tunnel and 1100 m for a 3 m wide tunnel (Fig. 1c). These situations, thus call for realigning of a tunnel through a better tunnelling media or a lower tunnel depth or both in order to reduce the anticipated support pressure.

4.2 Depth Effect on Tunnel Closure

4.2.1 Non-squeezing Ground Condition

For the study, Fig. 2 has been plotted using Eqn. 3. In Fig. 2, normalized tunnel closure, i.e., tunnel closure expressed as percentage of tunnel size has been taken on Y-axis. The figure has been drawn for good rock mass (N = 40) keeping support stiffness as 10 MPa. It can be seen in Fig. 2 that the normalised tunnel closure increases from 0.118 per cent to 0.198 per cent on increasing tunnel depth from 300 to 700 m. This rise of about 70 per cent of normalised closure is insignificant since the actual value is small. Therefore, it is implied that tunnel depth has less or practically no effect on tunnel closure in non-squeezing ground conditions.

4.2.2 Squeezing Ground Condition

Using Eqn. 4, Fig. 3 has been plotted for studying the effect of tunnel depth on tunnel closure in squeezing ground condition keeping support stiffness as 10 MPa and tunnel radius as 6 m. Fig. 3 shows plots for moderately good, fair and poor rock masses represented by the values of N as 10, 1 and 0.5 respectively. It can be seen in Fig. 3 that on increasing the tunnel depth from 300 m to 700 m the normalised tunnel closure increases from 1.55 per cent to 3.06 per cent for moderately good rock masses. This rise of about 100 per cent in normalised closure is significant from the point of view of support design. Further, it has been found out that in poor rock masses also the normalised tunnel closure increases by 100 per cent on increasing tunnel depth from 300 to 700 m.

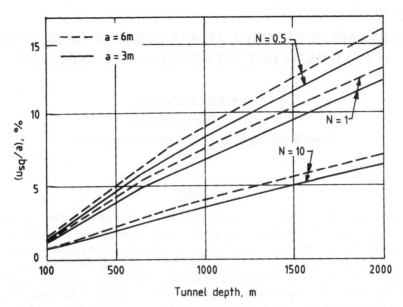

Fig. 3 : Variation of normalised tunnel closure under squeezing ground condition (u_{sq}) with tunnel depth

5.0 CONCLUSIONS

From the above study it can be concluded that

1. The tunnel depth has significant effect on support pressures and tunnel closures in squeezing ground conditions and practically no effect under non-squeezing ground conditions.
2. The effect of tunnel depth on support pressure is more in comparison to tunnel closure.
3. The depth effect on support pressure increases with deterioration in rock mass quality probably because the confinement decreases and the degree of freedom for the movement of rock blocks increases in larger tunnels.
4. The study would help the planners and designers to take decisions on realigning a tunnel through a better tunnelling media or through lesser overburden or both in order to reduce the anticipated support pressure and closure in tunnels.

ACKNOWLEDGEMENTS

Authors are thankful to the Director, Central Mining Research Institute for kindly permitting to publish the paper. The views expressed in the paper are those of the authors and not necessarily of the Institute to which they belong.

REFERENCES

Barton, N., Lien, R. and Lunde, J. (1974), Engineering classification of rock masses for the designs of tunnel supports, *Rock Mechanics*, **6**, pp. 189-236.

Goel, R.K. (1994). Correlations for predicting support pressures and closures in tunnels, *Ph.D. Thesis*, Nagpur University, India, p. 308.

Goel, R.K., Jethwa, J.L. and Paithankar, A.G. (1995), Indian experiences with Q and RMR systems. *Tunnelling and Underground Space Technology,* 10 : 1, pp. 97-109.

Goel, R.K., Jethwa, J.L. and Paithankar, A.G. (1995a), An empirical approach for predicting ground condition for tunnelling and its practical benefits, *Proc. 35th U.S. Sym. on Rock Mechanics*, USA, pp. 431-435.

PROBLEMS IN NUMERICAL ASSESSMENT OF SURFACE SETTLEMENT DUE TO SHALLOW TUNNELLING

PRABHAT KUMAR

Structural Engineering Division

Central Building Research Institute, Roorkee, India

SYNOPSIS

This paper shows that the analytical response of free surface to a shallow underground excavation depends upon the shape of initial stress field, which is employed in the analysis. The choice lies between gravity and uniform loads. The linearly varying gravity loading leads to an unrealistic finding. The outstanding problems associated with the research on shallow tunnelling are identified. It is anticipated that substantial tunnelling work will be needed in India in near future. An indigenous technology development will ensure full utilization of the available resources. Such an endeavour requires close co-operation between the research institutions and sponsoring / funding agencies.

1.0 INTRODUCTION

Removing some material from the underground medium creates underground structures. These are attractive because of less thermal variation, natural camouflage and automatic seismic disaster mitigation. Besides the space which is spared on the ground can be used for environment regeneration. Therefore, the underground space is ideally suited for defense applications, storage of essential commodities and rapid mass transport in large metropolitan cities. The hydroelectric power generation projects are usually located in hilly regions at higher altitude. These provide electricity, water for drinking and irrigation and help in flood routing. Most underground constructions usually suffer from surprises, which are discovered at an advance stage of construction. These unscheduled stoppages cause delay in the project completion and result in huge cost over runs.

In modern times, because of availability of large and fast computers and suitable numerical methods such as the finite element method (FEM), it is possible to exercise more control on the underground construction through analytical modelling. This provides an estimate of the medium response to the underground excavation. Yet there are several problems which have not been addressed by the existing research. Some such problems associated with the shallow tunnelling are described in this paper.

2.0 SHALLOW TUNNELS

The tunnels in which the depth of tunnel center (H) from the free surface is less than five times the tunnel radius (R) may be classified as shallow. The effect of tunnelling on the existing state of stress in the underground medium is confined to within H/R <= 5. As a consequence, an analytical study of shallow tunnel requires an explicit modelling of the free surface. In a real situation, the free surface contains the existing structures, which must not be damaged on account of shallow tunnelling. This implies that the surface settlements and the vibrations due to blasting and / or drilling must remain within the prescribed limits. In addition to safety of the existing structures, the services located at shallow depths like, water pipes, sewer lines and telephone and electricity cables must not be disturbed to avoid public inconvenience.

3.0 EXPERIENCE WITH SURFACE SETTLEMENT COMPUTATION

The free surface settles due to ground loss. In the analysis of tunnels, the excavation surface is to be made free of all shear and radial stresses, which come into action due to the initial stress field of the underground medium. The vertical initial stress is termed as gravity, which is contributed by the self-weight of the material. Its magnitude is proportional to the distance from the free surface. When the underground opening is deep or its vertical dimension is very small, the linearly variation of the vertical initial stress may be ignored and it may be approximated as a uniform field. It is of interest to know what difference the shape of the vertical stress field makes in the analytical derivation of the surface settlement. The following problem and its solution answer this question.

3.1 Geometry

Depth of tunnel center from free surface	$H = 750$ cm.
Radius of opening	$R = 200$ cm.
Unit weight of material	$\gamma = 2400$ kg $/ m^3$

3.2 Medium Properties

(a) Isotropic	Modulus of elasticity	$E = 34500$ kg/cm^2
	Poisson ratio	$\nu = 0.20$
	Shear modulus	$G = 0.417\, E$
(b) Anisotropy 1	$E_h = 0.6E$	$E_v = E$
	$\nu_h = 0.125$	$\nu_v = 0.375$
	$G_{vh} = 1.0\, E$	
(c) Anisotropy 2	$E_h = 0.6E$	$E_v = E$
	$\nu_h = 0.125$	$\nu_v = 0.375$
	$G_{vh} = 0.05\, E$	

The subscript h and v apply to the horizontal and the vertical directions, respectively. The material model b and c differ only in the shear modulus. A different analysis is performed for each of the above material model.

3.3 Loading

Initial (horizontal / vertical) initial stress ration = 1.5

Shape (a)   Linear variation with a zero value at the free surface

Shape (b)   Uniform variation corresponding to the magnitude at tunnel center

3.4 Numerical Model

The finite-infinite element mesh of the problem is shown in Fig. 1. It contains 226 nodes, 58 finite elements and 12 infinite elements. By virtue of symmetry only half of the problem is analyzed in a plane strain formulation.

3.5 Numerical Results

Figs. 2a and 2b show the results of present analysis. Figure 2c shows the numerical and experimental results obtained by Lee and Rowe (1989).

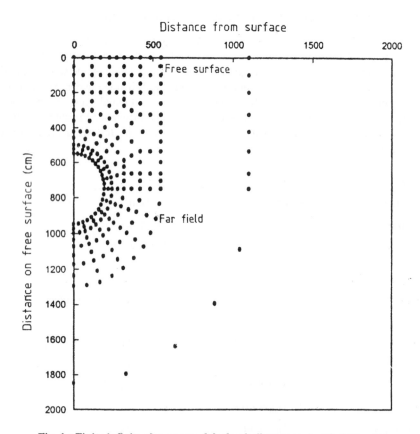

Fig. 1 : Finite infinite element model of a shallow tunnel with H/R = 3.5

4.0 DISCUSSION

Fig. 3a shows the results of a parametric study by Oteo and Sagaseta (1982) which employs truncation method to deal with the unbounded analysis domain and applies gravity loading. The bottom artificial boundary is initially placed very close to the tunnel bottom and then it is gradually moved away. Accordingly, the free surface deformation pattern changes from a settlement to heave. This feature is present in Fig. 2a because the infinite element of this study places the bottom boundary of truncation method at infinity. The surface heave may be unrealistic but it verifies the present study.

Fig. 3b is taken from a study by Lee and Rowe (1989), in which a uniform initial stress field is employed. It shows that the linear FEM analysis results underestimate the surface deformation. An anisotropic material model with a reduced shear modulus is more satisfactory. This conclusion is supported by Fig. 2b, which is obtained in the present study.

This study with adequate verification against the published results demonstrates the indigenous technology of underground excavation. However, the question about the variable or a uniform load to be used in the shallow tunnel analysis still remains open.

If there are no structures on the ground, then, in all possibilities, the tunnel will be constructed by a cut and cover procedure. If the free surface happens to be heavily built up, then, the weight of the existing structures is likely to be sufficient to suppress the linear variation of the gravity loading. Finally, the settlement associated with the tunnel construction may cause the arching action in the ground so that the load transfer due to gravity may not develop.

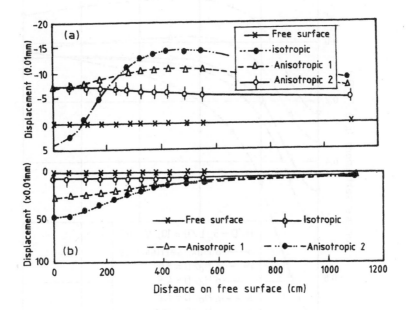

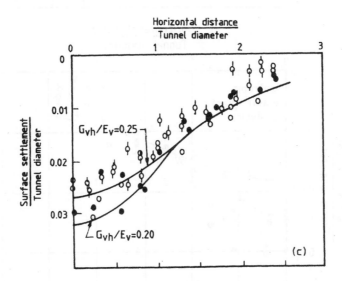

Fig. 2 : Results of analysis of a shallow tunnel with H/R = 3.5

     a. Under variable vertical initial stress field
     b. Under uniform vertical initial stress field
     c. Numerical and experimental results of Lee and Rowe (1989)

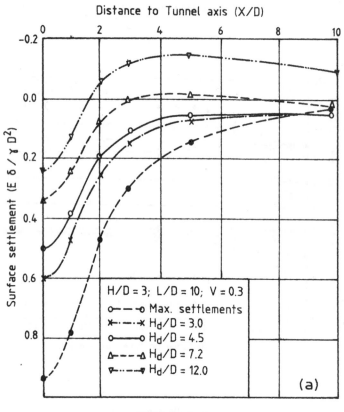

Distance to Tunnel axis (X/D)

Surface settlement (E δ / γ D²)

H/D = 3; L/D = 10; V = 0.3
o— —o Max. settlements
✕—·—✕ H_d/D = 3.0
o——o H_d/D = 4.5
△— — —△ H_d/D = 7.2
▽—··—▽ H_d/D = 12.0

(a)

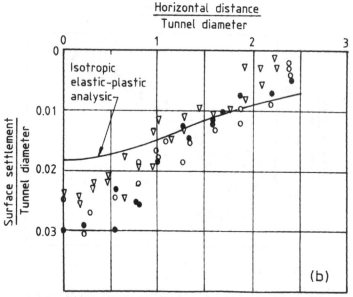

$$\frac{\text{Horizontal distance}}{\text{Tunnel diameter}}$$

Isotropic
elastic-plastic
analysic

$$\frac{\text{Surface settlement}}{\text{Tunnel diameter}}$$

(b)

Fig. 3 : Some published results on the shallow tunnels

    a. By Oteo and Sagaseta (1982)
    b. By Lee and Rowe (1989)

5.0 OUTSTANDING RESEARCH PROBLEMS

It is possible to extend the existing technology of modelling and analysis of tunnels according to the following.

(a) Medium response is very sensitive to the anisotropic characterization of the medium. An efficient and reliable method of anisotropic material parameter estimation is needed.

(b) Several new empirical methodologies to estimate the surface settlement are available. These have to be verified against the field data of Indian tunnels.

(c) New tunnelling method, earth pressure balanced shield tunnelling has been developed to ensure that the surface settlement remain within the prescribed limits. It is possible to include the earth pressure balancing action in the analysis.

(d) The problem of vibration transmission due to drilling and blasting associated with shallow tunnelling still require substantial effort. It is particularly relevant in the Indian conditions as most of the structures are designed without regard to seismic or tunnelling vibrations.

6.0 CONCLUSION

Although sufficient progress has been made in the indigenous know-how development on the modelling and analysis of underground excavation, there are several challenging problems, which are yet to be addressed in the present research. These outstanding problems are relevant in the national context. For a speedy progress it is necessary that the research institutions and the sponsoring agencies come together and cooperate.

ACKNOWLEDGEMENT

This paper is published with the permission of the Director of the Central Building Research Institute, Roorkee.

REFERENCES

Lee, K. M. and Rowe, R. K. (1989). Deformation caused by surface loading and tunnelling: Role of elastic anisotropy. Geo-technique, Vol. 39 (1), pp. 125-140.

Oteo, C.S. and Sagaseta, C. (1982). Prediction of settlement due to underground openings. Proceedings of International Symposium on Numerical Models in Geo-mechanics, Editor R. Dungar *et.al.*, pp. 653-659.

CHECKING ACCURACY OF NUMERICAL MODELS

RAJBAL SINGH

Central Soil and Materials
Research Stations, New Delhi, India

RAJINDER BHASIN   **AXCEL MAKURAT**

NGI, Norway

SYNOPSIS

The formulation of finite element method (FEM) and coupled finite element and boundary element method (FEBEM) used for the analysis of tunnel has been discussed briefly in this paper. The application of universal distinct element code (UDEC) is shown for tunnel problem. Further, the displacement and stresses predicted by coupled FEBEM and FEM analysis have been compared with analytical solution to check the accuracy of results. These results have further been compared with UDEC and UDEC with boundary elements. From this study, it is concluded that FEM and UDEC can be used for the analysis of tunnel excavation with little compromise on the accuracy. However, boundary element boundary may be used with both FEM and UDEC to increase the accuracy in the analysis.

1.0 INTRODUCTION

Many hydroelectric projects are being planned, designed and constructed all over the world. It is particularly so in the Himalayan region of Indian subcontinent. An important component of these projects is underground opening, be it for power generation or water conveyance from one place to the other. Many of these openings often pass through complex geological formations having different initial *in-situ* conditions. These complexities make it necessary to understand the influence of the geological conditions and the resulting states of the stresses on the ultimate behaviour of the openings.

The determination of the distribution of stresses and displacements around underground openings is one of the most important steps in the design of the openings. The magnitude of the secondary stresses developed after the excavation of opening is governed by a variety of factors such as the size of the cavity, the method of its excavation, the *in-situ* stress field, material properties and the length of period during which the opening is left unsupported. The different approaches for the analysis of induced stresses and displacements around underground openings of regular shapes are available in the literature. The application and feasibility of appropriate analysis approach can be judged by keeping in view all the methods.

In this paper, the formulation of FEM and coupled FEBEM used for the analysis of tunnel has been discussed briefly. The application of UDEC is shown for tunnel problem. Further, the displacement and stresses predicted by coupled FEBEM and FEM analysis have been compared with analytical solution to check the accuracy of results. These results have further been compared with UDEC and UDEC with boundary elements.

2.0 METHODS FOR ANALYSIS

The methods for the analysis of underground openings can be divided as :

- Analytical methods or closed form solutions
- Experimental methods.
- Numerical methods.

Till recently, the tunnelling engineers had an access to the analytical or closed form solutions (CFS) for the analysis and design of the openings. The analytical solutions are based on linear-elastic theory for homogeneous and isotropic geological media. For very limited cases, elasto-plastic solutions are also available. As such, these are useful only for preliminary calculations and for testing the finite element programme. Further, these solutions are available for regular shapes of openings such as circular, square and elliptical. Experimental methods are used to make equivalent material models for limited cases. However, for the analysis of underground openings of different shapes with geological media of the type commonly encountered in practice, these are inadequate and one has to resort to numerical methods.

The numerical methods can further be divided into the following categories :

- Finite element method (FEM).
- Finite element method with infinite elements.
- Boundary element method (BEM)
- Coupled finite element and boundary element method (FEBEM).
- Universal distinct element code (UDEC).

With the advent of digital computers, the numerical methods such as finite element method (FEM) and boundary element method (BEM) are increasingly used for the analysis of underground openings. FEM can be used for the analysis of any shape of underground opening incorporating practically any type of material behaviour of geological media and any complex boundary condition. In this method, the fixation of external boundary to represent infinite domain as encountered in the case of underground openings introduces approximation in the results. Also, a large domain is to be discretized and the preparation of data is tedious.

The use of BEM requires only the boundary of excavation to be descretized. BEM can be used for openings of any geometrical shape, but with limited zones of different geological media. This method is more accurate and less tedious than FEM in analysing a particular case of underground opening. In the case of opening where the material properties are likely to vary significantly near the face of the excavation, it is difficult to adopt BEM. For such cases, finite elements can be used near the face of excavations to take care of the complexities in material properties and boundary elements can be used away from the opening to take care of the infinite conditions of the media. This could be particularly useful in the case of underground excavation with boundary at infinity. The coupled finite element and boundary element method (FEBEM) is being used extensively for finding out accurate solutions.

The coupling of FEM and BEM was first proposed by Zienkiewicz *et al.* (1977) and Kelly *et al.* (1979). The application of FEBEM to linear elastic analysis of underground openings was discussed by Brady and Wassying (1981), Beer and Meek (1981), Varadarajan and Singh (1982) and Varadarajan *et al.* (1983, 1985). Beer (1983) applied FEBEM to visco-plastic analysis of mine pillar problem by using Drucker-Prager yield criterion. Singh *et al.* (1985) and Vardarajan *et al.* (1987) applied the FEBEM to elasto-plastic analysis of underground opening. Singh (1985) applied the method for elastic and elasto-plastic analysis of circular opening in layered rock media.

3.0 FINITE ELEMENT METHOD

The six basic steps involved in FEM (Desai and Abel, 1972) are as follows :

Step I   :   Discretization.

Step II   :   Selection of displacement model.

Step III :   Formulation of element stiffness matrices [Ke].

$$[K_e] = [B]^T [D] [B] \quad\quad\quad \text{... (1)}$$

where, [B] is the matrix, which relates the strain and displacement and [D] is the matrix of material
parameters.

Step IV   :   Assmbly of element stiffness matrices and introduction of boundary conditions.

$$[K] \{u\} = \{Q\} \quad\quad\quad \text{... (2)}$$

where, [K] is the total stiffness matrix, {u} is the displacement vector and {Q} is the load vector.

Step V :   Solution of unknown displacements.

Step VI :   Computation of elements strains and stresses from the nodal displacements.

The formulation of element stiffness matrix in step III requires material parameters as input data in finite element analysis. The accuracy of results of any analysis is dependent upon the material properties. It is, therefore, very important to evaluate material properties with as much accuracy as possible.

For any finite element analysis the input data required could be subdivided into three main classifications such as geometry of the structures and support condition, material properties and the loading to which the structure is subjected. For most of the analysis, *in-situ* stresses, shear strength parameters and modulus of deformation of rock mass are very important input parameters. In any computer method, output is garbage if input is not correct. Hence, it is very important step to check the results of numerical modelling with some available solution.

4.0 FORMULATION OF FEBEM

The finite elements are coupled with boundary element using direct formulation of BEM. The detailed procedure of coupling has been presented by Singh (1985) and it is briefly discussed here.

The equilibrium equation of the system in terms of total stiffness matrix, [K] is given by :

$$[K] \{u\} = \{Q\} \qquad \qquad \ldots (3)$$

where $\{u\}$ is the nodal displacement vector, $\{Q\}$ is the nodal load vector and [K] is total stiffness matrix of finite element and boundary element regions and given by:

$$[K] = [K_f] + [K_b] \qquad \qquad \ldots (4)$$

In which $[K_f]$ and $[K_b]$ are the stiffness matrices of finite element and boundary element regions, respectively, and are defined as :

$$[K_f] = [B]^T [D] [B] \qquad \qquad \ldots (5)$$

$$[K_b] = 1/2 [C] [U] [T] + ([C] [U] [T])^T \qquad \qquad \ldots (6)$$

where, n is the number of finite elements, [B] is the strain displacement matrix, [D] is the elastic matrix, [C] contains the collected area for all the boundary elements, and [U] and [T] are the matrices of the Kernel shape functions which are obtained by the procedure given by Watson (1979).

The boundary elements stiffness matrix (BESM) given by Eq. (6) is developed for full circular opening. Due to symmetry of geometry and loading, the BESEM is reduced corresponding to the quarter section in the present investigation by using condensation procedure given by Sharma *et al.* (1985). This results in reduced size of BESM. Then, this condensed BESM is added to the stiffness matrix, [K] of FEBEM. The further steps for the evaluation of displacements and stresses are same as in the standard FEM.

4.1 Computer Programmes

A computer programme FEBEM was developed for two-dimensional analysis of underground openings with plane strain condition. Simulation of one step excavation has been incorporated in the computer programme. The same FEBEM computer programme is used for FEM analysis also. The main computer programme of FEBEM has been developed in three stages for linear elastic analysis. In the first two stages, the computer programmes of FEM and BEM have been developed and tested for the analysis of circular opening by comparing the results with available closed form solutions. In the third stage, coupling programmes of FEM and BEM has developed the computer software of FEBEM. The FEBEM programme has been developed using 8-noded isoperimetric finite elements and 3-noded parabolic boundary elements. The programme has the provision of automatic generation of nodal coordinates, nodal connectivity of elements and loading on nodal points. The complete programme FEBEM is further developed to incorporate elasto-plastic analysis and layered rock media as discussed in detail by Singh (1985).

5.0 UDEC NUMERICAL MODELLING

The Universal Distinct Element Code (UDEC) is a two-dimensional numerical computer programme based on the distinct element method for dicontinuum modelling, UDEC simulates the response of discontinuous media (such as a jointed rock mass) subjected to either static or dynamic loading. The discontinuous media is represented as an assemblage of discrete blocks. The discontinuities are treated as boundary conditions between blocks; large displacement along discontinuities and rotation of blocks are allowed. Individual blocks behave as either rigid or deformable material. Deformable blocks are subdivided into a mesh of finite-difference element, and each element responds according to a prescribed linear or non-linear stress-strain law. The relative motion of the discontinuities is also governed by linear or non-linear force-displacement relations for movement in both the normal and shear directions.

UDEC has many built-in material behaviour models, for both the intact blocks and the discontinuities, which permit the simulation of response representative of discontinuous geologic, or similar, material.

The formulation and development of the distinct element method embodied in UDEC has progressed in last three decades beginning with the initial presentation by Cundall (1971). In 1985, Cundall and Itasca staff adapted UDEC specifically to perform engineering calculations on an IBM-compatible microcomputer. Now, UDEC can be installed on other types of computer workstations and mainframes.

UDEC is primarily intended for analysis in rock engineering projects, ranging from studies of the progressive failure of rock slopes to evaluations of the influence of rock joints, faults, bedding planes etc., on underground excavations and rock foundations. UDEC is ideally suited to study potential modes of failure directly related to the presence of discontinuous features.

The programme can best be used when the geologic structure is fairly well defined from the observation or geologic mapping. A wide variety of joint pattern can be generated in the model. A screen plotting facility allows the user to instantly view the joint pattern. Adjustment can easily be made before the final pattern is selected for analysis.

6.0 CASE STUDIES

6.1 Tunnel Analysis by FEM and FEBEM

When modelling infinite bodies (e.g. tunnels and underground cavern) or very large bodies, it may not be possible to cover the whole body with blocks due to constraints on memory and computer time. Artificial boundaries are placed sufficiently far away from the area of interest that the behaviour in that area is not greatly affected. It is useful to know how far away to place these boundaries and what error might be expected in the stresses and displacement computed for the areas of interest. A series of numerical experiments were performed on a numerical model containing a circular tunnel in an elastic material.

A two-dimensional analysis with plane strain condition has been carried out. The excavation of the openings simulated in single step. The discretizations of the circular opening for FEBEM and FEM are shown in Figs. 1 and 2, respectively. In FEBEM, the boundary between finite element and boundary element regions has been taken at a distance of four times the radius of the opening from the center of the opening, Varadarajan *et al.* (1985). The boundary has been fixed at eight times the radius of the opening in the case of FEM.

For the present comparison purposes, the excavation of the circular tunnel opening in a geological medium of the infinite extent was considered. The radius of the tunnel was taken as 1.0 meter. The geological medium was considered to be homogeneous and the behaviour was assumed to be linear-elastic. The modulus of deformation and Poisson's ratio were taken as 1 GPa and 0.25 respectively, for hydrostatic stress condition of .01 GPa.

85

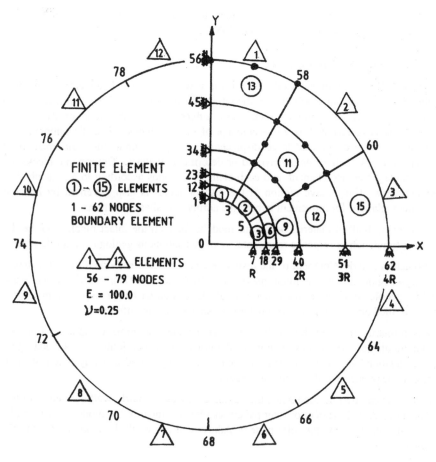

Fig. 1 : FEBEM discretization for circular tunnel

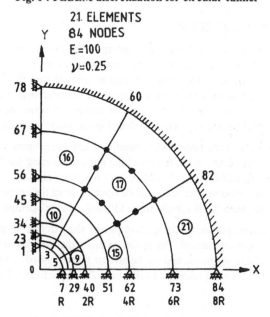

Fig. 2 : FEM discretization for circular tunnel

86

A comparison of displacements from FEM and coupled FEBEM with closed form solution are shown in Fig. 3. Based on this study, Singh (1985) concluded that boundary between FE and BE interface can be fixed at 4 times the radius of the tunnel while the boundary can be fixed at 8 times the radius of the tunnel in the case of FEM.

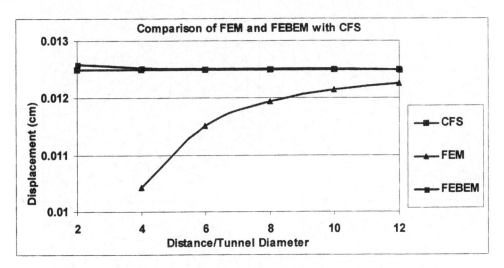

Fig. 3 : Fixation for boundary for FEM and FEBEM and comparison with CFS

6.2 Tunnel Analysis by UDEC

The circular tunnel model as used earlier by Singh (1985) and Varadarajan *et al.* (1985) was developed. Singh (1985) used a circular tunnel model for analyzing the same using Finite Element Method (FEM) and Coupled Finite Element and Boundary Element Method (FEBEM). The results of these analyses were compared with available Closed Form Solution (CFS) as given by Pender (1980). The same properties of rock mass were taken for the purpose of comparison. The boundary distance in the case of UDEC were also fixed at the distances of 2, 4, 6, 8, 10 and 12 times the radius of the tunnel. The discretization of UDEC is shown in Fig. 4.

The ratio of boundary distances and tunnel radius has been plotted against resulting displacement for CFS, FEM, FEBEM and UDEC as shown in Fig. 5. The results of CFS, FEM and FEBEM analysis are referred from Singh (1985) and Varadarajan *et al.* (1985). The UDEC analysis was performed during the research fellowship at NGI by Singh (1999).

It is seen from the results in Fig. 5 that the boundary fixation does not make much difference in FEBEM and the boundary between finite element and boundary element regions can be taken as 3 to 4 times the radius of the tunnel. However, the boundary must be fixed at a distance of 10 to 12 times the radius of the tunnel in the cases of FEM and UDEC. However, the boundary should not be less than 8 times the radius of the tunnel for having sufficient accuracy in the results.

6.2.1 *UDEC with Boundary Elements (UDEC-BE)*

The boundary element (BE) boundary condition is an artificial boundary that simulates the effect of an infinite extent of isotropic, linear elastic material. The direct element formulation as described by Sharma *et al.* (1984) has been coupled with finite element method in FEBEM and the same scheme is coupled to the distinct element scheme along the outer boundary of UDEC model. The distinct element region is embedded in the elastic domain. The region is first subjected to *in-situ* initial stresses. If a chance is made in the distinct element region such as an excavation, the continuity of excavation induced displacement and stresses must be satisfied at the distinct element-boundary element interface.

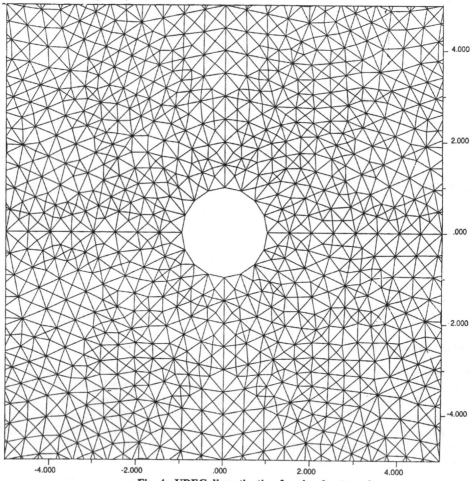

Fig. 4 : UDEC discretization for circular tunnel

In the UDEC formulation, boundary element nodes coincide with block corners (grid points) on the periphery of the distinct element domain. Determination of induced displacements at each boundary element node permits direct determination of induced nodal force by multiplying the nodal displacement vector and the stiffness matrix for the boundary element domain.

The nodal forces determined could then be applied to appropriate block corners located on the periphery of the distinct element domain in the subsequent iteration step. Thus, as the relaxation process continues, nodal displacements and forces are updated and used in the subsequent calculation cycle. Coupling between the two domains must ensure that displacements and traction's at all the nodal points are continuous across the interface. These conditions require that the region of non-linear material behaviour be confined inside the distinct element domain including the surface of joint sliding or separation.

The study shown by Singh (1985) is interesting to note that the boundary can be fixed at a distance of 8 R in FEM while it can be fixed safely at 4R in the case of FEBEM with a better accuracy in the resulting deformations and stresses. It can, therefore, be advantageous to use boundary element boundary in the case of UDEC since the code is having a provision for boundary element. The discretization for UDEC with BE boundary is shown in Fig. 6.

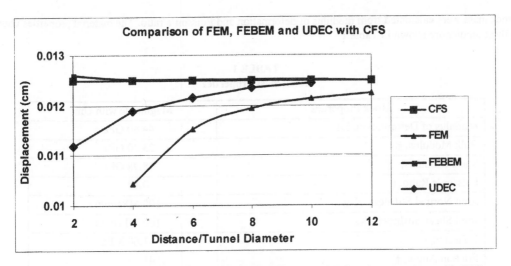

Fig. 5 : Fixation for boundary for FEM, FEBEM and UDEC and comparison with CFS

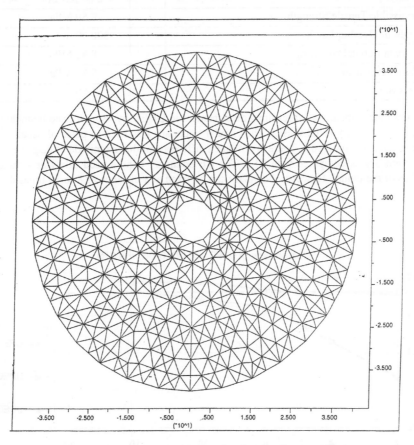

Fig. 6 : UDEC-BE discretization for circular tunnel

The rock mass properties were taken the same as used by Singh (1985) . The values of normal and shear joint stiffness were chosen for this model by trial and error increasing or decreasing in the model. About thirty models were run for fixing the final values of the normal and shear stiffness. Bulk modulus and shear modulus of rock mass were calculated from modulus of deformation and Poisson's ratio. The material properties chosen for these studies are shown in Table 1.

TABLE 1
Rock mass properties

Rock Properties	Magnitude with Unit
Modulus of Deformation, Ed	44.80 GPa
Bulk Modulus, K	23.30 GPa
Shear Modulus, G	18.98 GPa
Poisson's Ratio, μ	0.18
Joint Normal Stiffness, JKn	100 MPa/mm
Joint Shear Stiffness, JKs	100 MPa/mm
Cohesion, c	1.60 MPa
Friction Angle, ϕ^0	41
Density, γ	29.4 KN/m
Permissible Tensile Strength	8.50 MPa
Height of Overburden, h	300.00 m
In-situ horizontal stress, σ_h	8.82 MPa
In-situ vertical stress, σ_v	8.82 MPa
Stress ratio, K	1.00

Fig. 7 shows the comparison of resulting displacement in UDEC by using a fixed boundary (UDEC) and boundary element boundary (UDEC-BE). The circular boundary was used in place of usual squire boundary being used in most of the problems. There is not much difference in the magnitude of displacement in the case of using BE boundary as compared with CFS. The BE boundary can be used safely at a distance of 3 to 4 times the radius of the tunnel. However, boundary can be fixed at a distance of 8 to 10 times the tunnel radius in UDEC.

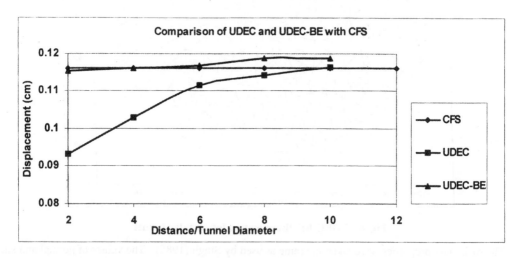

Fig. 7 : Fixation for boundary UDEC and UDEC-BE and comparison with CFS

6.3 Comparison among FEM, FEBEM and UDEC

Table 2 shows the comparison of accuracy for FEBEM, FEM, UDEC and UDEC-BE. To evaluate the accuracy of the results, the analytical solution for deep circular tunnel given by Pender (1980) has been used. The percentage errors are indicated with respect to closed form solution results as also shown in Table 2. In the case of FEBEM, the percentage error is very small and is in the order of 0.09 per cent for displacement. In the case of FEM, the error is 4.47 per cent, which is about 50 times higher than that observed in the case FEBEM. Similarly, the magnitudes of error in stresses predicted by FEM analysis are higher than those predicted by FEBEM. Thus, in all the cases, FEBEM gives more accurate results than FEM and this is due to the fact that the interface boundary between finite element and boundary element regions is fixed at four times the radius of the opening (Fig. 1) while it is fixed at eight times the radius of the opening in the case of FEM (Fig. 2).

TABLE 2
Comparison of stresses and displacements around a tunnel from different analysis

Sl.No	Points of Discussion	CFS	FEBEM	FEM	UDEC	UDEC-BE
1.	Displacements (Percentage error)	-0.01250	-0.012511 (0.09%)	-0.011949 (-4.47%)	-0.012010 (-3.92%)	-0.012510 (-0.08%)
2.	Major Stress (Percentage error)	1.69790	1.69750 (-0.02%)	1.64590 (-3.06%)	_____	_____
3.	Minor Stress (Percentage error)	0.30270	0.30270 (0.20)	0.29350 (-2.85%)	_____	_____

Similar conditions are applied between UDEC and UDEC-BE, which gives more accurate results than UDEC without boundary elements as shown in Table 2. The error in UDEC as compared to CFS is 3.92 per cent, which is about 50 times higher than the error of 0.08 per cent in UDEC-BE (UDEC with boundary element). The interface boundary between finite difference element of UDEC and boundary element regions is fixed at 3 to 4 times the radius of the opening while it is fixed at 8 to 10 times the radius of the opening in the case of FEM.

7.0 CONCLUSIONS

The following conclusions can be drawn on the basis of circular tunnel excavation analysis by FEM, FEBEM, UDEC and UDEC-BE :

The Finite Element Method (FEM) and Boundary Element Method (BEM) two well-established methods used for the analysis of underground openings.

➢ The BEM can be used advantageously for any shape of the opening. It is very easy to use and at the same time gives very accurate results. However, it is very difficult to incorporate discontinuities and complexity of rock behaviour in BEM.

➢ FEM may be used incorporating practically any type of rock media and rock mass bahaviour. Further, different opening shapes and sequence of excavation may be conveniently adopted. However, the extent of boundary distance to represent infinite domain and necessity of discretizing, a large domain introduced some approximation in the results. Furthermore, the preparation and checking of input and output data are tedious and time consuming.

➢ An efficient alternative is to utilize the advantages of both FEM and BEM by coupling them especially for the analysis of underground openings with significant modification of variation in rock mass properties near the tunnel excavation. Finite element may be used near the excavation surface for taking care of the complexities in the material behaviour and boundary element may be used away from it to take into account of infinite domain.

- ➤ From the analysis of circular tunnel excavation and comparison of the results as predicted by FEBEM and FEM analysis, it has been found that the FEBEM is more accurate and economical than FEM. Therefore, it is concluded that the coupled FEBEM can efficiently be utilised for the analysis of tunnel excavations. The interface boundary between finite element and boundary element regions is fixed at four times the radius of the opening while it is fixed at eight times the radius of the opening in the case of FEM.

- ➤ UDEC with boundary element boundary gives more accurate results than UDEC with fixed outer boundary. The interface boundary between finite different element of UDEC and boundary element regions is fixed at 3 to 4 times the radius of the opening while it is fixed at 8 to 10 times the radius of the opening in the case of FEM.

- ➤ The importance of rock mass properties as input parameter to numerical modelling has been shown by analysis. The change in material properties affects the results of analysis. The input properties must be determined properly during investigation of the project.

- ➤ From this study, it can be concluded that FEM and UDEC can be used for the analysis of tunnel excavation with little compromise on the accuracy. However, boundary element boundary may be used with both FEM and UDEC to increase the accuracy in the analysis.

REFERENCES

Barton, Nick (1991), "Geotechnical Design", World Tunnelling, November.

Beer, G. (1983), 'Finite element, boundary element and coupled analysis of unbounded problems in elastostatics', Int. J. Num. Meth, Engng., 19, p. 567-580.

Beer, G. and Meek, J.L. (1981), 'The coupling of boundary and finite element methods for infinite problems in elastoplasticity', Proceedings of the Third International Seminar Irvine, California, edited by C.A. Brebbia, p. 575-591.

Brady, B.H.G. and Wassyng, A. (1981), 'A coupled finite element method of stress analysis', Int. J. Rock Mech. Min. Sci & Geomech, Abstr., 18, p. 475-485.

Cundall, P.A. (1971), 'A computer model for simulating progressive large scale movements in blocky rock systems', Proc. of the Symp. Int. Soc. Rock Mech., Nancy, France, Vol. 1, Paper No. II-8.

Desai, C.S. and Abel, J.F. (1972), 'Introduction to the finite element method', Van Nostrand Reinhold Company, New York, p. 477.

Kelly, D.W., Mustoe, C.G.W and Zienkiewicz, O.C. (1979) 'Coupling boundary element methods with other numerical methods', in Banerjee P.K. and Butterfield R. (eds.), Development in Boundary Element Methods, Vol. 1, Applied Science Publications, London, Chap. 10, p. 251-285.

Pender, M.T. (1980), 'Elastic solutions for a deep circular tunnel, geotechnique XXX, 2, p. 216-222.

Sharma, K.G., Varadarajan, A. and Singh, R.B. (1985), 'Condensation of boundary element stiffness matrix in FEBEM analysis', Commun. Appl. Num. Methods, 1, p. 61-65.

Singh, Rajbal (1999), 'Report on research fellowship at NGI', submitted to Norwegian Geotechnical Institute (NGI), Norway, p. 55.

Singh, R.B. (1985), 'Coupled FEBEM analysis of underground openings', Ph.D. Thesis, Department of Civil Engineering , Indian Institute of Technology, New Delhi, India.

Singh, R.B., Sharma, K.G. and Varadarajan, A. (1985), 'Elastoplastic analysis of circular opening by FEBEM', Second Inti. Conf. on Computer Aided Analysis and Design in Civil Engineering, Roorkee, India, III, pp. 128-134.

Varadarajan, A. and Singh, R.B. (1982), 'Analysis of tunnels by coupling FEM with BEM', Proc. 4th Int. Conf. of Num. Meth. in Geomech., Edmonton, Canada, Vol. 2, pp. 611-618.

Varadarajan, A., Sharma K.G. and Singh R.B. (1983), 'Analysis of circular tunnel by coupled FEBEM', Proc. Indian Geotechnical Conference, Madras, India, VI, pp. 113-118.

Varadarajan, A., Sharma, K.G. and Singh, R.B. (1985), 'Some aspects of coupled FEBEM analysis of underground openings', Int. J. Num. and Analytical Meth. in Geomechanics, 9, pp. 557-571.

Varadarajan, A., Sharma, K.G. and Singh, R.B. (1987), 'Elastoplastic analysis of an underground opening by FEM and coupled FEBEM. Int. J. Num. Anal. Meth in Geomech.

Watson, J.O. (1979), 'Advanced implementation of the boundary element for two and three dimensional elastostatics', in Banerjee P.K. and Butterfield, R. (eds.), Development in Boundary Element Method, Vol. 1, Applied Science Publication, London.

Zienkiewicz, O.C., Kelly, D.W. and Dettess, P. (1977), 'The coupling of finite element method and boundary solution procedures', Int. J. Num. Meth. Engg. 11, pp. 355-375.

HIGHWAY TUNNELS AND UNDERGROUND STORAGE

UNDERGROUND STORAGE IN ZIMBABWE

ANDERS CLAESSON

Vice President (East)

Skanska International Civil Engineering A.B., Hongkong

SYNOPSIS

In 1989 Skanska International Civil Engineering was awarded a contract by National Oil Company of Zimbabwe (NOCZIM) to perform a complete feasibility study and construction of an underground strategic protected storage plant for storing of diesel, petrol and jetfuel.

As the area around the capital Harare, consists of good bedrock with a stable ground water table, Skanska adopted the Scandinavian storage method of unlined rock caverns.

During the last fifty years more than one hundred storage plants of this type have been constructed in Sweden for commercial as well as strategic use.

The advantages of underground storage are technical, environmental and economical.

At this time, there was an urgent need for economically viable strategic storage of fuel in Zimbabwe.

After conducting a series of geological investigations, Skanska could determine a cavern design that provided the most technical and economically viable solution.

When all financing was in place, the construction started in 1992. All excavation works were carried out by the drill and blast method, using heavy mechanized equipment. The plant is connected to the fuel import harbour at Beira in Mocambique, via a pipe line, and has now been in reliable operation since 1996.

It has contributed to a more efficient and safer operation of fuel distribution in Zimbabwe.

1.0 THE PROJECT

Zimbabwe – a land locked country, during the first ten years of independence, suffered from the impact of fluctuating oil prices and disturbances in the transport of products through their neighbouring countries (South Africa and Mocambique).

Consequently, the government decided to investigate, through NOCZIM, the possibility of storing petroleum products in a way that would provide security against logistic disturbances and have a stabilizing effect on the petroleum prices. An initial feasibility study prepared by others proposed storing diesel, petrol and jet fuel in buried concrete tanks.

Skanska International Civil Engineering AB was in 1989, after competition with a number of international contractors, awarded a turnkey contract for construction of a strategic fuel storage. The contract consisted of four different phases : evaluation, investigation, design and construction (Fig. 1).

2.0 PHASE 1

The work started with an evaluation of the concrete tank concept. The Scandinavian way of storage, in unlined rock caverns, had not yet been studied. Rock conditions in Eastern Africa are often good, even better than in Scandinavia. However, the weathering can sometimes be deeper in Africa, and therefore lead to more extensive initial investigations. In view of the existing geological conditions around Harare, Skanska made a comparison between concrete tanks and rock caverns. This study indicated considerable advantage, both with regards to cost and technical aspects, for the rock cavern concept. It was therefore decided, by NOCZIM, to continue to work on the cavern solution.

3.0 PHASE 2

The feasibility study, including geological investigations, consisted of seismic refraction measurement, in a number of different locations. These investigations resulted in the choice of a site, where a core drilling programme was performed, including hydro-geological measurements. The area chosen consisted of very high quality granite, although some minor fault zones had been detected.

Location :
15 km south-east of Harare, Zimbabwe

Client :
National Oil Company of Zimbabwe (NOCZIM)

Contract :
Design and construction of civil and electron-mechanical works

Value of Contract :
US$ 67 million

Financing :
BITS, Norsk Eksportfinans and local financing

Construction Period :
June 1992 – February 1996

Subcontractor :
Kvaerner Energy a.s, Norway

Fig. 1 : Underground fuel storage project, Zimbabwe

4.0 FIELD INVESTIGATIONS

One criteria for selection of possible sites was, a close distance to the railway and main road between Harare and Beira in Mocambique, which is the main harbour for the import of petroleum products.

The geological investigations were conducted in the following order :

- Visit to the sites, studies of geological publications, hydro-geological studies and resistivity surveys. These investigations were conducted on a number of locations. Further evaluation led to the choice of the present site and to more detailed investigations.
- Next step was to perform seismic refraction measurements, in order to determine the thickness of the overburden soil layer and possible fault zones. The result of the seismic investigation indicated, in principle, very good rock conditions.
- To verify these results and make a hydrogeological survey, a core drilling programme was established, containing six bore holes, to a depth of one hundred metres. The cores showed granite, with some inclined slabs of diabase. In the interface between the granite and the diabase, there were some layers of clay, of minor thickness.

The ground water was of good quality, with very small variations in the ground water level during the year.

5.0 PHASE 3

After analysis of the geological investigations and taking into account all necessary criteria set up by NOCZIM, regarding storage volumes and products, Skanska presented a conceptual design for underground storage plant in unlined rock caverns, in accordance with the Scandinavian concept. This concept is based on the following principles :

- The ground has to contain ground water.
- The product to be stored is lighter than water.
- The product is not soluble in water.

The advantages are :

- Minimal requirement of land area.
- Low maintenance cost.
- No risk of ground water pollution from oil leakage. The oil is contained under the ground water level. The ground water pressure is higher than the pressure from the product, thus preventing the product from leaking into the rock. For this reason the access tunnel used for excavation of the caverns is filled with water.
- Low risk of fire and explosion.
- Lower cost than steel tanks at volumes above 50,000 m³.
- Excellent protection against sabotage and acts of war.

More than one hundred storages of this type have been constructed in Scandinavia during the last fifty years, and Skanska has constructed around seventy different plants.

With this vast experience as background, Skanska started the design work with the purpose of presenting a turnkey proposal.

The chosen principle is that of a storage on fixed water bed, which means that the interface level between the product and seepage water in the bottom of the cavern, is kept at a constant level.

Recent experience from storing of modern cracked petroleum products, shows that contact between water and product shall be kept at a minimum.

Another factor that had to be studied, was the impact of the temperature. In Sweden, the underground temperature is + 8°C and in Zimbabwe + 20°C. The temperature can influence the risk for bacterial damage on the product, in the contact zone between water and product. This matter has been taken into consideration when designing the caverns.

A long-term (ten years) test is ongoing on site, carried out by the Swedish Defence Research Organisation (FOA).

The total storage volume required by NOCZIM was around 400,000 m³. This volume is divided into; 50 per cent diesel, 30 per cent petrol and 20 per cent jet fuel.

Diesel and petrol are stored in two caverns for each product, 20 m wide and 24 m high. Jetfuel is stored in two silos, 36 m in diameter and 44 m high. The caverns and silos are sealed off from the access tunnel with concrete barriers.

Silos are used for storing jetfuel in order to minimize the contact between leakage water and the product. In the silos, an "umbrella" of plastic sheet is erected in the ceiling, to avoid water dripping into the product.

Above the caverns there is a system of pipe tunnels, where all necessary pipes and electro-mechanical installations are situated.

For operational reasons there is also an underground "service area", 12 m wide, 6.5 m high and 115 m long. It consists of a control room HV transformer, switchgears, water handling, ventilation fans, workshop, storage, kitchen, changing room and emergency diesel generators with an internal water cooling system. The entire plant can be operated from the control room via a computerised control system. With the installed diesel generators, the plant can operate independently, without any power supply from the outside.

There is a strategic protection of all entrances by concrete barriers, making the underground part completely bomb proof.

There are also concrete barriers between the pipe tunnel and the service area, in case fire or explosion would occur in these locations.

The service area have a CO_2 system for extinguishing fire in electrical installations.

The rock cover for the plant is about 35 m (Fig. 2).

Underground storage in Zimbabwe.

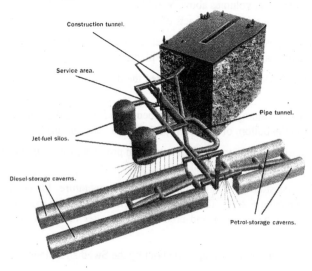

Fig. 2 : Underground storage in Zimbabwe

The control of water is an important part of the design and operation of an underground storage plant of this type.

One can distinguish the following main issues :

- *Ground Water*
  The ground water level is continuously checked through six bore holes around the storage area. This is carried out both during construction and operation.

- *Tunnel water for cooling of emergency diesel generators*
  Water from the top of the access tunnel is pumped through a pump shaft for cooling of the diesel generators. The cooling water outlet is at the bottom of the access tunnels, in order to have a thermal circulation.

- *Rainwater*
  The ramp leading down to the tunnel entrance will receive 1000 m³ during one hour of a 50 year intensity rain. The water flows down the access tunnel and is temporarily stored there and will be pumped to the surface through submersible pumps in the pump shaft in the "service area".

- *Water curtain system*
  Water curtains are drilled from the pipe tunnel between the caverns for different products, in order to prevent leakage and consequent mixing of products. The water curtain system maintains a constant over pressure in the drill holes. Ground water is collected at the surface and stored in a separate tanks underground. The water used for water curtains are completely separated from the rain water system, to avoid biological growth in the drill holes.

- *Water leakage into the caverns*
  All major water inflows into caverns are sealed by cement grout during the construction. The remaining leakage is collected in the pump pit and pumped to the surface oil separator before sent to the re-infiltration drains. The soil separator is designed according to API-norms.

- *Sewage water*
  Sewage water from the "service area" is pumped to a three-compartment septic tank above ground.

5.1 Storage of Jetfuel

Long-term storage has been used for a long time in Sweden, without affecting the products. On rare occasion, there has occurred development of sludge while storing jetfuel and heavy fuel oil. For this reason, the Swedish Defence Research Organisation (FOA) participated during the designing of the plant. They have conducted extensive microbiological research on how to handle this phenomenon. Their findings have been taken into account when designing the jetfuel storage.

6.0 PHASE 4

6.1 Construction

The construction work started in June 1992. Blasting works were completed in June 1994. Installation works started in January 1994 and were completed in June 1995. The whole plant was commissioned and handed over to the client in February 1996.

6.2 Tunnel

Drilling operations were fully mechanized and performed by two Atlas Copco Drill Jumbos no. 175 and 135. For the bench drilling in the silos, crawler drilling rigs ROC 601 and ROC 301 were used.

The size of the access tunnel was 60 m², in order to allow the lorries to pass each other.

Mucking was done by a number of wheel loaders of different sizes, depending on location. Transport of rock was performed by twelve heavy SCANIA trucks with rock body.

All shotcreting was carried out as wet shotcrete and applied by a Stabilator made hydraulic robot.

The rock support was performed with an "active design".

1. Four types of rock classes were defined.
2. Geological mapping was continuously made after each blast.
3. A preliminary classification of rock classes was carried out by the tunnel crew.
4. Rock support was carried out according to the principal design.
5. Every three months the designer visited site to assess all geological mapping and support work.
6. If considered necessary for the permanent works, additional support was carried out.

All areas with permanent installations were protected by one or two layers of shotcrete. For the grouting operations only cement was used.

6.3 Silos (Fig. 3)

The procedure for excavation of the silos :

1. Excavate the top dome. Special design of rock support using finite element analysis, made to ensure the protection of the 36 m diameter dome.
2. Access tunnel down to the bottom of the silo.
3. Small shaft excavated by long hole drilling between top and bottom of the silo.
4. Bench drilling in stages, with the blasted rock falling down the shaft, to be mucked out.
5. Reinforcement and grouting to limit water in leakage from the walls, carried out in connection with the bench blasting.

6.4 Caverns

The caverns were excavated by top heading and two horizontally drilled benches.

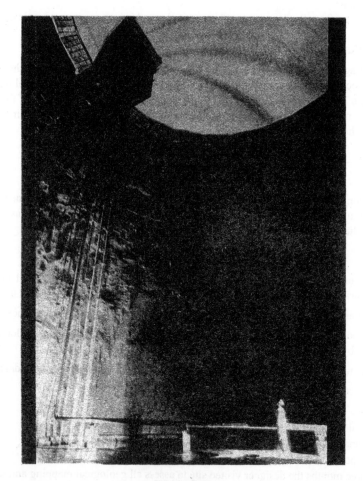

Fig. 3 : Soils

By using the above mentioned heavy equipment and transport of all excavated rock through a large access tunnel, it was possible to obtain a very cost and time effective rock excavation.

6.5 Shafts

Three shafts were made; (1) for pipe installations, (2) for diesel generator exhaust fumes, and (3) as an emergency exit. The shafts were constructed with long hole drilling and mucking out from the bottom.

6.6 Product Handling

The initial planning was to fill the storage by rail cars from a siding above the storage.

When construction of the storage was in progress, NOCZIM decided to extend the pipeline, from Beira port to the border town Mutare, all the way to the Harare terminal.

It was then decided to connect the pipeline terminal with the storage. This way, it was in principle possible to pump directly from a tanker in Beira, into the caverns 500 km away.

Distribution of products to consumers is done by rail.

The siding can take 20 rail cars, that can be filled by diesel, fuel and jetfuel at the same time.

6.7 Electro-mechanical Installations

Each cavern is connected to the pipe tunnel by a shaft. There is an explosion proof concrete deck at the top of the shaft with a number of casing pipes (Fig. 4).

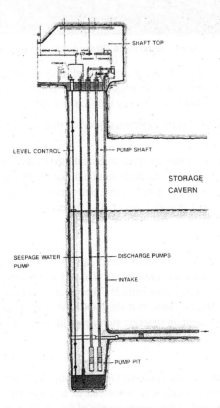

Fig. 4 : Disesel cavern pumping installations section (NOCZIM, Zimbabwe)

Through these pipes, submersible pumps are installed, for products and leakage water, down to the pump pit in the cavern.

There are also a number of control devices for product and water level, in order to obtain maximum security.

An intake pipe, including discharge pipes to the terminal, are installed.

The products pass through a water separation filter in the pipe tunnel, before reaching the distribution area at the terminal (Fig. 5).

Fig. 5 : Water separation filter in the pipe tunnel

One of the challenging aspects of M&E design, and overall project design, is to establish the appropriate level of technology. A level which minimises the possibilities of human error, yet, has enough sophisticated automatic technology and still is easy and cost effective to trouble-shoot and maintain.

Training of the future NOCZIM plant operators was carried out over a two-year period, prior to commissioning and handing-over.

Two months of training at plants of similar types in Scandinavia was followed by, two months hands-on operational training on a specially developed computer simulator for the designed plant. The future operational staff, worked integrally with Kvaerner's installations personnel, during the construction, testing and commissioning of the works, to enable a smooth hand-over of the plant.

7.0 COST

The total cost for the initial studies, the field investigations and the subsequent full turn-key contract comprising design, civil works construction, installation of E&M equipment commissioning and training of operational staff was 67 MUSD. This cost also included the pipeline between the pipeline terminal and the storage (3 pipes 3 km each). It should be noted that the NOCZIM plant is provided with maximum defence security measures to meet the requirement on strategic storage of different petroleum products.

8.0 FINAL COMMENT

Since the hand-over of the plant to the National Oil Company of Zimbabwe in 1996, fuel consumption has doubled in Zimbabwe. The plant, which was conceived as a long-term strategic storage, has become a short-term large-volume storage, critical to the steady supply of fuel to the country. The NOCZIM is a success story in terms of both design, implementation and utilization. The adopted turn-key concept resulted in a significant saving of time between the feasibility study stage and the construction phase and it gave the Owner a contract with no interface problems between the different kinds of works, a fixed programme for the implementation and a fixed price.

LARGEST-SIZED SLURRY TYPE SHIELD TUNNELLING THROUGH EXTREMELY SOFT GROUND : TRANS-TOKYO BAY HIGHWAY PROJECT

SHOUEI IKEDA   **MASAYOSHI OKAZAKI**   **TORU GOTO**

Civil Engineering Department, Shimizu Corporation

SYNOPSIS

The Trans-Tokyo bay highway, a 15 km highway linking Kawasaki with Kisarazu across the Tokyo bay, was completed in December 1997. For the construction of the 10 km under-seabed tunnel part of the highway, where as high as 0.5 MPa water pressure acts, the largest sized (outside diameter of 14.14 m) slurry shield tunnelling machines were employed along with several latest technologies which include the automatic segment assembly robot. The tunnel in the Kawasaki side slope passes through Ac1 layer, an extremely soft holocene clayey deposit. The Ac1 layer is so soft as to be called "mayonnaise layer", and has been anticipated to encounter various difficulties.

This report describes the outline of Trans-Tokyo bay highway project, latest technologies of automatizing for tunnelling, and discusses tunnel construction performance through Ac1 layer in which the excavating by shield machine was difficult.

1.0 OUTLINE OF THE PROJECT

Trans-Tokyo bay highway will be an ordinary toll road with a length of 15.1 km lying across Tokyo bay forming a straight-line linking the city of Kawasaki in Kanagawa prefecture to the city of Kisarazu in Chiba prefecture. When it have been completed and is integrated with the WANGAN Expressway and other highways to form a loop line encompassing the Tokyo region, it is expected to relieve traffic congestion, and play a big part in industrial activity.

It consists of both tunnel and bridge section. The Kawasaki city side is a 10 km shield tunnel, while the Kisarazu side consists of a bridge approximately 5 km long. One artificial island (Kawasaki man-made island) will stand at the corner of the tunnel section and another (Kisarazu man-made island) forms the link between the tunnel and bridge section. This project, which is the largest marine civil engineering project ever attempted, incorporates the most advanced technology at every stage of design and construction.

The overview of construction work and the typical structure of tunnel are shown in Figs. 1 and 2.

2.0 AUTOMATIZING FOR THE TUNNELLING

This project had some automatic equipments as latest technologies. This section, particularly, describes automatic segment assembling and intensive shield control system among them.

2.1 Automatic Segment Assembling

Segments had been assembled manually. In this project, as shown in Fig. 3, this system was introduced. It can handle segments of100 kN in weight, and assemble 11 segments into a ring of 13.9 m in outside diameter. It provides automatic and accurate control for gripping at higher speed.

PLAN

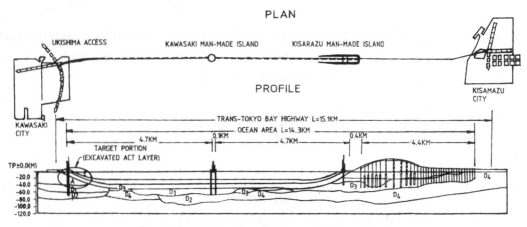

Fig. 1 : Trans Tokyo bay highway project outline

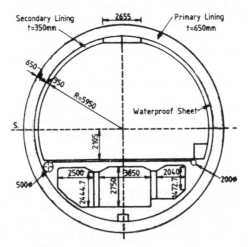

Fig. 2 : Typical structure of tunnel

This system consists of a segment erector robot, segment transport conveyor, segment transport hoist and segment transporter. The segments placed on the transporter are sent forward in sequence, and then to the conveyor by the hoist. The conveyor supplies segments to the erector robot after adjusting of the segments. The erector robot grips the segments, aligns it with already assembled segments, press the waterproofing seals on the connection face (segmental joint or circumferential joint), tighten the bolt, then return to initial position automatically.

2.2 Intensive Shield Control System

The system has been designed to intensively control and monitor the operation of the slurry shield tunnelling machine and associated equipment from the central control room in the site office.

3.0 EXCAVTION THROUGH AC1 LAYER

3.1 Outline of Excavation

As shown in Fig. 1, at the tip of the slope where improvement ended, a gradually increasing overburden (from 0.81D to 1.06D) would consist of the extremely soft Acl layer. Dubbed the "mayonnaise layer", which would have to be excavated for a length of 150 m. There were almost no reports of shield tunnelling through such a soft layer, so no methodology had been established for such work. Because of lack of a force to counter the buoyancy of the tunnel itself in this problem section, the safety factor might fall below 1.0. As a result, uplift of the tunnel was anticipated during excavation. In this difficult situation, the utmost care was required before and during excavation work.

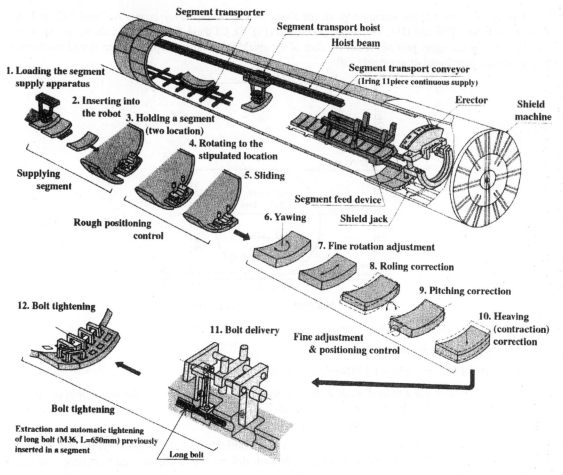

Fig. 3 : Conceptual diagram of the assembly process

Excavation of the target section had the following characteristics :

- Excavation of extremely soft Acl layer

1. Work organisation to allow continuous excavation (specifically, on-site shift system of work and constant manning by the staff of manufacturers involved).

2. Careful management of slurry pressure at the cutting race.

3. Management of physical characteristics of slurry (specific gravity and viscosity).

4. Feedback of seabed surface settlement to excavation work.

5. Feedback of automatic in-tunnel settlement measurements.

6. Management of shield machine load

• Countermeasure to buoyancy

Introduction of dead load into the tunnel (such as extremely thick steel H-beams or used rails) and its installation.

3.2 Detail of Soil Profile

During excavation at the leading edge of the slope, two holocene clayey layers, Ac1 and Ac2, and two Pleistocene layers, D1s and D1c, were to be penetrated (Fig. 4). Of these four layers, as shown in Fig. 4, the Ac1 layer lies at the upper part of the cutting face of the shield machine and represents the tunnel overburden and, as already mentioned, is extremely soft.

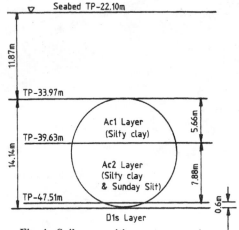

Fig. 4 : Soil composition at target point

The Ac1 layer has the following characteristics:

Ac1 layer (Yurakucho layer, silty clay)

$N = 0$; $\rho_t = 14$ kN/m^3; $E50 = 1.5$ MN/m^2; $\omega_n = 120\%$; $\omega_{n1} = 107\%$

- The layer has a natural water content of about 120%, which exceeds the liquid limit of 107%. It was therefore likely to be easily liquefied by the simple agitation caused by a cutter or the application of even a small external force.

- It has a failure strain of 5% - 10% with an accordingly high deformation coefficient. The layer was therefore expected to undergo significant deformation at the face due to changes in slurry pressure.

- The layer's low strength means that there is only a small difference between active and passive earth pressure, and this limits the controllable range of slurry pressure at the cutting face. A situation was expected where extremely tight control would be required for the following shield due to influence of excavation by the preceding shield.

3.3 Key Actions

3.3.1 *Stable Excavation*

Many ideas for ensuring the stable excavation of this layer had been under study ever since the work contract was awarded. Of these ideas, it was decided to implement the following :

(1) Management of slurry pressure at the cutting face

Since the Ac1 layer has a very low bearing strength, the allowable range of slurry pressure is small. Further, preliminary field experiments revealed the possibility of the most dangerous type of event that can occur during shield tunnelling; that is, blow-out following excessive pressure on the overburden. Consequently, whatever unexpected changes in pressure might occur, excess pressure over the overburden pressure was to be avoided at all costs.

The conclusion was reached that slurry pressure should be managed at the top edge of the shield so as to ensure safety of the Ac1 layer. Fig. 5 shows the set pressure was to be not below the active pressure without exceeding blow pressure (= overburden pressure) despite the active pressure (± 20 kPa). The set value at the section start point was 0.375 MPa at this top edge of the shield. Strict control was required to maintain the difference between the blow-out pressure and the active pressure within the fluctuation range (40 kPa).

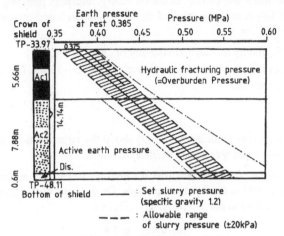

Fig. 5 : Slurry pressure management

(2) Measurement of seabed changes

To ensure early detection of signs of blow-out or collapse during shield work, it is important to install pore water pressure gauges in a mesh pattern on the sea bed to allow changes in the sea bed (heave and settlement) to be measured during the work. Since conventional pore water pressure gauges fail to provide reliable results, being affected by tides, atmospheric pressure changes, and variations in water temperature, we developed a special system comprising two standard hydraulic pressure gauges with anchors fixed in the supporting layer, a water tank set on a tower above the surface, and a mesh of 24 pore water pressure gauges linked by connecting pipes. This system is shown in Fig. 6. This allowed changes on the sea bed to be more precisely predicted from the difference between the readings given by the standard water pressure gauges and the pore gauges.

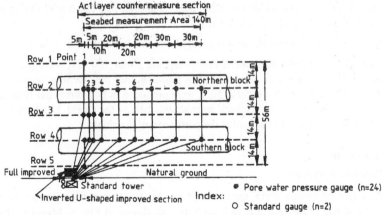

Fig. 6 : Arrangement of pore water pressure guages on the sea bed

Preliminary FEM analysis provided estimates of 3 cm heave as corresponding to blow-out pressure and 19 cm settlement corresponding to active collapse pressure. These values were used as guidelines throughout the work.

(3) Use of high quality slurry

If the slurry pressure at the cutting face exceeds the blow-out pressure, the ground may fracture hydraulically. One effective means of avoiding this is to increase the specific gravity and viscosity of the slurry. It is not possible to increase the specific gravity and viscosity without limit, however, since this would hinder excavations because there are certain limits on slurry transport pumping capacity. The following settings were therefore established for excavation of the target section based on previous experience with the inverted U-shaped improved section.

- Slurry feed: specific gravity of 1.24 and viscosity of 28 to 30 sec.
- Slurry discharge: specific gravity of 1.28 and viscosity of 35 – 37 sec. (the pump loading limit)
- Excavation rate: 15 mm/min.

3.3.2 *Buoyancy*

As noted, the small overburden (minimum 0.81 D) above the target section meant that an ample safety margin against buoyancy was necessary to ensure stable excavation.

Since the actual level of the ground surface (the sea bed) greatly affects resistance to buoyancy, the sea bed was surveyed in February 1994 to accurately map levels. On the basis of this survey data, it was calculated that there would be insufficient load to prevent buoyancy. (Fig. 7).

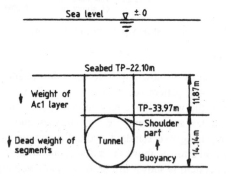

Fig. 7 : Schematic diagram for buoyancy calculation

Consequently, it was decided to bring additional load into the tunnel during excavation. The buoyancy safety factor during excavations was set at 1.0.

The heavy materials and equipment brought into the tunnel were as follows :

Extremely thick H-section steels, each weighing about 100 kN, were placed at a pitch of 750 mm as sleepers. Bundles of three 500 N/m rails were placed across the sleepers, with a maximum of 9 bundles per tunnel ring. 480 ingots, each weighing 50N, were produced in bolt boxes per ring. Ingots, each 8.58 kN, were produced and loaded into a work vehicle immediately behind the shield. Load absorbing jacks were attached to roundness retainers to utilize the load at the protrusion of the shield.

3.4 Excavation

3.4.1 *Excavation Process*

No major problems were encountered as the shield machine continued excavation in the target section at a speed of about 5.5 R (8.3 m) a day, as shown in Fig. 8. Minor problems did, however, arise, particularly with the machinery, over a length of about 80 rings (120 m) from the point of entry into natural ground (468th ring) to the 550th ring. In this area, an unexpectedly large quantity of ancient shells (fossils of large oyster shells, 20-30 cm long) was encountered.

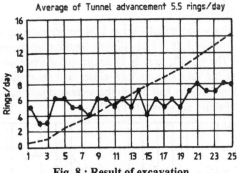

Fig. 8 : Result of excavation

To cope with this problem, crushing pumps were made ready prior to excavation. Despite this precaution, many malfunctions occurred in other parts of the slurry outflow lines.

The frequency of these problems led to considerable lost time as equipment and parts had to be changed, repaired, and improved. As a result, the daily advance was considerably slowed in the first half of the section. As excavation proceeded toward the latter half, where no fossils were encountered (after June 19), no pipe clogging occurred even with increased jacking speeds of up to around 20 mm/min., and the advance was stabilized at a daily rate of 7 to 8 rings.

3.4.2 Excavation Results and Discussion

(1) Slurry pressure at face

The slurry pressure set point was placed at the most critical location on the face, meaning the top edge of the shield. Then the mean of the hydraulic fracture pressure and the effective active earth pressure and pore pressure was used as the face slurry pressure (the effective active earth pressure and pore pressure +20 kPa). However, as the shield entered the natural ground (at the 468th ring), the amount of soil being excavated decreased suddenly. It was judged that this reflected a flow of some slurry into the fossilized shell layer, and so the set slurry pressure at the face was reduced to the effective active earth pressure and pore pressure + 10 kPa.

Excavation continued through the remainder of the section at this pressure. This set value was considered appropriate since the shield machine successfully penetrated the section without any problems of heave or settlement in front of the face, as indicated by sea bed measurement data.

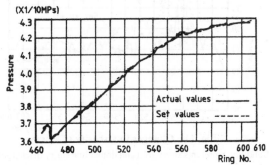

Fig. 9 : Face slurry pressure : set value and actural value

Figure 9 shows the graph of face slurry pressure, the actual slurry pressure closely followed this set value, meaning that ideal conditions were maintained. Fluctuations were within the range originally planned,

±20 kPa; to be specific, fluctuations were ±15 kPa during the slower excavation of the first half and ±10 kPa during the smoother excavation of the latter half.

(2) Specific gravity and viscosity of slurry

The specific gravity of the slurry was initially set at 1.24 before entering the Ac1 layer. It was gradually raised to the set value, and remained at 1.25 – 1.26, slightly higher than the set value, from around the 480th ring until the end of the section. The specific gravity of discharged slurry rose in accordance with changes in the feed, and was maintained at the set value of 1.28 close to it throughout the first half of the section. However, it rose to around 1.30 in the latter half due to the increased jacking rate and the proportion of sand. The changes in specific gravity of slurry are shown in Fig. 10.

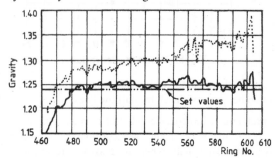

Fig. 10 : Changes in specific gravity of slurry
(Solid line represents slurry feed and dotted line slurry discharge)

Slurry viscosity was a crucial management parameter in the effort to prevent the Ac1 layer from fracturing. Past experiments have revealed that a higher viscosity gives more effective fracture prevention. Given the upper limit of slurry discharge viscosity (40 s), 28 to 30 s was chosen as the set value. As it turned out, the value gradually increased as the specific gravity rose in the inverted U-shaped improved section after remaining almost at the set value in the first half. At the peak, it reached 38 to 39 s, which was very close to the upper limit. This value was thought to be the maximum possible for the slurry to remain liquid, since uniform agitation was not possible in the preparation tank. The viscosity of the slurry feed fell due to influence of the increasing amount of sand in the cross-section, falling to 25 s and ultimately to 23 s. As the viscosity began to drop (after the 530th ring), very careful attention was paid to changes in sea bed measurements, and we were ready to add a thickener if large changes were seen. Fortunately, no such major changes occurred. The changes in viscosity of slurry are shown in Fig. 11.

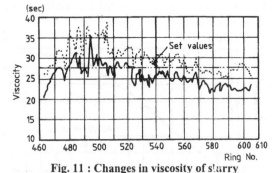

Fig. 11 : Changes in viscosity of slurry
(Solid line represents slurry feed and dotted line slurry discharge)

(4) Measurement of settlement of the sea bed

These measurements correspond to the ground level monitoring that is typically carried out during shield tunnelling work. In our case, connected – pipe pore water pressure gauges were used in addition to the usual sea bed measurement gauges.

112

The measurement turned out to be a remarkably effective means of understanding the tendency toward settlement or heave of the sea bed, despite variations in the data caused by swaying of the standard tower under the influence of wave motion. In particular, when changing set values, the data served as a good guideline for judgement as to whether the new value was appropriate or not.

The data show that there was almost no change in the sea bed before arrival of the face. Point 1 at Row 4 (borderline with the improved ground) underwent a settlement of about 3 cm after passage of the shield. Other than this, the generally observed behaviour was an heave of about 3 - 4 cm after the shield passed bv.

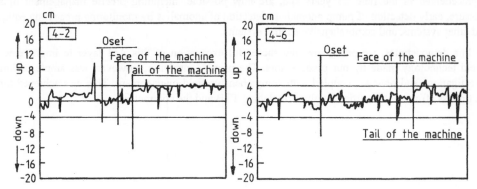

Fig. 12 : Time history of vertical displacement of sea bed

This heave began when the central part of the shield machine passed directly under the measuring point and was almost complete by the time the tail passed. The fact that there was almost no change before the arrival of the face may be interpreted as indicating the quality of management of the face slurry pressure and excavated volume.

A reasonable interpretation of the observed sea bed heave after passage of the shield, given the point in time when it began, is that it was caused by the influence of backfilling. The backfilling pressure (water pressure +0.05 MPa) as earlier mentioned corresponds to an initial effective stress, which means this pressure could have heaved the sea bed with ease. Regarding settlement, there was an initial fear that filling of void around the shield would be inadequate, but we deduce that filling around the shield was actually sufficient thanks to the effect of simultaneous filling from the shield machine.

4.0 CONCLUSIONS

In the early stages of the planning for this excavation work, various problems were anticipated while excavating the Ac1 layer. The problem layer is immediately below the sea bed and is characterized as extremely soft. Our success is excavating the tunnel through this layer is partly attributable to past experience with excavation of the Keiyo Line Haneda Tunnel, a case with many similarities. Of greater importance, however, are a range of advances made in shield tunnelling technology over the 25 years since the Haneda Tunnel was constructed. These are, for example :

1. Improvements to the tail seal :

    The development of the wire brush during the 1980's led to a remarkable improvement in the water sealing capabilities of the shield tail. In our work, four lines of wire brushes were used, and strict management of automatic lubrication between the wire brushes was implemented. This led to successful prevention of water leakage at the tail, which is one major drawback of the shield tunnelling method.

2. Improvements to backfilling :

    The improved tail seal promoted a transition from immediate backfilling methods to simultaneous backfilling. Further, two-part fillers which have a high strength in early stage have been developed since

113

around the middle of the 1980s. In our case, filling system from shield-mounted filling pipes was successfully achieved allowing early and ideal filling to be accomplished. This process was also a very effective way to prevent segment uplift.

3.   Improvements to information-assisted construction :

The advances in personal computers have made it possible to carry out calculations on site as rapidly and accurately as is possible with a conventional mainframe computer. Remarkable achievements, unprecedented in the field 25 years ago, are now possible, including precise management of slurry pressure, early detection of pump anomalies, detection of anomalies by monitoring processing equipment and other systems, and comprehensive control systems.

As this demonstrates, we partly owe the successful excavation of this Ac1 layer to improvements in shield technology made by our predecessors. The prime contributor, however, was ample preliminary investigations and reviews, coupled with comprehensive field management. This was achieved through the tireless efforts of all parties involved, including TTB and the contractors.

MARINE OUTFALLS PROJECT - ISSUES AND CHALLENGES

K. MUNZ   **G.R. HARIDAS**
Project Manager   *Technical Advisor*

Dyckerhoff and Widmann AG., Mumbai, India

SYNOPSIS

Mumbai-the commercial and industrial centre of India is acutely suffering from all amenities including housing, water supply, sanitation and transport, on account of rising population, which has touched nearly 12 million mark for an area of 603 sq km of the municipal limits of Brhanmumbai Municipal Corporation. Out of 2000Mld of sewage generated only 1.5% of which gets treated and rest all dumped into the sea without even primary treatment, causing widespread impairment of coastal water quality. The corporation finalised an integrated water management scheme under the World Bank Aid to solve the problem and one of the proposals covered Marine Outfalls, which is described here under. The proposal consisted excavating 2 nos of tunnels under the seabed at a depth of about 65 m going over 3.5 km to discharge the sewage after preliminary treatment to get a dilution of over 1 in 50 to ensure environmental friendly situation. The article deals with the excavation of the tunnels using a full face modern TBM with a segment erector with all necessary backup units, the problems faced during the excavation and the manner in which the same was solved. It highlights the role of Geology and site investigation, advance planning and the allocation of risks for proper and effective implementation of such hazardous projects.

1.0 GENERAL

Mumbai - the commercial and industrial centre of India, has witnessed a rapid growth of population during the past few decades. The 9.9 million, Population of 1991 census has now risen to nearly 12 million, reside within the Municipal limits of Brihanmumbai, with a land area of 603 sq km. The infrastructure development, however, has not kept pace with the population growth. As a result, the city suffers from shortage of all basic amenities including housing, water supply, sanitation and transport. Amongst these, the sanitation is perhaps the most neglected sector, with only 98.5% of 2000 Mld of untreated sewage generated is dumped into sea, resulting in widespread impairment of coastal water quality.

The integrated wastewater management scheme as finalised by the Brihanmumbai Municipal Corporation (BMC), divides the municipal region into seven drainage zones. For Colaba, Worli and Bandra, the plan provides for disposal by outfalls into coastal sea, with the other four drainage zones provided with aerated lagoons before disposal into the adjacent creeks. (Figs. 1 and 2)

2.0 MARINE OUTFALLS

This paper deals with Worli and Bandra - the two of the marine outfalls built to dispose of the wastewater into the sea after preliminary treatment. The salient design features of the scheme is to bore 2 tunnels nearly 30 m below the sea bed to a length over 3.5 km for releasing the wastewater under differential head, the details of which are provided in Table 1. The headwork for the outfall comprises of influent pump station, preliminary treatment facilities to screen coarse and fine grit and effluent pumping station to provide the motive force for wastewater disposal against the tide, when necessary.

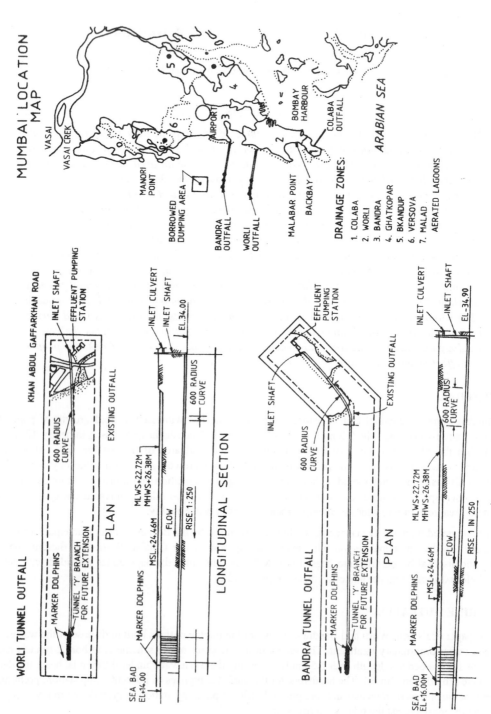

Fig. 1 : Drainage zones and general arrangement of outfall

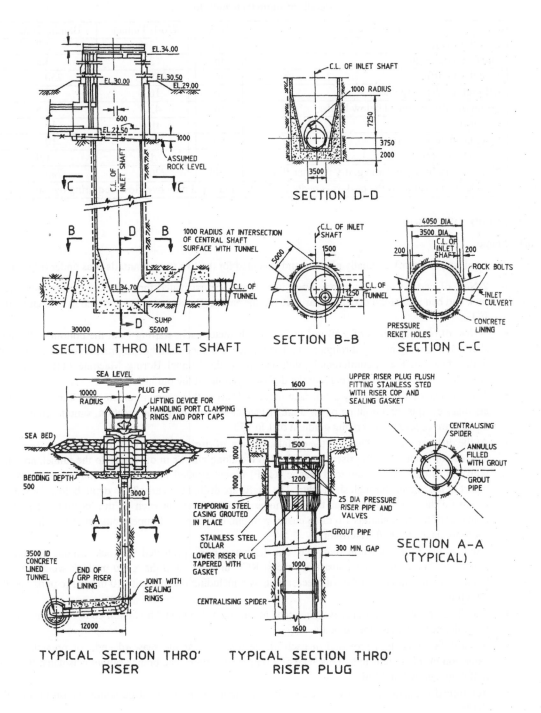

Fig. 2 : Details of inlet shaft and TBM assembly chamber at the start and riser diffuser structure at diffuser end

TABLE 1
Details of submarine outfalls

Particulars	Unit	Worli Tunnel	Bandra Tunnel
Length of Outfall	km	3.4	3.7
Finished Diameter	m	3.5	3.5
Length of Diffuser	m	250	250
Differential discharge head	m	16.4	18.6
Depth of diffuser tunnel below MSL	m	42.5	41.5
No. of risers	no	10	10
Riser head ports per riser	no	10	10
Riser Diameter	m	1.0	1.0
*Design Discharge under avg. dry weather flow	Mld	726	785
* Under wet conditions the peak is expected to be 2.5 times and may overflow, requiring by-pass.			

3.0 SCOPE OF WORK

3.1 Scope

The brief scope of work as finally planned and executed for the two tunnel outfalls is as under (Fig. 2) :

- 2-Nos. of vertical *in-situ* concrete lined inlet shafts are excavated through over burden, weathered rock and bed rock, for a depth of nearly 66 m, one each at Bandra and Worli, of 9.0 m O/D and 8.0 m I/D.

- 2-Nos. of outfall tunnels of 3.50 m I/D finished diameter each, starting from the bottom of the inlet shafts and passing mainly through rock. The tunnels are lined with bolted precast concrete segments with dowels in the circumferential joints except at the Tunnel Boring Machine (TBM) erection chambers, the backshunt and parts of the diffuser sections of the outfalls.

- The tunnels are excavated by a full face modern TBM with a segment erector, backup train and all ancillary equipment. Grouting of the annular face closely followed the erection, simultaneously. The tunnels are excavated to a rising gradient of 1 in 250.

- At the diffuser section of each of the outfalls, 10 nos. of riser shafts are constructed into the seabed in advance of tunnelling. The risers, spaced 30 m c/c along and 12 m off centre to the tunnel, are lined with 1.0 m dia., glass reinforced plastic (GRP) pipes, grouted into the rock and connected at the invert of the tunnels through short adits. The shafts are drilled to nearly 36 m depth below the seabed, to ensure their bottom is lower than the invert of the tunnel, at each of the locations. The outer diameter of the riser shaft is 1.6 m, to ensure a minimum 300mm of impervious grout fills the annular space to make the shaft watertight.

- Each of the riser discharge units has ten discharge ports protected by rock armour around the structures. These structures consist of precast RCC units of 7.0m dia x 1.0 m thick weighing ~ 80 tonnes. The maximum number of precast units is 8 including the unit with 10 discharge ports and the top plug.

- The diffuser section is progressively reduced at the end of the tunnel by *in-situ* concrete lining and in-fill concrete to increase the velocity.

- A short "Y" branch is provided in each of the tunnels near the seaward end, for possible future extension of the outfalls upto a distance of 7.0 km from shore. The branch tunnel is isolated from the outfall tunnel by bulkhead.

- Reinforced concrete inlet culvert at each outfall connects the effluent pumping station to the inlet shafts.

Besides the above main features, there are other relevant minor structures like – auxiliary, storm outlet culvert, bulkhead gates, penstocks, auxiliary outlets etc.

118

3.2 Contract

The contract was based on the FIDIC conditions and as such following the clause 12, covered the soil risk. Advance probing was a mandatory requirement.

4.0 GEOTECHNICAL SURVEY AND INVESTIGATION

4.1 Site Investigation

The Geotechnical investigation had been carried out in the period 1989-90. There were in all of 21 boreholes (17 vertical and 4 inclined) for Worli and 22 boreholes (17 vertical and 5 inclined) for Bandra alignments. They indicated tuff, volcanic breccia and basalt rock along the tunnel with very high RQD's (90 to 100%) and nearly zero fracture index. The permeability near tunnel levels were generally less than 3 lugeons (*i.e.* very low inflows). This signified massive rock structure with very few joints and low permeability. In general the rock strata in Mumbai region dips around 10° towards the West. Though the borelogs indicated very high RQD and low fracture index, the examination of cores from some samples exhibited the property of disking (parting of the rock core along several bedding planes). The long exposure could have attributed to the deterioration of the rock samples, due to the manner in which the cores had been handled in storing and transportation.

As the tunnel had to be excavated under the seabed, careful study of the furnished geological data was necessary. The detailed study of the said report revealed following :

4.2 Worli Outfall

The geological section showed 22% unweathered and strong basalt (rock) in various state with approximate 78% of the formation in tuff and tuff-breccia in all their compositions and states along the tunnel line. This formation could be classified as medium soft to moderately strong in strength. The high values of core recovery and RQD indicated the formation to be characterised with wider spaced fissures. There was no indication of existence of significant tectonic faults or fault zones. The measurement of lugeons recorded gave an indication of small to medium quantities of water ingress.

4.3 Bandra Outfall

The profile at this location revealed 18% Basalt and 3% Diorite rock in various states, all unweathered and very strong with low permeability. The proportion of the tuffs and tuff-breccias accounted for around 79%, with higher quantities of unweathered but weaker rock. The tuff was considered weak to moderately weak and the volcanic breccias moderately strong. Though generally the RQD values were greater than 90, at localised places they were 60~88%, which may not necessarily be resulting from faults. The rock formation was characterised by wider spaced fissures. There was, however, an indication of existence of a fault with strong shifting of the strata at around 3650 m station. This was evaluated with respect to the advance of the TBM, by making necessary provision for advance probing and the ingression of water. This was an indication that the water ingression in the Bandra tunnel could be slightly more than in the Worli tunnel.

5.0 SELECTION OF TUNNEL BORING MACHINE

The tender programme was based on the execution of both the outfalls by only one TBM capable of fulfilling all the tender requirements, in sequential order. The TBM had to be assembled and tested before the excavation and after the excavation dismantled removed to the second outfall for completion in the identical manner. The selection of the TBM was guided by the need to work under the geo-technical conditions described earlier, under the seabed, besides the erection of segments around 16~40 m behind the tunnel face. This was an indication of the need to permit relaxation in the *in-situ* stresses around the excavation, to ensure that the lining provided subsequently should not get overloaded by external stresses.

119

The TBM, thus, had to be an open type hard rock with a proven design and with a cutter-head of dismantlable type. After selecting two of the well known makes *viz.* Robbins and Wirth, the final choice was zeroed on to the Wirth machine TBS 111 400/450 E (Fig. 3). The salient details of this machine are as under :

Fig. 3

Machine type

- Cutter Head Dia...............................4.05 m
- Driving Power................................4 x 200 kW
- Torque with 0.95 efficiency ....1000 kNm
- No. of Cutters and their Size...…..............32 nos. and Disc dia. 431mm
- Limitations in Rock Strength….............300 MPa

Advance

- Advance Force............................…....8400 kN
- Boring Stroke.............................…....1200 mm
- Speed…...0 to 7.5 m/h
- Hydraulic Pressure...........................280 bars

Gripper System

- Grip Shield Nos...............................12
- Gripping Force.............................…..22,000 kN
- Hydraulic Pressure Max...................300 bars
- Surface Pressure @ Max. Force............~39 kp/cm²

Conveyor Belt
- Width...650 mm
- Belt Speed...................................2 m/sec.

Electrical Equipment
- Installed Total Power.......................~1000 kW
- Transformer Power.........................2 x 650 kVA
- High-Voltage...............................10, 000 V
- Operating Voltage.........................3 Phase, 380 V, 50 Hz,

Dimensions of the Machine
- Length......................................15 m
- Weight......................................200 t

This basic TBM is fitted with an advance probe drill, mounted 15 m behind, capable of drilling up to 120 m ahead of the cutter-head, to enable identification of strata, fissures/faults, water seams etc. so that necessary advance actions can be initiated to deal with the situations anticipated. In response to the Clients request, 1 no. of core drill is also installed to collect core samples up to 6 m ahead of the cutter head. Besides, this there are 2 nos. of vertical drills with articulated arms installed for rock-bolting purpose. The system enabled monitoring of water ingress by probing in different directions and then controlling the same by sealing with packers added by grout injection.

The TBM has been provided with the necessary segment installation equipment (L~19 m) trailing behind, with a comprehensive back up system of over all length of 120 m accommodating 16 nos. of wagons and a ramp. The erector is able to erect 2 nos. of rings each of 1.2 m from one position. Each ring consists of 8 nos. of precast concrete segments, of which the 2 segments to be erected at the top last are wedge-type with others cast in uniform size. The back up unit is well equipped with closed circuit TV for monitoring mining operation and guiding/correcting as necessary with the help of laser guided system. The mining operation is augmented by 2 nos. of trains pulled by diesel locomotives with 4 nos. of 9 cum muck wagons, 2 nos. of segment cars, 1 passenger car and 1 agitator car in alternative run. It carried a magazine accommodating 100 m length of ventilation ducts, which used to get installed at every 100 m of mining.

6.0 GROUND REALITIES

6.1 Previous Experience

The published reports on the tunnels excavated earlier for the water supply in Mumbai had revealed that the rock types to be predominantly volcanic breccia and basalt, with a small amount of tuffs generally containing varied thickness of breccia and shale layers. The compressive strength of rock was generally in the range between 17~70 Mpa and the E-value 0.07~1.5x10^{4}Mpa. There were layers of soft non-crystalline volcanic tuffaceous material along with belts of flaky laminated carbonaceous sedimentary shale and bands of volcanic ashes, clayey and sandy calcareous materials. Except at few locations, joints were compact. There were no rock falls, barring a few rock spallings on account of opening of bedding planes. At some local spots the seepage was around 6000 l/min. All these tunnels are lined with *in-situ* concrete after the entire excavations were completed, without any rock falls – an indication of long stand-up time.

6.2 Additional Bores

As the exploratory boreholes had been spaced at relatively large distance and there was a need for the additional bores to get a more decisive basic information. The contract provided for the additional bores. It was a mandatory requirement. Based on the data collected and that available in the tender an expected geological profile was plotted.

6.3 Additional Probing

It was admitted that the additional information could only be obtained by advance probing during the advance of the tunnel. A Tamrock HL 500S machine mounted on the TBM about 15 m behind the cutter-head for probing, which gave valuable information for the mining operations. An on-site geologist was assigned to record and monitor continuously the changes in the penetration rates, pressures applied, description of rock strata met, ground water inflows etc. The rock cuttings were helpful in identification of types of rock met, while penetration rates indicated the strength of rock. The comparison of the information so gathered with that carried out earlier along with the support measures undertaken there, helped to asses the support measures to be undertaken at the new location appropriately. It also helped in identifying the problematic zones in terms of ground stability, the location and order of water inflows expected etc., thus ascertaining its role in the mining operation.

7.0 PROBLEMS ENCOUNTERED

7.1 Rockfalls

The actual rock mass turned out to be the subaqueous lava flows consisting of compact, vesicular and amygdoloidal basalt, thick intertrappeann beds of volcanic breccia and tuff with intercalation of thin discontinuous shale constituted (Figs. 4 and 5). The formations exhibited 10° dip in the westerly direction. The intensity of rockfalls varied in different types, with no falls in basalt and diorite. The first large rockfalls were observed at Worli within the first 246 m from start and particularly in the crown portion for a height up to 1.5 m. above the tunnel periphery due to intercalation of shale. The major falls were in the layered tuff / volcanic breccia with shale intercalation or shale with minor tuff, with minor else where. The instability was mainly due to intersection of several joint sets in the crown below the bedding joints, dips in the formation, as also the weak rock conditions and *in-situ* stresses in the rock. Individual layers in these rock types varied from a few millimetres to over 0.6 m in thickness. These falls occurred in the form of sheets, slabs and blocks and were generally restricted to the crown portion, with some large rockfalls observed on the sides. At some places, weakness of rock led to the sinking of gripper pads of TBM.

7.2 Temporary Supports

With the first major unanticipated rockfall occurrence, the work suffered badly for nearly two weeks. The double arches of ISMC 125 beams had to be installed as supports at 0.6 to 2.4 m spacing as required, before proceeding. With stabilisation of rock by regular supports, only single/semi circular arches made from ISMC 100 channels were adequate as supports along with mesh. Subsequently, only steel straps (2 m x 15 cm x 1 cm) at 1.2 m spacing anchored through 2 nos. of swellex bolts were found to be adequate as supports. These swellex bolts come in 2 sizes standard and super, with shear loads at failure of 93kN and 245kN respectively. They are capable of accommodating 90~100% of their tensile strength under the shear loading. The standard bolt is of 32mm dia. in a folded condition and expandable up to 40mm by inflation. They are, however, anchored in 38mm dia., holes. As a thumb rule, the length of the bolt (Lb) in metres can be taken as under :

$$Lb \text{ in metres} = (2.0 + 0.15s)/ ESR$$

Where, s is the span of the opening in metres and ESR (Excavation Support Ratio) a number depending upon the excavation category, which varies from 0.8 to 5.0.

These swellex bolts came handy because of their ease of installation and assurance of immediate support. It is a versatile device in the form of folded tubular steel bolt when subjected to high water-pressure opens up to expand the bore hole slightly and inturn draw perfect locking all along the length to act as an anchor fastener. These straps were fixed in the crown of tunnel along with two swellex bolts (2 m long) and mesh. There were conditions where the support spacing had to be decreased to 0.6 m with additional supports provided on the sidewall portion. With the TBM's rapid rate of advance and confined workspace, a well-developed system of drilling and bolting is essential.

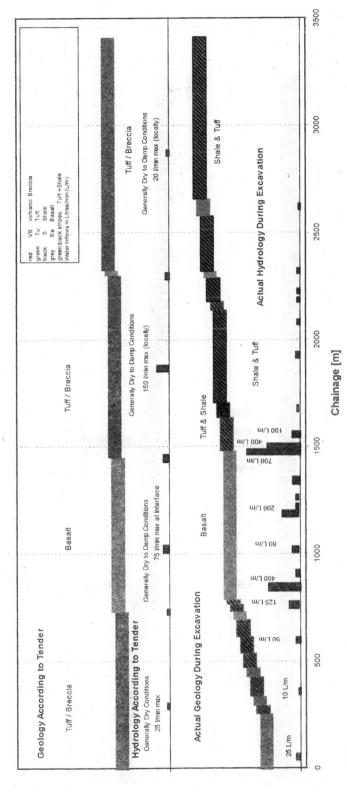

Fig. 4 : Comparison of geology and hydrology (tender and actual) for Worli Tunnel

Note: Water inflows shown are only those which were actually measured in some of the crown and sidewalls portions of tunnel. There were very often large inflows from invert and other inaccessible portions of tunnel where measurements could not be done. True inflows could be 2 to 3 times of those shown in above figure.

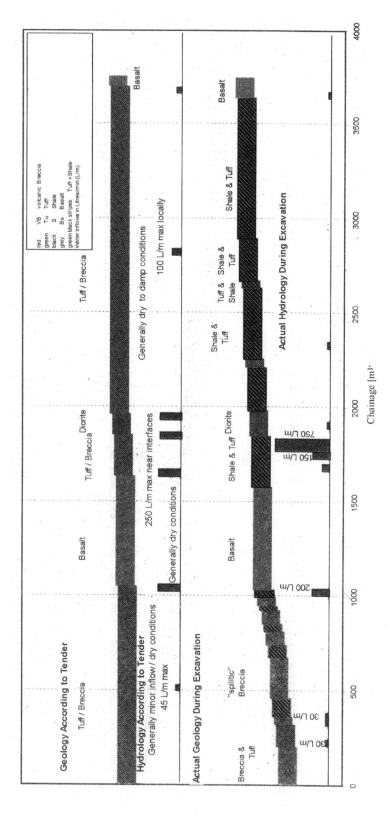

Fig. 5 : Comparison of geology and hydrology (tender and actual) for Bandra Tunnel

7.3 Clay Bearing Strata

Shale is a rock formed by consolidation of clay and these were sometimes found as seams interbedded with shale. The presence of clay often caused excavation problems, as it clogged the cutter heads, and several shifts were lost in cleaning up the cutter head. At one point 3 days were lost just on account of a major cutter-head blockage at Worli. As this clayey stratum was estimated to be about 40 m length, further mining operations were carried out without using the water sprinklers in the cutter head. Presence of water was found to aggravate the problem. This approach was found to be effective whenever shale or weak strata was encountered. Regular cutter head checks and cleaning became the need of the hour to prevent development of similar problem. The sinking of the erector sledge at the invert portion caused further delays in ring building. The erector had to be lifted up by means of jacks in such soft clay/rock portions of the tunnel.

7.4 Convergence of Tunnel

Yet another problem observed in shale rock or shale bearing strata was the convergence. This usually occurred in the tunnel roof section and sometimes also at the invert. The shrinkage of tunnel size was more noticeable in the vertical direction and found to be as much as 10 cm. The ring building had to be undertaken by jiggering at the heaved invert level to make room for placing rings. Similarly, there were difficulties in fixing the top segments of the ring. This demanded provision of closely spaced supports to minimise the intensity of convergence.

7.5 Water Inflow in Tunnel

Water inflows are expected phenomenon when mining under the seabed. There were differences in the flow characteristics between the Worli and Bandra tunnels (Figs. 4 and 5). At Worli basalt had large to very large (>125 l/min) inflows through vertical to nearly vertical joints or fracture zones. There were rarely inflows through bedding joints in basalt at Worli. At Bandra the water inflows within the basalt or diorite bodies were minor or moderate (through bedding or inclined joints). There were no fracture zones in these types of rocks at Bandra, like at Worli. However, large to very large inflows were encountered nearer to the interfaces of basalt or diorite with shale. Large inflows caused water accumulation at the invert level leading to delays in ring building. In order to facilitate ring building a barrier had to be built by using sandbags ahead of ring location, and the water accumulated pumped out. The salty water caused frequent electrical failures, despite use of high-protection rated cables. The seepage problem was tackled to some extent by using plastic sheets in the crown to divert the water along the tunnel sidewalls thus protecting the electrical cables as well as TBM equipment.

Continuous probing helped in identification of possible zones of water inflows ahead of the tunnel face. However there were instances when actual inflows were up to 3 times of those predicted from these probes. No grouting was called for water inflow up to 150 l/min. However, where inflows were much more than 150 l/min, generally 3 to 5 additional probe holes were called for and curtain grouting was resorted to with the help of packers. Using water to cement ratio 2:1 or 1:1 with about 2.5 % bentonite grouting was carried out. These holes were mostly in the upper portions of the tunnel. There was one instance in Basalt body at Worli where a total inflow from probe holes was nearly 1500 l/min. In this case 11 holes were drilled all around the tunnel periphery, and 20 cum. of grout had to be pumped, before we could control. However, the results were not always this good, as grouting in some cases helped in reducing the intensity of inflow and sometimes even diverting it to the other locations. In shale the water inflows were not localised to a few joints but spread through several bedding joints. No amount of grouting in this portion could help in reducing water inflows.

7.6 Problems Related to Hard Rock

There were frequent wearing out of cutter disks in hard rock like basalt and diorite due to the blockage of disks. A lot of dust pollution was also caused when mining through hard rock. Sprinkling water on the conveyor belt, at the time of mining controlled this. While probing in basalt rock, several drill rods and bits were lost due to breaking. The use of high tensile strength steel rods solved the problem.

8.0 DISCUSSIONS AND SUGGESTIONS

8.1 General

A successful tunnelling operation demands identifying and managing ground risks. It first calls for identification of areas to be considered to address the issue of risks and then deal with its management to increase the confidence of both the client and the contractor alike. It requires allocation of risks between parties, adoption of ground reference conditions and adoption and implementation of technology to suit the situation. During the construction, the partnering form of project management, stimulated with exchange of information and discussions, is called for. This generates a healthier work culture and atmosphere leading to continuous improvement in the production and quality of construction.

8.2 Role of Geology and Site Investigation

Geology has a very strong influence on the construction activity, particularly in the tunnel projects. Though the costs involved in carrying out site investigations are usually less than 2% of the project costs, misinterpretation may lead the projects to suffer severely when they overrun actual time and costs. In a project of the type under discussions, it is, therefore, important to have independent studies from experts in the geological field. The core samples need to be thoroughly examined by the geologist on the spot and fresh borelogs should be taken for confirmation when in doubt. It is preferable to spend a little more at the time and money during initial site investigation rather than face an irreversible situation later on, leading to cost and time overruns.

8.3 Advance Planning

A joint pre-and post–award planning needs to be drawn for the entire project to minimise the problems of risks arising during the construction so that it can minimise their effects. This would infuse confidence in both the client and the contractor to invest in the project, as each will be assured of being protected from the potentially large losses being incurred and hence be prepared to accept the challenge.

8.4 Allocation of Risk

Any project, and in particular a tunnel project, is required to be undertaken only after a detailed study of the basic requirements coupled with the ground conditions, giving due considerations to the cost and the time. No contractor is expected to have the benefit of knowing the ground conditions beforehand until he has plunged into it. He faces the unknowns and hence unanticipated risks leading to cost and time overruns, besides earning a bad name in the profession, for no fault of his. This calls for drawing first the jointly agreed terms and conditions for the execution of the contract between the parties, with known risks identified.

ACKNOWLEDGEMENTS

The authors wish to thank the Brihanmumbai Municipal Corporation and M/s Binnie and Partners (India) Ltd., in association with M/s Tata Consulting Engineers for providing all the co-operation in executing this challenging project. They also wish to place on record the services, devotion and commitments displayed by the young Indian engineers who mastered the mining technology to complete the second tunnel in record time, achieving a peak output of 555m of finished tunnel in a month, despite heavy odds.

THE STURE LPG STORAGE PROJECT

BENGT NIKLASSON
Senior Manager
Rock & Process Engineering

BJÖRN STILLE
Designer
Underground Structures

LARS ÖSTERLUND
Geologist

Skanska Teknik AB, Sweden

SYNOPSIS

In order to extract the lighter products from the incoming North Sea crude oil the Norwegian oil company Norsk Hydro in 1997 decided to build a LPG extraction plant at their crude oil terminal close to the city of Bergen in Norway. Included in this project was an underground LPG storage facility of 60000 m^3. Skanska Teknik was awarded the design contract from Raytheon Engineers and Contractors - the main contractor. Beside the design contract Skanska Teknik also signed the contract of cooling down the cavern since the design concept was to refrigerate the storage. A Norwegian contractor carried out all construction works, above and under ground.

The storage facility comprises an access tunnel, a water curtain, a shaft system and a storage cavern. The cavern is surrounded by a frozen zone in order to prevent leakage of LPG. Before the cavern was put into operation it was pre-cooled down to a temperature of -20°. After that the storage was inerted and the LPG was sprayed into the cavern at a temperature of -36°. The main concerns in the view of design was the sealing of the cavern, the effect of the water curtain system and the mechanical rock behaviour during the cooling down process.

1.0 INTRODUCTION

In the Øy-garden archipelago some 50 km NNW of the city of Bergen is the co-owned Sture Terminal for reception and dispatch of North Sea oil. The terminal showed in Fig. 1 is operated by Norsk Hydro. The oil is transported through a 113 km long pipeline from five oil fields in the North Sea. The oil is stored in five underground rock caverns with a total capacity of 1 million m^3 from where it is shipped out in tankers. About 120.000 m^3 a day is received from the North Sea fields. The terminal area holds two jetties admitting tankers of up to 300.000 DWT and with a loading capability of 2 x 11.250 m^3 per hour. Worth noticing in this context is that about 1% of the worlds crude oil passes through the Sture terminal .

During late autumn of 1997 a contract was signed for the construction of an underground, refrigerated LPG cavern for propane/butane extracted from incoming crude oil. In competition with two other groups of contractors, Raytheon Engineers and Constructors got the contract involving the rock cavern, the gas extraction plant, the plant for reception of cooling water, pipe-work, harbour constructions, pumps etc. Skanska Teknik was the consultant to Raytheon on the designing of the rock cavern and its refrigeration. The actual cooling down work was also executed by Skanska Teknik.

Skanska Teknik executed the preliminary studies, the geotechnical field investigation, and the detailed design of all underground construction work. In addition to this the storage has been chilled to -20 degrees by means of air cooling. This activity was completed in March 1999 and the cavern was then sealed. The next step was to inert the storage before it was filled with LPG. This further lowered the temperature to -35 degrees. Norsk Hydro took the facility into operation during autumn 1999.

Fig. 1 : The shaft top location at the Sture terminal

2.0   DESCRIPTION OF THE CONSTRUCTION SITE

The facility is situated in the southern part of the terminal area in conjunction with the south jetty. In the preliminary design, the intention was to drive the access tunnel to the gas storage from the existing underground cavern area situated further north. This would involve pumping dry the old tunnel, scaling and possibly carry out rock reinforcing before work on the new access to the LPG cavern could be started. Skanska Teknik instead proposed that an access tunnel could be built from surface level down to the LPG cavern as the topography of the area was suitable. This way there would be no need to open up the old installation. The access tunnel has a cross section of 40 m². Two water curtain tunnels branch out from the access tunnel. The underground cavern has an effective storage capacity of 60.000 m³. The roof of the cavern is at −29 m level. Fig. 2 shows the general layout of the Sture Storage facility. The gas will be stored at a temperature of approx. -35°C, and with an over pressure of 0,1-0,5 bar. The rock surrounding the cavern is frozen to avoid inflow of water or gas penetration from the cavern out into the surrounding rock. A temperate water curtain (with a pressure higher than the max 0,7 bar pressure of the cavern) is located above this area. This water curtain serves two purposes. One of filling any developing fractures in the frozen bedrock with water that immediately would freeze and seal of the fracture, and another of stopping the frozen zone from reaching the ground surface and resulting in permafrost.

3.0   GEOLOGY AND GEOTECHNICAL FIELD INVESTIGATION

The bedrock in the area mainly consists of various types of Precambrian gneiss over 700 million years old, Fig. 3. The gneiss is medium - to coarse-grained. The topography of Sture Island is dominated by a number of mountain ridges, formed by glacial erosion, running in a NNW direction. The rock stratification runs NNW dipping 20-50 degrees to the east. The main joint systems that can be observed in the area are striking between NW and NE, and dipping 70-90 degrees to the east. The number of fracture density is quite low - some 2-8 per cubic meter of rock.

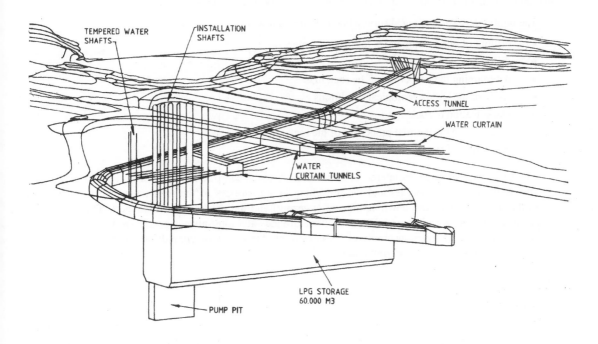

Fig. 2 : General layout of the Sture LPG storage

Fig. 3 : Precambrian gneiss

The major weakness zones/shear zones running across Sture Island in a N-S direction could sometimes be clay filled with a clay thickness of anything from 0,2 –2,0 m. More often these clays are of a swelling type. Water loss measurements show that the rock mass has a low permeability, $10^{-8} - 10^{-9}$ m/s. Groundwater flow could therefore mainly be expected in the N-S running, fairly sparse, joint systems.

The preliminary geo-technical survey involved the following :

- drilling of observation holes for ground water
- drilling of core holes
- determining the properties of the rock mass
- study of the hydro-geological conditions
- classification of rock mass
- determining the thermal properties of the rock mass
- measuring of stresses in the rock mass

There are two main areas of interest in a survey, prior to the construction of a frozen rock cavern for LPG gas. One is the thermal properties and the prevailing in situ stresses of the rock mass, because these data govern the design of the cool down equipment (freezing) and the degree of rock support. The other is the groundwater situation in and around the installation. Groundwater must always be present in, and around, the frozen rock to assure the impermeability of the cavern, but the inflow of groundwater to the cavern must be kept at a minimum as it affects freezing efficiency.

4.0 DESIGN WORK

4.1 General

Basis for the design have been the field investigation, site inspections, and the concluding geological engineering report for the Sture 4 oil storage project. Supplementary guidelines on construction, and specifications for execution, supervision, control etc. have been given by Norsk Hydro.

Three sets of documents have been produced to ensure well-defined and systematic construction solutions :

- Design specifications, description of constructions together with all required data, client's requirements, calculations and design solutions
- Technical specifications, technical descriptions, building specifications with all requirements
- Drawings and other plans describing the layout

A major part of the design work has been to define and examine the critical activities that have been identified in the preliminary phase of the project, *i.e.* gas tightness, maintaining a stable groundwater level and refrigeration stability.

4.2 Gas Impermeability

A groundwater model has been established using data from the geotechnical field studies and earlier experiences of the area. The rock mass is generally considered to be very low permeable with minor leaks along the essentially vertical fractures. No major aquifers in the area have been found, for which reason the area is considered sensitive to outflow and lowering of the groundwater level, *i.e.* it can take a long time to build up the groundwater pressure in spite of the high rainfall in the area. Rainwater will mainly run off as surface water.

In order to meet the client's requirements of a maximum water inflow of 40 m³/week, and to maintain the groundwater level, it was considered necessary with continuous probe drilling and monitoring of in and outflow of water. Limits for pre-grouting was set at 1 Lugeon during water pressure tests. In spite of the low inflow allowance, infiltration was at an initial stage considered necessary to maintain an acceptable groundwater level in the rock. However, this was never necessary to perform.

The following scenario has guided the design work in order to attain gas impermeability, *i.e.* assurance against gas leakage :

1. LPG leaks through a crack that has dehydrated.
2. LPG leaks through a crack in the frozen zone, the ground water pressure is too low to hold back the gas.

Based on the hydro-geological model and the requirements for gas tightness, it was concluded that a water curtain was necessary to include in the design. A search for literature on similar projects/problems resulted in works such as "Gas tightness of unlined hard rock caverns" (Kjörholt) and "Water curtains in a gas storage – an experimental study" (Söder). However, only very few underground frozen gas storage's have been built and experience is quite limited.

Controlling the limits of the frozen area sets special demands on the design of the water curtain. The water screen for the cavern was designed with the use of linked finite element method calculations, flow analysis and reference literature. During operation temperate water will circulate in the water curtain and the access tunnel.

Fig. 4 shows the extension of the frozen zone after 10 years.

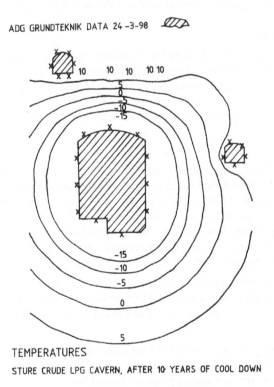

ADG GRUNDTEKNIK DATA 24-3-98

TEMPERATURES
STURE CRUDE LPG CAVERN, AFTER 10 YEARS OF COOL DOWN

**Fig. 4 : The plot shows the extension of the frozen zone after 10 years of cooling - bold line.
The access tunnel is filled with tempered water**

4.3 Rock Reinforcement

The designing of the underground rock reinforcement work was split into two parts: the access tunnel and the rock cavern.

The Q-method was used when designing the reinforcing programme in the access tunnel, where the rock was defined as :

Reinforcement Class I :   block size >0.6m, Q>4, selective bolting

Reinforcement Class II :   block size <0.6m, 1>Q>4, systematic bolting pattern

Reinforcement Class III :   block size <0.6m, partly with clay-filled fractures, 0,1>Q>1, systematic bolting pattern for roof and wall

Earlier projects in the area have encountered problems with swelling clays. Specially designed reinforcement programme were constructed at major clay zones according to the recommendations of Broms and Heiner in "Reinforcement of clay zones in rock cavern installations", 1979.

Concerning the cavern, it was necessary to do a careful study of the rock characteristics and how it was affected by freezing. For this reason a FEM calculation was performed linking thermal and rock mechanical characteristics to study the total stability during cooling down. The block stability of the rock mass was analysed for the main fracture orientations observed. The FEM calculations and the block stability analysis served as a basis for the design of the reinforcement work to be done. Rock reinforcement of the cavern roof was done as a systematic 2 x 2 m bolting pattern with a shot crete layer of 80 mm. The wall rock reinforcement was designed at site by Skanska Teknik in co-operation with the sub contractor. Only fibre-reinforced shotcrete was used in the rock cavern. All shotcrete was anchored with CT-bolts, fully grouted hot dip galvanised anchor bolts, of 4-6 m length.

4.4 Monitoring Programme

The groundwater levels and the effects of the excavation of the rock cavern are regularly checked in observation wells. Convergence gauging and optical measurements were carried out to verify rock-stress and total stability. A mini strain gauge extensometer was used to monitor the roof deformation during the cooling down phase. Temperature gauges were installed around the cavern in order to monitor the expansion of the 0° isotherm. During blasting work, adjacent installations was monitored using vibration measuring equipment.

5.0 ROCK EXCAVATION

Excavation has been done using conventional drilling and blasting techniques. Holes for blasting, probing, grouting and bolting were drilled using an Atlas Copco Rocket Boomer 353. An Atlas Copco Roc 742/642 HC was used for bench drilling of the upper bench. A Boart Aquamaster 300TS well-drilling unit with down-the-hole-hammer, was used for the water curtain holes. Rock excavation commenced in February 1998 and was finished in November the same year. Excavation and removal of muck, was done working North Sea shifts, *i.e.* 120 h/week. All cycle activities, rock excavation, drilling, charging, mucking and reinforcement work have been done by the same crew enabling full responsibility for the total process.

The access tunnel is 380 m long and has a cross section of 40 m^2, approx. 6,5 x 6,5 m. Two water curtain tunnels, each 30 m long and with a cross section of 38 m^2, branch out from it. A total of 21 water curtain bore holes have been drilled from the water curtain tunnels as a shield 14 m above the cavern roof. These bore holes have a diameter of 110 mm.

There are two entrances to the rock cavern from the access tunnel, one leading to the gallery level and the other to the upper bench. These tunnels have a cross section of 25 m^2 and are 25 m long. The reason for the smaller cross section is that the concrete sealing plugs are located here.

The rock storage has an effective storage capacity of 60.000 m^3, with the dimensions 118 x 21 x 29,5 m (L x W x H). The cavern is divided into a gallery, two benches and a pump pit.

The height of the gallery is 8,5 m, the upper bench 10 m, and the lower bench 11 m. Drilling of gallery and lower bench was done with horizontal drilling, while the upper bench was drilled with vertical holes. Blasting of the upper bench was done using pre-splitting along the contours, while the gallery and lower bench are blasted using conventional perimeter blasting technique. Drilling of the upper bench was done with a 51 mm drill bit in contour and helper bore holes and with a 76 mm drill bit in the stoping holes. A 51 mm drill bit was used for all bore holes in the gallery and lower bench.

The pump pit, 10 x 4 x 15 m (L x W x D), was excavated in 3 benches. Mucking was enabled by a clam shell bucket mounted on a hydraulic boom ranging 16 to 18 m down. Fig. 5 shows this special equipment.

Charging of the bore holes was done as follows :

contour holes – "gula rör", contour explosive in 22 mm diameter plastic pipes

helper holes – "vita rör", in 25 mm diameter plastic pipes

stoping and cut holes – ANFO, Ammonium Nitrate and Fuel Oil.

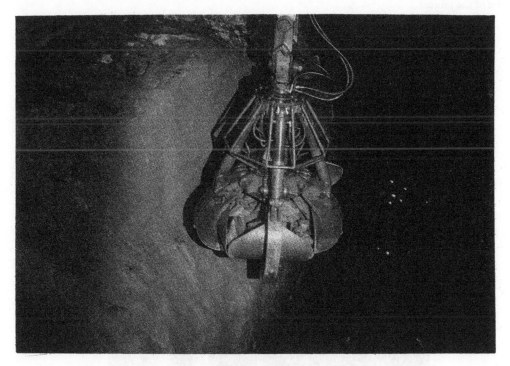

Fig. 5 : Shell bucket equipment for mucking of the 15 m deep pump pit

All rounds were initiated using Non-electric detonators.

For mucking a Caterpillar 980 using a side-dumping bucket was used. Iveco trucks, with a capacity of 10 m³, were used for transporting blasted rock.

Shafts for installation of pumps and monitoring equipment were drilled by raise-boring in order to obtain maximum drilling precision. A total of six shafts have been drilled at the eastern end of the rock cavern.

6.0 EXPERIENCES

6.1 Design

The set-up used for the designing of the project worked very well. A very fast start up of the project resulted in a forced design phase. Review proceeding was solved in a pragmatic way with representatives from the Client and Raytheon, the main contractor frequently visiting Skanska Teknik in Stockholm. In case of revisions, these were made during the ongoing meeting and approved drawing could be sent to Norway the very same day. Skanska Teknik has had a site manager in Norway as a link between designer and constructor, and to check sub-contractors work during time-wise critical construction phases. The design work has been using the "Active Design" philosophy where "Prognosis, Observation and Action" are the key words. The prognosis is based on calculations, the field investigation and experiences from the other underground projects on the Sture island and elsewhere. The observation phase comprises not only the results from the monitoring programme but from all observations including every day visual inspections at site. If observations indicate a discrepancy between the planned and actual some kind of action must be taken. This action is already planned and described in the action plan based on for example a risk analysis assessment where scenarios of unforeseen events are described. Skanska Teknik's site engineer has been a member of Raytheon's construction management team. This way of working tightly with the client and production people is essential for the success of the project.

Fig. 6 : The photograph shows the 29 m high cavern. The air cooling unit could be seen
on the right side of the cavern. Spray pipes are mounted in the cavern roof

6.2 Thermal Properties

Information from the geotechnical investigation has clearly pointed at the difficulties when it comes to certain testing methods for correctly describing the thermal properties of the bedrock. The samples have been taken from drill cores and been analysed by two from each other independent laboratories using different methods of analysis. Unfortunately the analyses have given different results, and in some cases not corresponded with values for the type of rock in question as described in reference literature.

6.3 Ground Water

In a rock mass with low permeability it is very important to monitor the groundwater pressure level during excavation. Sometimes this can be difficult to check through observation holes as the water only appears locally in those fractures that can carry groundwater. It is difficult to pinpoint these fractures with observation holes. One bore hole can show an unchanged water level, whereas the rock mass a few meters away can have a significantly lower groundwater level. A programme was therefore devised to measure the pore pressure in the tunnel and in the gallery of the cavern.

6.4 Excavation

Excavation presented no problems. It is worth observing that pre-splitting in the rock cavern and in the pump pit has been very successful and resulted in smooth walls.

6.5 Deformations

Only small deformations where monitored in the cavern roof during cool down. Expected deformations where less than one millimetre per m. Monitored where 0.1 mm/m.

6.6 Time Schedule

The cavern project was completed in 22 months. Approx. 10 months for rock excavation works including shafts and water curtain system. Actual cool down period was 4 months. The rock surface inside the cavern was at that time -25° and the 0° isotherm 4 m inside the rock. Calculated cool down time was in accordance with actual. No gas leakage has occurred.

6.7 Ground Freezing

The Sture Project together with a similar refrigerated LPG storage project in Sweden that was running parallel to the Sture Project has given Skanska Teknik very good experience in ground freezing technology. This has recently been utilised in tunnel projects in Sweden where ground freezing has been used to stabilise soil and clay material enabling tunnelling through parts with no rock cover.

HIGHWAY TUNNELS DESIGNED TO MEET INTERNATIONAL STANDARDS

D.B. POWELL **E. HANSON** **D. LEVERENZ**

Mott MacDonald, UK **Wilbur Smith and Associates, India**

SYNOPSIS

The Anik-Panjarpole Link Study carried out by Wilbur Smith and Associates (WSA) on behalf of the Mumbai Metropolitan Regional Development Authority (MMRDA) is part of a strategic plan to relieve congestion round the metropolitan area of Mumbai. A section of the proposed twin 3-lane highway from station 3+600 to 4+000 passes through a ridge of basaltic lavas and could be constructed either in cut and cover or tunnel. To develop a cost comparison, concept designs were undertaken for both options and the cost benefits, in terms of capital and whole life costs, assessed.

Mott MacDonald were commissioned by WSA to carry out the tunnel concept design. The request was for a tunnel that met current international standards. The AASHTO standards (1986, 1993) applied for highways design were used for defining the traffic and equipment gauges. The guidelines published by PIARC were used for the design of the ventilation, lighting and associated systems to meet normal, peak and emergency operational requirements. In addition the provision of a safe dry comfortable environment for the driver was given a high priority.

Installing mechanical and electrical systems, tunnel finishes and maintaining a dry drip free tunnel come at a cost. This paper reviews some of the technical aspects of the concept design and the cost implications of this type of construction. These costs are compared using local and international costs where appropriate. The costs in relation to fatalities are also assessed. These help place the provision of a safe driving environment into context.

1.0 INTRODUCTION

The Anik-Panjarpole Link (APL) is part of the Mumbai Urban Transport Project (MUTP), Phase II, a multi-modal project developed with World Bank assistance to bring about improvement in traffic and transportation in the Mumbai Metropolitan Region. This new dual 3-lane 5.2km long road will create a new link between the existing urban core and port areas and the rapidly developing areas to the north and east. It will provide an alternative route for heavy truck traffic that currently passes through exiting residential and commercial areas, (Fig. 1). Preliminary engineering for the APL was completed in early 2000.

On all projects, especially those in highly populated areas there are constraints. The corridor and vertical alignment chosen for the link avoids most of the existing infrastructures, including railways, power supply routes, industrial concerns and local housing developments. What was difficult to avoid was the prominent ridge located between stations 3+600 and 4+000 and the local village communities that are situated in and around this area, (Fig. 2). Earlier studies proposed an open-cut solution through the ridge while acknowledging that there would be a significant impact on the villages of Ashok Nagar, Sahyadri Nagar and Sautam Nagar, all of which have taken on a more permanent aspect over the years.

The open cut would be excavated in weathered and fresh basaltic rocks and would have a major impact on the skyline. In view of the topography and other environmental issues, particularly the impact on the local communities, the designers considered a tunnel option for this section. To assist in the design, a further phase of site investigations was carried out to better define the ground conditions and the mechanical properties of the basaltic lavas so that realistic estimates of the tunnel construction costs could be prepared.

Additional traffic studies were also performed during preliminary engineering to provide updated predictions of the number and mix of vehicles likely to use the link road. This information is important in order to develop the ventilation system. Recommendations for controlling exhaust emissions have a significant impact on the capital and whole life operation and maintenance costs of highway tunnels.

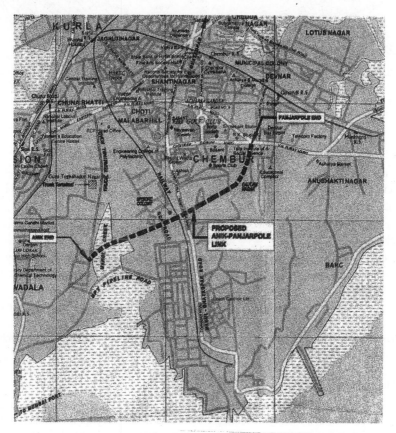

Fig. 1 : Project location

In addition, visits were made to two major highway tunnelling contracts currently under construction near Mumbai as part of the upgrading of NH4. These twin 4-lane tunnels are among some of the largest in the world in terms of span and length and offered a means of evaluating the behaviour of the basalts and the effectiveness of different support systems. They also provided valuable information for developing a reasonably accurate cost estimate for the Anik-Panjarpole tunnels. This was important since the cost of tunnel construction is high, usually of the order of 4 to 10 times highways constructed at grade and often 2 to 4 times the cost of cut and cover solutions. In this case, even with the grades adjusted to optimise the amount of excavation, the open cuts were deep. When other factors such as environmental impacts and aesthetics were considered a good case could be made for promoting a tunnel option.

2.0 PROJECT DESIGN BASIS

The following section discusses the design criteria and assumptions used in developing the tunnel concept design. In terms of cost, the civil works are the major component. However, in terms of operating safely and providing the driver with a comfortable environment, ventilation, lighting, finishes and other mechanical and electrical (M&E) systems, including management and control to maintain the tunnel environment, are important.

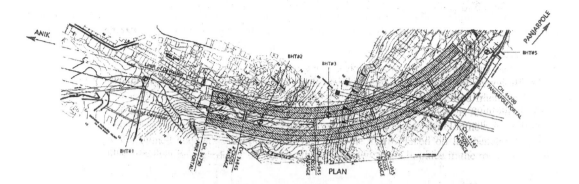

Fig. 2 : Horizontal alignment of the proposed tunnel section

The electrical and mechanical systems should be designed to ensure that the tunnels will operate safely and in a satisfactory manner at all times. The operating conditions that need to be considered in the design can be broadly categorised as :

Normal Operations - These exist when the traffic is free flowing. This mode of operation should prevail for the majority of the time.

Abnormal Operations - Congestion occurs when traffic is inhibited from flowing freely. This can occur as a result of maintenance being carried out, vehicle breakdown, extreme weather conditions (*e.g.* torrential rain), etc.

Emergency Operations - These will involve the intervention of emergency services (*e.g.* fire, medical, police).

The various systems that are installed are also required to react individually and in combination to meet all of these operating conditions.

2.1 Tunnel Alignment Options

The tunnel option allowed gradients to be kept to 0.90% or less, Fig. 3. This optimised the requirements for road drainage and, more importantly, reduces the emissions from heavy goods vehicles to low levels. This was an important advantage over the cut and cover option where much higher gradients, above 3 per cent, were required to keep the quantities of excavation to a minimum.

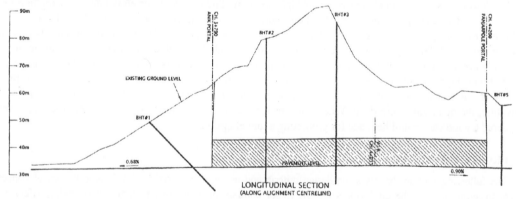

Fig. 3 : Proposed vertical alignment and tunnel length of 430m

2.2 Tunnel Cross-section

The tunnel cross-section is typically determined by the need to provide a stable excavation, sufficient space for installation of temporary and permanent support and the traffic envelopes necessary to provide the clearances for safe operation of a 3-lane tunnel, (Fig. 4). In addition, equipment gauges for the installation of ventilation fans, lighting, signage and traffic control and surveillance systems have to be specified. Space also needs to be provided for emergency walkways, ducting, tunnel finishes, niches, drainage and the road pavement.

As discussed later, excavation is the largest single cost item for tunnel construction and it is important to keep this to a minimum.

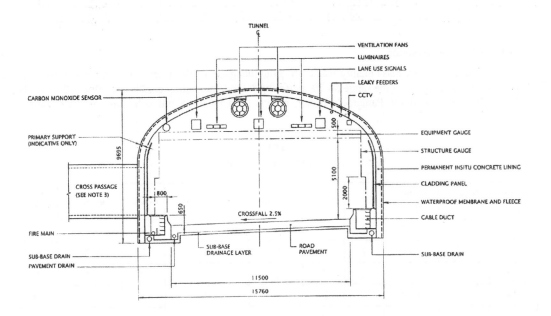

Fig. 4 : Tunnel cross-section and key design elements

2.3 Geological and Geotechnical Conditions

The site investigations established that the majority of the tunnel would be constructed in amygdaloidal vesicular basalts. These are relatively massive and impermeable when fresh and Rock Quality Designations (RQD's) are in excess of 70 per cent. The exceptions are a number of small isolated zones representing the contacts between different flows and narrow tachylitic basalts that have been subject to hydrothermal alteration. While further investigations would be necessary to locate any of these close to the tunnel, in general similar geological conditions on the NH4 Panvel Tunnel provided a very good medium for tunnel construction.

The geomechanical properties of the basalts are summarised in Table 1. The high strength-stress ratios (10 or more) indicate that the rock mass behaviour will be predominantly elastic and the excavations will be self-supporting providing that the potential for block failures is controlled.

Groundwater levels are close to the tunnel. Values of hydraulic conductivity were in the range 1×10^{-7} to $1 \times 10^{-8} ms^{-1}$. The structure of the lava flows and the low permeability values indicate that the tunnel is likely to be dry.

TABLE 1
Geomechanical properties

Borehole Number	Depth (m)	UCS (kg/cm^2)	PLI (kg/cm^2)	Dry Density (kg/cm^2)
1	35.70	693	-	2.81
1	42.35	-	13.72	2.50
1	68.50	715	-	2.63
1	72.40	-	37.69	2.78
3	22.96	297	-	2.53
3	32.95	397	-	2.60
3	42.80	273	-	2.54
3	59.30	415	18.86	2.70

2.4 Tunnel Support Systems

An assessment of rock mass quality was carried out using the RMR system of Bieniawski (1976), Table 2. Fresh basalts have values in the range 64-70 indicating "Good" tunnelling conditions. Where the basalts are weathered, for instance at the Panjarpole portal, ratings drop to 49-55 but still indicate "Fair" tunnelling conditions.

TABLE 2
Average rock mass ratings

Borehole Number	Rock Type	Depth (m)	UCS	RQD	Spacing	Discontinuity	GWT	Reduction	RMR
T1	W	0-18	2	13	8	26	8	-2	55
T1	F	18-95	7	17	10	28	8	-2	68
T2	W	0-1	2	3	5	26	15	-2	49
T2	F	1-50	4	20	10	30	8	-2	70
T3	W	0-18	2	3	5	26	15	-2	49
T3	F	18-75	4	17	10	27	8	-2	64
T5	W	0-6	2	3	5	26	15	-2	49
T5	F	6-25	7	17	8	28	8	-2	68

The primary support classes have been developed using the RMR recommendations, and combined with checks for specific block failures and the potential for slabbing in the crown where the contacts between flows are close to the tunnel perimeter. Rockbolts and shotcrete have been specified for a range of support classes in order to maintain a stable excavation, Fig. 5. As experienced during construction of the Panvel tunnel, most of the tunnel length is anticipated to require minimal support - mostly spotbolting and shotcrete applied locally.

To provide long term safety, low maintenance costs and a comfortable driving environment many highway tunnels employ either cast-in-place concrete or fibre reinforced shotcrete permanent linings. A majority of these, for operational and maintenance purposes, and particularly those that will be heavily used, prefer a cast-in-place concrete secondary lining and this was recommended for Anik-Panjarpole. The design requirements for the lining in hard rock conditions are relatively simple. For a self-supporting excavation that is stable, ground loads in the long term are small or negligible and hydrostatic pressures nominal provided that an invert drainage system is provided. A minimum concrete thickness of 300 mm was specified for ease of placement, Fig. 4.

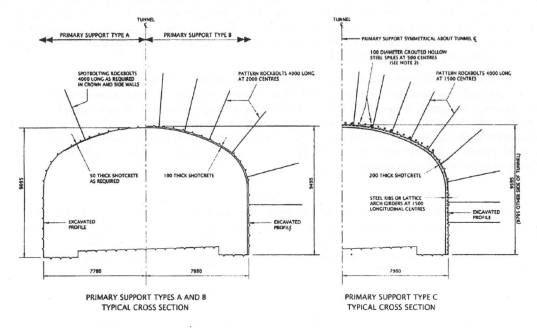

PRIMARY SUPPORT TYPES A AND B
TYPICAL CROSS SECTION

PRIMARY SUPPORT TYPE C
TYPICAL CROSS SECTION

Fig. 5 : Primary rock support types

2.5 Tunnel Ventilation

Tunnels longer than 300 m (approx.) require mechanical ventilation.

The tunnel ventilation system for the proposed tunnel was designed to:

- Dilute vehicle exhaust emissions to acceptable levels during all operating conditions;
- Control smoke in the event of a fire occurring in order to assist evacuation procedures and fire fighting;
- Ensure that air temperatures in the tunnels are kept at acceptable levels;
- Air quality at the portals and the regions surrounding the portals are acceptable.

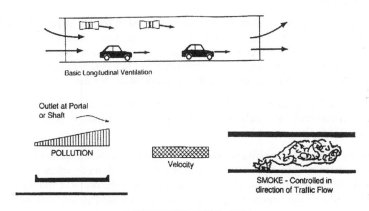

Basic Longitudinal Ventilation

LONGITUDINAL VENTILATION

Fig. 6 : Schematic of longitudinal ventilation

141

For the length of the tunnel, approximately 430 m, with the traffic flows given in Table 3, longitudinal ventilation is appropriate, Fig. 6. This is the most common type of ventilation system used for short tunnels since it places the smallest operating burden on the fans and does not require separate air ducts, making it the most cost-effective option available in terms of both capital and running costs. The design is based on the estimated values of peak and off-peak traffic flowrates, traffic mixes and air quantities discharged to the atmosphere at the portals. These assumptions provided a first estimate of the air quantities required in the tunnels for the range of operating conditions.

In its simplest form, longitudinal ventilation consists of a number of jet fans mounted in the tunnel crown which are used to compliment the natural piston effect of uni-directional traffic flow to maintain air quality in the tunnels at acceptable levels. In the event of a fire incident, longitudinal air movement is used to move the smoke in one direction such that evacuation can take place into the stream of fresh air. In a tunnel containing contraflow traffic, of the order of 3 per cent of usage for most tunnels, it is inevitable that smoke will be blown towards people on one side of the fire. Where this is the case the fan control system has to be set to switch on fans to give the best possible solution for smoke control.

TABLE 3
Projected traffic on APL, in number of vehicles per day in one direction, for years 2003, 2013, 2023

	2003	2013	2023
Cars	22687	36910	51910
Buses	3660	5953	8373
Light commercial	1830	2977	4186
Heavy commercial	8416	13693	19257
Total all vehicles*	36593	59533	83726

*Assumed peak flows 6% of the daily figures above.

This first stage of design enables the electrical design, civil design and environmental dispersion modelling to be progressed.

2.6 Tunnel Lighting

For reasons of road safety, tunnels over about 100 m in length, depending upon the tunnel geometry, should be illuminated and are designed to the same principles as ordinary road lighting. However, there are special features that need to be considered because of the confinement. This particularly applies to the transitions at the portals where vehicles move from daylight to a darker interior and vice versa. It takes time for the driver's vision to adjust and the interior lighting should allow for this.

BS 5489, Part 7 was used for designing the tunnel lighting. This standard is based on the work of the Commission International de l'Eclairage (CIE) and the Permanent International Association of Road Congresses (PIARC).

Safe driving demands that obstacles on the carriageway of the tunnel are recognisable at the appropriate stopping sight distance. The eye detects obstacles better and more quickly with increasing background luminance and with increasing luminance differences between the obstacle and the background. In this regard, luminance is defined by the amount of light from the surface carriageway and walls that is reflected into the driver's eyes.

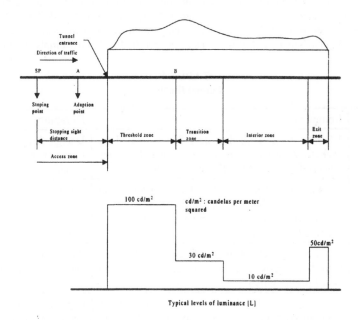

Fig. 7 : Longitudinal cross-section of one-way tunnel showing typical lighting levels

The luminance is influenced by the type of illumination and the requirements for transition zones, particularly at the tunnel entrance and exit, to cope with the difference in light levels between daylight and the tunnel luminance. The separate zones of illumination that need to be taken into account are indicated in Fig. 7, and typical levels of luminance are shown.

All of these zones need to reflect the driver's perceptions so that he is aware of other vehicles and able to cope safely with any emergencies.

The lighting design was undertaken with luminaires mounted from the tunnel roof as shown in Fig. 4.

2.6.1 *Control of Luminance*

The ambient level of luminance outside of the tunnels will vary from very high during some parts of the day, when the sun is at its brightest, to quite dull during stormy or cloudy conditions. In order to meet the ratio of external to internal luminance required by the Standards the number of luminaires switched on at any one time will require adjusting automatically. To achieve this, photometers mounted external to the tunnel are arranged to measure the portal luminance. Lighting on the approach roads to the tunnel should have a luminance of at least 1.5 cd/m squared and it is preferable to provide dark retaining walls and surrounds, to assist with the driver's ability to adapt to the sudden changes in luminance levels.

In the event of a loss of the mains power supply the lighting in the tunnel would switch off. A number of luminaires are designated for emergencies with a requirement for them to be uniformly distributed throughout the tunnel and connected to a maintained power supply for a period of time up to one hour to enable the tunnel to be evacuated in a safe and orderly manner. The level of illuminance provided under these conditions is usually in the order of 10% of the minimum daytime level of illuminance. The tunnel should not be used for traffic in these lighting conditions.

143

2.7 Tunnel Finishes

It is increasingly recognised that tunnels require not only carefully designed lighting but appropriate wall finishes to enhance driving safety. There is a wide range of finish options that could be used and some of these are listed in Table 4. A high level of luminance is usually required from the wall surfaces and reflectance values should be 0.6 or higher. The height of wall has been specified at 4 m to comply with this requirement.

TABLE 4
Tunnel finishes

Finish	Comments	Durability
Heavy duty vitreous enamel panels	1.5 mm thick flanged plates. Panels sealed by gaskets. Any choice of colour - usually beige. Good reflectivity. Matt finish preferred.	Design life of 100 years plus. Low maintenance costs. Good impact resistance.
Light duty vitreous enamel panels	0.4mm thick flanged plates. Panels sealed with gaskets. Any choice of colour - usually beige. Good reflectivity, matt finish preferred. Stiff backing boards required.	Design life of 100 years but low impact resistance and corrosion possible due to hairline cracks in enamel.
Cement reinforced fibre board	Composite panels. Sealed with strips and lighter than VE panels. Matt finish, light so backing boards not required.	Poor impact resistance. Lower design life due to warping.
Painted concrete	Good quality finish required. Best to use mineralised paints for durability. Tend to retain dirt and therefore need frequent cleaning.	High maintenance costs, needs repairs/repainting at regular intervals.
Ceramic tiles	Not usual for large tunnels. Good reflective properties, tough and easily washed.	Good durability but often become dislodged and cracked.
Coated steel panels	Normally used as external cladding for buildings. Tend to retain dirt so hard to clean.	Colour changes affect coatings and repainting required.

Experience on other projects worldwide suggests that in the long term heavy duty vitreous enamel (VE) panels have the highest capital cost but very low whole life costs and these are recommended. High gloss textured finishes should be avoided because of reflection and the inherent glare problems from daylight or vehicle stop lamps. The ideal texture is semi-matt, possibly combined with a noise absorbing porous surface.

2.8 Traffic Control and Surveillance Systems

One of the prime tasks of a traffic management scheme is to organise the flow of traffic in an orderly and safe manner. Several means of achieving this are currently in use within tunnels :

- Speed control signs
- Traffic lane control signals
- Variable message signs on the approaches to the tunnel
- Traffic loops built into the carriageway
- Closed circuit television for surveillance of operations.

2.8.1 *Traffic Loops*

Traffic loops are set into the surface of each of the running lanes and give information regarding traffic volume, traffic speed and traffic congestion. Sets of loops are installed at discreet intervals. The loops are connected to traffic counting software to give the information required by the end user.

2.8.2 *Closed Circuit Television*

Closed circuit television has now become an essential tool in the monitoring not only of traffic but of incident management. Closed circuit television cameras linked to an operator's control desk are used to give traffic management information both inside and outside of the tunnel, and are used as an aid to control incidents. Closed circuit television cameras of the pan tilt and zoom type are mounted outside of the tunnel on high masts to give a general view of the traffic conditions at the tunnel portals and on the walls within the tunnel to give a view of the general flow of traffic.

The use of pan tilt and zoom cameras is particularly beneficial to tunnel applications.

3.0 PROJECT COSTS

3.1 General Considerations

In general, tunnel design and an appropriate level of services is relative to the level of usage. There are many examples worldwide where sophisticated M&E systems are not installed, *e.g.* Norway, where traffic volumes are very low and the cost of providing state of the art ventilation, lighting and management and control systems is not cost-effective. However, for environmental and safety reasons, the degree of sophistication, particularly on the more highly technical M&E end of design, has increased due to a better understanding of fire life safety issues and the benefits of prevention of road accidents in terms of cost.

A breakdown of costs for 3-lane highway tunnels with a full compliment of services is summarised in Fig. 8. There will be some variation for individual projects depending on factors such as the choice of linings, ventilation system, finishes, lighting, operational and maintenance requirements, but, in general, these are typical. Civil costs for tunnels longer than about 500 m will be about 90 per cent of the capital cost of a project and excavation is the largest single item at about 30 per cent. M&E costs are of the order of 10-15 per cent although for relatively short tunnels such as Anik-Panjarpole this is likely to be much higher.

When looking at whole life costs, operations and maintenance are also an important cost element. These could be as high as 2-3 per cent of the capital cost per annum and this is significantly higher than the 1 per cent typical of highways at grade. These are direct costs that can be assessed when looking at the design elements of tunnels. What is much more difficult to assess, and often not taken into consideration, are the direct and indirect costs that accrue from an increase in the number of accidents that occur if safety concerns are not addressed fully. Issues such as poor visibility, poor lighting or poor ventilation take on a much greater significance when looking at these costs.

A recent review by Department of the Environment, Transport and Regions (1998) that provided an evaluation of the benefits of preventing road accidents indicates a cost saving of £1,207,670 per fatality in the UK. This assessment looked at lost output, human costs and the costs of medical services. As a proportion of the construction costs this is quite high. For the most part, accident levels in tunnels due to careless driving are low because drivers tend to travel more slowly and take more care. Nevertheless, there will be a correlation between the level of services that affect driver safety and the number of accidents.

The recent problems experienced in the Mont Blanc Tunnel in France and the Tauern Tunnel in Austria highlighted the risk of fires in tunnels. Data from PIARC shows that a fire occurs approximately every six million vehicle miles and there is almost no correlation with accident frequency. For the most part, highway tunnel fires have been rare but the actual cost of fatalities and repairing damage on these projects is much higher than £1.2 million per person. The cost of the repairs for the Mont Blanc Tunnel (estimated at approximately £100 million) and lost revenue (estimated at £10 million per month) both have to be factored into the benefit of avoiding such events through the use of proven fire life safety technology. This raises issues such as whether longitudinal or semi-tranverse ventilation is more effective in terms of smoke control, the main concern in preventing fatalities. There appears to be no consensus on this at present but semi-transverse ventilation adds approximately 15 per cent to construction costs.

For these reasons, and recognising that there is always pressure to keep the capital costs of projects low, the level of usage of the Anik-Panjarpole tunnels in metropolitan Mumbai was considered high enough to justify the full range of services listed in Section 2. For a tunnel 400 m in length, with expected high traffic flows, particularly of heavy goods vehicles which have a higher fire rating (10-20 times that of a passenger car depending on the cargo) good lighting is important and longitudinal ventilation for smoke control a necessary requirement.

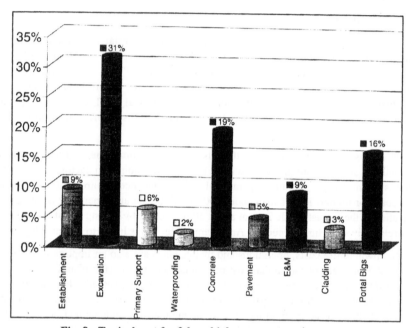

Fig. 8 : Typical cost for 3-lane highway construction

3.2 Tunnel Cost Estimate

A breakdown of the construction costs for the Anik-Panjarpole Tunnel is given in Fig. 9. This shows a similar breakdown to Fig. 8 but there are differences. The M&E costs as a proportion of the capital cost estimate are much higher than 10 per cent. This partly reflects the much shorter tunnel length and also the difficulty of identifying and specifying local products when preparing the estimate. Portal building costs are much lower since a simple control building only is required for longitudinal ventilation. Adjusting the M&E costs to reflect fully local costs would result in most cost items corresponding reasonably well with the values given in Fig. 8.

It is not particularly useful to look at the overall cost of construction in India compared to other countries. The cost differential is high because of the much lower labour, plant and material costs. In countries such as the UK and Hong Kong, where 3-lane tunnels have been constructed, the cost of construction are very similar, *i.e.* of the order of £25,000 per linear metre. By comparison, the predicted cost for the Anik-Panjarpole is £9550 and the costs provided by the contractor for the Panvel tunnel, not including M&E, were approximately £3300 linear metre The large differential between the Anik-Panjarpole and Panvel tunnels is certainly related to the difference in length, but also the level of M&E

systems and the provision of tunnel finishes with a full secondary concrete lining. However, the £3300 quoted by the constructors for the Panvel tunnels is remarkably low considering that these are for 4-lane tunnels approximately 1100 m in length.

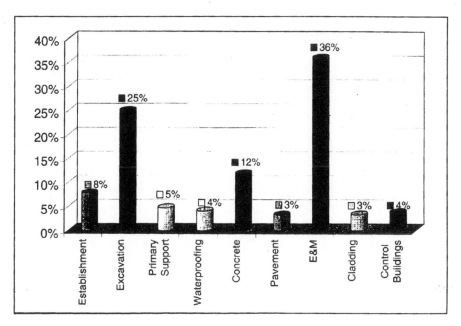

Fig. 9 : Construction costs for Anik-Panjarpole tunnel options

The provision of fire life safety systems such as those included in the Anik-Panjarpole tunnels comes at quite a high cost. It would be useful to carry out a study of the benefit of accident prevention in India. If it is assumed that the UK cost of 3-lane tunnels is an order of magnitude greater than the cost per linear metre of construction in India, this would suggest that the cost benefit of preventing accidents is approximately Rs. 8 million. This value seems unrealistic when labour costs and other factors are considered. A lot more effort and information is required to develop values so that informed engineering decisions and judgements on the provision of tunnel services and fire life safety systems in particular can be made.

ACKNOWLEDGEMENTS

This work has been carried out as part of a project study for the Metropolitan and Mumbai Regional Development Authority (MMRDA). MMRDA has been the key agency in formulating the MUTP project, through which significant improvements in pedestrian subways, roads, bus transport and surburban railway services are being undertaken and their kind permission to publish this information is gratefully acknowledged.

REFERENCES

AASHTO, 1986, 1993. "Guide for Design of Pavement Structures." Washington, USA.

PIARC (Permanent International Association of Road Congresses).

Bieniawski, Z.T. 1976. "Rock Mass Classification in Rock Engineering." Proc. Symposium on Exploration for Rock Engineering, Johannesburg, Volume 1, pp. 97-106.

Department of the Environment, Transport and the Regions. "1998 Valuation of the Benefits of Prevention of Road Accidents and Casualties." Highways Economics Note No. 1:1998.

METRO TUNNELLING

CAIRO METRO LINE 2 - CONSTRUCTION PROBLEMS
AND THEIR SOLUTIONS

A.J. BURCHELL
Pacific Consultants International (PCI), New Delhi, India

SYNOPSIS

The 11 km phase I section of Cairo Metro Line No. 2 involved 8 km of underground construction involving large diameter Tunnel Boring Machines and deep cut and cover station boxes. The saturated sands of Cairo make any underground construction difficult and problems arose in the early parts of the construction which brought about design and construction changes to reduce risks to the latter part of the works. This paper outlines the problems encountered ranging from ground treatment failures at stations, difficulties at tunnel break-ins and break outs at the stations through to loss of ground at the face of the slurry TBM which in one case caused an overlying structure to collapse. The lessons learnt from these problems are described and recommendations given for minimising the risks in underground construction in saturated soils.

1.0 INTRODUCTION

Cairo is the largest city in Africa with a population estimated at 15 million and growing rapidly. The Egyptian Government embarked on a heavy rail metro system in the 1980's and Line 1, built by cut and cover techniques, was completed in 1989. Line 2 commenced construction in 1993 under a contract awarded to Interinfra of France with the civil works joint venture led by Campenon-Bernard and was constructed in two phases (Fig. 1). The first phase totalled 11km and runs underground for most of its length on the east bank of the Nile. Bored tunnelling was introduced for the first time on the metro using a single 8.4 m diameter tunnel to carry both metro tracks. Stations remained as cut and cover but are deeper than the Line 1 stations to give the required cover to the bored tunnel and to allow Line 2 to pass under Line 1 at Mubarak and Sadat stations (Fig. 2).

Fig. 1 : Cairo metro route plan

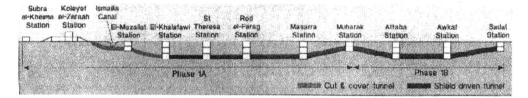

Fig. 2 : Line 2 phase 1 longitudinal profile

Ground conditions in Cairo are predominantly water bearing sands which are approximately 60 to 90 m thick along the route of the metro. The soil profile consists of 2 to 4 m of fill underlain by a clay deposit 4 m to 9 m thick with sand and silt lenses. Below this lies the alluvial sand laid down by the Nile as it meandered across the valley over the annals of time. The deposits are by no means uniform and lenses of gravels and large cobbles were encountered at depth within the sand. Due to the promixity of the river Nile, ground water levels remain at 3 to 4 m below the surface throughout the year.

2.0 STATION CONSTRUCTION

The station boxes are typically 150 m long; 20 m wide and 25 m deep. The main design concern was how to form a stable cofferdam in the water bearing sands within which to excavate down to 25 m depth and cast the station structure. The concept adopted followed the principles used for the cut and cover works on Line 1. This involved 50 m deep diaphragm walls and a 7 m thick grouted plug at the base of walls covering the complete station area. This plug was designed to reduce the permeability of the sand to 1×10^{-6} m/sec and was located at 50 m depth to enable the weight of soil above the plug the resist the uplift pressure (Fig. 3). Furthermore the stations were divided into 3 compartments by transverse slurry cut-off walls.

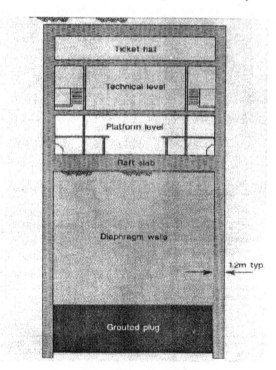

Fig. 3 : Typical station arrangement

The plug was created by drilling and installing tube a manchette grout pipes on a 1.4 m grid. Each pipe was fitted with sleeves or grout holes at 0.5 m spacing throughout the 7 m depth of the plug. There were approximately 2000 grout pipes per station and 13 sleeves or injection points for each grout pipe. The grouting system was first of all to use cement-bentonite to fill any larger voids followed by a sodium silicate soft gel to give the required permeability. Grouting pressures and volumes were monitored by computerised equipment and re-injection carried out where pressures were low. On completion of the grouting for a particular box, a pumping test was carried out and if successful, excavation was allowed to commence.

Following a successful pumping test result for each of the 3 boxes of the station, construction commenced in that box with the roof slab cast directly on formwork placed on the ground and then excavation continuing to the technical slab. This was then cast and followed by excavation to the base slab with two intermediate rows of struts which could only be removed after completion of the base slab. Corbels were cast along the diaphragm wall / base slab junction to counteract the uplift forces on the base slab.

3.0 TUNNEL CONSTRUCTION

The bored tunnels were constructed using two 9.4 m Herrenknecht Slurry Tunnelling Machines (TBMs). Each machine was assembled inside the completed station box then tunnelled to the next station and broke into a prepared tunnel eye. The TBM was then pushed through the intermediate station and relaunched to the next and final station where it was removed. The TBM launching points or break-ins were carried out within a block of jet grout with additional slurry walls and a soft gel block outside the slurry walls. The sequence of operations was to complete the jet grout block and other works during station excavation and then to demolish the diaphragm wall after probing through the wall to verify the integrity of the grout. At the TBM launch points (break-in) a rubber gasket or seal was bolted onto the wall to fit tightly around the TBM body as the machine moved out of the station and to contain the slurry pressure. The TBM arrival points or break outs were designed as a single jet grout block of 15 m length in order to have the TBM buried in the jet grout before it entered the station (Fig. 4).

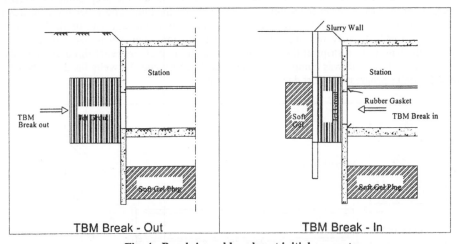

Fig. 4 : Break-in and break-out initial concepts

A further complexity was the requirement to connect the main tunnel to annex structures located between stations. These annex structures provided for low point sumps and ventilation equipment and were small diaphragm wall boxes connected by a short adit to the main tunnel. As with the break-ins and break outs, this short adit was designed to be constructed by hand in a complete block of jet grouting.

4.0 PIPING INCIDENTS

In order to excavate the station safely, the integrity of the deep, 7 m thick grouted plug was essential. The grouting results were overall very good with over 90 per cent of the boxes achieving or exceeding the desired permeability of 1×10^{-6} m/s. However, piping incidents occurred at 3 stations, Mazallat, Sadat and Abdeen. In the case of Mazallat, excavation was approaching the basè slab when a single isolated leakage zone was observed at the excavation level with water welling up through the sand at a rate of several cum / hr. Sandbags were piled over the soft, leakage area and an injection pipe pushed easily down into the piping area. Polyurethane and cement / silicate grouts were injected directly into the piping zone and effected a seal which allowed excavation to proceed.

The incident at Abdeen station was considerably more serious. The eastern box in the station had been excavated successfully without problem but failure of the pumping test in the western box delayed the start of excavation while regrouting took place. The rapid tunnelling progress meant that the TBM was ahead of programme approaching Abdeen from the east and so the decision was taken to carry out the break out preparation and receive the TBM into the station where lenghty refurbishment to the cutterhead could be carried out. This all proceeded without incident and the TBM arrived and was pushed into the station in October 1995. Meanwhile in the western box the technical slab and been cast and excavation was proceeding down to the base slab. Suddenly in November 1995 a 'fountain' of water and sand appeared at the excavation level in quantities that were estimated at over 100 m^3/hr and there was concern that this would erode the plug and lead to a catastrophic failure (Fig. 5). The option of flooding the station to balance the external water pressure was no longer possible because the TBM had junctioned from the east and indeed any uncontrolled inflow would now cause considerable damage to the TBM and lead to flooding of the adjacent Attaba station where the second TBM was under erection. The immediate action was to backfill the western box with any available material adding weight to the plug and at the same time build a bund wall to separate the east and west boxes. Pumps were also mobilised together with drilling equipment. When approximately 5 m of backfill had been placed over the majority of the area of the western box, the piping subsided but did not stop altogether however stability was sufficient to investigate the cause and plan remedial measures. Additional deep wells were installed both inside the outside the box and peizometers placed immediately outside the wall to detect any drawdown that would indicate the location of the defect in the plug. Based on this data a pattern of additional ground treatment was carried out inside the box, above the existing plug level together with injection of polyurethane and cement-silicate grouts from the outside of the box. This succeeded in reducing but not stopping the leakage. The next measure taken was then to activate the deep pumping wells installed around the outside of the box to lower the water pressure acting on the plug. The water pressure was reduced by effectively 5 m head which further reduced the leakage to a manageable limit. Excavation then recommenced but instead of excavating the whole box to formation level it was decided to tackle this in sections and cast the base slab in strips at the same time maintaining pressure relief through the wells outside the box. Only after the base slab was complete and the corbels cast could the wells inside the box be finally sealed. In total 5 months progress was lost due to this incident.

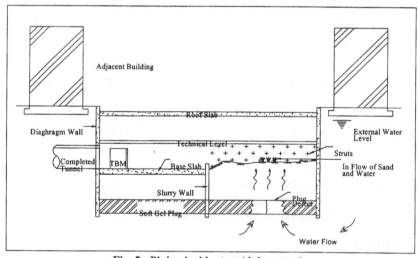

Fig. 5 : Piping incident at Abdeen station

These piping incidents highlight the need to very carefully evaluate all grouting parameters and regrout where concerns are felt with regard to low pressures or volumes. In addition contrigency drilling and grouting equipment with polyurethane or similar fast setting grouts should be available to enable any problems to be dealt with as quickly as possible.

5.0 BREAK-OUT / BREAK-IN PROBLEMS

The station-tunnel connections proved to be a major area of risk and several problems arose. The initial design for the break out was for a block of jet grout to act as a self supporting zone to allow the tunnel opening in the diaphragm wall to be formed by demolition of the wall ready to receive the TBM.

In August 1994 during the demolition of the diaphragm wall at Rod El Farag station in preparation for arrival of the TBM a large flow of water suddenly appeared washing in quantities of sand and causing a depression to rapidly form on the ground surface totalling some 150 m³. The excavation plant inside the station was mobilised to push large quantities of sand up against the leakage area and concrete was placed into the hole on the surface. Drilling rigs were also mobilised and polyurethane grout injected below the level of the jet grout plug. This grout effectively flowed with the water for a short time and then gelled within the jet grout mass blocking the leakage paths. Following this systematic pattern of cement and silicate grouting was carried out to stabilise the area. On later examination the leakage had occurred due to a 'window' in the jet grout mass where one of the jet columns had deviated from the vertical position.

In October 1994 a similar incident occurred at Masarra station. This time the results were more serious and the depression on the ground surface caused a 1000 mm water main to burst. This caused large quantities of water to enter the station and the decision was taken to allow the station to flood in order to stabilise or balance the external water pressures.

These two incidents led to a detailed review of the break out design and the contractor proposed to supplement the jet grout block with a perimeter slurry wall and a base plug of sodium silicate grout. These additional measures were carried out at both the Rod-El Farag and the Masarra break outs before continuing and completing the demolition of the diaphragm wall eyes. From then on, all future break outs were first constructed with a slurry wall and base plug, followed by a pumping test to demonstrate the integrity of the slurry walls and plug and then finally carrying out the jet grout block (Fig. 6). With these measures in place no further incidents arose at the break outs.

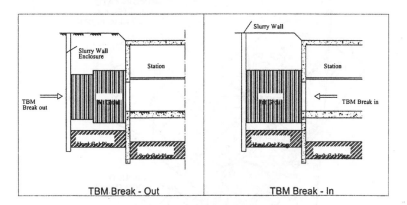

Fig. 6 : Revised TBM break-in and break-out designs

In addition to the break out problems, both TBMs caused minor collapses as they commenced the break-ins. In each case the cutterhead of the TBM had penetrated the jet grout and the slurry wall but face stability could not be maintained in the soft gel treated section. The cavity resulting on the surface was quickly filled with concrete and then tunnelling resumed with hardly any loss in progress. It was considered that the grout tubes left in the ground for the soft gel had in some way disturbed the sand at the face as the cutterhead tried to cut into them and that this mechanism triggered instability. These problems brought about a redesign of the break-in concept in a similar manner to the break-out and no further problems arose.

6.0 LEROY BUILDING COLLAPSE

The Leroy building is a large masonry 3 storey building built in the early 1900's and situated over the line of the tunnel very close to the end wall of Attaba station. A plan view is shown in Fig. 7. After considering various options for protecting the building it was agreed to evacuate the residents, to prop the façade arches and windows and then to proceed with the tunnelling works. Cracking was expected and it was planned to repair this after tunnelling before allowing the residents to return. The close proximity of the building to the end wall of Attaba station meant that the revised break-in design with slurry walls and a base plug that was developed after the August and October 1994 incidents could not be utilised. Instead the proposal was to carry out a block of jet grout with a mineral grout base and to perform much of this grouting using inclined drilling. This was carried out, verification holes were drilled through the station end wall and then the diaphragm wall successfully broken out to form the tunnel eye. At the same time the TBM was being erected and was finally ready to commence tunnelling in mid October 1995. Even with a rubber seal around the tunnel eye to prevent the slurry at the TBM face from running back into the station, it is always necessary to build the slurry face pressure up gradually so that as the TBM enters virgin soil it is at the correct or full pressure to maintain stability in the sand. Unfortunately there were defects in the mineral grout causing full water pressure to act on the face of the TBM before full slurry pressure was achieved. This caused loss of ground at the face which in turn led to the collapse of a large brick built culvert situated above the TBM and below the buildings foundation. The effect was then a sudden loss of the slurry pressure and full collapse of ground into the cutterhead chamber (approx 150 m³). Almost immediately this collapse reached the building foundations and several of the internal masonry columns and arches collapsed. The building façade fortunately remained intact and to the press and the public there was no apparent evidence of a problem.

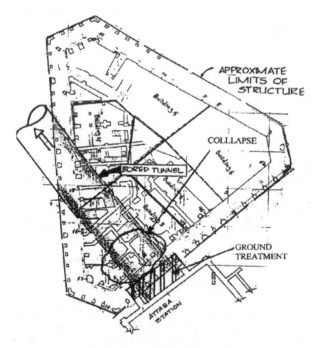

Fig. 7 : Leroy building layout

156

To recover the situation, limited demolition of the building above the disturbed zone was carried out and then the exposed cavity filled with lean concrete. Some grouting was also undertaken from the surface to fill any residual voids. 5 weeks after the collapse the slurry pressure was reapplied to the face of the TBM and apart from a small area of leakage on the surface the pressure held and the TBM proceeded through the disturbed zone, under the remaining length of the building without further incident.

7.0 TBM PROBLEM

The overall performance of the two TBM's was very good with a peak progress rate of 28.5 m/day and average rates of 14m/day. In addition ground settlements were very low averaging 13 mm and the finished tunnel quality and watertightness met the exceedingly high specification requirements.

The main problem encountered was one of boulders. These had not been identified in the site investigation and occurred as cobbles / boulders of around 200 mm but occasionally larger. During tunnelling from Masarra to Mubarak a boulder layer was encountered but tunnelling progressed reasonably well until one boulder got struck at the entry to the slurry pipe. This completely stopped operations and required the slurry to be removed from the face and replaced by compressed air. Men entered the face and after a few hours removed the boulder. At the same time an inspection was made of the cutterhead which showed that the boulder had caused some damage and repairs were needed. While these repairs were in progress the sand face supported now by compressed air became unstable. It started to dry at the top and raveling occurred. The men were quickly evacuated and the chamber refilled with slurry but this was not in sufficient time to prevent the ground collapsing and a cavity formed on the road above. Fortunately the contractor reacted quickly and the affected section of road had been cordoned off. The cavity was filled with concrete and the drive recommenced with the repairs completed the following day.

A second layer of boulders occurred between Attaba and Abden. This time a complete section of cutterhead picks were damaged and the sand had started to abrade the face of the cutterhead itself. Again emergency repairs were carried out and despite limiting the compressed air working to 24 hours, once again the face became unstable and a collapse occurred leading to a cavity in the road. The intervention period was subsequently reduced and no further events took place. When the TBM was brought into the station box at Abdeen it was found that the steel plate on the face of the cutterhead had worn through and exposed the support frameworks behind.

These problems with boulders brought about a review of the cutterhead design which was slightly modified to include more openings. In addition a 'stone crusher' was subsequently fitted at the bottom of the chamber in front of the slurry outlet pipe so that cobbles coming through the openings in the cutterhead could be crushed and then safely disposed of through the system.

8.0 CONCLUSIONS

The construction of the Cairo Metro Line 2 posed several major challenges to both designer and contractor. Inevitably problems arose but these were dealt with in an expedient and professional manner. The piping problems that occurred represent a low percentage of problems with the grouted plug concept of construction is sands. Time and effort must be spent on verification of grouted plugs and in analysing all grouting records to try to identify areas of potential weakness for regrouting. Contrigency measures such as polyurethane injection should also be on standby. Break-ins and break-outs must be treated with caution and reliance on a single barrier of ground treatment is not recommended. The developed concept of a slurry wall box with base plug as one line of defence and then filled with jet grout as a strengthening measure is advised in granular soils. The possibility of boulders and cobbles should be carefully evaluated in the design of tunnelling machines and if entry into the face of the tunnelling machines is necessary then this should be in short deviations compatible with the stand up time of the ground under compressed air support. It is of great credit to the contractor that the construction problems were overcome and the project completed and opened to the public ahead of time and with a quality in the completed tunnel works that is of exceptionally high standard.

REFERENCES

Bellarosa A. 1999, Construciton of Cairo Metro Line 2, Proceedings of the Institution of Civil Engineers, UK, Civil Engineering Volume 132 No.2.

Richards D.P. 1997, Slurry Shield Tunnels on Cairo Metro, Proceedings, Rapid Excavation and Tunnelling Conference, Las Vegas, USA 1997.

Campo D.W. 1997, A review of the grouting on Line 2 of the Cairo Metro, Proceedings, Rapid Excavation and Tunnelling Conference, Las Vegas, USA 1997.

Burchell A.J. 1994, Stations and Tunnels on the Cairo Metro, Tunnels and Tunnels and Tunnelling, Middle East Special Edition.

PROSPECTS OF UNDERGROUND FACILITIES FOR DELHI

A.K. DUBE

Emeritus Scientist

Central Road Research Institute, New Delhi, India

SYNOPSIS

Megacities like Delhi are prone to congestion and with growing population pressure the existing infrastructural facilities are strained to maximum. Lack of space on surface becomes a serious problem for any upgradation on creation of any facility like roads, pipelines, telecom and power network and places for monitoring and control of civic facilities. More cars add confusion to road traffic and need to be tackled. In the developed countries the authorities took initiative to construct facilities down below to lessen pressure on surface which is simply not available. Traffic bypass tunnels, parking lots, pedestrian crossings conducts for power, telecom cables, water and sewage mains, storages for various utility goods water etc. can definitely be constructed underground. This may lot more ease the pressure on surface and provide comforts to all users. About 40 per cent of Delhi is founded on Aravali rocks which may be host to numerous useful civic utilities. What are the prospects and what can be done has been narrated in the text. It may provide food for thought for policy planners, technocrats and conscious citizens to conceive some worthwhile schemes and get them constructed for the common good.

1.0 INTRODUCTION

Delhi, the national capital territory is the home of about 140 lacs persons. They live and work here and move about with in the urban area for livelihood and industrial activities. The area is getting congested faster and no space is left for infrastructural development on surface. There is need to look for other avenues and one of them is to construct under ground civic facilities. Some of the facilities like underground road bypass, parking lots, service tunnels, underground storages of oil and gasses, water, cold storages and waste disposal can be considered. In several advanced countries, numerous of the above mentioned facilities do exist and many more are under construction. The ground realities, the need for infrastructure development and the availability of Aravali system rocks in some of the Delhi areas are quite lucrative. In this paper an attempt has been made to assess the development of some such facilities in Delhi area.

2.0 SCENARIO IN DELHI

The urbanisation of India is growing at a faster pace and the megacities are most attractive places for migrants. It is estimated that by the year 2010 more than 50 per cent of Indians would be living in cities. We very well know that every year 5 lacs persons settle in Delhi making it more congested. Unplanned growth leaves no space for civic facilities for sustaining growing population. The roads are shrinking due to encroachments, movement of vehicles slowed down due to increasing numbers of vehicles, no space for laying of pipelines for water supply and sewage disposal, telecommunication and power cables, parking of vehicles and many other social needs. Dehlities get less water and there is a shortfall of about 20 per cent and it is growing very fast. Waste disposal and storages for petroleum products pose a serious risk to the sprawling city.

Delhi is on the banks of river Yamuna but it's share in river water is low as negligible. During raining season the river swells and floods low lying areas of the city and during other seasons the water is very less and it becomes unsuitable for human consumption due to unrestricted flow of industrial and domestic effluence.

About 40 per cent of Delhi is rocky and very good Aravali rocks are exposed in many parts of the city, specially the ridge. These rocks are mostly quartzite which are competent and are considered as good tunnelling media. The rocks classified as fair to good are easy to excavate and can be supported economically. Fig. 1 gives the exposures of these rocks and depths from surface at which they can be found. Most of the rockey areas of Delhi are devoid of water as there are no or fewer acquifers. The problem of loss of ground water because of underground construction, therefore, shall not be a major concern. If ground water exists in some pockets, it can be protected well by providing appropriate protection measures within the tunnel system.

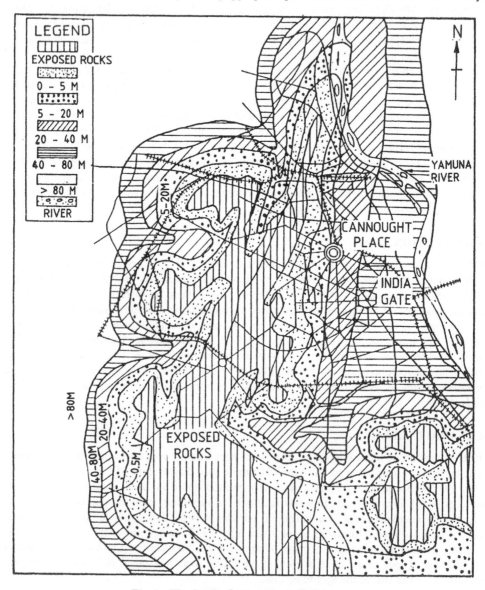

Fig. 1 : The depth of quartzites in Delhi area

3.0 PROSPECTIVE UNDERGROUND FACILITIES

As already stated many of the infrastructural facilities can be built underground where there is no space available on surface. In many countries the following facilities are operational :

1. Underground traffic by passes and parking lots.

2. Underground passages in airports, city centres etc

3. Underground metros

4. Underground service tunnels

5. Underground storages for food, chemicals, oil, gas and water

6. Nuclear waste repositories

7. Strategic underground facilities

In India, we can start a process of at least data, collection, concept development and preparing feasibility reports of some useful facilities. This percusitive may reduce the construction time considerably because we may have some data available to us to begin with. In our country we built numerous hydro-electric projects involving tunnels and underground caverns. The experience has demonstrated that the construction was very much delayed and there had been considerable cost escalation. This was primarily due to lack of data, careful study and appropriate strategy to execute a project with speed. Our country is a fast growing country and a major economic and strategic power in the South-East Asian region and in future can afford numerous underground facilities. In the following paragraphs more details are given:

4.0 UNDERGROUND TRAFFIC BY-PASS AND PARKING LOTS

Delhi has a very congested and heavily built up area in many places. There is no place for road or pedestrian crossing. Under such circumstances an underground facility would be most suitable. Delhi is a historical city having numerous historical and archaeological monuments. The usual fly overs obstruct the vision around such monuments and are objected to by heritage conscious citizens. In many historical cities of the world, fly overs are not permitted nearer to old monuments and some may be fine for Delhi. Since long, a need had been felt to have a traffic bypass from Sarai Kale Khan to Dr. Zakir Hussain Marg. The area has Humayun Tomb, railway tracks, a nursery and forested area. It can be an ideal site for a tunnel bypass. Fig. 2 shows such a setting. The area probably may have Indo-Gangatic alluvium. These alluviums are typically of gravels, sands, silty clays with remains of animals and plants. Yamuna is a new river system and had a shifting tendency. Hence the deposits show significant local variations. Due to being in the vicinity of perennial river Yamuna there may be ground water close to surface. The underground bypass may provide a direct access to Dr. Zakir Hussain Marg from the vicinity of Nizamuddin Bridge. It will shorten the distance by about 3.5 km if a 1.5 km long tunnel bypass is made from the ring road opposite Nizamuddin Bridge and Dr. Zakir Hussain Marg.

The Nehru Place area is a most suitable place for underground parking lot. At present vast area in front of buildings is used for parking lots and is very much congested and environmentally degraded. It is possible to construct underground parking system for cars and numerous buses which are lined up along roads increasing the congestion. The Kailash Hills, just behind Nehru Place, can accommodate all the vehicles in underground parking system. Enough space can be created for car repairshops and shops for general merchandise which are at present sold on pavements of central vista making the place very congested. The area vacated by presently assigned parking space can be used for beautifying, the place by planting flower beds, shrubs and trees. The approach to various buildings can be through underground corridors which will save the users from inclement weather which they face when approaching their work place from the parking lots. It is possible to prepare an efficient system which may serve the needs of the area. Other attractive locations are Rajendra Place and business area of Vasant Kunj, having good rock exposures.

Fig. 2 : Key plan of study area with proposed tunnel alignment

5.0 UNDERGROUND PASSES IN AIRPORT, CITY CENTRES

In an airport men and materials are moved to the aircraft from the terminal buildings. It is mostly done by mobile vehicles which cross the operational area frequently and cause congestion in a very high security place like an airport.

The Palam Airport area is situated mostly on the Aravali rock exposure and it can be possible to construct a network of tunnels for ease of movement of men and material in the technical area. It is possible to plan and design a system of tunnels and underground space to provide better passenger services. Underground network is energy efficient for air conditioning and can be built in places where there are numerous constraints on surface. At some airports in the west there are tunnels for transfer of cargo.

6.0 UNDERGROUND METROS

Underground metros are means for fast commuter traffic in congested metropolises. In Calcutta it is operational and in Delhi it is already under construction. It may help greatly in easing commuter problems.

7.0 UNDERGROUND SERVICE TUNNELS

Water mains, sewage transport pipelines, telecommunication cables, power cables etc. are to traverse the city. They are usually housed in conduits built in dug up trenches. Any scheme to alter them or to increase their capacity, they are to be dug up again. It creates scars a roads, pavements and cause traffic stopagge or congestion when the construction work goes on. All this can be avoided if the service tunnels are made in congested areas to carry service utilities. These tunnels can be designed and built as per the needs of the situation. Underground substations, telephone exchanges and control and monitoring systems can also be housed in underground space specially built. Such service tunnels will help greatly in easing the congestion on surface and may make the city servicing more efficient.

8.0 UNDERGROUND STORAGES FOR FOOD, CHEMICAL, OIL, GAS AND WATER

Old and abondoned mines had been a good place for storage of french beverages and wines for many years. In India there are good prospects for such storages in many old mining fields. Cold storages which are very economical (up to 60 per cent energy saving) can be housed in abondoned mines or freshly excavated space can also be created to serve special purpose.

Storage of oil and gas are very common in advanced countries and such storage is a very well developed technology. Swedes were the pioneer in this field and at present there are over 400 such facilities for oil and gas. With the increasing use of LPG and LNG in industry and transport, a need has been felt to create underground storages for these gases. The LPG can be liquified under hydrostatic pressure of water deeper inside a cavern at least 150 m deep. The water acts both as sealant as well as pressurising media for the gas. In India first LPG underground storage is at present under construction at Visakhapatnam and this technology had been found to be very economical and safe for bulk storages. In-ground storages for LNG had been constructed in Japan in good numbers and they go very well with the current refrigerated LNG bulk liners ferrying LNG from far off lands. This business is growing and it may succeed in India also. There is a world wide trend to discourage the use of pressurised storages for gases and in our country we may have to adopt this technique for our bulk requirements of gases. It is heartening to know that South Korea and Zimbwabe have built underground storages for products and they found the experience rewarding.

Water is going to be scarce during the twenty first century and India is already facing serious problems in many places and Delhi is one of them. The ground water as well as some water from Yamuna with the blessings of neighbouring states is the only source of water for about 140 lacs population of Delhi. There is a shortage of 20 per cent and this gap may grow in the years to come. To cover up the shortfall Delhi should have its own source of water. Delhi Government can think about storing the flood waters of Yamuna in underground space. The Aravali rocks which are very good tunnelling media are spreaded over 40 per cent of Delhi territory. It is possible to develop underground storage facilities in phases. The water during floods is no body's property and it does not violate any river water sharing agreement. Any storage on surface needs larger tract of land which is not available. Secondly, the surface storage of water breeds insects and a good quantity of water evaporates. The underground stored water is free from any loss or the environmental issue. It is possible to design a most suitable and economical underground system in view of a vast experience available in India about underground construction. The underground system need not be lined to effect economy. The water seeping through rock cracks may help in enriching the ground water table which is going down very fast.

9.0 NUCLEAR WASTE REPOSITORIES

Since last 40 years nuclear energy development grew very fast. Super powers developed nuclear weapons also. Both the energy and weapon programmes generated huge wastes. Low level and intermediate level wastes are tackled easily, however the high level wastes are problematic. So far no solution had been found and the research is being done to find a technology for underground disposal of nuclear waste. There are stringent specifications for underground waste repository. It should have a proven stability for 10,000 years, the repository should confine and contain the waste for very long period of time. The science of rock mechanics in at present unable to provide answers to many of the issues ensuring confinement and no chance of migrations of neucloids by air water or contact through rocks.

In India we are at present looking for appropriate sites and Delhi is not a candidate for such a possibility in the near future but may consider construction of nuclear power plant to become self sufficient in power.

10.0 STRATEGIC UNDERGROUND FACILITIES

Delhi is the capital of India and houses the highest civil and millions authority. In view of this it is important to have strategic facilities for safe abode for top functionaries in war times. These abodes should be interconnected by sub-surface passages. The city had to be protected against air attacks, hence there had to be numerous shelters where people can shift temporarily.

The offensive and defensive weapons can be housed underground to ensure safety and its effectiveness during war times. Military hardware may also to be stores safely and underground storages are considered to be very appropriate.

It may be worth mentioning here that during the Vietnam war the Vietnamese fought the American forces very effectively with the help of underground shelters and passages. The Americans are now perfecting underground war fare in numerous underground facility developed by them in Nevada test site. During Kargil war it is believed that the Pakistani forces entrenched themselves in underground bunkers connected by narrow tunnels. It is a good lesson for India to learn and start developing underground strategic facilities. Delhi has lots of potentialities for defense facilities.

11.0 CONCLUSION

The matter of development of underground utility facilities in Delhi is an open issue and needs debating and drawing appropriate conclusions. There is thus a need to have a forum for this purpose. The Delhi Govt. may appoint an expert group to study the problem thoroughly and to prepare a plan for action.

ACKNOWLEDGEMENTS

Author is thankful to Mr. A.K. Gauba, Sr. Stenographer, Director's Office, Central Road Research Institute, New Delhi for preparing the paper neatly.

DESIGN STRATEGIES AND PLANNING FOR FIRE AND LIFE SAFETY IN DELHI METRO

A.K. GUPTA

Rail India Technical and Economic, Services (RITES)
New Delhi, India

ABHAY BAKRE

Delhi Metro Rail Corporation, (DMRC)
New Delhi, India

SYNOPSIS

Delhi will get its first underground metro by March 2005. In view of recent major fire tragedies in Delhi, which resulted in casualties, the issue of fire and life safety is highly important and sensitive to the people of Delhi. This issue becomes much more critical in an underground public place like a metro station and tunnels due to the confined space and larger public presence. The only other underground metro system ever constructed in India at Calcutta has witnessed a few fire incidents in recent years. The general consultants to Delhi Metro Rail Corporation(DMRC) have developed an elaborate design and planning criteria for the fire and life safety in Delhi Metro. This is based on the need for providing safe and quick evacuation of passengers in the event of fire emergencies. These criteria generally follow the guidelines issued by the National Fire Protection Association(NFPA).

This paper describes the basic principles adopted in the design and planning of underground stations, including assessment of occupant load and exit capacity, and arrangements of emergency exits. It also describes the basic design principles of tunnel ventilation and smoke extraction system including emergency evacuation of passengers from affected trains.

1.0 INTRODUCTION

The Delhi Metro Rail Corporation has been entrusted with the responsibility of constructing Mass Rapid Transit System in Delhi. The main objective of the project is to build a world class Metro - a vehicle to promote dignity and discipline in the city. This cannot happen without building a safe and efficient transit system. Thus one of its major concerns is to ensure that satisfactory fire precaution and safety features are incorporated in the system design such that the lives of the passengers will be safeguarded in the event of fire.

Delhi Metro is only the second metro in India after the Calcutta Metro. It is observed that all local building regulations and fire protection code from National Building code of India - 1983 (Amendment No. 3, 1997) are not meant for an underground transit station. Presently all passenger stations and terminals of air, surface and marine public transportation services are to be covered under Group D, 'Assembly Buildings', of National Building Code of India. But provisions under this category are basically meant for any building or part of a building where number of persons not less than 50 congregate or gather for amusement, recreation etc. A mass rapid transit station deals with mainly large quantity of rapidly moving passengers who seldom gather except at platform for a few minutes while waiting for a train. Hence, except for some general aspects, there are fundamental differences in the concept of evacuation facilities and nature of occupancy between an ordinary building and a transit station.

The conceptual layouts of the underground stations at the Detailed Project Report (DPR)(5) stage for Phase-I of the transit system were largely based on criteria adopted by Hong Kong MTRC but electrical and mechanical services including fire detection and control were largely based on local codes. The General Consultants while preparing detailed design and planning criteria for the Delhi MRTS, came up with recommendations for adopting NFPA 130 – 1997 as the basic guideline for the design of fire safety and tunnel ventilation system. This standard specially deals with fire protection and life safety in fixed guideway transit system. Subsequently it was also decided that the station layouts should also comply with NFPA 130 to achieve a compatible design in all respects.

The provisions of NFPA 130 were originally based on the criteria adopted in North American Transit Systems, which were very lightly loaded and based on luxurious space for passengers. Hence when NFPA 130 was adopted in station design, it yielded significant increase in station sizes than those envisaged in DPR which was due to large trainload of 2388 passengers, high frequency of train with 2 minutes headway, and huge 60,000 passengers peak hour peak direction traffic (PHPDT). Hence other mass transit systems designed on NFPA 130, such as, Singapore MRTS and Bangkok MRTS were studied and their salient features were picked up for incorporation in Delhi MRTS design. A thorough study was undertaken by the General Consultants to find out techno-economic effect on station design in view of the various important parameters of NFPA-130, Singapore and Bangkok MRTS designs and additional operating and safety procedures to be adopted. This concluded into the final tender design and planning criteria.

2.0 GENERAL STATION LAYOUT

The first phase of Delhi MRTS consists of about 11 km of underground tunnel with 10 underground stations. Five stations; Vishwavidyalaya, Old Secretariat, Civil Lines, Connaught Place and Central Secretariat will be side platform stations and remaining five stations; ISBT, Delhi Main, Chawri Bazaar, New Delhi and Patel Chowk will be island platform stations. ISBT and Connaught Place stations are interchange stations having interchange with above ground rail corridor and future East-West corridor respectively. Depth of concourse level is normally about 7.0 m from ground level whereas that of platform level is normally about 12.3 m from ground level. Except Chawri Bazaar all other stations will be constructed by cut and cover method. Chawri Bazaar station will be constructed by tunnelling method. Depth of platform level at Chawri Bazaar and Connaught Place stations is likely to be 20 m from ground level.

3.0 MEANS OF EGRESS AND EVACUATION OF STATION

3.1 Mean of Egress

Stations are having 185 m platform length with a central concourse above, the length of which varies from 100 m to 185 m from station to station. Normally staircases are provided for down movement and escalators for up movement at stations. Exiting facilities during emergency consists of these normal staircases and escalators as well as additional emergency staircases provided near the ends of station platform. In underground stations escalators keep on moving upwards under a guaranteed power supply and help in speedy evacuation of station in emergency. One escalator is discounted in evacuation capacity calculations as required by NFPA- 130. Under fire or emergency, passengers should be able to see the exit routes, which are the same routes that are normally used for ingress and egress, and use them to come out of the station in the quickest possible time without creating panic or a stampede. All staircases and escalators are uniformly laid along the concourse and platform to provide least travel distance on the platform to reach an exit. It also helps in spreading the passengers and providing alternate exit routes. Fig. 1 shows a typical station layout.

Most of the stations have two unpaid areas at concourse level, which are wide apart and leads to separate entrances/exits to stations through subways. This provides an inbuilt fire separation on the concourse with two egress paths. Each platform is provided with atleast one emergency staircase which, by-passes the concourse passenger area to provide an alternate route of evacuation in case of a fire at the concourse, which may block some of the normal entry/exit routes. All the exit routes will be provided with fresh air supply through fire rated ducts.

Time is the main criterion in the planning of evacuation of passengers from the station. Escalators, staircases, and fire gates provided in the station are not only capable of handling traffic flow at peak period, they are also designed to ensure that passengers are able to leave the station within a specified time frame.

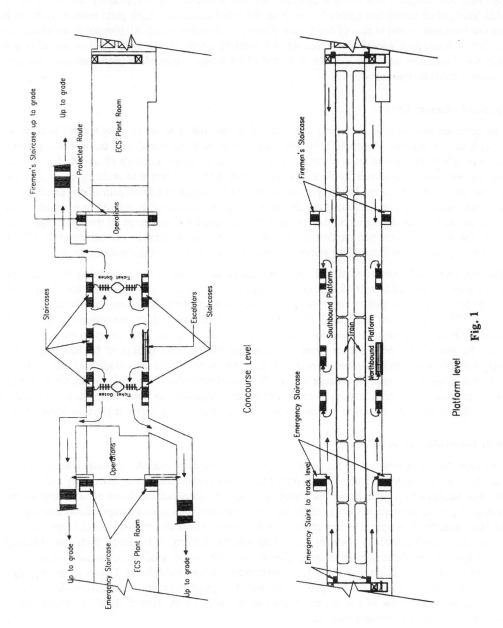

Concourse Level

Platform level

Fig. 1

167

The time frame set by NFPA 130 is 4 minutes for evacuation of the 'station occupant load' from the station platform and 6 minutes for evacuation from the most remote point on the platform to a point of safety (at-grade). This has been adopted in underground stations.

All non public areas of stations where only a small number of DMRC staff will be working, are to be provided with fire compartmentation, fire detection and alarm system. All offices and plant rooms are to be separated from public circulation spaces by two-hour fire rated separation. Each plant room is going to be a separate compartment so that in case of fire at such places, it can be contained. All non-public areas are also to be provided with adequate escape routes leading to a point of safety. Where only one direction of escape is available, travel distances would be less than 20 m and where escape is possible in more than one direction the travel distance could be less than 40 m.

3.2 Station Occupant Load

The occupant load used in the emergency exit calculation noted above is composed of two parts: the detraining load (*i.e.* number of people being discharged from arriving trains) and the entraining load (*i.e.* number of people waiting for trains at platform). For each train direction, a calculated trainload is derived by dividing the peak hourly trainloads by 50 to get peak minute train load and then multiplying by 2 times the headway to allow for one missed headway. Detraining load is subject to a maximum of the crush load of a train of 2388 passengers. For computing the detraining load, the calculated trainloads in the peak direction only are assumed to enter the station in the normal traffic direction and discharge all their passengers. In the operation plan, the train on the non-incident line is required to by-pass the station without stopping. Following trains will be regulated from entering the station and thus would not increase the detraining load.

The entraining load for terminal and wayside stations is calculated by multiplying the peak minute passengers entering the platform by 6 or 2 times the headway, whichever is higher. The entraining load for interchange stations is calculated by multiplying the peak minute passengers entering the platform by 2 times the headway. Thus one missed headway is allowed for delayed services in the computation of entraining load at platform. Above computation is done for peak flow direction. In the off peak direction entraining load is calculated by multiplying peak minute passengers entering the platform multiplied by headway, thus not considering any delayed services in the off peak direction. Above provisions are unique for Delhi Metro. This is based on arriving at an equitable balance among chances of emergency occurring during the peak minute flow, chances of delay occurring in peak direction and volume of passengers handled at stations so that a truly balanced design is achieved. A sample calculation for a typical Delhi Metro station is given in Annexure 1.

3.3 Exit Time Calculation

Exit capacities are calculated on the basis of 558.8 mm wide exit lanes. A fractional lane exceeding the width of 304.8mm is counted as half a lane. A stopped escalator is considered as an emergency exit of 1.5 lane capacity, the escalator step width being 1m. Capacity of stairs and stopped escalators per exit lane for the up direction is 35 persons per minute (ppm) and for the down direction is 40 ppm. Passages, doors and gates have per exit lane capacity of 50 ppm. The fare gates are in the form of retractable barriers and the capacity per gate is 50 ppm. For each sidewall in passages 1/2 a lane and on platform 1 1/2 lanes from platform edge is discounted from the available number of lanes. Due to these NFPA 130 guidelines, normally staircases have been provided in the integral multiples of 1 lane width, the minimum width being 3 lanes for normal ingress/ egress of passengers and 2 lanes for emergency staircases. Design of staircases and escalators is done initially for normal peak minute flow of passengers in and out of stations with desired 'level of service' and later checked for emergency exit times to be achieved.

Evacuation time from the platform can be calculated by dividing the station occupancy load by the exit capacity available from platform to concourse. Exit capacity beyond the concourse through fare barriers and entrances to station is normally provided in excess or equal to that provided from platform to concourse. If at any point such exit capacity is lesser, passengers have to wait at such point of exit and the waiting times are added to the walking time through the longest exit route to a safe place, which is normally at ground level, outside the station. Thus, platform exit time when added together with the time taken in travel through the longest exit route from platform exits to the ground should be less than or equal to 6 minutes.

In such calculations the walking speed is taken as 61 m/minute on level, 15.24 m/minute in up direction and 18.3 m/minute in down direction on staircases and stopped escalators. Escalator speed has been taken as 0.65 m/second and its capacity has been taken as 135 passengers/minute.

Besides calculating the evacuation times, the ratio of escalator lanes and stair lanes is also checked to ensure that escalators do not account for more than half of the units of exit at any one level.

3.4 Firemen's and Emergency Staircases

In addition to the staircases and escalators provided for the normal flow of passengers, enclosed stairs are provided at one or both ends of the station. One of these enclosed stairs is designated as the firemen's stair, whereas the others (if any) are termed as emergency stairs. Emergency stairs either lead to a point of safety at-grade or discharge to the passages leading to the normal entry/exit of stations. For security reasons, doors to the enclosed emergency stairs will be normally locked. During an emergency, locking device to the doors of the enclosed stairs will be released remotely and the exit signs switched on to invite passengers at the platform to use them. As a safety feature, the lock and light can be automatically released/switched on by the activation of the station fire alarm system or upon the loss of power. The door will also be equipped with a break-glass plunger to de-energize the locking device manually if there is a need to do so. Emergency staircases are so designed that service staff can use them during normal operation as service stairs.

Every station is to be provided with a firemen's staircase which will be a totally independent staircase for the exclusive use of firemen and station operational staff. The staircase will exit at ground level to open air and it will incorporate a flood proof landing with a lockable door. Except at ground level, the staircase will be protected by a smoke lobby and fire rated doors. This stair will be a protected route to gain access to the station control room and each principal level of the station. Additional stairways have been identified for the use of fire fighters, where access from the Firemen's Access Stairway to certain parts of the building is obstructed by tracks, or in other ways. It is to be provided with leaky coaxial cables for use by the emergency services for communication, an auxiliary fire indication panel, station drawings in a cabinet and a telephone at street level within the staircase enclosure. Firemen's stairs are not taken into consideration for calculating the emergency exit capacities for the station.

4.0 TUNNEL VENTILATION AND ENVIRONMENT CONTROL SYSTEM

Underground MRTS tunnels utilize significant quantities of electrical and mechanical energy resulting in inevitable waste heat due to train equipment, station equipment and other plant and equipment. It combines with lots of body heat, vapour and carbon dioxide released by a crowded environment with lots of passengers. Ventilation and air-conditioning are the principle means by which underground metros are able to remove their waste heat, maintaining passenger comfort and the elimination of obnoxious gases. In addition to providing comfort under normal condition, tunnel ventilation is also designed to control and extract the smoke and hot air from inside the subway to the atmosphere.

Tunnel Ventilation System being designed for the Delhi Metro has following principle functions:

- Providing an acceptable environment in the tunnel, station trackway and in the stations for passengers, staff and equipment.

- Controlling smoke movement by achieving "critical velocity" and preventing the phenomenon of backlayering in the event of fire.

- Providing a non-polluted tenable air environment in the emergency egress path to the passengers and staff for their safe evacuation to a safe place on ground.

Tunnel Ventilation System for Delhi Metro consists of Tunnel Ventilation Fans (TVFs), Trackway Exhaust System (TES), Smoke Exhaust Fans (SEFs), Tunnel Booster Fans (TBFs), Dampers and Ventilation Shafts. A schematic diagram showing these equipment is shown in Fig. 2.

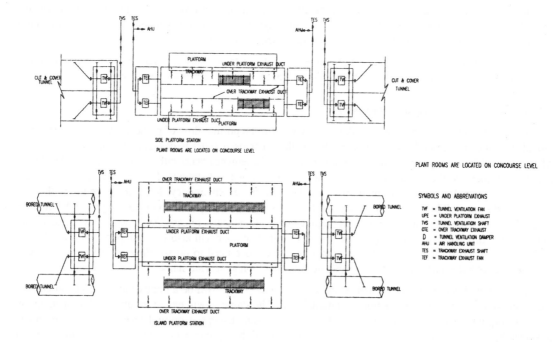

Fig. 2

4.1 Tunnel Ventilation Fans

At each end of stations, beyond the station platform, but within the overall station box, two longitudinal large diameter, and reversible TVFs are to be installed at the concourse level. Each fan is connected at one end to one tunnel through opening and damper at tunnel crown location. The other ends of the two TVFs are connected to a common ventilation shaft. A by-pass damper is also provided for regulating draught relief during normal piston effect operation. This draught relief route connects ventilation shaft to the tunnel, bypassing the TVFs.

4.2 Trackway Exhaust System

At every station two sets of ducts are provided for extracting heat and smoke from each trackway and the train standing on it. One set of duct, Over Trackway Exhaust (OTE), is provided just above the track mainly for extracting released heat from the air-conditioning units on trains. Another set of ducts, Under Platform Exhaust (UPE), is provided under the platform to extract heat and fumes released by the vehicle motors, brakes and other under-car equipment. OTE and UPE are connected at each end of the station to two Trackway Exhaust Fans (TEFs). The ducts and fans alongwith associated dampers and sound attenuaters make the Trackway Exhaust System. The combined efficiency of the system should be at least 65 per cent. The efficiency is defined as the ratio of the train heat directly captured by the system before it can mix in the trainway to the total heat released by the train as it dwells in the station.

The two TEFs at each end of station are located adjacent to TVF room at the concourse level. Other ends of these TEFs are connected to Trackway Exhaust Shaft and Air Handling Units(AHUs) through dampers. TEFs in operational mode may be either supplying return air to AHUs or sending smoke/hot air to Trackway Exhaust Shaft. TEFs may be stopped for controlling tunnel fires.

4.3 Tunnel Booster Fans

Two Tunnel Booster Fans(TBFs) are to be installed for each trackway at the crossover location. Their operation is important in moving air in the desired direction through the crossover. The fans are to be mounted on walls of tunnel. These TBFs are required during congested/emergency operations during which both TBFs in the incident tunnel should be operated in a direction consistent with the flow of fresh air. During emergency tunnel operations, both in the incident tunnel and the non-incident tunnel, TBFs will be energized. The operation of TBF in the non-incident tunnel will be against the flow of fresh air and the pressure resistance so developed helps fresh airflow through the incident tunnel.

4.4 Tunnel Ventilation Shaft

One Tunnel Ventilation Shaft(TVS) and one Trackway Exhaust Shaft(TES) has been provided at each end of stations. Each of them connects to the two TVFs or two TEFs respectively at one end and leads to the atmosphere at the ground level. Estimated cross sectional area are 20 sq.m for TVS and 12 sq.m for TES. Terminal air velocity from TES in areas where public would be exposed to the airstream, will not exceed 2.5 m/second, otherwise it will not exceed 5.0 m/second.

4.5 Station Environment Control System

Every underground station of Delhi Metro will be provided with a station air conditioning system, which is integral to the ventilation system. It consists of Supply Air Ducts(SADs) and Smoke Exhaust Ducts(SEDs) at concourse level and SADs at platform level. SADs supply cool air received from AHUs or fresh air from Supply Air Shafts. Trackway Exhaust System under normal conditions delivers return air from platform level. During fire emergencies at station platform, SADs at platform level work as Smoke Exhaust Ducts. There are two Smoke Exhaust Fans at each end of stations (four per station) which are connected to SEDs at one end and Trackway Exhaust Shafts at other ends. These fans extract smoke from the concourse level and platform level during fire emergencies at stations.

4.6 Normal Operating Condition

The Tunnel Ventilation System adopted for Delhi Metro is known as 'Closed System'. During normal operating conditions ventilation is achieved due to the piston effect as trains move through the tunnels. This gives the cheapest mode of ventilation. During summer and hot rainy season in Delhi, this operating condition will yield highest operational efficiency with only station air conditioning in operation, and ventilation shaft dampers in closed condition. Fresh and cool air fed into the station will be recirculated through Trackway Exhaust System to the Station Air Conditioning System. Relief of train piston air will be through by-pass shafts that permit the air to be exchanged between the two running tunnels. During temperate outside conditions such as, spring, autumn and in winter or anytime the station air conditioning plant is shutdown, the ventilation shaft dampers will be opened and the TES can discharge directly to atmosphere.

4.7 Congested Operating Condition

During congested operating conditions delay disrupts scheduled operation of trains and results in the idling of one or more trains in various tunnel segments or stations for extended periods of time. During such an incident, the train service regulator will halt as many subsequent trains as possible in stations, rather than in tunnels where vehicle air conditioning may fail due to higher air temperature. The design of Delhi Metro's system is based on the logical course of only one train being permitted in a ventilation zone. When a train is forced to halt midway in the tunnel and air temperature rises in the tunnel, TVFs will be activated in a push pull mode. In each ventilation zone TVFs would be operated so that a steady flow of fresh air is passed over the idling train. To generate push pull effect, one TVF of each ventilation zone will be operated in supply mode and one TVF in exhaust mode so as to move air through the tunnel in the direction train travels.

4.8 Computer Simulation

Tunnel ventilation will be designed using a Subway Environment Simulation (SES) computer programme. This will be in accordance with the Subway Environmental Design Handbook. The SES computer programme version 4 will be used to model the subway environment. The computer simulation programme will model the tunnel sections and all the stations as well as train performance and civil design data to aid in the design and verify the adequacy of all systems during normal, congested as well as emergency conditions. The different types of fire models will be simulated to arrive at the correct design of the ventilation system. The train fire load will be assumed as 20 MW. The small fire load in the station area will be taken as 1 MW. The heat release and smoke release rates including the smoke visibility criteria will be derived as per the guidelines of NFPA-130. The performance of the ventilation and smoke extraction system will be validated using Computational Fluid Design(CFD) techniques.

5.0 MEANS OF EGRESS AND EVACUATION OF TUNNELS

In the event of detection of fire in a train, which is running midway in tunnel between two stations, the first priority will be to take the train non-stop to the next station in the earliest possible time. However, there will be incidents in which a train becomes disabled in the tunnel and it becomes necessary to evacuate the passengers. In such cases, passengers will leave the train by a ramp at the front or back end of the train and will walk along the trackbed under the guidance and control of authorized, trained staff or other authorized personnel as warranted under an emergency situation. Passengers will be evacuated to a safe place to the ground level in the quickest time through a safe and protected route, usually by the walkway to the nearest station and then through the station to the ground.

NFPA 130 requires emergency exit stairways to be provided throughout the tunnels, spaced so that the distance to an emergency exit is not greater than 381m. It also provides for cross passages in lieu of emergency exit stairways. In Delhi MRTS, the trainways are to be constructed in the form of two separate tunnels with bored method of construction and separated into two trainways by a minimum of 2 hr rated fire walls in cut and cover method of construction. Cross passages are to be provided between separated trainways spaced not farther than 244m apart. In determining the requirement of these exit facilities, the emergency stairs located at ends of each underground station are counted as emergency exit stairs from tunnels. All such openings in cross passages will be protected with fire door assemblies having a fire protection rating of 1 1/2 hours with a self-closing door. The ventilation system will provide fresh air to the evacuating passengers during this period.

6.0 EMERGENCY SMOKE EXTRACTION FROM STATION AND TUNNEL

6.1 Fire in Train Stranded in Tunnel

As soon as a fire is detected in a train and the train becomes disabled in the tunnel, emergency smoke extraction and fresh air supply system of tunnel ventilation will be activated remotely in the affected tunnel / trainway. If the fire is towards the front end of train then, fresh air is supplied in the tunnel from the rear station end and smoke extraction is started from the approaching station end. If the fire is towards the rear end of train then, fresh air is supplied in the tunnel from the approaching station end and smoke extraction is done from the rear station end. This is done remotely from control room on receipt of information about exact location of fire in train from driver, or other authorized personnel in the train or trackway. This is done to ensure a constant supply of non-contaminated air to the majority of passengers on the train. However when the fire is in the middle of a train then fresh air can be supplied from either end.

The vehicle is designed to provide adequate duration of fire resistance from a fire either in the underframe area or above the roof, which results in disabling of the train. Hence in all such cases passengers would be able to move inside the train from one coach to other coach. Under the guidance of the driver or any other authorized staff, train passengers would be directed to come out on the desired end of the train. This is possible by means of emergency doors of 1100mm width provided at each end of the train in the driving cab. Passengers would be alighted on the trackbed through a ramp that would come out of the emergency door. Once all passengers are moved to the end of the train, they are further directed towards the nearest station through the trackway, from where they will be evacuated to the ground.

In such cases Trackway Exhaust System at the two stations bordering that incident tunnel section may need to be de-energized. The operation of TEFs, even on the downstream, or smoke filled, side of tunnel fire may be hazardous if smoke is drawn into an otherwise protected station. The first Tunnel Ventilation Shaft (on the exhaust side of a tunnel fire) is expected to remove all of the smoke generated by a vehicle fire in a tunnel ventilation zone. This confines the smoke flow and thus, the incident area, to one ventilation zone. The TEFs on the upstream, or smoke-free, side of a tunnel fire must be de-energized during tunnel fires. Fig. 3 shows a typical schematic diagram for such an evacuation.

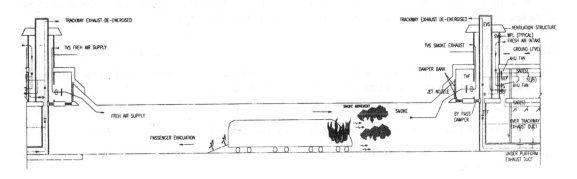

Fig. 3

In the event of failure of the tunnel ventilation to provide a sufficient amount of non-contaminated air, passengers from the disabled train will be directed to move to the adjacent trainway through the nearest cross passageway. A non-contaminated environment will be provided in the non-incident trainway through proper tunnel ventilation. In the contaminated trainway, the tunnel ventilation will help in controlling smoke in the vicinity of the passengers. From the non-contaminated trainway, the passengers will be evacuated safely to a nearby station, from where they will be finally brought to the safe ground level. During such emergency evacuation, train services will remain suspended in the non-contaminated trainway.

6.2 Fire in Train Stranded at Station

If the affected train is standing on the station trackway, the objective of emergency ventilation is to remove smoke coming out from vehicles in the most localized manner. Trackway Exhaust System is operated in open mode to extract smoke from top and bottom of vehicles and dispose through Trackway Exhaust Shafts in open air. Supply Air Ducts provided on platform are also switched into Smoke Extraction Mode and smoke will be extracted by SEFs and disposed through TESs. Supply of fresh air may be done through the tunnels by operating TVFs at adjacent stations in supply mode. Thus Trackway Exhaust System operates in concert with TVFs during a fire in the station.

7.0 OTHER DESIGN FEATURES

Apart from the design features dealt above in detail there are other important design features which are worth mentioning here. These are an elaborate fire protection and life safety in stations consisting of Fire Detection System, Automatic Sprinkler System, Fire Hydrants and Fire Extinguishers. The design also includes the fundamental fire risk reducing methods like using non-combustible fire resistant or fire retardant materials in the construction of the stations and tunnels; limiting combustible loads in trains; reliable power source backed up with emergency supply; and sophisticated system of station supervision like closed circuit TV; communication and train signalling etc.

8.0 CONCLUSION

In summary, the Delhi MRTS has been designed to incorporate the required fire prevention and life safety measures which are capable of effectively detecting and suppressing fire at an early stage. Should the fire situation warrant emergency evacuation to safe areas, the evacuation process can be safely completed within a very short time by means of the escape routes provided.

A set of comprehensive emergency procedures and disaster management plan will be formulated by the operation department to deal with different emergency situations and fire scenarios. In all the cases a comfortable and safe environment will be maintained for the passengers inside the subway.

With all these measures, it is believed that the Delhi Mass Rapid Transit System will be observed as one of the best designed system worldwide with respect to fire and life safety.

REFERENCES

Outline Design Criteria for 'Station Architectural Works' and 'Tunnel Ventilation Works' for underground metro corridor, October 1999, Delhi Metro Rail Corporation.

NFPA-130, 1997, Fixed Guideway Transit Systems, National Fire Protection Association.

Transportation Report, June 1999, by General Consultants to Delhi Metro Rail Corporation.

Subway Environmental Design Handbook, US Department of Transportation, Urban Mass Transit Administration.

Detailed Project Report, 1995 For Integrated Multi-Modal Mass Rapid Transport System For Delhi (Modified First Phase) by RITES.

National Building Code of India, 1983.

SAMPLE CALCULATIONS

For illustration purposes an intermediate side platform station(Civil Lines) has been taken.
Crush Trainload = 2388 passengers
Train service interval = 2 minutes

For Northbound Platform	For Southbound platform
TRAFFIC ESTIMATES FOR YEAR 2021;	
Morning peak Peak hour link load 15275 passengers	Morning peak Peak hour link load 16223 passengers
Evening peak Peak hour link load 15502 passengers	Evening peak Peak hour link load 12979 passengers
Morning peak Peak minute 3 passengers entering station station load 20 passengers exiting station	Morning peak Peak minute 23 passengers entering station station load 3 passengers exiting station
Evening peak Peak minute 3 passengers entering station station load 21 passengers exiting station	Evening peak Peak minute 18 passengers entering station station load 3 passengers exiting station
STATION OCCUPANT LOAD	
Morning Peak Detraining load $=15275 \div 50 \times 2 \times 2 = 1222$ passengers Entraining load $= 3 \times 6 = 18$ passengers Station occupant load $= 1222+18 = 1240$ passengers Evening Peak Detraining load $=15502 \div 50 \times 2 \times 2 = 1240$ passengers Entraining load $= 3 \times 6 = 18$ passengers Station occupant load $= 1240+18 = 1258$ passengers Critical station occupant load = **1258 passengers (Evening peak)**	Morning Peak Detraining Load $= 16223 \div 50 \times 2 \times 2 = 1298$ passengers Entraining load $= 23 \times 6 = 138$ passengers Station occupant load $= 1298+138 = 1436$ passengers Evening Peak Detraining load $= 12979 \div 50 \times 2 \times 2 = 1038$ passengers Entraining load $= 18 \times 6 = 108$ passengers Station occupant load $= 1038+108 = 1146$ passengers Critical station occupant load = **1436 passengers (Morning peak)**
EMERGENCY EXIT CAPACITY CALCULATION	
At Civil Lines station, open stairs 3 numbers of 1.8 m width are provided at southbound platform and 2 numbers of 1.8 m width are provided at northbound platform. 1 number 1 m wide escalator is provided at northbound platform. At concourse level 2 fare barriers are provided at the two ends of common paid concourse. Each fare barrier consists of 3 ticket gates and 1 emergency gate of 1.2 m width. From concourse to ground there are 3 staircases each having 4.5 m width.	
Exit capacity available with 2×1.8m stairs = 2×3 lanes $\times 35$ ppm = 210 ppm Exit capacity available with escalator = 0 ppm (Due to 1 escalator discounted) Total exit capacity available = 210+0 = 210 ppm	Exit capacity available with 3×1.8m stairs = 3×3 lanes $\times 35$ ppm = 315 ppm Exit capacity available with escalator = 0 ppm (Due to no escalator provided) Total exit capacity available = 315+0 = 315 ppm
At Concourse	
Exit capacity through fare barrier Ticket Gates = 6×1 lane $\times 50$ ppm = 300 ppm, Emergency Gates = 2×2 lanes $\times 50$ ppm = 200 ppm, Total exit capacity = 500 ppm.	
TEST NUMBER 1	
To evacuate station occupant load from station platform in 4 minutes or less.	

Evacuation time = station occupant load÷platform exit capacity $= 1258 \div 210 = 6$ minutes. Hence, exit capacity should be increased by providing emergency stairs. Exit capacity required at platform =1258÷4 =315 ppm. Therefore additional capacity in the form of emergency stairs required = 315–210 = 105 ppm. Therefore emergency stairs required = 105 ÷ 35 = 3 lanes = 1.8 m wide staircase. Platform evacuation time with emergency stairs = 4 minutes	Evacuation time = station occupant load ÷ platform exit capacity $= 1436 \div 315 = 4.6$ minutes. Hence, exit capacity should be increased by providing emergency stairs. Exit capacity required at platform = 1436÷4 = 359 ppm. Therefore additional capacity in the form of emergency stairs required = 359–315 = 44 ppm. Therefore emergency stairs required = 44÷35 = 2 lanes = 1.2 m wide staircase. Platform evacuation time with emergency stairs = $1436 \div (315+70) = 3.7$ minutes

TEST NUMBER 2

To evacuate station occupant load from the most remote point on the platform to a point of safety in 6 minutes or less.

Walking Time For Longest Exit Route

Platform to concourse, T1 = 5.3 m÷ (15.24 m/minute) = 0.35 minutes.
On concourse/subway, T2 = 75 m ÷ (61 m/minute) = 1.23 minutes.
Concourse to grade, T3 = 7 m÷ (15.24 m/minute) = 0.46 minutes.
At grade, T4 = 5 m÷ (61 m/minute) = 0.08 minutes.
Therefore total walking time = 0.35+1.23+0.46+0.08 = 2.12 minutes.

Waiting Time At Fare Barrier

Occupant load at fare barrier = occupant load at platform (evening peak) + proportion of normal waiting passengers at southbound platform arriving through open stairs and escalators - emergency stairs capacity at southbound platform = 1258+18×2×9÷11–105×4 = 1258+30–420 = 868 passengers. Waiting time at fare barrier = 868÷ (fare gate capacity) - 4 minutes=868÷500-4=1.74-4=0 (no waiting time).	Occupant load at fare barrier = occupant load at platform (morning peak) + proportion of normal waiting passengers at southbound platform arriving through open stairs and escalators - emergency stairs capacity at southbound platform = 1436+3×2×6÷9-70×3.7 = 1436+4 –259 = 1181 passengers. Waiting time at fare barrier = 1181÷ (fare gate capacity) -3.7 minutes=1181÷500-3. 7=2.36-4=0 (no waiting time).

Waiting Time At Concourse Exit

Exit Capacity from concourse to ground
= exit capacity of 3 × 4.5 m stairs - exit capacity of emergency stairs opening directly in the subway.
= 3 × 8 lanes × 35 ppm - (3 lanes +2 lanes) × 35 ppm
= 840×175 = 665 ppm.

= (occupant load at fare barrier) ÷ (exit capacity concourse to ground) - 4minutes = 868÷665-4 = 1.3-4 = 0 (no waiting time)	= (occupant load at fare barrier) ÷ (exit capacity concourse to ground) – 4minutes = 1181÷665–3.7 = 1.8–4 = 0 (no waiting time)

Total Exit Time

= Platform evacuation time + total walking time + waiting time at fare barrier + waiting time at concourse exit = 4+2.12+0+0 = 6.12 minutes. This is higher than 6 minutes. Hence at northbound platform emergency stairs must be increased to 4 lanes from 3 lanes provided from test 1.	= Platform evacuation time + total walking time + waiting time at fare barrier + waiting time at concourse exit = 3.7+2.12+0+0 = 5.82 minutes. This is lower than 6 minutes hence o.k.

RENEWING THE URBAN INFRASTRUCTURE : THE REBUILDING OF THE SOUTHFIELD SEWER

AWNI QAQISH
Assistant Director

BHARAT DOSHI
Head Engineer

Detroit Water & Sewerage Department
Detroit, Michigan, USA

JEROME C. NEYER
Chairman

KEITH M. SWAFFAR
Executive Vice President

NTH Consultants, Ltd.
Detroit, Michigan, USA

SYNOPSIS

The Southfield sewer is a 3.05m (10 ft.) diameter pipe that conveys combined sewage from the City of Detroit's northwest regions to the wastewater treatment plant. The sewer was constructed, in tunnel, in the 1920'ó. When originally constructed, the sewer was located beneath a surface street. However, in the 1960's, a major urban freeway was constructed directly over the sewer. That freeway is now one of the most heavily travelled roads in the State of Michigan. In 1992, a detailed inspection of the tunnel revealed several areas of distress that required remedial action. Within one reach, having a length of about 66.96 m (200 ft.), the project team concluded that the most appropriate remedial method was abandonment of the existing barrel and its replacement with a bypass tunnel. Based on the sewers repair history in this area, the owner elected to include several other previous repair areas and construct a by-pass sewer having a length of approximately 1127.76 m (3700 ft.). Because the sewer was located directly beneath the travel lanes of the freeway, the connections from the existing sewer to the bypass tunnel needed to be made directly beneath the existing travel lanes. However, due to the freeway's use, MDOT would permit only weekend closures of the road. These constraints required the development of an innovative top-down construction approach that required only weekend road closures while providing appropriate protection to personnel working beneath the freeway.

1.0 INTRODUCTION

The Southfield sewer is a 3.05 m (10 ft.) diameter tunnel carrying combined sewage from northwest Detroit to the City's wastewater treatment plant. The tunnel was originally constructed in the 1920's in soft ground ranging from a full face of soft to medium clay to a face condition consisting of wet sands and silts overlain by clays. When originally constructed, the sewer alignment was located beneath Southfield road, a surface street. However, in the 1960's the surface street was replaced by the Southfield freeway, a major urban highway, which is one of the busiest freeways in Michigan. When the freeway was constructed, the vertical alignment of the road was changed dramatically to allow the freeway to cross beneath major surface street intersections along its alignment. Consequently, following construction of the freeway, the ground cover above the tunnel was reduced from generally 10.57 m (35 ft.) to as little as 4.57 m (15 ft).

Because of the freeway construction, a detailed inspection of the sewer revealed a reach of approximately 60.96 m (200 ft.) in length, located directly below the freeway lanes, which was severely distressed. Failure of this section of the sewer would result in the closing of one of the busiest roadways in the state as well as the potential backup of sewage into homes and spilling into nearby surface waters. As such, repair of the distressed section required development of a procedure that would not jeopardize the conveyance of flow through the existing sewer nor impede travel on the overlying freeway.

Based on these constraints, the DWSD and NTH team developed a repair concept that entailed the mining of a new 3.05 m (10 ft.) diameter by-pass sewer away from the traffic lanes. The final by-pass length of 1127.76 m (3700 ft.) was selected based on the location of the observed distress as well as the proximity of previous repair areas to the distressed section.

The new by-pass section was connected to the existing sewer, up stream and down stream of the distressed area, using a "top down" construction approach, that required only minimal freeway closures during weekend periods.

The "top-down" procedure called for the freeway to be closed only briefly on several weekends. During these periods, the contractor removed the pavement in the area of the connection and excavated a shaft approximately 3.05 m (10 ft.) deep. A soldier pile and lagging bracing system was then installed, temporary bridge abutments constructed and a "lid" of prestressed concrete beams placed over the shaft to span the excavation and support the traffic load. The balance of the excavation and connection of the bypass then took place below the "lid" while traffic drove over the shaft. When the connection had been completed, the freeway was again closed briefly on a weekend. During this closure, the "lid" was removed ; the shaft support system removed and the shaft filled with cement stabilized fly-ash. The pavement was then restored and the freeway reopened. Following completion of the by-pass, the existing distressed section of sewer was abandoned in-place using cement stabilized fly-ash grout placed from the surface.

The project was completed on time and within budget. There were no unanticipated freeway disruptions and the sewage flow was never interrupted. The sewer and the freeway are now ready for many more years of service.

2.0 SEWER CONSTRUCTION AND HISTORY

The Southfield sewer alignment extends over a distance of 16.09 km (10 miles) within the cities of Detroit and Dearborn, Michigan. The sewer diameter is generally 3.05 m (10 ft.) however there are reaches within the system where the diameter varies from as small as 1.83 m and 22.80 cm (6 ft. 9 inches) up to 3.66 m 15.24 cm (12 ft. 6 inches). The sewer was originally built over the period between approximately 1927 and 1930 The sewer was constructed in tunnel using handmining techniques at depths on the order of 12.19 m (40 ft.). Although the available records do not indicate the specifics of the primary earth support system used, other tunnels mined in the area during the same period used a variety of systems. These consisted of steel liner plates, concrete blocks, ribs and lagging or a combination of these. Over the last decade, construction activities within two reaches of the sewer have uncovered evidence of both steel liner plate and pre-cast concrete block primary liners. Following excavation and placement of the primary lining system, a secondary liner consisting of 40.64 cm (16 inches) of non-reinforced concrete was placed within the tunnel.

In 1929, prior to completion of the project, three portions of the sewer collapsed. One of these sections was located near the site of the current distress. Based on the available documentation, repairs to the sewer at that time were made in open cut with steel sheet piling used below an initial precut excavation. Once the collapsed portion of the barrel was reached, repairs were facilitated using a combination of cast-in-place concrete, shotcrete and bricks. Near the current repair location, the sewer diameter was reduced to 2.80 m (9.2 ft.) within one 80.77 m (265 ft.) reach because of previous repair activities.

The Southfield freeway was constructed over the sewer's alignment during the period from 1961 to 1964. Construction of the freeway entailed depressing the road surface at all major street crossings. The excavation to achieve the required highway grades at major intersections resulted in the ground cover over the sewer being reduced to as little as 7.62 m (25 ft.) in some areas.

Shortly after the completion of freeway construction in 1966, settlement of the pavement surface at the Warren exit ramp (near the current repair area) was noted. Soil borings performed at that time revealed a significant void beneath the pavement surface. Subsequent inspection of the sewer indicated significant damage of the lining in the area requiring immediate repair. As such, the barrel was stabilized by the placement of concentric brick rings within a 36.58 m (120 ft.) reach. These repair activities resulted in a reduced pipe diameter of 2.44 m (8 ft.) within the damaged section.

3.0 CONDITION ASSESSMENTS

Major inspections and condition assessments of the Southfield sewer have been conducted over the last 40 years. The first major condition assessment of the facility occurred during 1959 and 1960 immediately before the overlying freeway construction. Subsequently, a full-scale inspection was also performed in 1966. In addition to these full inspections, other more limited inspections were performed in 1976, 1979 and 1984. The most recent major inspection resulting in the by-pass construction detailed herein was performed in 1992.

3.1 Previous Inspections

The sewer inspections conducted in 1959/1960 and again in 1966, were performed by DWSD working with the Wayne County Road Commission. Review of inspection reports from this period indicates that during the 1959-1960 inspection, flow levels in the sewer were such that the inspecting team was able to walk the entire length of the sewer. However, reports from the 1966 inspection indicated that the inspecting team encountered flow depths of 1.22 m to 2.13 m (4 ft. to 7 ft.) to 7 feet requiring the use of rubber rafts to inspect much of the alignment. Since the 1966 inspection, high flow levels in the sewer have been a concern for each inspection. As such, each of the partial inspections performed by the owner between 1976 and 1990 were performed during low flow periods. This usually resulted in the inspections being performed on Sunday mornings from 4 AM to 7 AM.

Review of the available inspection documentation indicates that during the last detailed inspections, a depressed area located near Paul and Whitlock Streets was identified. This reach of sewer presented significant difficulties to the inspection teams that used rubber rafts for the survey in 1966. Local legend tells that members of the inspection team had to hang from the sides of the raft while one team member laid on his back and pushed on the tunnel crown to navigate the raft through the depressed area. The depression may in fact date back to the previous repair activities performed prior to the tunnel being placed into service. Over the next 26 years, partial inspections of the sewer including observation and monitoring of the depressed area were conducted by DWSD on an annual or biannual basis.

3.2 1992 Inspection

In response to DWSD's concerns about the condition of the entire Southfield sewer system, a joint venture of NTH and Walbridge Aldinger was retained to develop and implement an inspection programme for the entire Southfield sewer system. Because of historic high flow levels within the sewer, the initial engineering activities consisted of a review of the available system data and development of a hydraulic model to evaluate flow conditions and flow control alternatives. Using this approach, the engineering team was able to model, design and install pumping and flow control within the sewer to enhance the inspection without risking an uncontrolled backup of the sewer. A temporary pump station was constructed and installed immediately north of the depressed area to allow this section to be dewatered. In addition, flow control devices were placed in the tunnel at 4 locations. In general, the flow control devices used as a result of the study consisted on segmental bulkheads supported by a ring anchored to the existing secondary liner. The ring and segmental stop logs were designed and fabricated in pieces small enough to be carried into the tunnel from the nearest manhole.

Following completion of the flow control evaluation, the inspection was conducted over a period of about 4 months with inspection activities being performed on Sunday mornings throughout the winter and spring as weather conditions permitted. When the inspection was finally completed, the 16.09 km (10 miles) of sewer had been divided into 16 segments, each of which was inspected in detail by a team of 5 individuals. Each inspection team was composed of staff from both DWSD as well as the NTH/WA joint venture. Within the inspection team, individuals were given specific asignments during the inspection such as team leader, photographer or data recorder.

Based on the inspection, the sewer was generally found to be in good condition. However, 10.16 cm (four reaches) within the system were identified as being in poor condition. Two of these distressed areas, identified as the Schoolcraft and Dayton reaches exhibited significant concrete deterioration and the presence of voids extending through the concrete. Conceptual repair techniques that were developed for these reaches included patching the voids and placing new concrete in eroded areas. In addition, two other reaches were noted to have extensive cracking and spalling. These two areas, identified as the Paul/ Whitlock and Dover reaches exhibited signs of significant distress and as such, the stability of the existing lining was in doubt. Because of these concerns and the presence of the overhead freeway, the owner elected to install temporary internal support to the tunnel to prevent a potential collapse. As such, W4 x 13 steel ribs were installed at 1.22 m (4 ft.) on center, throughout the limits of the noted distress.

Concurrent with the installation of steel ribs, alternatives were developed for the rehabilitation of the severely distressed sections. Within the Dover repair area, the selected repair alternative consisted of 6-inches of shotcrete placed outside the existing support ribs within a 34.75 m (114 ft.) reach of the tunnel. Using this repair procedure, a minimum sewer diameter was 2.44 m (8 ft.) was maintained throughout the repair area. Other activities required to facilitate this repair procedure included construction of a new access shaft, within a greenbelt area, south of the repair area as well as temporary flow control to keep the area dewatered during shotcrete placement.

Within the Paul / Whitlock area, the severely distressed section was estimated to extend over a distance of 60.96 m (200 ft.). Within this area, the inspection had reveled not only the presence of severe cracking, but also that the vertical grade of the sewer had been severely impacted by settlement of the last 60+ years. Measurements made at the time of the inspection indicated that within the distressed section, the invert of the sewer was depressed as much as 0.91 m (3 ft.). At several locations, a probe rod could be pushed through the concrete secondary liner to the existing primary liner.

Due to the level of distress noted, three repair alternatives were developed and presented to the owner. These consisted of one repair alternate within the current horizontal alignment and two by-pass repair options. The in-line repair alternate called for the open cut removal and replacement of the existing distressed section of sewer. Due to the presence of the freeway, efforts were made to minimize the extent of the repair to minimize the impact on the overlying roadway. This option entailed replacement of approximately 60.96 m (200 linear ft.) feet of sewer. Because this option would have required closure of the northbound traffic lanes, it was concluded that any open cut replacement was not practical. As such, attention moved to the construction of a by-pass sewer around the distressed area. Two by-passes, short and long, were evaluated. The short by-pass concept called for the mining of a new tunnel having a length of approximately 152.4 m (500 ft.). The long by-pass concept entailed mining a new tunnel over an alignment of at least 609.6 m (2000 linear ft.). In considering each alternate, the owner eventually chose the long by-pass concept because it remediated other areas within the sewer in which the condition of the concrete was questionable or which had been repaired in the past. When all of these factors were considered, the final length of the by-pass section extended over an alignment of approximately 1127.76 m (3700 linear ft.).

4.0 BY-PASS DESIGN AND CONSTRUCTION

Once the internal support ribs had been installed and the preferred repair option selected, the design team prepared plans and specifications for the construction of the by-pass tunnel, tapping of the existing sewer and abandonment of the existing distressed reach. The concept eventually selected consisted of tapping the existing sewer at the north terminus of the by-pass, beneath the right northbound travel lanes. The by-pass would then be extended to the adjacent service drive where the alignment would continue to the south. At the south terminus of the alignment, the by-pass would reconnect with the existing sewer again, beneath the existing freeway surface. In preparation for the new construction, important elements of the design included evaluating the existing subsurface conditions along the new alignment and the anticipated flow levels within the existing sewer. In addition, because the sewer at the tap locations was located beneath the freeway travel lanes, a solution needed to be devised which would permit the required construction without impacting the freeway users. The solution adopted to resolve this problem was to utilize "top down" construction techniques for the project elements located beneath the freeway travel lanes. This concept entailed the construction of temporary bridges over the tap locations to permit construction activities to proceed while traffic flows were maintained overhead. Fig. 1 shows the general configuration of the north tap from the by-pass tunnel to the original sewer.

4.1 Top-Down Construction Concept

The "top down" construction method required the northbound lanes of the freeway to be closed for short periods. Working with the Michigan Department of Transportation (MDOT), the design team arranged for several weekend freeway closures. In general, the freeway was closed from Friday evening, after rush hour, to Monday morning, before rush hour. The concept developed for the project was for the contractor to use this time, approximately 54 continuous hours, to build foundations for temporary bridges to span the excavation as well as to start initial shaft construction activities. Once the foundations were constructed, precast bridge beams were placed over the excavations and temporary pavement placed to restore the road surface.

Now that the freeway closure issues had been addressed, the contractor was able to continue his work under the freeway without concern for the traffic above. The project contractor, JayDee Contracting, made the connections between the original sewer tunnel and the new by-pass using a series of flumes to carry sewage through the shaft area while the connections were being made. Since the Southfield sewer flows full during storm events, the shaft and flume system was designed to allow flooding of the shaft during these events. Thus, the flow in the sewer was not impeded by the construction.

Construction of the by-pass tunnel itself was relatively straight forward except for the need to connect a number of existing lateral sewers to the new tunnel. There were also several storm drains on the freeway itself that drained into the original Southfield sewer. These had to be rerouted to the by-pass tunnel. Connection of the lateral sewers to the by-pass tunnel required construction of shafts at the intersection of the sewers that flowed into the Southfield sewer. Once again, it was necessary to maintain flow in both the lateral sewers as well as the main tunnel during construction of the connections.

Once the by-pass had been completed and the connections made at the two shafts in the freeway, the original sewer was backfilled with a flowable fill consisting of cement, flyash and water. The shafts within the freeway were removed during weekend shutdowns and backfilled with compacted fill. A new concrete pavement was placed over the area of the shafts and traffic was free to role over the repair area on Monday morning.

The quality of the design was proven by the completion of the project ahead of schedule and within the budget established by DWSD. There were no spills of sewage into the Rouge river and no backups of sewage into the nearby basements. Furthermore, there were no disruptions of traffic during the weekdays. The Southfield sewer is now in the best shape it has been in since it was built nearly 70 years ago.

4.2 Sub-surface Conditions

During the conceptual design phase, a thorough geotechnical investigation was performed along the proposed alignment. The investigation consisted of drilling of borings at the locations of major structures and shafts as well as at intervals of approximately 152.4 m (500 ft.) along the proposed tunnel alignment. In general, the sub-surface conditions encountered along the proposed alignment were fairly uniform consisting of fill materials underlain by native cohesive soils. The cohesive soils are in turn underlain by native granular materials that extend to the explored depth. A sub-surface profile section depicting the conditions encountered along the alignment is presented on Fig. 2.

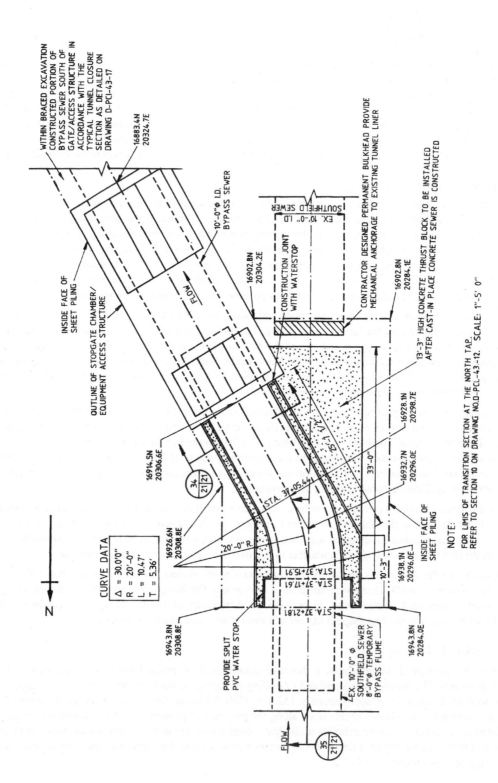

WITHIN BRACED EXCAVATION CONSTRUCTED PORTION OF BYPASS SEWER SOUTH OF GATE/ACCESS STRUCTURE IN ACCORDANCE WITH THE TYPICAL TUNNEL CLOSURE SECTION AS DETAILED ON DRAWING D-PCI-43-17

16883.4N
20324.7E

INSIDE FACE OF SHEET PILING

OUTLINE OF STOPGATE CHAMBER/ EQUIPMENT ACCESS STRUCTURE

10'-0"Φ I.D. BYPASS SEWER

FLOW

16902.8N
20304.2E

CONSTRUCTION JOINT WITH WATERSTOP

SOUTHFIELD SEWER

EX. 10'-0" I.D

CONTRACTOR DESIGNED PERMANENT BULKHEAD PROVIDE MECHANICAL ANCHORAGE TO EXISTING TUNNEL LINER

16902.8N
20284.1E

13'-3" HIGH CONCRETE THRUST BLOCK TO BE INSTALLED AFTER CAST-IN PLACE CONCRETE SEWER IS CONSTRUCTED

16914.5N
20306.6E

34
21 21

25'-1 1/2"

16928.1N
20298.7E

33'-0"

16932.7N
20296.0E

INSIDE FACE OF SHEET PILING

STA. 37+05.44

16926.6N
20308.8E

20'-0" R.

STA. 37+15.91

16938.1N
20296.0E

10'-3"

STA. 37+17.61

CURVE DATA

Δ = 30.0'0"
R = 20'-0"
L = 10.47'
T = 5.36'

N

16943.8N
20308.8E

PROVIDE SPLIT PVC WATER STOP

STA. 37+21.81

EX. 10'-0" Φ SOUTHFIELD SEWER 8'-0"Φ TEMPORARY BYPASS FLUME

16943.8N
20284.0E

FLOW

35
21 21

NOTE:
FOR LIMIS OF TRANSITION SECTION AT THE NORTH TAP,
REFER TO SECTION 10 ON DRAWING NO.D-PCI-43-12.   SCALE: 1"-5' 0"

Fig. 1 : North bypass connection plan

182

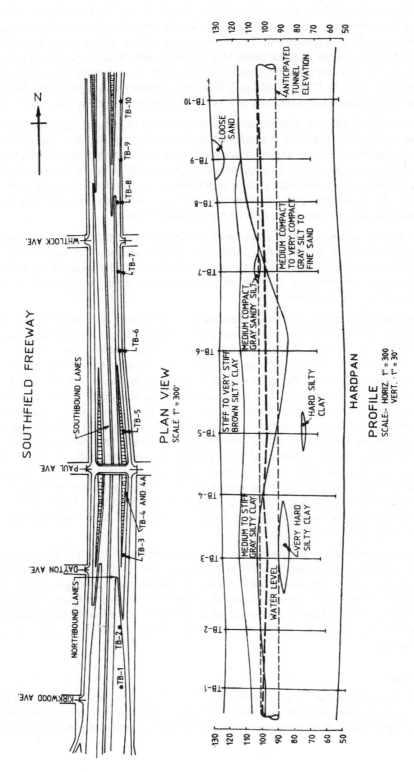

Fig. 2 : Southfield sewer generalized subsurface profile

183

As indicated on the profile section, the upper native cohesive soils generally consist of stiff to very stiff brown silty clay. This material typically has shear strengths on the order of 9.760 T/sq m (2000 10 sf) to 12.20 T/sq m (2500 psf) and extends to depths ranging from about 3.05 m to 5.18 m (10 ft. to 17 ft.) along the alignment. The brown cohesive clays are in turn underlain by gray silty clays which are somewhat softer, ranging in consistency from medium to stiff. The vertical extent of these soils is somewhat variable along the alignment extending to below the proposed tunnel bore near the middle of the alignment and terminating above the proposed tunnel at the north and south ends of the alignment. Testing of these materials indicated shear strengths on the order of (4.88 T/ sq m (1000 psf). Based on laboratory testing performed during the project, these materials can be classified based on the Unified Classification System as low plasticity CL materials.

Beneath the surficial cohesive soils, the native materials consist of granular deposits of medium compact to very compact gray silt to fine sand. As indicated on the profile section, these soils extend beneath the proposed tunnel invert throughout the alignment. Standard penetration tests on these materials performed in the field indicated SPT resistances ranging from approximately 30 to nearly 80 blows. Laboratory testing on these soils indicated the materials are somewhat variable in gradation with classifications ranging from SM to SC or ML. Based on SPT resistances, depth of overburden and gradation characteristics, these materials were assigned angles of internal friction varying from 33 degrees to 40 degrees.

Based on the sub-surface conditions encountered, it was apparent that face conditions during mining of the by-pass tunnel would be quite variable. Within the northern and southern most reaches, anticipated conditions would consist of a full face of clay. However, within the central portions, the face was expected to consist entirely of silt and fine sand. Consequently, within the transition zones, a face condition of clay overlying sand could also be anticipated.

During the geotechnical investigation, groundwater measurements were taken during the drilling of each test boring. In addition, after the completion of drilling, pneumatic or open standpipe piezometers were installed in most of the borings. Based on the groundwater data obtained during the investigation, it appeared that the phreatic surface along the tunnel alignment was located between elevations 105 and 103. These elevations corresponded to the upper quarter of the proposed tunnel excavation. Because of the granular materials within the proposed face elevations along the alignment, control of groundwater and running ground were major concerns during the upcoming mining and shaft construction activities. Therefore, a field pump test was conducted to better define the hydraulic properties of the granular materials that would be encountered during construction. The pump test involved the installation of a pumping well within the southern portions of the alignment. The test was conducted by pumping from the well and monitoring the draw down within the nearby piezometers. Based on the field pumping test, the transmissivity, hydraulic conductivity and storativity of the aquifer were estimated to be 0.31 million litres/m, 22638 litres/m^2 (25,000gpd/ft, 556gpd/ft^2) and 0.00015, respectively. This data was subsequently provided to the contractors with the bidding documents such that they could estimate appropriate dewatering systems prior to submitting their bids.

In addition to groundwater, explosive and toxic gases were measured at each of the test boring locations. Concentrations of methane gas ranged from 0.1 to 2 percent of the lower explosive limit (LEL) within the tunnel zone and as high as 38% of the LEL outside the tunnel zone. Hydrogen sulfide gas was also encountered at most of the boring locations. In general, hydrogen sulfide concentrations ranged from 0.1 parts per million (PPM) to 0.8 PPM. Toxic gas levels as high as 21.8 PPM were encountered outside the tunnel zone within one of the borings. Based on this data, the tunnel was classified as "potentially gassy" based on federal OSHA requirements.

4.3 Shaft Design

The shafts in which the new by-pass sewer was to be connected to the original Southfield sewer were constructed in the travel lanes of the freeway. Since freeway closure was only an option for short periods on weekends, the "top-down" method of construction was adopted. The system required minimal closure of the freeway, which could be accomplished by closing the freeway at 11:00 PM on Friday and reopening it at 5:00 AM on Monday. There were three periods of weekend closure to construct the two shafts and two periods of weekend closure to remove the shafts and restore the pavement.

During the first closures, the pavement was removed and soldier pile support beams were installed to a level below the invert of the sewer. The soldier piles were used for support of the surrounding ground as well as to support the temporary bridge abutments. Each of the two shafts was approximately 18.29 m (60 ft.) long parallel to the tunnel and 9.14 m (30 ft.) wide were 418.29 m. After the support beams were installed, the shafts were excavated to a depth of approximately 3.05 m (10 ft.). Pile supported bridge beam supports were also constructed at the ends of each shaft. Precast prestressed concrete box beams were then seated on the supports and a wearing surface of asphaltic concrete placed over the beams. Concrete median barriers were placed along the edge of the shaft nearest the edge of the pavement so that no cars would drive on the shoulder and come too near the shaft. The general configuration of the temporary bridges is shown in Fig. 3. Fig. 4 shows temporary bridge over the access shaft and Fig. 5 shows a typical access shaft construction.

Once the shafts had been covered with the box beams and the asphalt surface course, excavation of the balance of the shaft was accomplished by equipment working below the travelled freeway. The contractor used low-headroom equipment to place the lagging boards behind the previously installed soldier beams. After the top of the original tunnel had been exposed, an opening was made in the sewer to permit bracing to be installed within the sewer. This precaution was mandated by the specifications to prevent damage to the old sewer until it was ready to be abandoned.

The shaft support system was then extended beyond the edge of the pavement so that a connection to the by-pass tunnel could be made in open cut. After mining and lining of the by-pass sewer had been completed, flumes were installed in the old sewer to divert low flows through the area of the shaft. An entire section of the old sewer was then removed to allow transfer of the flow to the by-pass sewer. During this process, there were several instances of high flow in the sewer. These flows were handled by allowing the shaft to flood and then allowing the overflow to return to the Southfield sewer when the flow subsided.

Removal of the box beams and the shaft support system went quite smoothly. Good planning on the part of the contractor and excellent cooperation among the team members, including MDOT, led to reopening the freeway ahead of schedule after completion of the removal operation.

4.4 Tunnel Construction

Once the shafts to tap the sewer had been constructed, the by-pass alignment was continued within a braced excavation to the adjacent freeway service drive. At this location, a conventional open face tunnel-boring machine was launched. Due to the anticipated granular soils located below the groundwater table, the contractor installed a series of dewatering wells along the alignment well in advance of tunnel construction activities. The dewatering system was very effective not only in dewatering the granular materials through which the tunnel was mined, but also in controlling methane and hydrogen sulfide gas from entering the tunnel.

During mining, the contractor used a primary earth support system consisting of steel rings and wood lagging. Due to the presence of fine granular soils and a relatively high groundwater table, a geo-filter fabric was placed behind the primary lining to prevent the migration of fines into the tunnel, should the dewatering system fail. Once the entire tunnel was mined, the contractor commenced to cast the secondary concrete liner. Because of settlement associated with ground loss beneath the original tunnel, controlling of cracks and the potential for the infiltration of soil fines was considered essential for the long-term performance of the by-pass sewer. As such, the secondary concrete liner was designed to be 35.6 cm (14 inches) in thickness with inner and outer reinforcing steel for bending resistance and crack control. In addition, to preclude groundwater infiltration at joint locations, all construction joints were required to contain water stops.

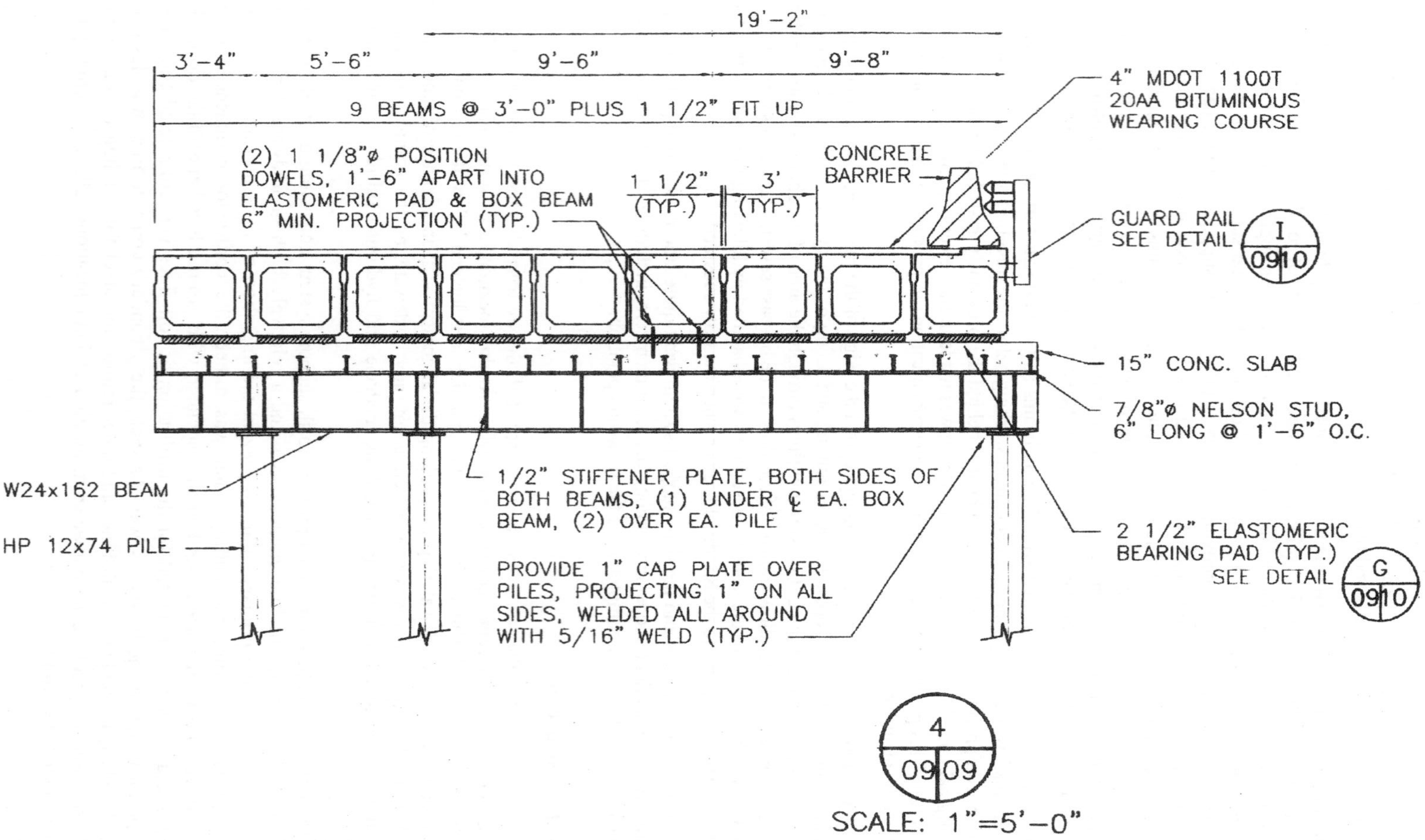

4" MDOT 1100T 20AA BITUMINOUS WEARING COURSE

GUARD RAIL SEE DETAIL $\dfrac{I}{0910}$

15" CONC. SLAB

7/8"⌀ NELSON STUD, 6" LONG @ 1'-6" O.C.

2 1/2" ELASTOMERIC BEARING PAD (TYP.) SEE DETAIL $\dfrac{G}{0910}$

CONCRETE BARRIER

(2) 1 1/8"⌀ POSITION DOWELS, 1'-6" APART INTO ELASTOMERIC PAD & BOX BEAM 6" MIN. PROJECTION (TYP.)

1 1/2" (TYP.) FIT UP

3' (TYP.)

3'-4"   5'-6"   9'-6"   9'-8"

19'-2"

9 BEAMS @ 3'-0" PLUS 1 1/2" FIT UP

1/2" STIFFENER PLATE, BOTH SIDES OF BOTH BEAMS, (1) UNDER ℄ EA. BOX BEAM, (2) OVER EA. PILE

PROVIDE 1" CAP PLATE OVER PILES, PROJECTING 1" ON ALL SIDES, WELDED ALL AROUND WITH 5/16" WELD (TYP.)

W24×162 BEAM

HP 12×74 PILE

$\dfrac{4}{0909}$

SCALE: 1"=5'-0"

Fig. 3 : North abutment section

186

Fig. 4 : Temporary bridge over access shaft

Fig. 5 : Typical access shaft construction

4.5 Flow Control

Since the Southfield sewer carried large flows during rain events, it was necessary to develop a plan to control the flow during the construction process. During periods of dry weather, the low flows of sanitary sewage were diverted into other sewers by means of a series of half-high dams, placed within the 3.05 m (10 ft.) sewer. These temporary dams were installed at McNichols Road and Tireman Avenue to divert the flow to the Northwest Interceptor and the Hubbell Sewer, respectively. When wet weather led to high flows, the dams were overtopped and flow continued down the Southfield Sewer.

When the connections were being made between the by-pass tunnel and the original sewer, flumes were required to pass the flows through the work areas. The flumes were half round sections of steel pipe that could be readily installed, redirected, and removed as the work progressed. The design of the junction chambers for the connections to the by-pass included provisions for forming the flumes for the final configuration into the final concrete structure.

Incorporated into the design of the by-pass tunnel were several control structures. The structures were designed to allow the owner to shut off flow in sections of the sewer for inspection or maintenance activities. The design also provided for four equipment access structures. These structures allow the owner to insert construction equipment into the tunnel for cleaning and for any other repairs that might be required in the future.

Near the southerly end of the by-pass tunnel, MDOT operated a pump station that took storm runoff from the freeway and pumped it into the original Southfield sewer. When the by-pass was constructed, it was necessary to modify both the intake and discharge lines for the pump station since the original sewer was being abandoned. The team of MDOT, DWSD, Contractor and NTH considered several options for keeping the pump station operational. They arrived at a solution that sent the discharge to one of the lateral sewers that intersected the Southfield sewer and constructed a specially reinforced intake tunnel that actually passed through the concrete liner of the by-pass tunnel. During the construction of these modifications to the pump station, several heavy rainfalls were experienced and the pump station was able to handle them quite well.

The flow control system also provided for flows higher than normally experienced. This was accomplished by permitting sewage to flow into the construction shafts whenever the flow exceeded that which could be handled by the flumes. When this happened, work was suspended until the flow subsided and the shaft cleaned up and work resumed. This concept proved quite effective in that there were no backups of sewage into basements and no diversion of flow to the adjacent Rouge river.

4.6 Construction Considerations

As with all construction projects, the contractor is the person responsible for making all the Engineer's plans really work. For this project, the contractor was Jay Dee Contractors, a firm with considerable talent and expertise in the underground construction field. Once the contract was awarded to Jay Dee, they became part of the team committed to getting the job done without problems.

The design team decided to leave the details of the shaft design to the contractor, subject to review by the owner and the designer. When Jay Dee was selected as the contractor, they developed some preliminary designs for the shaft support system. These designs were the subject of several meetings between the designer/owner team and the contractor. Eventually, the shaft designs were finalized with the input of all affected parties, including MDOT.

As with most underground construction projects, situations arose as the construction proceeded which were not anticipated by the designer or the owner. When these situations arose, the entire team of designer, owner, and contractor met to brainstorm potential solutions to the problems being encountered. The result of this teamwork during the construction process was that there were no unresolved issues. This allowed construction to proceed without delays and with no claims for extra money or extra time by the contractor.

Regular weekly meetings of all parties were held throughout the progress of the work. These meetings enabled the team members to be current on progress and on conditions actually being encountered as the work progressed. All parties felt free to discuss any issue at these meetings. This spirit of trust among the team members was an important factor in the success of what could have been a very contentious project.

5.0 SUMMARY

Collection and treatment of sanitary sewage is one of the keys for providing a healthy environment to the people of Detroit (or to the people of the world, for that matter). This process is complicated by the fact that many older urban sewerage system are designed to collect and transport both sanitary sewage and storm flows. The Southfield sewer is one such system. Built more than 70 years ago, it had served the people of Detroit well. However, its continued service life was jeopardized by one section of distressed tunnel. To replace the entire 16.09 km (10 miles) long Southfield system would have been cost prohibitive. By replacing only the distressed area, the citizens of the region were saved that cost.

Our society also depends upon our system of urban freeways to move about. Disruption of the freeway would have had extremely adverse effects upon the thousands of people who daily travel the Southfield freeway. By devising a system that allowed the traffic to flow over and the sewage to flow under, the cost and aggravation of a freeway closure were eliminated. By scheduling short periods of closure on weekends, the daily traffic flow was never interrupted.

One of the challenges facing our society today is how best to deal with the essential infrastructure which was built many years ago. This project demonstrates that when owners, designers and contractors work together, much of the ageing infrastructure can be saved at a fraction of the cost of replacing it. Thus, the replacement of a section of the Southfield sewer can, in a sense, be considered as a demonstration project for future efforts to save the existing infrastructure.

Had the repair not been accomplished in an expeditious manner, there is a real possibility that sewage would have had to be diverted to the Rouge river in the event of a failure. Thus, the successful completion of the project eliminated the risk of failure and the resultant environmental damage to the Rouge river ecosystem.

Sewers are things that very few people ever see or even think about. But they do expect them to work all the time. Freeways are things that most people see regularly and often think about. They also expect freeways to get them where they want to go when they want.

PRE-CONDITION SURVEY AND MONITORING OF THE STRUCTURES LIKELY TO BE AFFECTED DURING TUNNELLING OF DELHI METRO

MANGU SINGH   KAMAL NAYAN

Delhi Metro Rail Corporation, New Delhi, India

SYNOPSIS

One of the major concerns in urban underground projects is the effect of ground settlement on the foundations and structures in the proximity of the project. The alignment of underground North-South Corridor of Phase I of Delhi Metro passes through heavily built-up area of Old Delhi, namely, Chawri Bazaar area. Hard rock, slightly weathered to highly weathered, is likely to be encountered during tunnelling in this area. The tunnelling is proposed to be done with the help of closed face Tunnel Boring Machines. In order to assess the effect of tunnelling on existing old buildings, a detailed pre-condition survey of the buildings has been carried out and all the buildings have been categorized on a scale from 0 to 5. The effect of ground settlement during tunnelling and the effect of vibrations due to Tunnel Boring Machine operation have been worked out based on Numerical Modelling. The combined effect of existing condition of a particular building, settlement and vibration has been superimposed and a final classification showing the degree of vulnerability to damage due to tunnelling has been done. The present paper gives a detailed account of pre-condition survey and proposed monitoring system during construction phase.

1.0 INTRODUCTION

To cater to transport needs of fast increasing population of Delhi, Government of India approved Phase I of the Delhi Mass Rapid Transit System (MRTS) in 1996. Phase I of the system includes an 11 km long underground Metro Corridor. The Metro Corridor traverses under the old walled city being one of the most densely populated parts of Delhi. (Fig. 1)

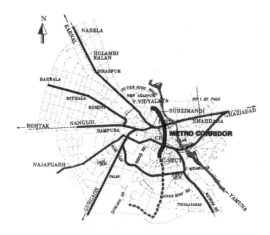

Fig. 1 : Route plan of Delhi MRTS showing metro corridor

2.0 ROUTE AND GROUND CONDITION

The chosen route for the Metro Corridor is essentially north-south running between Vishwa Vidyalaya in the north to Central Secretariat in the south. Ten stations are located along the underground section. The route follows the major roads in the northern part between Vishwa Vidyalaya to ISBT and from Patel Chowk to Central Secretariat in the southern part, whereas under the central area between ISBT to Patel Chowk the alignment runs below a large number of existing structures.

Ground conditions vary along the route and several geotechnical studies undertaken (Gahir, 2000) indicates that sandy silty clay to hard quartzite rock are likely to be encountered during underground construction. Bedrock is available at variable depth.

3.0 PROXIMITY TO BUILDINGS/STRUCTURES

The proximity of buildings/structures along the route varies considerably. In the northern part and southern part, where the route follow the roads the buildings are generally 10 to 15 m away from construction perimeter of underground structures/tunnels. The alignment between New Delhi to Patel Chowk falls mostly below open area or railway tracks and most of the buildings in this area are quite away from the construction perimeter. The section between Delhi Main to New Delhi through Chawri Bazaar lies below a large concentration of old buildings of two or three stories, with condition of their foundation and structure frames being in a very vulnerable condition.

4.0 GEOLOGY AND METHOD OF CONSTRUCTION

Because of the proximity of a large number of buildings in the stretch between Delhi Main to New Delhi, detailed soil investigations were carried out in whatever open space was available along and in the vicinity of alignment. The investigations indicated that the alignment in this area passes mainly through highly to completely weathered quartzite and ortho-quartzite with thin bands of schists at isolated locations. However at some reaches the degree of weathering of quartzite was relatively low and rock could be classified as slightly to moderately weathered.

Keeping in view the ground conditions along the alignment in this area and space constraints it has been planned to construct the tunnels between New Delhi to Delhi Main by means of dual mode rock Tunnel Boring Machines, so that they can operate in closed mode in the event that completely weathered zones are encountered or fissures of Delhi silt are found extending down into the rock mass. Chawri Bazaar station lying between the two stations will be a bored tunnel station and will be constructed by enlarging the two running tunnels to platform size and connected by passenger adits and ventilation adits. (Fig. 2)

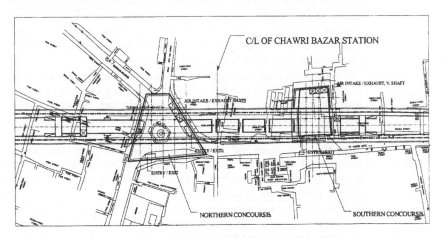

Fig. 2 : Layout of Chawri Bazaar station

191

5.0 NEED FOR CONDITION SURVEY

The buildings along the alignment between New Delhi to Delhi Main mainly belong to the Old City era and they were mostly built in the early part of 20th century. With time, due to commercial needs unauthorized construction has taken place in the form of additional structures, projections to the existing buildings without adequate consideration to the capacity of the supporting structure and good engineering practice. No records of the construction in this area were kept and buildings have also not been well maintained because the residents are generally tenants and not the owners of the buildings. As such it was felt that vulnerability of the buildings should be assessed before undertaking major tunnelling works underneath. It was therefore decided that features of the existing buildings *i.e.* pre-construction condition of the buildings should be recorded and the effect of likely surfaces subsidence and vibration due to tunnel construction should be studied in order to assess the suitable construction methodology, strengthening measures and monitoring system during construction.

6.0 METHODOLOGY FOR CONDITION SURVEY

A condition survey of buildings was undertaken by first defining the zone of influence *i.e.* the limits or the boundary on surface between which the buildings are likely to be affected due to tunnelling activity along the alignment. This was decided based on the proposed alignment, contours of Chawri Bazar station layout and depth of tunnel. All the buildings falling fully or partially in the zone of influence were identified. The physical condition of each building was then undertaken by visual inspection. Mapping of all major defects and photographs of existing signs of distress such as cracks, dampness, spalling, wall bulging, missing bricks, exposed rebars, deteriorating timber elements, settlement, tilts, poor structural supports were undertaken. After this exercise all the buildings were rated from 0 to 5 based on Burland *et.al.*, (1977) Classification of Visible Damage to Buildings (Table 1) and all these details were systematically recorded based on the layout of the buildings along the alignment.

TABLE 1
Classification based on physical condition of buildings (Burland *et al.*, 1977)

Category of damage	Degree of damage	Description of typical damage
0	Negligible	Hairline cracks of less than about 0.1mm.
1	Very slight	Fine cracks, which can easily be treated during normal decoration. Damage generally restricted to internal wall finishes. Close inspection may reveal some cracks in external brickwork or masonry. Typical crack widths up to 1 mm.
2	Slight	Cracks easily filled, redecoration probably required. Recurrent cracks can be masked by suitable linings. Cracks may be visible externally: some external repointing may be required to ensure weather tightness. Doors and windows may stick slightly. Typical crack widths up to 5mm.
3	Moderate	The cracks require some opening up and can be patched by a mason. Repointing of external brickwork and possibly a small amount of brickwork to be replaced. Doors and windows sticking. Service pipes may fracture. Weather tightness often impaired. Typical crack widths are 5 to 15 mm or several greater than 3 mm
4	Severe	Extensive repair work involving breaking-out and replacing sections of walls, especially over doors and windows. Window and doorframes distorted, floor sloping noticeably. Walls leaning or bulging noticeably, some loss of bearing in beams. Service pipes disrupted. Typical crack widths are 15 to 25 mm but also depend on the number of cracks.
5	Very severe	This requires a major repair job involving partial or complete rebuilding. Beams lose bearings, walls lean badly and require shoring. Windows broken with distortion. Danger of instability. Typical crack widths are greater than 25 mm but depends on the number of cracks.

The building condition survey records were made available to tenderers for the Metro Corridor contract and shall be made available to the design and build contractor who will be responsible for tunnelling work so that he is fully aware of the condition of the buildings along the route and is therefore able to do any necessary modifications to his construction methodology and ensure any necessary features in the Tunnel Boring Machines. In general, the details were recorded in the following manner to effect easy identification of the condition of a building.

(i)   Key-plan showing the location of the building concerned and its outer dimensions with respect to tunnel alignment and its surrounding building and lanes.

(ii)   Details of building mainly its use, construction features including foundations details, probable age of the building.

(iii)   Map of defects along with photographs.

(iv)   Classification of condition of building.

The survey work was conducted for a length of 1250 m along the alignment and involved approximately 1200 buildings. The details of the condition survey work have been compiled in a very systematic manner so as to ensure easy retrieval of details with respect to any building so surveyed. For each building one folder has been created which contains key plan showing location of building, building details, map of defects and their photographs. These folders have been compiled in areawise binders in 25 volumes. The index of each volume contains list showing building number in ascending order.

7.0 RESULTS OF CONDITION SURVEY

In general almost all the buildings in this area showed one sign of distress or the other. Most of the buildings had brick masonry structure with 1 to 1.5 m deep spread footings as foundations. Roof slabs mostly consists of stone slabs supported on wooden beams however in several buildings these were replaced by T-Section concrete beams. Although the buildings were originally designed for one or two floors, more floors have been added to most buildings and often these new upper floors project beyond the lower floors with projected portions supported on columns. In several cases, load-bearing walls have been removed or reduced in thickness resulting in redistribution of stresses. A common site in the area is of deteriorating wooden beams in the roof slabs resulting in sagging floors and a lot of buildings also have dampness. The most common types of defects, because of these reasons, observed are as follows :

7.1 Cracking of Masonry

This is the most common and widespread form of defect in the area. Cracking occurs due to movement within the masonry construction which could have been caused by one or several of the reasons mentioned above, such as inadequate foundation, redistribution of loads, sagging roof slabs resulting in loss of restraint for the walls, deteriorating wooden beams leaving voids in the walls and therefore causing movements and cracking etc.

7.2 Bulging of Walls

This can be seen in several buildings. This may be attributed to movement within the masonry and loss of restraint to the walls due to the reasons mentioned above under 'Cracking of Masonry', or additional may be due to differential subsidence and settlement of the foundation.

7.3 Settlement

Several buildings show differential settlement, which could be due to the loads in excess to what the foundation can safely transfer to the ground. Addition of floors may have contributed to this, together with unevenly spread foundation loads.

After analyzing the condition of the 1200 buildings in this area it was found that degree of damage to most of the buildings falling within zone of influence are in 'moderate' to 'severe' condition. 86 per cent of the buildings falls in the category 3-5 on a scale of 0-5 and 24 per cent of buildings fall in the category 4-5.

8.0 SETTLEMENT AND VIBRATION EFFECT ON BUILDINGS

As the condition of most of the buildings was very poor, studies were conducted to assess the impact of settlement and vibration effects. These studies were done by means of numerical modelling using the software Fast Lagrangian Analysis of Continua (FLAC). This analysis took into account non-linear material model based on Mohr-Coulomb failure criterion, large strain, effective stress analysis including analysis of support such as tunnel lining. The settlement analysis also took into account the effect of excavation due to the second tunnel. In order to take into account varying degrees of weathering and depth of overlying soil along the alignment, the section was divided into five segments based on the depth of tunnel and soil and rock profile. Using the subsidence data as arrived from numerical modelling the horizontal strains in the buildings were calculated and using this data the likely effect on the buildings was assessed. The relationship between the estimated tensile strain induced in the building and the potential damage is given in Table 2 below.

TABLE 2
Relationship between category of damage and limiting tensile strain (Boscardin and Cording, 1989)

Category of damage	Normal degree of severity	Limiting tensile strain (%)
0	Negligible	0-0.05
1	Very Slight	0.05-0.075
2	Slight	0.075-0.15
3	Moderate	0.15-0.3
4 to 5	Severe to Very Severe	>0.3

The subsidence analysis indicated that the buildings in these area are not likely to undergo tensile strain more than 0.075 per cent thereby limiting the increase in damage to the structure to "Very Slight" only.

Analysis was also done to assess the effect of vibration on the buildings due to the operation of the Tunnel Boring Machines. The analysis produced the likely peak particle velocity on the structure. This was compared with the ISO proposals which suggests that peak particle velocity on a load bearing element at ground level should not exceed 3 - 5 mm per second as a threshold, 5 - 30 mm per second for minor damage. The German DIN 4150 specifies a limit of 4 mm per second for buildings in poor condition. The study indicated that the peak particle velocity of vibration at surface structures lie well within limits prescribed above.

After conducting the condition survey, settlement and vibration analysis, the buildings were grouped based on vulnerability taking into account all the three parameters and it was decided that necessary precautions shall be taken for all the buildings which are likely to be vulnerable based on these parameters.

9.0 PRECAUTIONS ADOPTED FOR DETAILED DESIGN AND CONSTRUCTION PHASE

Even though the settlement and vibration analysis carried out indicates that the impact of tunnelling will be very insignificant all necessary precautions shall be taken to ensure the safety of the buildings and their occupants. Necessary provisions have been made into the contract documents to take care of these matters. It has been stipulated in the contract documents that during the design stage the contractor will undertake a three-stage analysis of the effect of tunnelling on the buildings. First, he will undertake an independent building condition survey and supplement the survey already done. He will then carry out a settlement assessment of all these buildings and then conduct building structural surveys for those buildings, which are likely to undergo large horizontal strains due to tunnelling works. He will then carry out necessary strengthening and protection measures to such building which are likely to be damaged as a result of large horizontal strains. Only after these steps have been taken will the contractor be permitted to undertake the tunnelling works. During the construction stage the contractor shall provide an extensive instrumentation scheme to monitor the various parameters likely to result in damage to buildings. The proposed instrumentation scheme shall ensure safety during and after the construction by providing early warning of any excessive or under ground movement of adjacent structures. This instrumentation scheme shall take into account existing ground conditions, location and condition of existing buildings, proposed method of construction and type of equipments proposed for construction work. Instrumentation shall comprise mainly tilt meters, inclinometers, extensometers, precise settlement markers, load cells, crack monitoring instruments etc. The frequency of monitoring, location, method of data recording and report generation system for each type of instrument shall be based on construction methodology and condition of adjacent structures. All the parameters associated with these instruments shall be identified during the design stage itself and then certain review / alarm levels shall be decided so that necessary remedial measures can be taken when such levels are reached. This concept of review levels has already been used in the Singapore Metro and has been found to be effective in safeguarding the buildings lying in the zone of influence of tunnelling activity.

Monitoring of structures shall commence sufficiently before construction commencement and shall continue until such time when movement due to the works has stopped or slowed sufficiently to be deemed negligible.

10.0 CONCLUSION

Success of any project depends greatly on public support and good will. Delhi Metro Rail Corporation, as its corporate philosophy, insists that construction activities should not endanger public. As such all necessary precautions and measures are being taken so that at all times the structures likely to be affected during tunnelling are well protected and a constant monitoring of these structures shall be done during construction.

REFERENCES

Boscardin, M.D. and Cording, E.G. (1989). Building response to excavation-induced settlement. Journal of Geo Engineering. ASCE, USA.

Burland, J.B. Broms, B. and De Mello, V.F.B. (1977). Behaviour of foundations and structures - SOA Report, Session 2, Proc,. 9th Int. Conf. SMFE, Tokyo.

Gahir, J. (2000) Initial Geotechnical Consideration for the Delhi Metro - International Conference on Tunnels and Underground Structures, Singapore, Balkema.

ENVIRONMENTAL PROTECTION IN DELHI MASS RAPID TRANSPORT SYSTEM (MRTS) PROJECT

S.A. VERMA

Rail India Technical and Economic Services, New Delhi, India

SYNOPSIS

The implementation of the Delhi Mass Rapid Transport System (MRTS) project which was long overdue, is a culmination of the plans of the Government of India and Delhi State Government to provide a fast, safe, cheap and environmentally friendly means of transport to city commuters. This project is an environmentally challenging project in many ways. There are a large number of environmental controls, which are to be observed during the construction and operation of any major development project. This paper analyses the rationale for the project in as much as it relates to environmental benefits that will accrue to the citizens of Delhi once the project is commissioned. It also focusses on the important factors of environmental controls during construction specially when this project is being constructed in an era when there is much more awareness in India of environmental issues and their impact on community health. Furthermore, it is to be implemented in the heart of a mega-city, which is also the country's capital. The ambient standards and statutory requirements that have been in place for quite sometime now, are planned to be effectively executed through various control mechanisms to be provided in the contract. The project is intended to serve as an example for projects of similar nature being planned in other cities of India.

1.0 INTRODUCTION

In Delhi, which has a population of about 13 million, only buses meet the mass transport needs. These are overcrowded and unreliable with long waiting periods at bus stops. Inadequate mass transport, improved socio-economic levels and the considerable increase in motor vehicle production have significantly resulted in an alarming increase in personalised vehicles. The vehicular population of Delhi, which is already very high, is growing at the rate of 0.3 million additional vehicles per year. This is matched by the high population growth. The growth of low capacity personalised vehicles is the increasing cause of acute road congestion, parking problems, fuel wastage, road accidents, and air and noise pollution. Transport is a major source of environmental pollution. Vehicular traffic accounts for 70 per cent of air pollution in Delhi. The need for a rail based mass transport system, therefore, can hardly be over-emphasised.

2.0 ENVIRONMENTAL RATIONALE

Rail based mass transport system is much less polluting than its road based counterpart. For example, road travel emits approximately twice the amount of carbon dioxide than rail for the equivalent passenger miles. Furthermore, rail has the capacity to carry a significantly higher number of passengers per area than road. A two track urban rail can carry upto 60,000 people per hour in each direction, whereas a two-lane road can only accommodate 6,000 each way in the same time. A double line rail uses an average of 12 m of land in width while a three-lane motorway uses an average of 47 m. Energy requirement per passenger kilometer is one-fifth vis-à-vis the other system. Finally, a rail based system it is faster, more reliable, safe and comfortable than a road based one. There are thus immense social and environmental benefits that would accrue as a result of the project. A rail-based MRTS system for Delhi has been recommended over a road-based system because it is non-polluting and environmental friendly system has a higher carrying capacity and requires less space to carry same number of passengers. Since the requirement of buses is estimated to decrease, as 1 million commuter trips will now be catered to by the Delhi MRTS, this will result in savings in cost of buses, and fuel. Lesser buses will reduce congestion that will translate into a lower air and noise pollution on Delhi roads.

3.0 THE PROJECT

With a view to reducing the problems of Delhi commuters, Government of India (GOI) and Government of National Capital Territory of Delhi (GNCTD) have launched the Multi-Modal Mass Rapid Transport System (MRTS) for Delhi. The present proposal of a modified first phase of the Delhi MRTS project will comprise a network of 11 km underground (Metro) along with 44.30 km of elevated/surface (Rail) corridors. It will have 45 stations in all. An estimated sixty thousand passengers per hour per direction will be carried by MRTS, Delhi with a headway of three minutes. The first section, elevated/surface corridor of 8-km length is planned to be commissioned by March 2002. The project is planned through one of the most thickly populated areas of Delhi as shown in Fig. 1. Metro corridor connects Delhi University with Central Secretariat via ISBT and Connaught Place. The rail corridor links Shahadara with Nangloi and Trinagar with Barwala. The rail corridor is partly elevated and partly at grade. A portion of this corridor, between Shahadara and Tis Hazari, will be opened to public in less than two years from now.

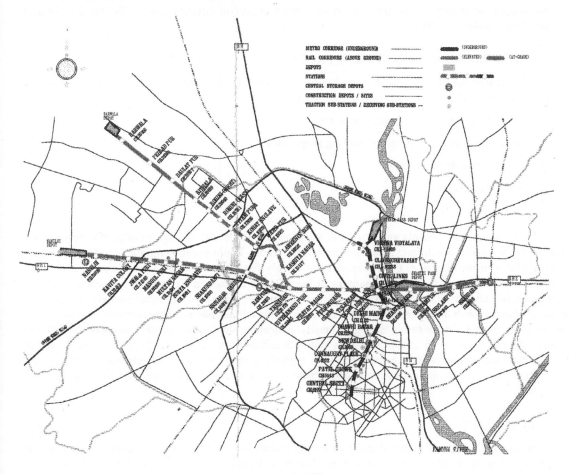

Fig. 1 : Delhi MRTS project

4.0 ENVIRONMENTAL CONTROLS DURING CONSTRUCTION

While it is easier to anticipate impacts and plan mitigation for the three phases of project cycle namely project design, siting and operation, it is relatively difficult to predict construction related impacts in view of the number of imponderables involved. Moreover, construction impacts being temporary in nature have not been given the attention they deserve in India. As awareness in India is growing on negative long-term construction related impacts, environmental design and control criteria have become a focus of increasing attention of environmental planners, statutory government enforcement agencies and the public alike.

The MRTS Delhi is likely to throw up a number of environmental challenges during construction since it is planned underneath or in close proximity to areas housing the state legislature, numerous government and private offices, commercial areas and residential complexes. It was therefore imperative to anticipate critical environmental issues and develop project specific environmental criteria for this project. To this extent Delhi MRTS Project can be considered a pioneer in applying construction related environmental controls. Infact, the corporate mission statement of DMRC clearly enunciates that construction activities should not lead to environmental degradation and should not inconvenience or endanger public. A detailed Environmental Impact Assessment conducted by Rail India Technical and Economic Services (RITES) also recommended design of control measures for impact mitigation during construction and operation of the MRTS system. For this project, an Environment Quality Manual has been prepared for guidance and compliance of the contractors, which list measures for environmental control as also the monitoring requirements of these parameters, during construction of the project. Thus, the main requirements to control environmental pollution shall be met through contractual obligations.

5.0 ENVIRONMENTAL MANAGEMENT PROCESS

An Environmental Management Process has been designed for this project to contain negative impacts during construction. The main elements of this design are :

- Establish environmental control measures
- Monitor compliance through monitoring programme
- Event contingency planning
- Environmental audit
- Environmental reporting

The environmental controls and standards are proposed to be met through Employer's Requirement (ER) on Environment, which form an integral part of the tender. Supplementing the ER is an Environmental Quality Management Manual (EQM) prepared exclusively for the project and which is applicable to all contracts under the project. The main purpose of the EQM is to make the contractors aware of DMRC's environmental concern. The manual also establishes guidelines for the environmental controls for guidance of the contractor. Such an Environmental Management approach has numerous advantages. It will provide database for impact determination and give timely indication should any control measure fail. It will monitor the effectiveness on control measures and help initiate remedial action if required. Finally it will help in determining contractor's compliance with regulatory requirements. Thus environmental controls have been established for air quality, noise and vibration, water quality, solid waste, health, landscape and aesthetics.

6.0 AIR QUALITY

Construction activities that may degrade air quality include excavation, material handling and stockpiling, vehicular movement, wind erosion in unpaved work areas and equipment usage. Control measures have been devised for each of the activity as outlined below.

6.1 Fugitive Dust Emissions

All necessary precautions shall be taken to minimise Fugitive Dust Emissions from operations involving excavation, grading, clearing of land and disposal of waste. Emissions of Fugitive Dust from any transport, handling, construction or storage activity shall not be allowed to remain visible in atmosphere beyond the property line of emission source. Dust generating material shall be transported in closed containers or covered trucks; loaded and unloaded in closed systems. At each construction site, the contractor shall provide storage facilities for dust generating materials that shall be of closed containers/bins or, wind protected shelters or, mat covering or, walled or, any combination of the above. The dust producing excavated material shall be loaded/placed in a manner that will minimise dust production. The material shall be stabilised each day and wetted. During dry weather, dust control methods shall be used continuously especially on windy, dry days to prevent any dust from blowing across the site perimeter. Water sprinklers shall be made available at all times for dust control use. Dust control activities shall continue even during any work stoppage. The contractor shall design and implement his blasting techniques so as to minimise dust generation. The contractor shall water down construction sites as required to suppress dust, during handling of excavation soil or debris or during demolition.

6.2 Pollution due to Equipment

Construction equipment shall be designed and equipped to prevent or control air pollution. Evidence of such design and equipment shall be maintained and made available for inspection by employer. In case the equipment or methods of working cause serious air pollution, then these shall be inspected and remedial proposals drawn up and implemented. Remedial measures shall include use of additional/alternative equipment or maintenance/modification of existing equipment. Transport vehicles and other equipment shall conform to emission standards fixed by the Statutory Agencies of Government of India and by the local Government from time to time. Necessary periodical checks and remedial measures including replacement, if required, shall be undertaken, so as to operate within permissible norms. The internal combustion engine powered vehicles used in the project shall be maintained periodically.

6.3 Dumping Sites

No earth, rock or debris shall be deposited on public or private right of way. The dumping sites shall be identified in advance and all excavated materials shall be dumped in the same. The temporary dumping areas shall be maintained at all times until the excavated material is re-utilised for back filling.

6.4 Impact Monitoring

Monitoring of Suspended Particulate Matter (SPM) at each construction site throughout the duration of project shall be carried out under Environmental Air Monitoring and Control Plan prepared for the Project. This plan specifies the duration, frequency and locations of air sampling, calibration requirements and manner of reporting the results of such an impact-monitoring programme so as to be useful for the project.

6.5 Air Quality Limits

Central Pollution Control Board, New Delhi has established National Ambient Air Quality Standards for various parameters. The Standard as given in Fig. 2, below for Limit Levels of SPM in ambient air may be followed in estimating the pollution level caused by contractor's activities.

Pollutant	Time weighted average	Sensitive area $\mu g/m^3$	Industrial area $\mu g/m^3$	Residential area $\mu g/m^3$
Suspended	Annual	70	360	140
Particulate matter	24 Hours	100	500	200

Source: Air (Prevention and Control of Pollution) Act, 1981

Fig. 2 : Air quality limits

7.0 NOISE

Noise shall be considered as an environmental constraint in the design, planning and execution of the works. All appropriate measures shall be taken to ensure that work carried out by the contractor, whether on or off the site, will not cause any unnecessary or excessive noise which may disturb the occupants of any nearby dwellings, schools, hospitals, or premises. The noise level reduction measures include that all powered mechanical equipment used in the works shall be effectively controlled, using the most modern techniques available including but not limited to silencers and mufflers. Acoustic screens or enclosures shall be constructed around any parts of the works from which excessive noise may be generated. Noise generated by work carried out whether continuously or intermittently during day and night time shall not exceed the maximum permissible noise limits. Necessary measures shall be taken to reduce the noise levels and thereafter maintain them at the appropriate levels, which do not exceed the statutory limits. Such measures may include without limitation the temporary or permanent cessation of use of certain items of equipment.

7.1 Noise Control Requirements

The contractors shall comply with applicable codes, regulations, and standards established by the Central and State Government and their agencies. To the extent required, the contractor shall use noise reduction measures listed below to minimize construction noise emission levels :

- Truck loading, unloading, and hauling operations shall be scheduled so as to minimize noise impact near noise sensitive locations and surrounding communities.

- Locate stationery equipment so as to minimize noise impact on the community.

- Use only well maintained plant at site, which should be serviced regularly.

- Plant and equipment known to emit noise strongly in one direction should, where possible, be oriented in a direction away from noise sensitive receptors.

- Silencers and mufflers on construction equipment should be properly fitted and maintained.

- Number of plant and equipment operating in critical areas close to noise sensitive receptors should be reduced.

- Equipment should be maintained such that parts of vehicles and loads are secure against vibrations and rattling.

- Physical separation between noise generators and noise receptors should be maximised.

- Grading of surfaced irregularities on construction sites should be undertaken to prevent the generation of impact noise and ground vibrations by passing vehicles.

- Scheduling work shall be undertaken to avoid simultaneous activities that generate high noise levels.

7.2 Noise Monitoring

A noise monitoring and control plan has been prepared for the project. It requires that the contractor include full and comprehensive details of all powered mechanical equipment, which the contractor proposes to use during day and night time, and of his proposed working methods and noise level reduction measures. The appropriate parameter for measuring construction noise impacts shall be the equivalent A-weighted sound pressure level (L_{eq}) measured in decibels (dB).

7.3 Noise Level Limits

A limit for construction noise is based on the existing ambient noise levels at the noise sensitive sites adjoining the construction sites. The noise levels at noise sensitive locations shall not exceed L_{eq} plus 10-dB (A) or more above existing ambient pre-construction noise levels for continuous construction activities. Where there are no ambient noise measurements, the construction activities shall be limited to levels measured at a distance of 75m from the construction limits or at the nearest affected building, whichever is closer, as given in Fig. 3. The groundborne noise levels within building structures due to tunnel boring machines and any other underground and tunnelling construction activities shall not cause interior noise levels to exceed the levels given below as measured in the inside of the affected noise sensitive structure : (a) Residential : L_{max} 55dB(A), (b) Commercial : L_{max} 60dB(A).

Land use	Maximum noise levels-L_{max} dB(A)	
	Day Time	Night Time
Residential	80	65
	at all times	
Commercial	85	
Industrial	90	

Source: Environmental Quality Management Manual of DMRC

Fig. 3 : Allowable construction related noise

8.0 WATER QUALITY AND MANAGEMENT OF LIQUID WASTE

The main sources of water pollution are the surface run-off from construction activities, drainage and dewatering operations from tunnels, sewage effluent and spillage of oil and waste.

8.1 Surface Water, Water Courses and Temporary Drainage

The contractor shall ensure at all times that the existing stream courses and drains within, and adjacent to the site are kept safe and free from any debris and any excavated materials arising from the works. It should be ensured that chemicals, earth, bentonite and concrete agitator washings are not deposited in the watercourses. They should be suitably treated and effluents and residue should be disposed off in a manner approved by local authorities. All water and waste products arising on the site shall be collected and removed from the site through a suitable and properly designed temporary drainage system and disposed off at a location and in a manner that will cause neither pollution nor nuisance. All precautions shall be taken for the avoidance of damage by flooding. Adequate precautions shall be taken to ensure that no spoil or debris of any kind is pushed, washed, falls or deposited on land adjacent to the site perimeter and carried to the water source. In such an event, the same shall be immediately removed and the affected land and areas restored to their natural state. Any mud slurry from drilling, tunnelling, or any underground construction or grouting, etc., shall not be discharged into the drainage system unless treatment is carried out.

8.2 Ground Water

Due to lowering of ground water levels and contamination of ground water, the discharge of water from the project site to outside, is not to be allowed, especially when the project site is in city limits. Any water obtained from dewatering systems installed in the works must be used only for construction purposes and should not be subsequently discharged into the drainage system. If not used, it shall be recharged to the ground water at suitable acquifer levels. The requirements of the Central Ground Water Board for discharge of water arising from dewatering shall be complied with.

8.3 Waste Water and Oil Waste

A wastewater drainage system shall be provided to drain wastewater into the sewerage system. The wastewater arising out of site office, canteen or toilet facilities constructed in the site shall be discharged into the sewers after obtaining prior approval of the agency controlling the system. Suitable oil removal / interceptors shall be provided to treat oil waste from workshop areas, etc. All measures shall be taken to prevent discharge of oil and grease during spillage from reaching drainage system or any water body through spill prevention and control measures.

8.4 Sediment Migration

Necessary steps shall be taken to prevent soil particles and debris from entering the wells or water discharge points by use of filters and sedimentation basins.

8.5 Water Quality Limits

The Indian government legislation and other state regulations in existence, in so far, as they relate to water pollution control and monitoring shall be complied with. In principle, Schedule VI of Environmental Protection Act as applicable to inland surface water shall be followed for compliance.

9.0 MANAGEMENT OF SOLID WASTE

9.1 Site Clearance

All precautions shall be taken to avoid any nuisance in the project area or site or locality. This shall be accomplished, wherever possible by suppression of nuisance at source. The work place should be free of trash, garbage, debris and weeds. It should be ensured that refuse containers are provided and used properly so that rodents, flea and other pests are not harbored and attracted.

9.2 Scrap and other Waste Material

The contractor shall remove waste in a timely manner. Scrap, trash, garbage and waste material shall be removed and disposed off at landfill sites after obtaining approval of Conservancy and Sanitation Engineering Department of Municipal Corporations for their disposal. Burning of wastes is prohibited. The contractor shall maintain and clean waste storage areas regularly.

9.3 Hazardous Waste

If encountered or generated as a result of contractor's activity, then waste classified as hazardous under the "Hazardous Wastes (Management & Handling) Rules, 1989" and chemicals classified as hazardous under "Manufacture, Storage and Import of Hazardous Chemicals", shall be disposed off in a manner in compliance with the procedure given in the rules aforesaid.

9.4 Food Related Waste

At site, metal or heavy-duty plastic 'Refuse Containers' with tight fitting lids for disposal of all garbage or trash associated with food shall be provided. The containers shall not have openings that allow access to rodents. Specific locations should be designated for consuming food and snacks to prevent random disposal of food related waste.

10.0 TRANSIENT WORKER SETTLEMENTS

The contractor shall be responsible for providing temporary accommodation for the work force/ labour and shall provide facilities to the labour residing in the area, in accordance with latest labour by-laws issued by local governments. The contractor shall prepare a "Workers' Plan" that will define and provide for the arrangement of the requirements of labour; like fuel, water supply, sanitary discharge, primary health, demobilization and de-encampment of work force, and the establishment of liaison with police and fire departments.

11.0 ENVIRONMENTAL AUDIT AND REPORT

It is intended to undertake environmental performance reviews through an Environmental Audit Programme. It will assess the effectiveness of the contractor's efforts in implementing mitigation measures and compliance with environmental standards. A checklist of environmental requirements associated with specific activities within the construction programme, shall be prepared. Audit at regular intervals shall be carried out against the criteria of compliance with procedures, environmental incidents, exceedance, accuracy of record keeping and effectiveness towards environmental management activities of the contractor. Under the reporting system designed for this project, the contractor is required to submit report on his activities on a monthly basis. The report shall contain results of various environmental monitoring programmes and their interpretation, measures taken for abatement/mitigation of impacts and action taken on environmental complaints, investigation and follow-up.

12.0 CONCLUSION

Awareness of environmental consequences during major construction projects has existed even though nothing much has been done in mitigation; probably due to their temporary nature and short duration of time involved. However, large-scale construction projects take 5 to 10 years to construct. Hence, if not acted upon effectively, construction related impacts have the potential to mould public opinion against the project resulting in decreased public cooperation, goodwill and sympathy and may put the project many years behind schedule. The DMRC has thus made a decision to provide for environmental control measures, which are executable through contractual means. This paper has discussed the pollution abatement measures for construction related impacts of this project. The statutory agencies are yet to come up with standards for construction related environmental impacts. The environmental protection requirements discussed in this paper could form the basis on which a detailed environmental design procedure could be developed. The techniques as discussed in this paper could serve as an example for adoption in other MRTS projects, which are planned in near future in some of the major Indian cities.

REFERENCES

Central Pollution Control Board, 1997, Pollution Control Acts and Rules issued thereunder, New Delhi.

Environment Protection Act, 1986.

Air (Prevention and Control of Pollution) Act, 1981.

RITES, 1995, Final Report on Environmental Impact Assessment for Delhi MRTS, New Delhi.

Kalyanasundaram T. and Verma S.A., 1999, Environmental protection requirements during construction of major projects, Proc. of Seminar on Construction Technology for new millennium, Indian Building Congress, Chennai.

Kalyanasundaram T. and Verma S.A., 2000, Transport Policy and Environmental Considerations, Proc. of Conference on Urban Transportation and the Environment CODATU IX, Mexico.

ENVIRONMENT CONTROL SYSTEM IN DELHI METRO

B.P. VERMA

Rail India Technical and Economics Services (RITES), New Delhi, India

SYNOPSIS

In an underground metro system, the Environment Control System (ECS) is inescapable to provide a tenable environment to the users and to offset the adverse effects of large amounts of heat being released from train equipment, braking, train air conditioning, and station equipment, etc. The presence of an ECS becomes even more important and vital to effect safe evacuation of passengers in case of an emergency like fire.

This paper presents the feature of the ECS proposed for the underground section of Delhi Metro.

1.0 INTRODUCTION

Delhi has a transportation infrastructure primarily in the shape of state owned bus services and privately owned motor vehicles. Delhi has the largest number of vehicles, much larger compared to any other metro city in India. The roads are totally inadequate to handle the present peak traffic and traffic jams are a common scene.

The Delhi Government appointed RITES to prepare the feasibility report for a suitable alternative transit system. RITES in their report proposed a total 195.8 km section of metro network to meet the transportation requirements of 2021. The government then established the Delhi Metro Rail Corporation (DMRC) in 1995 to undertake the challenging task of designing and constructing a metro system. At present about 55.3 km metro network is proposed to be completed by March 2005 in the first phase.

Delhi metro has two type of sections (a) on the surface or elevated from ground level, and (b) underground. The environment control system is required for the underground section which is about 11 km long from Vishwa Vidyalaya to Central Secretariat. In India, the only working metro system is in Calcutta in which the environment control aspects have also been given consideration in system design. However, the Delhi metro system shall be an improved system in respect of ECS on account of the adoption of new system philosophy, latest technology for air-conditioning equipments, mechanical ventilation provisions for plant rooms and inclusion of emergency handling measures like smoke extraction system for stations and tunnel ventilation system for tunnel.

2.0 NECESSITY OF ENVIRONMENT CONTROL SYSTEM

The underground structures in a metro system are isolated from the natural environment / atmosphere and therefore a tenable environment is to be created artificially to support human life. The ECS, therefore, is required for the following reasons :

(a) To maintain uninterrupted supply of oxygen to the passengers and staff.

(b) To remove large amount of heat released from train equipment, braking, station equipment, lighting, etc., and create comfortable environment for the passengers.

(c) To remove other harmful gases / fumes from batteries and heat from various equipment under operation in plant rooms.

(d) To ensure safety of human life at stations or in the trains in case of emergencies like stalling of trains, fire, etc., and providing safe evacuation path to the passengers.

(e) To effect smoke and fire control.

3.0 SYSTEM PHILOSOPHY – OPEN VS. CLOSED SYSTEM

3.1 Calcutta System in Brief (open system)

In the Calcutta metro, the air is collected from the atmosphere, pre-cooled in air washeries, cooled in Air Handling Units (AHU) and supplied to station area. Shafts are provided for collection of fresh air from atmosphere or to exhaust hot air to atmosphere. The cooled air picks up heat gains of various sources. Part of this air is permitted to exit to the atmosphere from the station and the balance flows into the tunnel. This air flow cools the tunnel environment and goes out to atmosphere from shaft provided in the tunnel between two stations. Since this system is open to the atmosphere at both ends, it is called 'open system'. In this system, hence, both stations and tunnel are air-conditioned and therefore coaches are not required to be air-conditioned and there is no separate tunnel ventilation system. The open system, however, lacks a conceived planning for handling emergencies, specially in case of fire, as there are no specific measures dedicated to this purpose (Fig. 1).

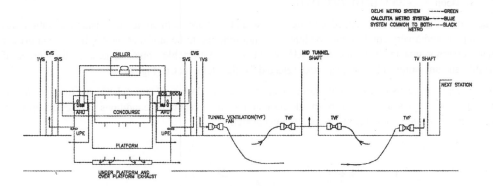

Fig. 1 : System arrangement - Calcutta metro vs. Delhi metro

3.2 Delhi Metro System (Closed system)

The Delhi metro system differs from the Calcutta system. Basically cooled air is supplied to the station and after picking up heat gains of various sources, is collected and returned to the AHU to be cooled again and for further recirculation to the station area. Only 10% fresh air is added to meet the biological and living requirements of the human body. Separate tunnel ventilation fans are placed at both ends of each station to induce sufficient air flow in the tunnel to remove smoke and heat away from passengers in case of fire in the tunnel. Reversible fans are used to have flexibility of inducing air flow in forward or reverse direction so that safe evacuation path can be provided to the passengers by moving smoke opposite to the direction in which passengers are moving for evacuation. As the air is collected and recirculated, the dampers in the exhaust shaft are normally closed and system therefore is known as a 'closed system'. In addition there is a smoke extraction system proposed at stations for smoke management in case of fire.

4.0 ENVIRONMENT CONTROL SYSTEM COMPONENTS

An environment control system for underground metro consists of the following components :

(a) Air-conditioning for station area : Comfortable environment for passengers.

(b) Mechanical ventilation for plant rooms : Heat extraction from plant rooms.

(c) Smoke control and extraction system for station area : Smoke management at the station for safety in case of fire at station.

(d) Tunnel ventilation system : Safe evacuation of passengers in case of stalling of trains or fire inside the tunnel.

5.0 DESIGN CRITERIA FOR STATION CONDITIONS

The Delhi climatic conditions have a wide range of variation from extreme cold to extreme hot. The average temperature during summer, winter and rainy seasons are as under :

	Max.	Min.	
Summer	36.7^0C	22.8^0C	The temperature in summer, however, goes up to 46°C
Winter	24.25^0C	9.3^0C	
Rainy	34.05^0C	24.15^0C	

The outside climatic conditions, selected for the design of the ECS are taken as :

Summer   - 43^0C (DB), 28^0C (WB)

Monsoon   - 35^0C (DB), 29^0C (WB)

The station conditions are decided keeping in view that a person is likely to stay inside the station only for about 2 minutes and a very comfortable conditions are not required to be aimed at. In ASHRAE comfort zone, a condition corresponding to 29^0C and 65% RH which is in comfortable zone but near to the border of the comfortable and uncomfortable zone, has been selected for the Delhi metro (Fig. 2).

As the ambient temperature is low during winters, the circulation of atmospheric air through the station shall be adequate to pick up the heat released and maintain the acceptable conditions. Therefore provision of heating arrangements are not made in air-conditioning design.

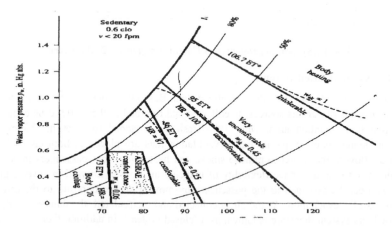

Fig. 2 : Ambient air temperature T_a, ^{0}F comfort and discomfort diagram

6.0 CLOSED SYSTEM FOR DELHI METRO

Initially in the RITES report, an open type system was proposed for Delhi metro. However, in a review by the General Consultants to DMRC, adoption of closed system was recommended as the closed system has many advantages. These are identified in Fig. 3 in respect of design philosophy :

		Open system	Closed system
1.	Flexibility of operating in various modes	Operation possible only in one mode	Operation possible in different modes as per the climatic conditions
2.	Economy	Bigger size of air-conditioning (AC) plant leading to high space requirements and operating cost	The AC plant is much smaller as tunnel is not air-conditioned and recirculation of cooled air during summer season, leading to less operating cost and space requirements
3.	Inclusion of emergency handling philosophy	The emergency handling capabilities of the system are limited	Emergency handling possible at station and tunnel to a greater extent
4.	Compatible to present technology		The system can be provided with platform screen doors at a later stage

Fig. 3 : Comparison of open and closed system

In view of above benefits, a closed system has been selected for the Delhi metro project. The basic designing of the integrated system shall be done using the 'Subway Environmental Simulation Programme (SES) developed by the 'Department of Transportation', USA.

7.0 BENEFITS OF CLOSED SYSTEM

There are various mode of operations possible in case of closed system which result economy in operation. The details are as under :

(a) Operation in winter - The ambient temperature is low and atmospheric air is sufficiently cold. In winter, therefore, the air is not cooled (Chillers not operating) and is directly circulated through the station. This picks up the heat gains inside the station and warm air is exhausted direct to the atmosphere from the station. In this case air is not recirculated and this mode is known as 'free cooling mode'. This provides economy in the operating cost as chillers and associated equipment are not required to be operated (Fig. 4).

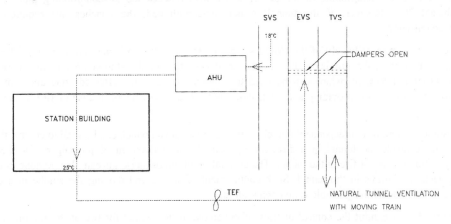

Fig. 4 : Closed system in cold climate

(b) Operation in summer - The Fig. 5 shows the operation of a closed system in summer season when ambient temperature is high.

The air is cooled in AHU and circulated in the station area. It picks up the heat released from different sources and the warm air is collected and returned to AHU for cooling again. The temperature of this return air is lower than the summer atmospheric temperature which would have been cooled in AHU in case of open system. Therefore, less cooling efforts and capacity required in case of closed system and the size of the air-conditioning plant is smaller. This gives added advantages of less operating cost and less space requirements.

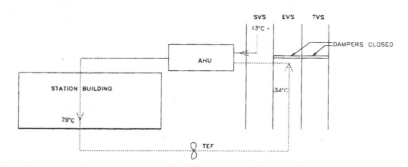

Fig. 5 : Closed system in hot climate

(c)   In the open system there is a concern about maintaining comfortable temperature inside the tunnel, thus tunnel is also cooled. However, in the case of the closed system, the benefit of the low sub-soil temperature is availed, especially in summer season, instead of tunnel cooling. The tunnel is surrounded by earth which is at a fairly constant temperature lower than the atmospheric temperature. If the tunnel is not cooled, the tunnel air temperature rises on account of heat released from train and there will be a heat flow from tunnel air to the soil. The heat transfer is more as the tunnel air temperature rises. In addition a flow of cooled air from the station into the tunnel with moving trains assists in maintaining the temperature. Therefore tunnel air temperature tends to stabilize after some temperature rise. The tunnel temperature, in case of Delhi metro under normal working condition is likely to be around 40°C.

This avoids air-conditioning of the tunnel which also reduces the air-conditioning plant capacity substantially. However, if the tunnel is not air-conditioned, the coaches are required to be air-conditioned.

(d)   The most significant benefit of the closed system is its capability to handle emergencies. The closed system facilitates the provision of a smoke control system in the station area including the zone pressurization concept, which can be used to contain smoke to the effected zone and remove the smoke from there for providing safe evacuation path to the passengers in case of fire in the station area.

In addition there is an independent tunnel ventilation mechanism which can be used to control the flow of smoke inside the tunnel in a direction opposite to the movement of passengers for their safe evacuation in case of fire in the tunnel. The circulation of atmospheric air can also be done in case of congestion of trains in the tunnel. Such containment of smoke and moving the smoke in a desired direction is not always possible is an open system.

Figures 6 to 9 exhibit the control of flow of the smoke inside tunnel for fire at the station or in the tunnel.

(e)   The latest metro systems are being built with platform screen doors (PSD) to separate the platform from the train. This reduces the air-conditioning requirements substantially and also makes easier to handle fire cases at stations. The PSDs do not permit train noise to enter into the station and also protect passengers from the possible danger of getting within the proximity of moving trains while they stand on the platform.

An open system can not be provided with PSDs whereas a closed system can conveniently be provided with PSDs,. Delhi metro has kept this provision in the system design such that PSDs can be provided in future.

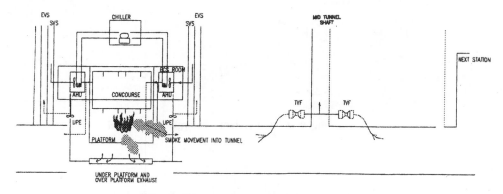

Fig. 6 : Open system – emergency at station

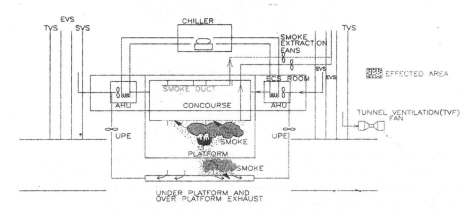

Fig. 7 : Closed system – emergency at station

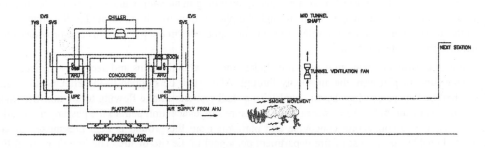

Fig. 8 : Open system – emergency in tunnel

8.0 SYSTEM ARRANGEMENTS FOR DELHI METRO

(i) There are three shafts provided at either end of the station namely, air supply shaft, exhaust shaft and a tunnel ventilation shaft for fresh air supply, air exhaust to atmosphere and for tunnel ventilation. The shaft arrangements are shown in the Fig. 10.

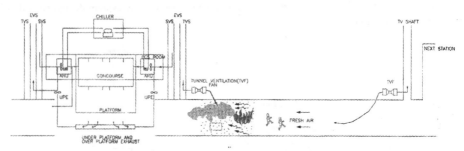

Fig. 9 : Closed system – emergency in tunnel

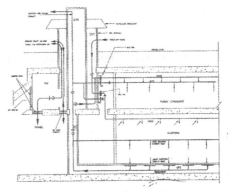

Fig. 10 : Shafts at each end of a station

(ii) The underground stations are in two floor arrangements. The first floor under the ground is the concourse where tickets shall be available and further one floor down is the platform. The chiller and associated equipment plant rooms are located above the ground level. Whereas the AHU, ventilation fans, smoke extraction fans and tunnel ventilation equipment are located in the plant rooms at concourse level. Trackway exhaust fans are installed for collecting the station return air and feeding to AHU for cooling and circulation.

(iii) Separate ventilation fans of adequate capacity operating in air supply and exhaust mode are provided for continuous circulation of atmospheric air through the plant rooms which are located at the concourse level and the platform. This removes the heat released from the equipment in the plant rooms and provides a tenable environment for the maintenance staff.

(iv) Smoke control and extraction system includes smoke extraction fans and necessary dampers. In case of fire at concourse or platform, the air supply through AHU shall be stopped to the effected floor and the other floor shall be kept pressurized by air supply. This shall restricts smoke flowing to the non-fire floor and also forces air to flow through the stairs/escalators to ensure safe evacuation path to the passengers. In addition the firemen's stair case connecting concourse and platform from the ground is kept pressurised to keep it smoke free and enable easy access of fire department personnel to the site of fire. The line diagram at Fig. 11 shows the ECS system components at the station.

(v) On the platform, two return air ducts namely, over track exhaust (OTE) and under platform exhaust (UPE) are proposed. This shall enable extraction of about 65% of heat released from the train in station area directly.

(vi) The tunnel ventilation system comprises, tunnel ventilation fans placed at both ends of the station box, on the concourse level. In normal operations, the ventilation provided inside the tunnel due to flow of air caused by the piston effect of moving trains, is considered adequate. However, in case of any eventuality of

210

tunnel temperature rising, due to stalling of trains or fire, these tunnel ventilation fans at two ends of the tunnel segment between two stations, shall be operated in supply and extraction mode so as to cause continuous flow of air inside the tunnel to move smoke away from the evacuating passengers. The air flow can be induced in both forward and reverse direction as the fans used are reversible.

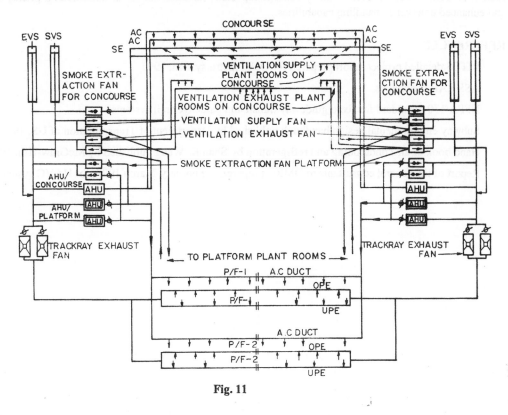

Fig. 11

The operation of ventilation fans and damper positions are shown in Fig. 12.

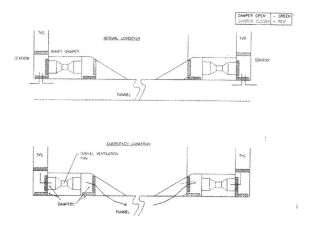

Fig. 12 : Tunnel Ventilation Operations

9.0 CONCLUSION

The Environment Control System (ECS) is inescapable for an underground metro system. ECS are of mainly two types *i.e.*, 'Open System' and 'Closed System' and the latter has been adopted for underground section of Delhi metro because of many benefits especially, smaller size of air-conditioning plants, economy and enhanced emergency handling capabilities.

REFERENCES

RITES detailed project report prepared for Delhi metro system, 1995.

ASHRAE handbooks.

NFPA 130 - standard for fixed guideway transit and passenger rail systems.

Subway environmental design handbook published by Department of Transportation of US, 1997.

Hand book of air-conditioning and refrigeration by Shan K. Wang published by McGraw-Hill, 1993.

Report of the general consultants to DMRC prepared in February-March 1999, for HAVC system.

TUNNELLING IN COMPLEX GEOLOGICAL SETTINGS

CASE STUDIES OF RAILWAY TUNNELLING ON UDHAMPUR —KATRA SECTION

RAKESH CHOPRA
Chief Engineer (Const.)

SANJEEV KUMAR LOHIA
Deputy Chief Engineer

PRAMOD SHARMA
Deputy Chief Engineer

ACHAL KHARE
Deputy Chief Engineer

Northern Railway, Udhampur, J&K, India

SYNOPSIS

Indian railways are constructing a rail link between Jammu and Baramulla. The work has started in the Jammu-Udhampur and Udhampur-Katra section. This paper brings out the difficulties being encountered in tunnelling through the sector of Himalayas as case studies in the Udhampur-Katra section. These deal essentially with the geological, tunnel design and selection aspects. The paper tries to bring out the consequential delays and shortcomings for this type of mountain ranges and the need for more research in the area. Authors have expressed their personal views in this paper.

1.0 INTRODUCTION

Udhampur-Katra section is the first leg of prestigious and challenging Udhampur-Srinagar-Baramulla Rail link project covering a total length of 287.0 km. The project envisages construction of a new single line broad gauge railway with maximum speed potential of 100 km. per hour with ruling gradient limited to 1 in 100. The maximum degree of curvature has therefore been restricted to 2.75^0. This has necessitated provision of 82 tunnels totalling to 89.6 km. out of the total route length of 287 km with the longest tunnel being of 10 km across Pir Panjal range. Besides 82 tunnels, there are 105 number major bridges and the longest and highest railway bridge is across river Chenab with span of 1000 m at a height of 390 m above bed level as against 79 km of tunnelling in 760 km involved in Konkan Railway. This project is the most difficult new railway line project undertaken on Indian subcontinent. The terrain through which the line is to be constructed is the most difficult ever encountered in the history of railway construction as it passes through young Himalayas which are full of geological surprises and numerous problems. The alignment of the project is shown in Fig. 1. This project once completed will not only fulfill the long cherished dreams of the people of J&K but also be an unparalleled engineering marvel. Presently the work has been taken up only on Udhampur-Katra section and Quazigund-Srinagar -Baramulla section of the project.

Udhampur-Katra section is 25 km long, but it involves construction of 8 tunnels, aggregating to a total length of 10 kms and 9 major bridges. Out of 8 tunnels, two are more than 2 km. long viz. Tunnel No. 1 (3.2 km) and Tunnel No. 3 (2.5 km). Since extensive tunnelling was involved through quite fragile geology, M/s NHPC were engaged as consultants for geological mapping and tunnel design while M/s RITES were engaged for geo-technical investigations.

2.0 CASE STUDY 1-GEOTECHNICAL INVESTIGATION AND PIT FALLS

The tunnels in Udhampur-Katra section fall in the Siwalik Group and Pleistocene to recent deposits. Tunnel No. 1 is expected to encounter unconsolidated overburden material consisting of boulders, cobbles and pebbles at both ends coupled with high water table constituting an extremely poor strata for tunnelling while the middle portion is expected to traverse through siltstone and sheared clay stone beds of upper Siwalik. Tunnel No. 2 is 1.4 km long and is expected to pass through terrace deposits and in some length through compacted sand. The terrace deposit consists of an heterogeneous mixture of boulders, cobbles and pebbles in a matrix of silt, with water table well above the tunnel level, constituting an extremely poor tunnelling media. The strata is as depicted in the Photo 1. Tunnel No. 3, 4 and 5 would be excavated within the middle Siwalik formation, with main rock types being thickly bedded, moderately soft, sparsely jointed sand stones and sheared clay stones, silt stones. Remaining tunnels are expected to pass through dolomite scree with mild cementation. Thus geologically a considerable length is to be tunnelled through extremely poor tunnelling media.

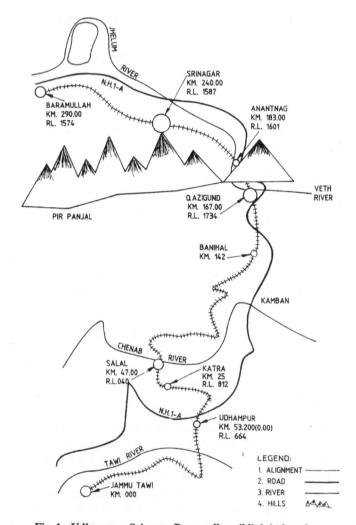

Fig. 1 : Udhampur, Srinagar Baramulla rail link index plan

Geo-technical investigations included detailed geological mapping by M/s NHPC, based on which a minimum investigation programme was chalked out. Systematic seismic refraction studies were conducted using 24 channel seismograph with 5 point pattern in selected stretches of tunnel alignment, to determine the seismic velocities of soil, overburden and underlying bed rock. This was done to determine their thickness and characteristics for design of tunnels. Seismic refraction survey was carried for all the tunnels in a length of 7.47 km. This was followed by confirmatory core drilling of 17 nos. bore holes (BH) with a view to assess the

216

Photo 1 : Exhibit boulder strated soil. Which is likely to be encountered at Udhampur end of Tunnel No. 2

physical and lithological characteristics of the overburden material and underlying bed rock. Permeability tests were conducted in bore holes in overburden and bed rock strata to assess the porosity and fractured condition of the strata as well as indirectly assessing the groutability of the strata. Exploratory pits were also dug out at locations of tunnels 1 and 2, to take undisturbed samples of overburden strata for conducting tests.

The geology of the Tunnel No. 1 on both ends was predicted as quite alarming. On Udhampur end of the tunnel, geological traverse predicted overburden material comprising boulders and cobbles mixed in red ferruginous soil from chainage 2200 to 2375. The geophysical scanning of the area showed seismic velocity of the order of 1100 m/sec with layer thickness of 9 m at Ch. 2200 m and 60 m at Ch. 2357.5 m. Fig. 2 shows chainage diagram.

Accordingly, it was interpreted to be bouldery strata from Ch. 2200 to 2425 m. This was also confirmed by two bore holes at Ch. 2225 m which showed an overburden comprising boulders set in sandy/silty matrix upto formation level. BH at Ch. 2355 m also showed sub rounded and angular rock fragments, sand, stone, quartzite etc., Set in silty matrix upto formation level. However, the work of open excavation on this end is in progress. This has revealed a clayey soil with pebbles and occasional occurrence of boulders in consolidate mass at chainage of 2280 m upto 12 m below the ground level.

On the Katra end of Tunnel 1 also, geological traverse predicted overburden material (terrace deposit comprising boulders, cobbles, pebbles in a matrix of sand and silt) form Ch. 4170 to Ch. 5300. As per seismic profiling, soft sand stone/silt stone was predicted upto Ch. 4800. Between Ch. 4800 to 5300 the strata was interpreted

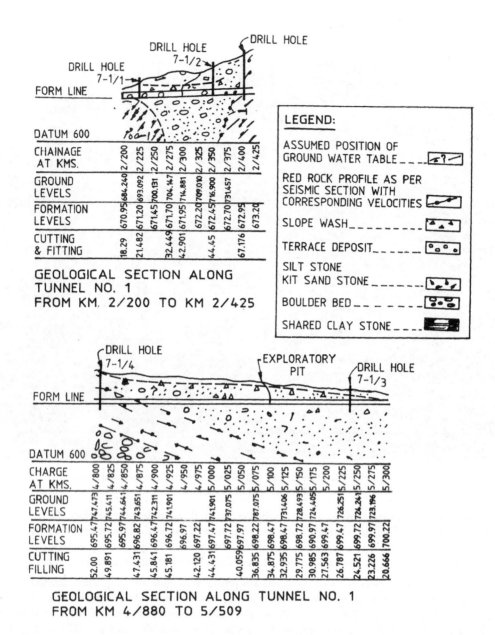

Fig. 2 : Udhampur-Katra rail-link (B.G.) geological section along Tunnel No. 1

as overburden. The confirmatory bore hole at Ch. 5215 showed pebbles, cobbles and boulders set in silty matrix with core recovery of 3 to 8% only. However, during excavation, the strata has been found to be of very soft sand rock, clay/slit with pebbles. Though because of presence of pebbles, it is difficult to extract a core, but the strata has sufficient in situ strength to make tunnelling comparatively easier as compared to the overburden type of strata. Thus, instead of having long lengths of cut and cover portion involving open cuts upto 42 m height, portals on Udhampur end and Katra end can be established earlier and tunnelling can be commenced.

From the above, it can be seen that in Himalayas, pitfalls can occur even after detailed geo-technical investigations. This has led to slow progress, frequent design changes, frequent dialogues with the geologists and a longer time to be spent by field engineers at the site.

3.0 CASE STUDY 2 — SUPPORT CLASSES AND THEIR APPLICATIONS

Five different classes of support measures have been suggested by the consultant[1] based on South African Council for Scientific and Industrial Research, Geo-mechanics classification system proposed by Bieniawski and Norwegian Geological Institute Tunnelling Quality Index System proposed by Baston. The five classes and Good rock, Fair rock, Poor rock, Very poor rock and Overburden.

The Rock Mass Rating (RMR) as defined by Bieniawski, 1989[2] depends upon the following parameters :

a) Uniaxial compressive strength of intact material
b) Rock quality designation (RQD)
c) Spacing of discontinuity
d) Condition of discontinuity (Roughness)
e) Ground water conditions and
f) Orientation of discontinuity (Dip and Strike)

The value of RMR depends largely on RQD, spacing and condition of discontinuities. In field, while calculating RMR a lot of subjectivity is involved because of the following reasons :-

(i)   The strata at various places consists of alternate bands of sand stone/silt stone/clay stone. The thickness of bands varies from few cm to 1-2m. The uniaxial compressive strength thus varies at very close intervals of even on pull of 2 to 2.5 m.

(ii) The RQD is difficult to be determined for even one pull. The strata and the fracture pattern varies at different heights on the same wall, between two walls and between walls and crown portion. If the complete area is divided into grids of 1.5 m by 1.5 m, the RQD of each grid square will be widely varying. Which value should be considered needs to be addressed ?

(iii) Spacing and condition of discontinuities is a very subjective assessment (Photo 2). Though some of the discontinuities are clearly discernible, but most of the time, it is very difficult for the field engineer to identify the fracture patterns and discontinuities. Thus even with best efforts, the RMR values remain quite subjective. The assessment of the same strata by different persons is therefore likely to result in different classifications. In such circumstances, which support class is to be provided remains an open ended issue.

**Photo 2 : Tunnel – 5 UDM end. Photo indicate highly fractured nature of rock.
No conclusion regarding fracture pattern can be established**

Furthermore, the RMR does not take into account the following site conditions, which can affect the strength and deformation characteristics of jointed rocks mass :

- As per Deere (1967) the spacing of joints slightly lesser than 10 cm will give zero percent RQD while a spacing of slightly more than 10 cm would give 100% RQD leading to deceptive assessment of the overall quality of rock mass. Number of joints per metre or joint frequency would definitely affect the UCS of the rock mass.

- The orientation of joints with respect to axis of loading would affect the strength of the rockmass but is not reflected in RMR system.

Thus the use of RMR system in field is not only highly subjective but also very difficult to be applied on day to day basis specially in young Himalayas where the geology changes at every metre. It can at best be used for preliminary design support system. There is, therefore, a need for a more simpler and practical approach for day to day use in the field.

4.0 CASE STUDY 3 — IMPROVEMENT IN LAGGING DESIGN

The design envisages placement of 50 mm thick RCC laggings behind steel ribs which is the ideal location for placement of lagging (Fig. 3). However, while doing back filling with placers, a lot of difficulty was faced in the field as the laggings used to fall due to vibration. As such the back fill upto spring level had to be done manually (metre by metre) by placing lagging in small heights only. This was a very time consuming activity. This was hampering the progress badly and was taking about 30 hrs. per metre including backfill concrete. To tackle the problem following designs were considered and trials conducted.

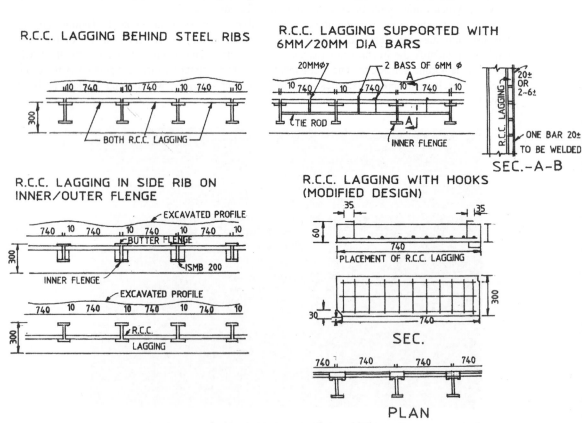

Fig. 3 : Improvement in design of RCC lagging

220

(i)   **Placement of laggings inside the ribs on the inner flange:** This concept is depicted in Fig. 3. However this would have reduced the lining thickness to only 100 mm rendering it impossible. Lining is essential for encasing the ribs and providing long term stability. Thus this would have necessitated increasing the section, making the whole proposal very uneconomical.

(ii)   **Placement of lagging inside the ribs on the other flange:** This concept is depicted in Fig. 3. This design was followed in Loktak and Reva project. However, this would have required the placement of wedges between the inner flange of the ribs and the lagging to keep them in position. Thus, this is also not a time and cost effective solution.

(iii) **Placing two 6 mm dia reinforcement bars behind ribs:** The arrangement is shown in Fig. 3. in this two 6 mm dia reinforcement bars were placed behind the ribs and welded at bottom the laggings were placed between the bars and the ribs. The laggings were also tied with binding wire to tie rods between the ribs. The bars were provided from bottom to one metre above spring level. However 6 mm dia bars proved to be quite flexible and laggings used to bulge outside.

(iv)   **Placing one 20 mm dia reinforcement bar behind ribs in the centre:** This arrangement considered providing 20 mm dia. single reinforcement bar between ribs at the back from bottom to one metre above spring level and placement of RCC laggings between the reinforcement bars and the ribs as shown in the Fig. 3. This was quite a sturdy arrangement and reduced the time of placement of lagging land backfill concrete from 30 hrs. to 20 hrs per metre. This design was though time effective but was not cost effective as additional reinforcement bar was required.

(v)   **Providng hooks on the lagging itself:** In this the design of the RCC lagging was modified by providing a hook on either side of the lagging at the time of casting itself by bending upwards one of the reinforcement as shown in the Fig. 3. The two diagonals corners are also tapered to the exact requirement of 200 mm lagging, so as to avoid entanglement of corner at the time of placement. With this design, the placement of lagging has become a very easy job, reducing the time of placement and backfill concrete to 12 hrs. per metre. This system has thus proved to be quite time effective and at the same time cost effective as well as no additional reinforcement is required.

5.0 CASE STUDY 4 — SECTION AND DESIGN OPTIMISATION FOR CONSTRUCTION EASE

The tunnel cross section initially advised by the consultant and followed on Udhampur-Katra section is shown in Fig. 4. Certain difficulties were being faced from construction points of view in adopting this cross section. These are described below :

5.1 Difficulties

(i) Deterioration of tunnel base due to slushing without zero level concrete

As seen from the Fig. 4, the base of the side drains is lower than the base of bottom lining in central portion of the tunnel. Thus, it is not possible to lay zero level concrete at one level with this profile. Therefore, the advantage of laying zero level concrete during construction phase could not be derived. As a result the tunnel base without any concrete was getting slushy due to ingress of water as well as movement of heavy machinery, affecting the tunnel progress.

(ii) Requirement of additional blast for wall beam trenching and side drain

The experience gained so far while working on tunnels 4 and 5 indicates that heading with full face blast is possible in the strata presently encountered. However, one additional blast was required to be taken for trenching on the sides due to the wall beam and side drain at lower level. This, being an additional activity, affected the progress of tunnelling.

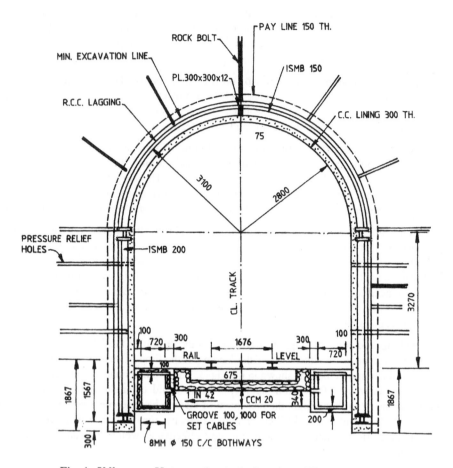

Fig. 4 : Udhampur Katra section typical section of T-5 in poor rock

(iii) Additional time required for fixing of spring level beam

Normally spring level beams are needed when the arch ribs are required to be provided at a closer spacing *i.e.* the arch ribs spacing is lesser than the spacing of the wall posts. In the present case of tunnelling on Udhampur-Katra section the spacing of arch ribs and walls posts is same. Therefore, the provision of spring level beam could be dispensed with. The erection of beams at spring level was taking about 3 hrs.

(iv) Difficulty in movement of gantry for concrete lining

The base of the tunnel being at two different levels, the movement of gantry for concrete lining would have got restricted due to trenching for the side drains.

(v) Infringement in the movement of hose pipe of the concrete pump due connection plates at crown

The arch rib is being fabricated into two parts to be connected through bolts with connection plates at the crown. While doing the concrete lining in arch portion above spring level, flexible hose pipe of concrete pump is required to be moved laterally for concreting the left as well as the right portion of the arch. The size of the hose pipe being 100 mm, minimum opening of about 125 mm is needed to allow the lateral movement of this pipe near the crown. However, the connection plate at the crown was permitting only 50 mm. Keeping in view the difficulties highlighted above, the tunnel cross section was modified as show in Fig. 5 with following features.

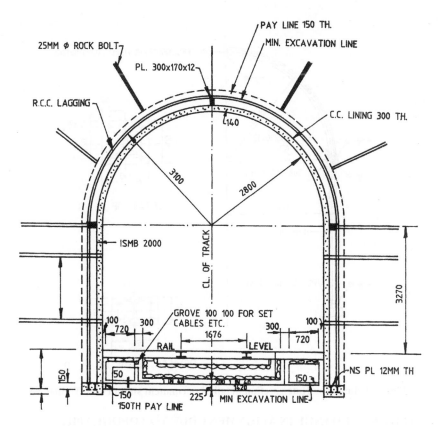

Fig. 5 : Udhampur Katra section revised typical section of Tunnel No. 5 in poor rock

5.2 Improvement/Modifications

(i) Provision of zero level concrete with common base

The base concrete earlier being done in one stage at two different levels (*i.e.* the central portion and the drain) is now being done in two stages as follows :

(a) First stage concrete at one base level with depth varying from 225 mm to 150 mm from centre towards end thereby providing a cross drainage gradient of about 1 in 40. This could ensure speedy laying of zero level concrete during construction stage as the tunnel progresses. Thus the problem of tunnel base getting slushy could be controlled to a large extent. The provision of cross gradient in the first stage concrete ensured better side drainage resulting into a clear and relatively dry central portion for easy movement of the machinery. The level of first stage concrete has been fixed in such a way so as to cover the full depth of the wall beam for the entire base width of the tunnel. Thus it will act as a strut between two wall beams giving additional structural advantage to compression member.

(b) Provision of second stage concrete of uniform 200 mm thickness in central portion (excluding side drain) ensured adequate depth of the side drain below formation level.

(ii) Modification of permanent steel support system

Since the spacing of wall post and arch rib is being kept same, the provision of spring level beam was dispensed with thus reducing the time for permanent support erection as well as effecting economy.

The plate connection near the crown was modified as shown in Fig. 6 to allow for adequate clearance for the movement of flexible hose of the concrete pump.

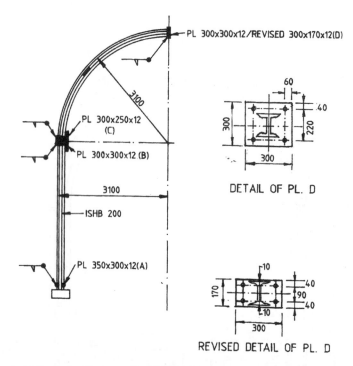

Fig. 6 : Udhampur Katra section detail of permanent support for Tunnel No. 5

6.0 CASE STUDY 5 — CHANGE IN ALIGNMENT DUE TO TOPOGRAPHY

6.1 The site is located in village Jakhrian of Udhampur district of Jammu and Kashmir state along Udhampur-Katra Rail Link Project. From chainage 0 to 1545 alignment is in cutting with a maximum cutting of 13 m at 1545. The cutting has been done and it indicates large variation in strata. The strata predominantly has boulder/gravel with matrix of sand, silt and clay; medium hard to hard shale with silt stone intercalation. The open cuts in the following stretch have slope failures on LHS of alignment.

1. Ch. 1127-1152.   2. Ch. 1162-1212
3. Ch. 1245-1262.   4. Ch. 1388-1452
5. Ch. 1470-1489 .

The main reasons for the failure as assessed area as follows :

1. Presence of soft, fragmentary and weathered shale and silt stone.
2. Alignment is parallel to strike of rocks.
3. Poor shear parameters of strata.(c = 0.2, ϕ = 0)
4. Lubrication of bedding planes during rains.

As the past experience in the adjoining patch of open cuts is not satisfactory, the stretch 1545-1825 was studied for provision of cut and cover tunnel.

6.2 Geology between Ch. 1545-1825

Entire alignment is in cutting Fig. 7 with a maximum depth of 32 m at Ch. 1750. To ascertain the strata two bore holes were drilled at Ch. 1700 (31 m deep) and Ch. 1750 (31 m deep). Bore log of both locations indicate :

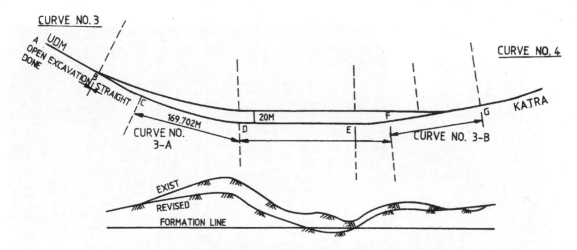

Fig. 7 : Plan showing shifting of alignment from km 1545.00 to 2180.00

(a) Presence of overburden *i.e.* boulder/gravel in a matrix of sand silt, and clay in top 11 m. In the strata full water loss was noticed.

(b) Below the overburden, highly fractured bands of clay stone, silt stone and sand stone are available with a average core recovery and average RQD in 20 m. Strata as under.

Sl. No	Boring location	Average care recovery (%)	Average RQD (%)
1.	1700	82	37
2.	1750	68	23*

* As per Bieniawskis[2] classification system (RMR) if RQD is between 0-25, rock quality is very poor and as per approximate general Tunneler's description it is a crushed rock.

However it may be mentioned that the strata in the area changes very frequently as visible from the open cutting upto 1545. (Photo '3'). Thus, as such no conclusion regarding the strata type can be drawn with sufficient degree of certainty.

6.3 Technical Details Regarding Original Alignment (1545 to 1825)

The cutting in this stretch varies from 13-32m. As such the cover available is 5 m to 24 m. For construction of tunnel a minimum 16m cover is desirable (from 2D consideration). This cover is available between Ch. 1700-1800. Thus if we go ahead on this alignment then the technically possible construction scheme is cut and cover from 1545 to 1700, 1800 to 1825 and tunnel from 1700 to 1800. The anticipated problems associated with this scheme are :

(A) For construction of cut and cover from 1545 to 1700, 1800 to 1825 open cutting required is 13 m to 24 m deep.

 (a) Open cuts will not be stable due to presence of overburden strata and steep slope of hill LHS of alignment.

 (b) Possibility of hill slope failure.

(B) For construction of tunnel from 1700 to 1800.

 (a) Geology shows presence of highly fractured bands of clay stone, silt stone and sand stone.

 (b) Dip is inclined towards the alignment which will have large over-breaks.

Owing to these problems change in alignment was considered.

225

Photo 3 : Exhibit the open geology at Ch. 1.400 to 1.500 geology shows variation of strata *i.e.* shale band and boulder strated soil. Depth of cutting is around 13 mtrs

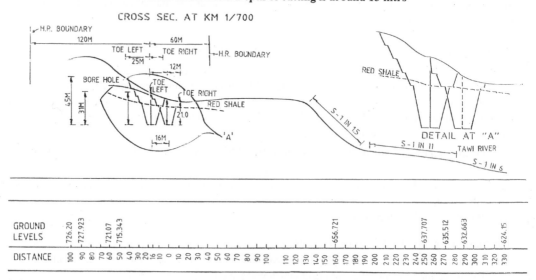

Fig. 8 : Cross Section at km 1/700

6.4 Technical Details Regarding Revised Alignment

The revised alignment (Fig. 8) is located on the RHS of the original alignment with a maximum shift of 20 m. The cutting involved varies from 13 m to 24 m. 2D cover is not available in almost the entire length as such proper cover for tunnelling is not available. However cut and cover construction becomes comparatively easier along this alignment in view of the following :

1. Cutting gets reduced from a maximum of 32 m from original alignment to 24 m in the revised alignment.
2. As the alignment shifts away from the hill slope possibility of un-stability of hill slope reduces.

In view of the above it is felt that construction of cut and cover along the revised alignment between Ch. 1545-1800 is a better technical option. However few precautions required at construction stage as under :

1. The open excavation in the stretch for cut and cover be done carefully and in short length preferably not more than 20 m at a time and before proceeding further the excavated patch is concreted and backfilled.
2. The work should be done in dry season.

7.0 CONCLUSIONS

1. The case studies reveal the uncertainties that prevail in the system of investigations.
2. As a result there are delays expected which lead to cost over runs.
3. Assessment of work, time schedule, cost for the purpose of tendering is not accurate.
4. Changes in the planning is necessary during the various stages of work progress.
5. Design needs to carry field past base experiences, as otherwise changes for construction ease or safety lead to delays and costs.
6. There is need for practical field methods for support classifications.
7. This is more difficult than even Konkan Railway Project, where there was 79 km of tunnelling in 760 km of length.

REFERENCES

Final geotechnical report on Rail Tunnels for Udhampur-Katra rail link project by NHPC June' 1999 and interim geotechnical report on Rail tunnels for Udhampur- Katra Rail Link Project by NHPC, Oct.' 1977.

Hoek, E., Kaiser, P.K., Bawden, W.F., (1995) Support of Underground Excavation in Hard Rock. Chapter 4, pp. 27-47.

Final Report on Geophysical Studies and Confirmatory Drilling of Boreholes for proposed Tunnels along Udhamplur-Katra Rail Link project by RITES' June' 1998.

BIBLIOGRAPHY

Barton, N. Lien, R. and Lunde, J. (1974), Engineering Classification of Rockmasses for the Design of Tunnel Support, Rock Mechanics, Vol. 6, No. 4, 1974, pp. 189-236.

Bienaiawski, Z.T. (1976), Rock Mass Classification in Rock Engineering, Proceeding of Symp. on Exploration for Rock Engineering, Johnannesburg.

Mal Barna, B.D. (1996), Tunnelling in Post Siwalik Boulder Conglomerate in Jammu Area : A Geotechnical Evaluation G.S.I. Spl. Publication 21 (2), pp. 97-100.

T. Ramamurthy. (1994), Classification and Characterisation of Rock Mass, Proceeding of Workshop on Tunnelling India '94, pp. 1-11.

MINING MECHANISM FOR TUNNELLING

P.K. PAWAR

Executive Engineer

**Arphal Canals Division
Karad, Distt.- Satara, Maharashtra, India**

V.V. GAIKWAD

Chief Engineer (Specified Project)

**Maharashtra Krishna Valley Development
Corporation, Irrigation Department, Pune
Maharashtra, India**

SYNOPSIS

Conventional method of tunnelling includes drilling/blasting with jackhammers in combination with airlegs and mucking by tippers. The paper describes the mining mechanism used for excavation of Arphal Tunnel of length 16.63 km. The discussion includes, various equipment's used, comparison of mining mechanism with conventional method, and various site observations for this methodology. Energy consumption per cum, for mining mechanism is less as compared to conventional method of tunnelling with vertical shaft. The method is ecofriendly causing less pollution during tunnelling operation as compared to the conventional method. Cost analysis for present project site indicates that the mining mechanism is costlier by about 15 to 20 per cent as compared to the conventional method.

1.0 GENERAL

Kanher dam of gross storage 10.10 TMC (286 MCM) has been constructed across the river Venna and is a part of Krishna Irrigation Project, in Maharashtra, India. The left bank canal of this dam when crosses Krishna river at km 20, the canal has been named as Arphal canal. Total planned length of this canal is 205 km. The alignment of express Arphal canal from km. 86 to 131 was passing through forest land as well as through steep slopes. Also in this length there was no irrigation by Arphal canal. Considering above difficulties, alternative alignment for canal with a tunnel of length 16.63 km between km 86 to 102 of Arphal canal was planned to carry 15 cumecs of discharge and is called Arphal tunnel (Fig. 1). This alternative alignment reduced the main canal length by 30 km. It will help to irrigate about 16758 ha of land in Sangli district of Maharashtra state. The Arphal tunnel is one of the biggest tunnel on canal in Maharastra, India.

The starting point of this tunnel (*i.e.* portal-I) is at Rajmachi in Satara district and end point (portal-II) at Hingangaon in Sangli district. The tunnel in general passes through compact basalt and volcanic breccia type of rock. As per the original planning, tunnel was to be excavated with conventional method of drilling/blasting using jackhammer, tippers etc. with three adit of 3.5 km length. Various alternative methodologies were considered during the bidding stage of this work and the methodology considering mining mechanism for tunnelling was considered for construction of this tunnel. The methodology considered construction of two inclined shafts each 365 m and 425 m. length instead of 3.5 km long 3 adits for excavation of main tunnel. Overall quantity saving for tunnelling was about 53,500 cum.

2.0 METHODOLOGY OF CONSTRUCTION

The entire length of 16.63 km has been planned to be excavated through portal-I (P-1) and portal-II (P2) and four faces through two inclined shafts S1 and S2, as shown in Fig. 1. The excavation of tunnel through portal P1 and P2 is by conventional method. Excavation of tunnel through shaft S1 and S2 is being done with use of state of the art mining machinery as discussed below.

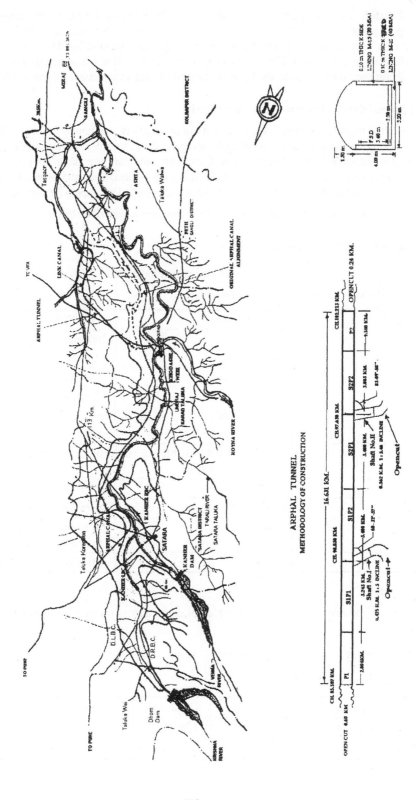

ARPHAL TUNNEL
METHODOLOGY OF CONSTRUCTION

Fig. 1 : Krishna irrigation project index plan

2.1 Construction of Inclined Shafts S1 and S2

Inclined shaft S1 has been planned with a slope of 1 in 5 with total length of shaft about 425 m and for shaft S2, slope is about 1 in 5.46 with length of 362 m. These shafts have been excavated by normal drilling and blasting with jack hammers in combination with airlegs and mucking by 0.6 cum capacity excavators directly loading on to skip handled by 7 ton pneumatic winch (Clark Champman). The excavated muck thereafter was discharged on to tipper, truck, by mechanical tipping arrangement for onward disposal.

2.2 Main Tunnel Excavation through Inclined Shafts

Drilling in the main face tunnel is being carried out by single Boom Jumbo Boomer 281 of Atlas copco made. The mucking operation is then carried out by using 0.6 cum capacity excavators directly discharging on a stage loader loading the 3 cum. Granby Mine Cars. The mine cars are handled by 8 ton locomotive to discharge the excavated muck on the conveyor installed in each inclined shaft at 5 m below bed level of tunnel for transportation to surface bunker. The muck thereafter is transported by tippers for disposal.

Various components of mining mechanism for tunnelling are as described below :-

2.3 Specification of Mining Mechanism

(Courtesy : Kvaerner Cementation India Ltd., Mumbai)

1. Boomer 281- Atlas Copco :- Main Components

(a)	Rock Drill	Rotation speed	0-300 RPM
(b)	Feed	Hole depth	3000 mm
		Feed force	12.5 kN.
(c)	Boom	Feed extension	1250 mm
		Boom extension	1250 mm
		Feed roll over	360
		Max lifting angle	+ 50 to −30
(d)	Power pack	Axial pump with variable displacement for impact and feed constant flow gear pump for rotation	
(e)	Carriers	Engine :	42 kW at 2300 RPM
		Max Speed :	10 km/hrs
		Braking system :-	All hydraulic with parking and emergency brake
		Steering	Articulated power steering.

2. Stage Loader

Rail mounted portable conveyor through which blasted muck is transported from face by excavator to Granby Mine Car.

3. Granby Mine Car :- Side tilting type capacity – 3 cum

4. Loco :- 8 T battery operated loco.

(a)	Hauling capacity	8 loaded Granby Mine Car
(b)	Maximum speed	14 km^1 hr
(c)	Average speed	10 - 12 km^1 hr

5. Belt Conveyor

(a)	Capacity	160 T^1 hr
(b)	Belt speed	2.05 m per sec
(c)	Belt width	900 mm
(d)	Troughing angle	30
(e)	Spacing of carrier roller	1 m
(f)	Spacing of return roller	1 m
(g)	Motor	100 HP, 415 V, 1440 RPM star delta

3.0 TIME CYCLE

3.1 Time Cycle for Drilling Operation of Main Tunnel with Jumbo

For the main tunnel of cross section 26.3 sq.m. Burn cut drilling pattern having 54 no. drill holes 41 mm diameter, 3.6 m length along with one no. empty hole of 76 mm diameter, 3.9 m length has been considered. With rate of penetration of Jumbo drill as 1.8 m to 1.5 m drilling time required for drilling entire profile is about 2.10 hours and adding 50 per cent more for setting time, total time for drilling operation is about 3 hours 20 minutes/30 minutes.

3.2 Charging, Blasting and Defuming:- 2 Hours

3.3 Time required for Mucking

Ex. 100 has been considered for loading with its theoretical capacity as 108 cum/hr. With the use of mine cars for transportation, stage loader, for practical purpose loading at 40 cum/hr if considered, it requires about 3 hours and 30 minutes, adding 30 minutes for setting etc., the mucking time is approximately about 4 hours.

Thus for one complete cycle, total commutative time for each face will be about 10 hours, adding 1 hour for miscellaneous work such as grading, pipe extension etc, total cycle time is approximately 11 hours. Thus with this methodology, two blast in each face can be easily achieved giving a production rate of 6 m x 25 days = 150 m/month.

Comparison of important parameters when drilling carried out by conventional method and Jumbo (Boomer) are described in Table 1.

TABLE 1

Description	Conventional	Jumbo
Hole size	32 mm	41 mm
No. of holes	65 nos.	54 nos.
Powder factor	1.0 kg/cum	1.4 kg/cum
Drilling rate	250 mm/min	1800 mm/min
Drilling length	2.7 m	3.6 m
Pull	2.1 m per cycle of operation at 16 hrs/cycle	3.0 m per cycle of operation at 12 hrs/cycle
Progress	3.0 m per day of 2 * 12 h shift per face	6.0 m per day of 2 * 12 h shift per face

4.0 ENERGY CONSUMPTION

The mining mechanism for tunnelling could be compared in terms of energy consumption as described below.

Conventional Method			Modern Technique	
By Jack Hammer			By Jumbo Drill	
Drilling	2 *Compressor	180 kW	1 Jumbo Drill	54 kW
Loading	Common method : (Ex.100)			
Transportation	2 * Winch	276 kW	2 * Loco + Loader	110 kW
To surface	2 * EOT	30 kW	1 * Belt	90 kW
Transportation to dump yard		Common	Method (Tippers)	
Total Energy		486 kW		254 kW
Output per face	90 m^3		180 m^3	
Energy consumption / cum		5.4 kW/m^3	1.4 kW/m^3	

Thus energy consumption per cum for mining mechanism is about 1.4 kW/cum while in case of conventional method with vertical shaft the energy consumption is about 5.4 kW/cum. Thus in terms of energy utilisation mining mechanism is energy conservative. The methodology is also ecofriendly, as could be seen from the comparison of ventilation parameters vide Table 2.

TABLE 2

Chainage	Ventilation parameter			Remarks
	C_O (PPM)	NO_2 (PPM)	O_2 (%)	
86/690	10	0	20.8	At Portal No.I, Drilling is in progress
86/675	40	0.2	20.6	(conventional method)
86/120	40	0.2	20.4	
97/190	0	0	20.7	At Shaft No.II, Boomer is working, drilling
97/165	0	0	20.6	is in progress (mining mechanism)
97/145	0	0	20.7	
98/040	4	0.7	20.5	At Shaft No.II near stage loader mucking is
98/050	7	0.1	20.4	in progress (mining mechanism)
97/600	1	0.1	20.6	
At face	44	1.5	20.4	At Portal No.II, mucking is in progress
101/750	38	3.6	20.4	(conventional method)

Mining mechanism for tunnelling involves use of hydraulically and electrically operated Jumbo Boomer, battery operated loco, mine cars and stage loaders, ventilation conditions inside the tunnel are uniform and are very good as compared to the conventional method as could be observed from Table 2.

5.0 COST ANALYSIS OF MINING MECHANISM FOR TUNNELLING

Table 3 indicates the comparison of cost of tunnel excavation with conventional method and using mining mechanism with boomer for drilling. Tunnel excavation using mining mechanism is costlier by about 15 to 20 per cent (approx.). Comparison of the various components of the cost indicate that labour input for mining mechanism is minimum and is less by about 20 to 35 per cent and thus indicates that method is mechanised one involving less labour input. The major component, which makes the system costlier, is the drilling and blasting cost. Blasting cost as compared to the conventional method is more due to high powder factor that is required, to get a good crushed muck for easy transportation by the conveyor belt. Cost analysis indicated in Table 3 is for the present project situation, with mining mechanism considered to be used for 13 km. of tunnel excavation.

Considering the high capital investment for various mechinery and plant as discussed vide para 2.3, the method is economical/useful for longer tunnel. With this methodology and for tunnel on open canal with restricted width of 5.00 m, concrete side lining for the tunnel is possible in parallel to the tunnel excavation activities such as drilling, mucking , etc. In case of a Arphal tunnel side lining with a lining height of 4.0 m is being executed for length of 8 m at each face simultaneously with tunnel excavation and both tunnel excavation and side lining activities are running close to parallel.

Maximum achieved monthly progress for mining mechanism is 135 m/month and average monthly progress achieved is 115 to 120 m/month while for tunnelling with vertical shaft maximum monthly progress could be about 75 m. Thus the method is very useful for speedy excavation of tunnel with simultaneous concrete side lining for tunnel of minimum width of 5.00 m.

TABLE 3
Cost analysis for tunnel excavation through portal, vertical shaft, and incline shaft

Sr. No.	Component	Through Portal (At entry and exit of tunnel)	Through vertical shaft	Through incline shaft (Mining Mechanism)	Remark
1.	Labour	83.05	101.69	65.83	As per minimum wages.
2.	Drilling and blasting				
	Machinery	54.68	53.03	119.49	
	Material	218.60	244.68	344.30	
	Total	273.28	297.71	463.79	
3.	Pipe line charges	32.50	23.75	18.80	For mining mechanism only water line and no air line.
4.	Mucking				
	Machinery	129.77	216.85	252.68	
	Rail line	0	0.00	13.76	
	Total	129.77	216.85	266.44	
5.	Ventilation charges	119.23	115.30	78.06	
6.	Total cost including other charges	799.00	936.00	1096.00	

6.0 OBSERVATION AND EXPERIENCE WITH THE MINING TECHNOLOGY FOR TUNNELLING

From the actual observations indicated in Table 4. It is seen that, the average cycle time for tunnelling through inclined shafts using mining mechanism is in the range of 13 to 14 hours giving average monthly progress of the order of 115 to 120 m. Increase in cycle time as compared to the theoretically predicted cycle time of 11 hours as discussed in para 3.0 is due to fractured, strata with broad joints, giving more boulders in the muck, requiring hand breaking and secondary breaking.

In this methodology, for transportation of muck over the conveyor boulder size needs to be of 30 kg. (30 cm long). If oversize boulders are met, they are required to be hand broken. As discussed in Table 1 initially the tunnel excavation was started with 54 no. of holes and power factor of 1.4 kg/cum. but muck received included large number of over size boulders which were required to be hand broken . Also with the restricted width of 5.0 to 5.5 m storage and handling of these over size boulders cause lot of trouble, hence to get good crushed muck of desired size number of drill holes were increased gradually upto 90 and powder factor upto 2.0. Thus explosive consumption is very high as compared to conventional method. Average progress of excavation through incline shaft - I is about 126 m while through shaft II it is about 113 m. At shaft - II rock met is jointed porphyritic basalt with some green in fillings. It exhibits vertical, horizontal, block, oblique and at some places columner jointing and hence muck received after blasting includes large quantum of over size boulders and thus affecting the mucking process and thereby giving less monthly progress. Such problem has not been observed at shaft - I. At shaft - I, rock observed is of breccia type at crown while it is massive below springing giving a good progress. As can be seen, from powder factor values, for tunnelling through shaft they are highest and are basically to have a good crushed muck, so as to avoid big boulders. Thus explosive requirement is very high as compared to the conventional method, and contributes to about 22 per cent more cost as compared to conventional method (in Table 4).

TABLE 4
Arphal tunnel (km 86.00 to 102.00)

Sr. No	Description	Unit	Portal-I	Portal-II	S1P1	S1P2	S2P1	S2P2
1.	Length executed	m	1379	1211	458	448	516	547
2.	Average monthly progress	M	118	131	127	126	114	113
3.	Average cycle time	Hrs.	12	12	13	14	13	14
4.	Average pull achieved	M	2.2	2.7	3	3	3	3
5.	Average powder factor	Kg/ Cum	1.2	0.98	1.9	1.9	2.02	2.02
6.	Overbreaks	%	3.00	4.79	5.86	5.71	4.68	4.18
7.	Geological conditions		Volcanic breccia	Hydro thermally altered basalt	Jointed basalt	Jointed basalt	Jointed porphyritic basalt	Jointed porphyritic basalt

Water required for drilling needs to be clean water for hydraulic operation of boomer and requirement is of the order of 85 lit/min. Various other maintenance problems involved are tensioning of conveyor belt and belt for stage loader once in a month by specilised agency. Since the method involves use of electrically operated loco, boomer and stage loader, it requires a electrical line of 3.3 kVA. Due to this while removing under cut by secondary drilling/blasting the electrical line is required to be shut down. Thus operation of undercut removal is little bit difficult as compared to the conventional method.

7.0 CONCLUSION

The method is a mechanised one, involving less labour input. Mining mechanism for tunnelling being mechanised technology, has an advantage of reduced accident rates. With the use of this method for the present project situation probable period of completion has been reduced by about six months and thus will help in accruing the irrigation benefits earlier. The method is ecofriendly and helps in speedy tunnelling causing less pollution during tunnelling. Thus even though the per cum cost of tunnelling is slightly more considering various advantages the method is superior to the conventional method.

TUNNELLING THROUGH ROCK COVER OF MORE THAN 1000 M — A CASE STUDY

G.S. PUNDHIR
Geologist

A.K. ACHARYA
Geologist

A.K. CHADHA
Sr. Manager

Geology Department, Nathpa Jhakri Power Corporation Ltd., Jhakri, Himachal Pradesh, India

SYNOPSIS

Tunnelling has always been a major challenge, more so when being done in the Himalayas. Apart from the geological risks and challenges which are an integral part of underground excavation, the location of structure is an important criteria for the experience of tunnelling through high cover of more than 1000 m has been of problems of high stress like distressing with spalling, exfoliation, rock bursting and other instability condition. The 27.3 km long head race tunnel of Nathpa Jhakri Project which is located in the lesser Himalayas passes through a varied geology of gneisses, schists, amphibolite, pegmatite and quartzite and through a rock cover as low as 9 m and as high as 140 m. The present paper is a case study dealing with the experiences of tunnelling through a cover more than 1000 m.

1.0 INTRODUCTION

The Nathpa Jhakri hydroelectric project is being constructed on the river Satluj in the Shimla and Kinnaur districts of Himachal Pradesh (India) having an installed capacity of 1500 MW (6 x 250). The project has many unique features like, the worlds largest underground desilting chamber (525 m long 16.5 m wide x 27.5 m depth), the worlds largest head race tunnel (10.15 m dia and 27.3 km long) and one of the worlds deepest surge shafts (21.6 m dia and 301 m deep), apart from a 60.5 m high concrete gravity diversion dam and an underground Power House Complex.

The 27.3 km long 10.15 m long finished dia head race tunnel is under construction since 1994. The construction is carried out at 13 faces from 7 adits. HRT passes through a variegated cover ranging from 9 m to 1450 m, with a maximum length having an average cover of ± 500 m. About 1850 m of HRT passes through a cover of more than a kilometer. Thin stretch of HRT from RD. 10900 to 12750 (Fig. 1b) have been excavated through two faces, one faces, about 2943 m from Nigulsari downstream side while remaining 3431 m from Wadhal upstream, side the cover over the HRT in this part consists mainly of rock and few meters of overburden and two nallas crossing at RD 14212 and 3360.

The dominant rock types exposed in the project area are a variety of metamorphic rocks like gneiss, schistose gneiss, schists, quartzites intruded by basic intrusions of amphibolites and acidic intrusions like pegmatite's and quartz veins belonging to the Jeori-Wangtu Gneissic complex of pre-cambrian age.

The part of the HRT under study consists of mainly gneiss, Augen gneisses, Schistose gneisses and quartz mica schists with thin bands of biotite schist intruded by the basic and acidic bodies (Fig. 2). In these reaches of high joint concentration little or no effect of stress/spalling observed.

Stress, spalling, slabbing, rock bursting, rock fall, buckling of rock bolt plates in the rock mass behaviour has been challenge to the construction people. The less supported tunnel reaches shows more deformation of the rock mass as compared to the additionally supported reaches of the tunnel. In most of the cases, the rock cover above the schistose visual observation and experience has helped in predicting the stress problems. Some experiences have been gathered by observing behaviour of different constitution of rock support installation in different reaches.

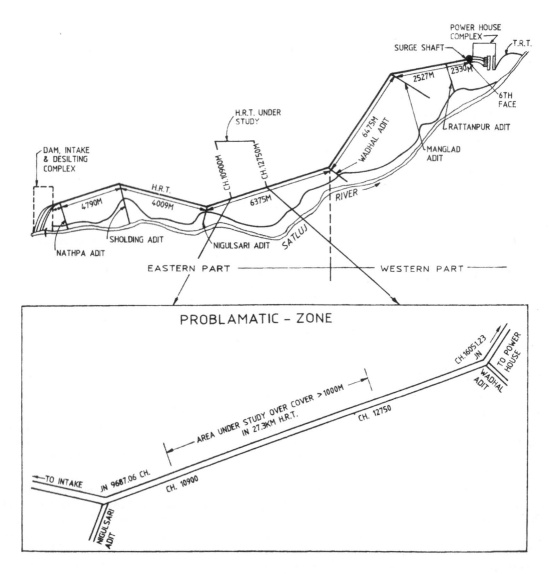

Fig. 1a : Layout plan of Nathpa-Jhakri Hydroelectric Project with location of HRT under study
Fig. 1b : Problamatic-zone

2.0 GEOLOGY

The rock exposed in and around the project area comprises of a variety of metamorphic rocks like gneiss, schistose gneiss, schist, quartzites and basic intrusions (amphibolites), granite and pegmatite. These unfossiliferous rocks belong to the Jeori Wangtu Gneissic complex of Pre-Cambrian age. These are surrounded by rocks of the Jutog Group, Salkhala Group and Rampur Group separated by thrusts of fault.

The project, in respect to the Geology can be divided into two parts, the eastern part and the western part (Fig. 1a). The eastern part of the project included the dam, desilting chambers and a part of the HRT (16 km) and the rock types encountered are predominately gneiss with quartz mica schist bands and having acid (granite, pegmatite and quartz veins) and basic (amphibolite) intrusion. The western part comprises of the remaining part of the HRT (11.3 km), surge shaft and the Power House Complex. The rocks exposed in this part are mainly quartz mica schist with gneiss band at places and quartzites, these are again intruded by basic and acidic intrusions.

236

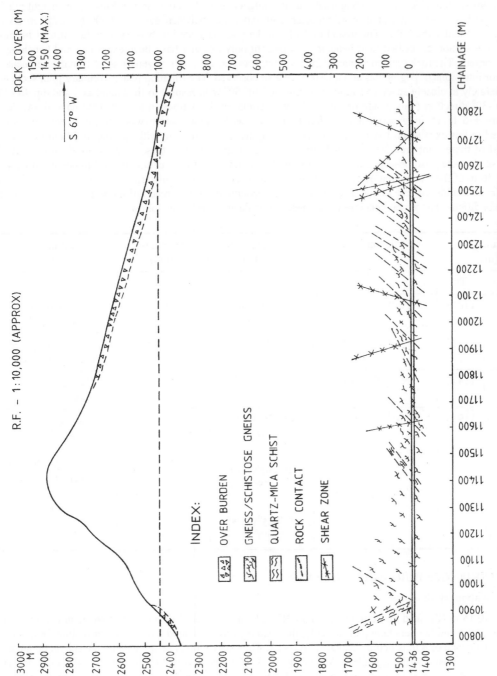

Fig. 2 : Longitudinal geological section between chainage 10900 to 12750 m with rock cover exceeding 1000 m

237

The area under study lies in the eastern part and the rock types exposed in the part of the HRT are gneiss, schistose gneiss and quartz mica schist with amphibolite, pegmatite, and quartz intrusions. The study area lies between two adits Nigulsari and Wadhal adit (Fig. 1b) and have been excavated into two parts. From Nigulsari the excavation was done 2900 m and from Wadhal the excavation was done 3432 m. The tunnel is aligned $N71^{\circ}E$ from Wadhal side and from Nigulsari side it is aligned $S67^{\circ}W$ in this stretch. Most of the height falling in this zone have been excavated from Nigulsari adit and it lies between the Ch. 10900 to 12750 m d/s of Nigulsari adit under the HRT. The tunnel of 10.15 m dia was excavated in two parts in the form of heading (75% of the section excavated) and benching. At present only heading have been excavated and benching is still in progress. The rock type encountered in this reach was gneisses, with augens at places, schistose gneisses and minor intrusions of pegmatites and quartz were also observed. The foliation planes (primary joint sections) are moderately to closely spaced striking at an angle of 30^{0}-50^{0} with respect to the tunnel axis and dip 40^{0}-70^{0} in N 10°-30° direction (valley side or right side) with some local vriations (n to NW dips) have shown long continuity + 5 to 15 m. Apart from the foliations, the steep dipping south westerly joints are commonly associated in this excavation having continuities ranging from 2 m to 12 m with spacing of 30 cm to 150 cm and most of these joint are clean. Apart from these clean joints some of these joints show shearing, having transverse character also observed which have same frequency but of long continuity and frequently observed after 25-30 m of the excavation of the tunnel and contain clay or chloritic material and granular material most of the time. In addition five major shear zones were encountered in this reach of the tunnel which had poor ground conditions. The general occurrence of these major shear zones are as follows :

Chainage (m)		Strike direction	Dip amount and direction	Thickness (cm)	Fillings
From	To				
11003	11013	$N40^{0}W$-$S40^{0}E$	$S50^{0}W/80^{0}$	200	Clay gouge and crushed rock
11321	11333	$N40^{0}W$-$S40^{0}E$	$S50^{0}W/75$-80^{0}	20-50	Chlorite/clay gouge and crushed rock
11472	11478	$N35^{0}W$-$S35^{0}E$	$N55^{0}E/75^{0}$	45-50	Clay gouge and crushed rock
11920	11934	$N40^{0}$-$45^{0}W$-$S40^{0}$-$45^{0}E$	$S45^{0}$-$50^{0}W/75^{0}$-80^{0}	75-80	Clay gouge and crushed rock
		$N40^{0}$-$45^{0}W$-$S40^{0}$-$45^{0}E$	$S45^{0}$-$50^{0}W/70^{0}$-75^{0}	25-30	Clay gouge, chlorite and crushed rock
		$N25^{0}$-$30^{0}W$-$S25^{0}$-$30^{0}E$		50	-do-
			$S60^{0}$-$65^{0}W/75^{0}$-80^{0}		

3.0 EXPERIENCES DURING TUNNELLING

3.1 Excavation Methodology

The 10.15 m (finished dia) circular shaped HRT is being excavated of horse-shoe shape in two parts *i.e.* the heading part and the benching part. Initially, the heading was excavated 2 m below spring line for a height of 7 m and a bench of 4 m is being excavated presently. The excavation is being done by the conventional drill and blast method using drilling jumbos. An average of 3 m of advancement of heading was achieved per day with rock support comprising of rock bolts, shotcrete, wiremesh and ribs as per site requirements being installed.

3.2 Support Installed

Presently rock bolts have been installed which are 25 mm dia, 4.15 m long, grouted. These are installed in a staggered fashion of 1.5 m x 1.5 m, which averages to around 12-15 rock bolts in a row above the spring line. This is the primary form of support in the HRT for good to fair rock conditions. Plain shotcrete (20-50cms) has also been provided in some reaches for the poor to very poor rock conditions, the primary form of support is ribs sets at 1.0 to .75 m C-C spacing and back filled with M40 concrete.

3.3 Observed Behaviour of Rock during Excavation

The different rock mass has behaved differently with advancement of the tunnel. The gneiss and schistose gneiss which are massive to less jointed act as brittle medium giving cracking sounds and shows deformation by spalling. When two or moe joint sets are present in this rock there has been rock falls without any noise. The quartz mica schist rock on the other hand can take some plastic deformation before spalling and gives no sound on deformation. Heavy spalling has been observed on the contact of these two different rock types because of the difference in their elasticity modulus with gneisses taking the major part of the load, and hence are more prone to spalling. In some limited reaches where the rock mass is highly jointed, there is little or no stress related problems.

The orientation of the foliation planes which are the primary joint sets has played a very important role in deciding the amount of support to be provided and also the locations for the support foliation planes making an angle of 30-40^0 with respect to the tunnel direction and their dip towards the valley side or right side of the tunnel has resulted in directing the maximum stresses of the rock cover to the right crown walls of the tunnel and hence almost all of the spalling, open joint planes, rock falls etc. have taken place on the right crown and wall of the tunnel during the excavation of these reaches. It is because of their uniformity, that predictions related to high stresses became easy for the tunnelling crew and thus the amount and type of support to be installed also came in easy, with the result that more emphasis was laid on the supporting of the right crown and wall in comparison to the left crown and wall. In gneiss and schist gneiss the rock mass showed sign of stressing ever after 10 or 15 days or after about 30 m of tunnel advancement. The quartz mica schist rock mass showed signs of crumpling on the right arch after the tunnel advancement of 10 m only, when the rock cover exceeded 1300 m. But with lesser rock cover the crumpling effect was delayed for a longer period of time and with quite a lot of tunnel advancement (100 m or more). In all the major five shear zones encountered in this part of the HRT which were provided with rib support and later on back filled with M40 concrete, there was no sign of movement/buckling in the ribs.

The visible signs of stress under the influence of high rock cover are being summarized below.

3.3.1 *Crackling Sounds*

Were heard time and again by the tunnelling crew and by the authors when the rock cover exceeded 1300 m. *i.e.* between Ch. 11250 to 11350 which has been predominantly excavated in more or less massive to competent gneiss and schist gneiss.

3.3.2 *Mild Rock Bursts*

This phenomenon was again observed between Ch. 11250 to 11350, *i.e.* in the gneiss and schistose gneiss rock. It was in the form of ejection of sharp edged thin slabs of rock from the tunnel periphery. The size of the rock columns liberated were 5 to 25cm in thickness and usually half a square feet to one square meter on surface.

3.3.3 *Rock Falls*

The majority of the rock falls took place on the right crown and in the competent rock mass like the gneiss and schistose gneiss, whenever one or two joints set is intersected with the foliation planes these rock falls took place after a time gap of 10-30 days from the excavation date and ranging in size from 1 m to 3 m. These falls have been confined between Ch. 11255 to 11660 with the maximum no. of rock falls recorded between Ch. 11300 to 11400.

3.3.4 *Open Foliation Planes*

This is the most common phenomenon observed on the right arch and wall of the tunnel in which the foliation planes were dipping. Although rock bolts are installed immediately in every cycle of the excavation, there has been visible sign of opening up of foliation planes on the right crown and wall only. This has been true for both the competent and the less competent rock masses and is more pronounced between Ch. 11300 to 12350, these openings range from a few mm to about 2cm.

3.3.5 *Spalling*

This is an again a very common phenomenon observed during the excavation of this part of the HRT and is more confined towards the right arch and wall. Visible signs of openings and later detachment of rock slabs along the periphery (particularly right side) of the tunnel have been taking place time and again. The depth of the spalling in relation to the drilled profile varied from 20cm to 200cm. The reaches where the thickness of spalling is at its maximum is attributed to the presence of thin (10-15cm) biotite schist/shears (<10 to 20 mm) in the main rock mass, which has resulted in variable rock strengths.

3.3.6 *Rock Bolt Plate Buckling/Breaking Away*

This phenomenon was confined in some reach between Ch. 1600 to 2000 on the right arch only. An instance or rock bolt plates breaking away was observed between Ch. 10866-10882. Where spalling had taken place in the gneissic rock.

4.0 MEASURES TAKEN

In reaches where there was deformation in the rock bolts plates of 4m, length 6m long rock bolts were installed in juxtaposition with the 4 m long rock bolts, mainly on the right arch. These reaches also showed a stability with no further sign of stress related problem. A 30 to 50 m thickness of plain shotcrete has been sprayed in conjunction with rock bolt where there has been openings (1-2 cm) in the foliation planes. This has again given stability to the rock mass. In the whole of the excavated reach, the tunnel have been supported with 4 and 6 m long rock bolts with plain shotcrete.

The reaches where heavy spalling took place on the right arch, the rock bolt density was increased along with application of 20 to 50 cm thickness of plain shotcrete which gave the rock mass stability. No further spalling was observed in these reaches with the passage of time after the installation of heavier support which includes longer bolts.

5.0 CONCLUSION

After passing through this zone in the heading section, it has been observed by the authors that contrary to common belief that tunnelling through high rock cover are prone to problems of high stresses does not seem to be true in the area under study and it can be said that the orientations of the discontinuities, their spacing along with litho units, has a very vital control over the extent of problems to be faced during tunnelling under such conditions. The orientation of major principle stress is also one of the determinant factor in the behaviour of tunnel under high rock cover. It was also concluded that the problem of high rock can be reduced to certain extent if the supports are installed within shortest possible time so as to arrest any initial deformations which has been proved in certain stretches in the area under study.

TUNNELLING IN HIMALAYAN GEOLOGY—A CASE STUDY OF A RAILWAY TUNNEL

JAGDIP RAI

General Manager

J&K Project Cell, IRCON International Ltd., New Delhi, India

1.0 INTRODUCTION

Tunnelling in Himalayas has been a challenging task for engineers, - be it for roads, for water carriage or for Railways. Himalayas being in nascent age geologically, engineers find hostile conditions and unpredictable problems while tunnelling in it. Indian Railways have taken upt the herculian task of connecting Kashmir valley with the rest of the country with broad gauge railway line. The railway line would connect Jammu Tawi with Baramulla passing through Udhampur-Katra-Quazigund and Srinagar (Fig. 1). The total length of tunnelling between Jammu to Quazigund (as beyond Quazigund the Kashmir valley starts) is approx. 99 kms.

Work has been taken up from Jammu to Udhampur, Udhampur to Katra and from Quazigund to Baramulla leaving Katra to Quazigund portion which will be taken up later on.

Jammu-Udhampur is the first leg (approx. 53 kms.) on this railway link. There are 21 tunnels measuring approx. 10 km between Jammu to Udhampur. The alignment passes through unstable geological formations and highly undulating and difficult hilly terrain of Shivalik ranges. The strata involved in tunnelling is highly varying consisting of conglomerate, clay stone, sand stone with water under moderate pressure at places makes the work more difficult.

2.0 SITE DESCRIPTION

The total length of the tunnel No. 8 is 2288 m. With portals at km 15.696 and 17.984 (km. 0 being at center line of Jammu Tawi Station building). In addition, there are two cut and covers of 17m and 156m length each on Jammu end and Srinagar end respectively. The maximum over burden is 243.8 m at 17.400 km. There has been no possibility of any adit or shaft. L-section of the tunnel is given at Fig. 2.

The approaches for either end of tunnel no. 8 are one of the most difficult in the whole project. Tunnel passes through very tough and nearly inaccessible terrain of very soft rock formation encountered with underground water. The strata comprises mainly of very weak to moderate sand stone with dull grey clay stone at few location. The work of T-8 is very tedious owing to its accessibility extremely difficult. There is no direct approach road. There are frequent land slides during heavy rainfall, causing closures of roads, resulting in less availability of working season of about 7 months against the full calendar year.

The shape of the tunnel is D-shaped with the supports of RSJ 150 x 150 x 12mm @ 80cm to 100cm spacing depending on the standup time and quality of rock encountered. A typical cross-section sketch of T-8 is enclosed as Fig. 3.

3.0 BRIEF HISTORY

It is evident from this paper that the agencies who did not have a detailed preplanning, foresight, appreciation for eminent and unforeseeable problems and engineering etiquettes could not be successful and finally failed.

The work of construction of tunnel 8 was awarded to agency A in November 1988. This agency could make only approach roads and 15m of tunnelling when the contract was terminated in November, 1993 *i.e.*, after 5 years. Agency A failed primarily because it could not make a good and efficient approach to tunnel site free of any incumberance.

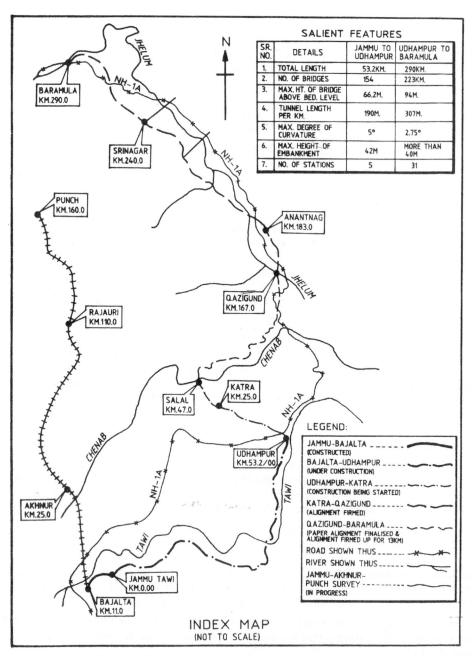

SALIENT FEATURES

SR. NO.	DETAILS	JAMMU TO UDHAMPUR	UDHAMPUR TO BARAMULA
1.	TOTAL LENGTH	53.2KM.	290KM.
2.	NO. OF BRIDGES	154	223KM.
3.	MAX. HT. OF BRIDGE ABOVE BED. LEVEL	66.2M.	94M.
4.	TUNNEL LENGTH PER KM.	190M.	307M.
5.	MAX. DEGREE OF CURVATURE	5°	2.75°
6.	MAX. HEIGHT OF EMBANKMENT	42M	MORE THAN 40M
7.	NO. OF STATIONS	5	31

LEGEND:

JAMMU-BAJALTA (CONSTRUCTED)

BAJALTA-UDHAMPUR (UNDER CONSTRUCTION)

UDHAMPUR-KATRA (CONSTRUCTION BEING STARTED)

KATRA-Q.AZIGUND (ALIGNMENT FIRMED)

Q.AZIGUND-BARAMULA (PAPER ALIGNMENT FINALISED & ALIGNMENT FIRMED UP FOR 13KM)

ROAD SHOWN THUS

RIVER SHOWN THUS

JAMMU-AKHNUR-PUNCH SURVEY (IN PROGRESS)

INDEX MAP
(NOT TO SCALE)

Fig. 1 : Jammu-Udhampur-Srinagar Baramula rail link

The work was again awarded to an agency B in September 1994. This agency although had constructed other smaller tunnels in the project earlier also could not make desired progress due to following :

- poor accessibility
- poor drilling mechanism followed
- poor drainage
- inadequate machinery
- poor working environment
- lack of engineering etiquettes

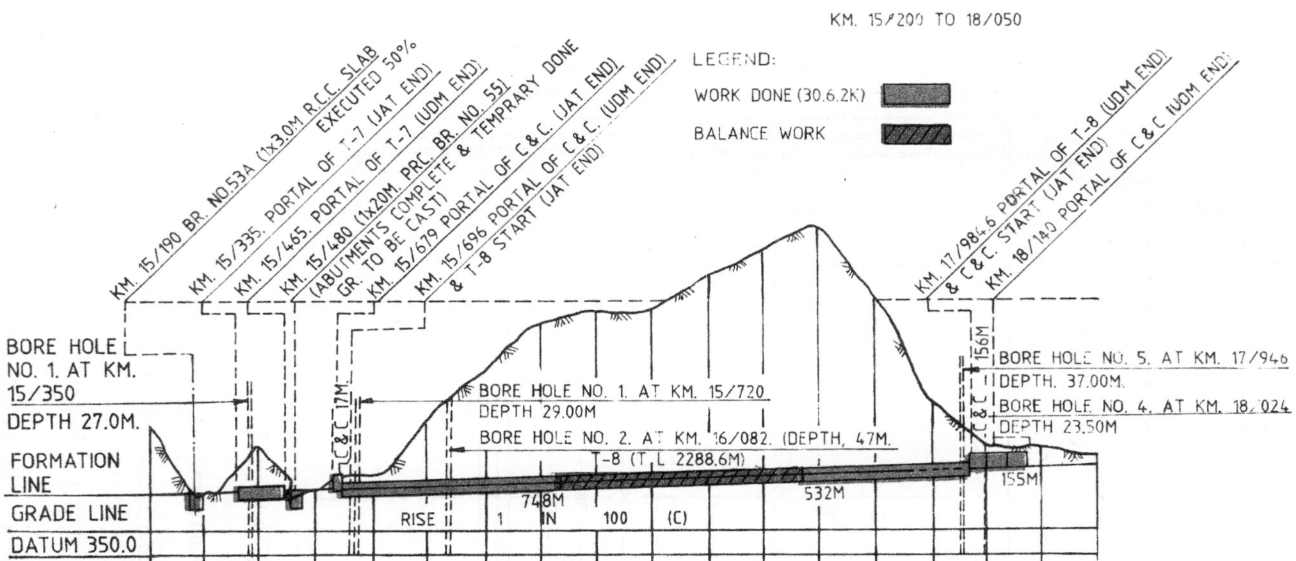

KM. 15/200 TO 18/050

LEGEND:
- WORK DONE (30.6.2K)
- BALANCE WORK

BORE HOLE NO. 5. AT KM. 17/946 — DEPTH 37.00M
BORE HOLE NO. 4. AT KM. 18.024 — DEPTH 23.50M
KM 17/98.6 PORTAL OF C&C START (JAT END)
KM 18/140 PORTAL OF C&C (UDM END)
155M   156M

BORE HOLE NO. 1. AT KM. 15/720
BORE HOLE NO. 2. AT KM. 16/082. (DEPTH 47M)
DEPTH 29.00M
T-8 (T.L 2288.6M)
532M   748M
RISE 1 IN 100 (C)
KM 15/696 PORTAL OF C&C (JAT END)
KM 15/679 PORTAL OF C&C (UDM END)
C&C T-8 COMPLETE & TEMPRARY DONE
T-8 START (JAT END)

BORE HOLE NO. 1. AT KM. 15/350 — DEPTH 27.0M
KM 15/790 BR. NO.53A (1x30M R.C.C. SLAB EXECUTED 50%)
KM 15/235 PORTAL OF T-7 (JAT END)
KM 15/465 PORTAL OF T-7 (UDM END)
KM 15/480 (1x20M) PRC BR. NO.55) ABUTMENTS COMPLETE & TEMPORARY GR TO BE CAST)
RISE 1 IN 748M
C&C 17M

	15/000	200	400	600	800	16/000	200	400	600	800	17/000	200	400	600	800	18/000	200	400
CUTTING & FILLING	68.277	4.559	44.794	1.508	66.130	90.405	99.926	143.268	150.181	148.20+	184.659	202.233	243.842	161.561	65.770	19.219	16.233	10.118
FORMATION LEVELS	407.846	409.846	411.759	413.759	415.759	417.759	419.759	421.759	423.759	425.759	427.759	429.759	431.759	433.759	435.759	437.759	439.586	441.452
GROUND LEVELS	476.123	405.287	456.553	415.267	481.889	508.164	519.685	565.027	573.940	573.963	612.418	631.992	675.601	595.320	501.529	456.978	455.809	451.570
DISTANCE IN KMS.	15/000	200	400	600	800	16/000	200	400	600	800	17/000	200	400	600	800	18/000	200	400

FORMATION LINE
GRADE LINE
DATUM 350.0

Fig. 2 : Northern railway Jammu-Udhampur rail link project L-section of tunnel no. 7 & 8 and connected works

243

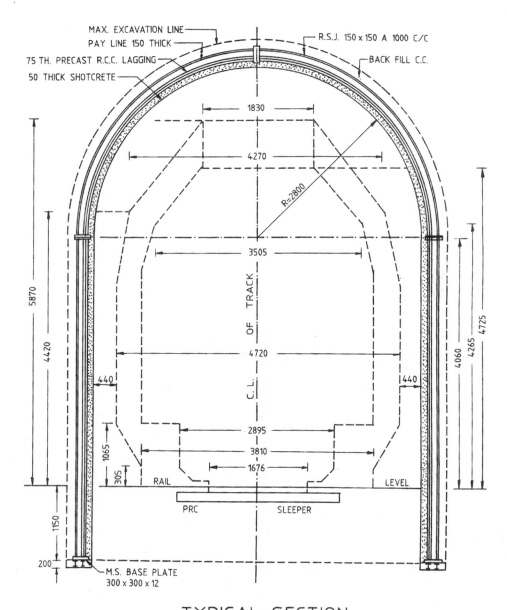

TYPICAL SECTION

Fig. 3 : Northern railway Jammu-Udhampur rail link (B.G.) typical section of tunnel no. 8

As a cumulative effect of above factors agency B also failed to give desired progress and the work was terminated in May, 1997 when progress was only 31 per cent.

Subsequently an indepth study was done examining various reasons for failures and for finding out ways to increase the pace of construction of tunnel.

It was considered to change the tunnel support system to reduce the time of construction as well as for better economy. Arguably tunnel supports using RSJ, RCC lagging, with back filling is an obsolete practice now-a-days, not followed in most part of the world. Consultations with various expert organizations such as CMRI, NIRM and WAPCOS were held and option for going in for NATM based support system with rock bolts and SFRS (steel fibre reinforced shotcrete) was considered. It is noted that the tunnel support being followed for this tunnel was designed by M/s WAPCOS based on Terzaghi's method. M/s WAPCOS however had also proposed tunnel support with rock bolts based on other methods *i.e.*, Bieniaswki, rock support interaction analysis and finite element analysis (detail given at Annexure 1). Tentative design with NATM was got done afresh from CMRI. This had the following distinct advantages :

(i)   Rapid and regulated progress

(ii)   Smaller tunnel profile

(iii)   Less cycle time

(iv)   More economical

(v)   Improved crack resistance

This option was however later on dropped keeping in view that no railway tunnel has been constructed in Himalayas with such type of support system and hence it was considered to go ahead with conventional rock support system, but with a representative section of 50 m only with NATM duly instrumented for creating data for further decision making for constructing other tunnels with NATM.

The work was awarded to the agency 'C' in February, 1999 and the work is in progress.

4.0 PROBLEMS AND SOLUTIONS

Tunnelling is a highly specialized work involving large number of uncertainties such as strata to be encountered, faults, presence of water/cofined acquifers and entrapped gases etc. Efforts are made to assess these factors. Yet, inspite of this the planning goes haywire and mishaps do take place involving loss of manpower, machinery and time. Thus it can be well appreciated that tunnelling is a hazardous activity faught with grave risk and uncertainties.

The single yard stick for evaluating the progress of any tunnelling work is the cycle time *i.e.*, time taken to complete one cycle of operations to get a desired progress. Table given at Annexure 2 shows the cycle time achieved before and after taking corrective measures. It is noted that the cycle time now is approx. 32 hrs. for 2 m progress as against approx. 72 hrs. taken earlier for the same progress. Thus the monthly progress is 85 m to 90 m now as against only 35 to 40 m earlier. This has been possible only after dedicated in depth study of problems and having practical solutions thereon.

A brief of important problems met with and solutions are given below:

4.1 Approachability

This has been a significant factor, not taken seriously by contractors and hence resulting in their failure. Figure 4 (a) and (b) showing various approaches. Initially when the work was awarded to agency A, the tunnel No. 6 was not through and hence there was no access to T-7 (which is part of T-8 contract) and T-8 (JAT end) through tunnel no. 6. The contractor with great difficulty made a long detour of approx. 4 km with steep grades and yet could only carry maximum load of 2.5MT. At the tunnel face of T-8 (Jammu end), there was no working space available as there was a flowing nallah adjoining. It was impossible for the agency to work in rainy season and in fact once some of the equipments were even washed away. Similarly, on Srinagar end of tunnel 8 the contractor constructed a katcha approach road. A cutting of approx. 250 m long and 12 to 15 m deep was required to reach the tunnel face. As there was a nallah adjoining, the water used to get accumulated in excavated parts and contractor was engaged in only bailing out water. Thus the agency spent most of the time in gaining an access to work site. It constructed actual tunnel of approx. 15 m only before the contract was terminated after a lapse of 5 years.

The work was then entrusted to agency B in 1994. By this time the accessibility from Jammu side to Jammu end of tunnel no. 7 was available through tunnel no. 6 by constructing an approach along toe of hills upto tunnel face. A diversion was made by cutting a hillock to prevent water of the nallah coming towards T-6 (Srinagar end) and the space in between T-6 and T-7 was filled up. Also an access was made along the nallah to Jammu end of T-8, abandoning earlier approach road constructed by agency A. This agency also failed miserably as during monsoons, the nallah was flooded and the work was stopped. It did not construct the vital link *i.e.*, bridge no. 55 to have direct accessibility to T-8 through T-7.

245

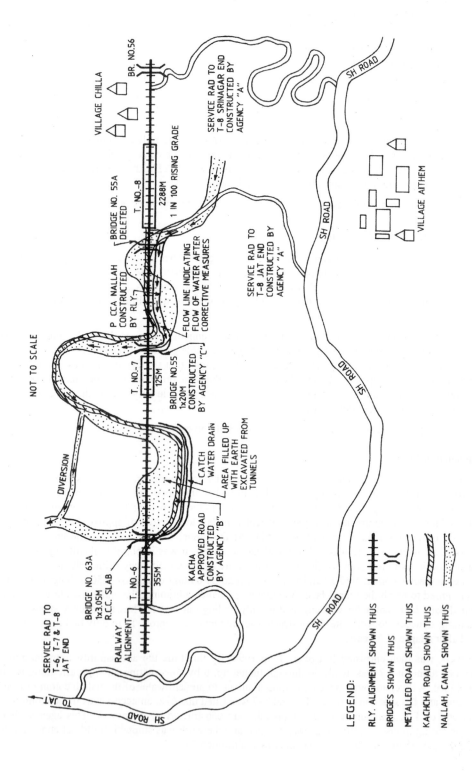

Fig. 4a : Sketch showing approaches to tunnel portals T-6, 7 and 8

Fig. 4b : Approaches Ports to T-7, T-8

On Srinagar end of T-8 a diversion was made by agency B and hence could divert the water of the nallah, however, as the tunnel was in rising gradient the seepage water alongwith the rain water was collected at the tunnel face during monsoon. The contractor was thus busy in bailing out water rather than doing actual tunnelling work most of the time. The work awarded to agency B was also terminated after 3 years in May, 1997.

The problem of accessibility was then examined in detail with a scientific over view. The Jammu end of the tunnel T-8 was made an all weather access by lining and proper guiding the nallah and proposing construction of bridge no. 55 immediately after the award of the work to the new agency. After the work was awarded the bridge was constructed and at present there is uninterrupted access to T-8 Jammu end. As regards Srinagar end of T-8, sumps were made at every 100 to 150 m interval to collect seepage water and pumps were provided in series to pump out the water into a bigger sump near the tunnel portal. From the sump a pipe line was laid with negative gradient of 1 in 400 to drain out the water nearly 300 m away into a gorge. Fig. 5 showing the layout. It is thus noted that the contractor now has uninterrupted access to the face of the tunnel and there is no problem of water accumulation.

It is further mentioned that as the tunnel is on rising gradient all along and that a period of more than 1.5 years elapsed before fixing agency C, the tunnel was totally flooded with rain/seepage water when agency C, reached at site. The depth of the water at the portal was nearly 4 m and at the face it was right upto the crown level. The new agency had to pump out this water which was made much easier by the already constructed underground pipe line laid in negative gradient (Fig. 5).

4.2 Drilling and Blasting

4.2.1 *Drilling*

The drilling of holes was done earlier with conventional jack hammers. This was a time consuming activity as there was a lot of dust and it was difficult to have easy access at the higher levels. To mitigate this problem it was considered prudent to use mechanised drilling. A conscious decision was taken to use drilling jumbos. Drilling jumbos with two booms drastically reduced the time for drilling operations. As there was no earlier experience of railway engineers at site some problem with drilling jumbos were also faced.

(i)   The dia of the hole made by jumbo was of 42 mm size which was further getting wider because of high pressure of water jet and very soft strata. This problem was tackled by reducing the water pressure to minimum.

(ii) The size of explosive cartridge used was 25 mm as against 42 mm dia hole. As a result there was no efficient blasting. The size of the cartridge was hence changed from 25 to 40 mm.

4.2.2 *Blasting*

The explosive used earlier was NG based. After discussions with experts and careful study of various explosives (distinct advantages of emulsion based explosive over NG based explosives, powder based explosives and slurry based explosives have been listed out by Mr. Jagdish and Mohd.

Nabiullah in their paper "storage characteristics of commercial explosives in Minetech" Volume 18 page 50 to 55 May-June 1997), it was found that this NG based explosives has distinct disadvantages over emulsion based explosive. Field tests for both NG based and emulsion based explosives were conducted and it was noticed that although emulsion based explosives were similar in strength but had distinct advantages over NG based in respect of defuming time, hazardousness, giddiness etc. After a careful study, the type of explosive was changed from NG based to emulsion based.

4.3 Ventilation and Lighting

It was noted that tunnel as well as the tunnel face was not well lighted and ventilated to give a harmonious working environment. To improve ventilation the ventilation pipes were properly sealed and blowers were provided at intermediate places. In addition, the tunnel was well lighted with tube lights after spacing of 25 to 30 m to give a friendly atmosphere and a feel of openness to the workers. This enhanced the productivity of the workers engaged in working inside the tunnel.

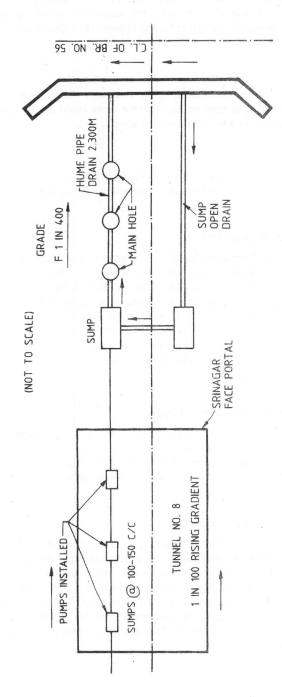

Fig. 5 : Arrangement for draining out seepage and rain water from T-8

249

4.4 Demucking

As the time of defuming was reduced drastically effort was made to speed up demucking. It was noticed that it was difficult for the tippers to turn round and take the excavated earth/rocks to the dumping place as the space within the tunnel was not adequate. As a result when a tipper was to move out, all other vehicles on way also had to go out. This problem was solved by providing larger size trolley refuges *i.e.*, 6 m wide instead of 3 m, which could work as shelters/turning places for the tippers. Thus, the time of demucking as well as usage of the machinery was greatly improved.

4.5 Erection of Ribs

It was noted that the lot of time was put in by the contractor for carrying the ribs inside and providing lagging. It was proposed to use a rib erector, however the same could not be used. A local machinery was developed to raise arch portion of the ribs and facilitate easy erection.

4.6 Backfilling

Earlier the backfilling was being done with a placer and with the concrete supplied by concrete mixers. To improve upon this batching plant was installed and a concrete pump was used, thus reducing the time of back filling.

4.7 Chimney Formation at Srinagar End

While agency B was doing drilling and blasting at Srinagar end of tunnel face, a large chimney formation 8 m high occurred and there was heavy seepage of water. It was noted that there was a small pond (baouli) at the top of the hill which got empty after formation of the chimney. agency B made arangements to tackle the chimney, however, unfortunately its contract was terminated. The chimney formation further widened with the passage of time and became deeper upto 15 to 16 m high. As there was heavy accumulation of water inside tunnel, no action could be taken immediately to plug the chimney. It was after the work was awarded to the new agency and after pumping out accumulated water and muck, that chimney formation was tackled using wider size arch ribs and forepoling technique followed by back filling and grouting.

5.0 FURTHER IMPROVEMENTS

5.1 Usage of Latest Technology for Tunnel Support

It is noted that although it was planned initially to go in for NATM based rock support for the entire length of the tunnel, however, later on it was decided to have only a representative length of 50 m and for the balance to go for conventional support system. Efforts should be made to construct other tunnels between Udhampur to Quazigund with the experience gained by instrumentation of representative tunnel using latest techniques and methodology.

5.2 Use of Road Headers for Tunnel Excavation

In the process of excavation by drilling and blasting we first disturb the rock mass equilibrium badly and then make arrangements for supporting rock mass. It is noted that in the Shivalik ranges of the Himalayas in this region the strata is primarily soft to very soft sand stone with at some places layers of shale. The strata is also water bearing and there is mild to heavy seepage (which increases during rainy season). Under such conditions it is considered prudent to have excavation with minimum disturbance to the rock mass. As such the usage of road headers for tunnel excavation for other tunnels on this alignment can be explored. BEML are making road headers upto 3.3 m dia only, however there are foreign manufacturers who make road headers for suiting railways cross-sections.

Support Requirements

S. No.	Method of Analysis	Design Parameter	Probable Support System	Recommended Support System
1.	Terzaghi	Rock load	Steel set RSJ. 180 x 150 @800/c/c	Steel set RSJ 180 x 150 @800/c/c
2.	Bleniawski	RMR-Rock class	25	25
	Ch 15.8 to Ch 16.1	31 Poor	Rock bolts spaced 1.0 to 1.5m plus wire mesh and 50mm shotcrete	Rock bolts spaced 1.0 to 1.5m plus wire mesh and 50mm shotcrete
	Ch 16.1 to Ch 17.7	52 Fair	250 3m long grouted rock bolts spaced 1m to 1.5m c/c plus 50mm shotcrete	250 3m long grouted rock bolts spaced 1m to 1.5m c/c plus 50mm shotcrete
	Ch 17.7 to end	62 Fair	same	same
3.	Rock-support interaction analysis Ch 15.8 to Ch 16.1		Steel set RSJ 150 x 150 @800 c/c	Steel set RSJ 150 x 150 @800 c/c
	Ch 16.1 to Ch 17.4	Support pressure 0.25 kg/cm^2	Rock bolt 250-6.0 m long @2.0 m c/c	Rock bolt 250-6.0 m long @2.0 m c/c
	Ch 17.4 to end	Support pressure 0.25 kg/m^2	Rock bolt 250-6 m long @2.0 c/c both ways	Rock bolt 250.6 m long @2.0 c/c both ways
4.	Finite element analysis Ch 15.8 to Ch. 16.1	Displacement	Rock bolt 250-3 m long @1.0-1.5 m c/c both way	Rock bolt 250-3 m long @1.0-1.5 m c/c both way
	Ch. 16.1 to Ch. 17.4	Displacement	Rock bolt 250-6 m long @2.0 m c/c both way	Rock bolt 250-6 m c/c both @2.0 m c/c both way

Reduction in Cycle time   (for 2m progress)

S.No.	Activity	Old Cycle (time in hrs.)	Revised cycle (time in hrs.)
1.	Marking of profile	4	2
2.	Drilling, loading and blasting	16-20	6
3.	Defuming	2-3	0.5
4.	De-mucking	12-14	6
5.	Rib erection and providing lagging	16	10
6.	Backfilling	18	8
7.	Total	$\simeq 72$	32

Monthly progress for two faces with 2-3 days for doing bed concrete with the improved system came to 80 to 90 m / per month as against 30 to 35 m / per month earlier.

TUNNELLING EXPERIENCES ON JAMMU-UDHAMPUR RAIL LINK

S.R. UJLAYAN

Chief Admn. Officer/Const.

Northern Railway

VINAY TANWAR

Dy. Chief Engineer/Constn.

Udhampur

SYNOPSIS

Jammu-Udhampur Rail Link (53.4 kms) forms a part of the prestigious and the most challenging Jammu-Udhampur-Srinagar Baramulla (340 kms) Railway Line connecting Jammu, the summer capital of the J&K State with Udhampur the district and Indian Army's Northern Commands Headquarter. Upon completion, it will be an engineering marvel and will supplement the transportation needs along with the National Highway, encourage tourism, ease army movements and thrust industrial development. The alignment traverses along the river Tawi on left bank and crosses over to the right bank at Manwal. Track traverses the domain of Shiwalik ranges of young Himalayas which is highly undulating and difficult hilly terrain. Construction of railway line envisages 85.22 lac cum of earthwork and rock cutting; 21 tunnels with total length of 10.680 kms, longest tunnel being 2.445 kms. and 158 bridges with spans up to 102 m (in prestressed concrete) and 154 m (in steel) and pier heights up to 68 m above bed. Out of 21 tunnels, 15 tunnels totaling to 8.612 kms have been completed and remaining are in progress. Indian Railways, with a rich history in tunnelling since 1889 when first tunnel commenced at Bhor Ghat, will conquer the mighty and unpredictable Himalayas on JURL with Broad Gauge line. Inadequate knowledge of strata in Himalayas makes tunnelling an extremely complex, arduous, hazardous and painfully slow work. This paper deals with the challenges faced in tunnelling, the problems encountered and the methods by which these were solved on Jammu-Udhampur new rail line project.

1.0 INTRODUCTION

1.1 Jammu-Udhampur rail link is the most difficult B.G railway project ever undertaken. The line criss-crosses numerous nallahs with highly undulating and unstable terrain of the Shiwalik Ranges of young Himalayas. The alignment of the line is as shown in Fig. 1. The strata through which tunnelling has been done and is further expected consists of :

Conglomerates	-	Compact/loose, massive/ or isolated boulders.
Clay stone	-	Blocky and seamy.
Silt stone	-	Blocky and seamy.
Sand rock	-	Soft compacted sand.
Sand stone	-	Ordinary and hard.
Boulder studded soil	-	Loose and compact
Loose mass	-	Deposited after failure of hill slope

1.2 The tunnels pass through unconsolidated or poorly consolidated sediments with rocks of upper/middle/lower Shiwalik and Muree formations. The alignment is also encountered by Jindrah thrust and the regional structure of Mastgarh anticline. In the initial stages of planning, sub soil investigations were carried out with isolated boring. However, the strata actually met with is highly varying in nature. In fact no boring is sufficient to foretell the nature as there are surprises even in few meters. The water encountered especially in sandstone, clay stone, and silt stone, makes tunnelling not only challenging, but hazardous too. The rocky strata is associated with faults, fissures and dips in different directions with weak strata in between even within one cross-section. The rock encountered is 'poor' to 'fair' with Rock Mass Rating of 32 to 62; the free standing time of 7 to 200 hours; the density of rock is in range of 2.4 gm/cc to 2.7 gm/cc and the unconfined compressive strength from 100 to 236 kg/cm². The Geological Survey of India have been associated right from the beginning. The support system in all the tunnels consists of thick web ISHB 150 x 150/200 x 150 spaced from 60 cm to 100 cm c/c. A typical cross-section of the tunnel is as shown in Fig. 2.

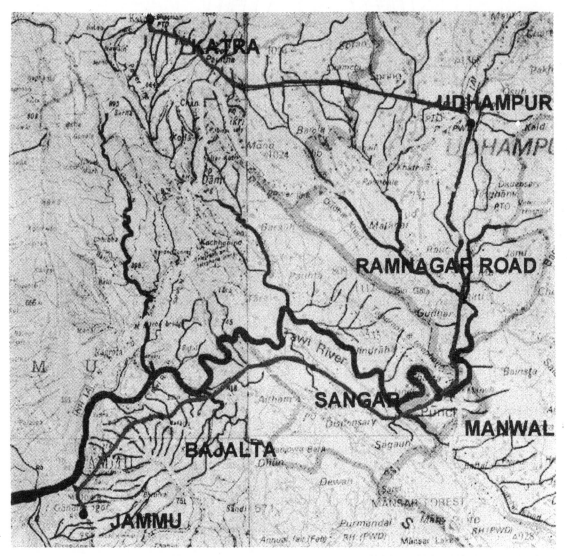

Fig. 1 : Jammu Udhampur katra rail project

2.0 METHOD OF TUNNELLING

2.1 Key factors, determining the method of tunnelling, are the size and shape of tunnels, equipments available, conditions of geological formations and extent of supports needed. Initially on JURL, the use of Tunnel Boring Machine (TBM) was considered, but this was given up after consultation with various agencies.

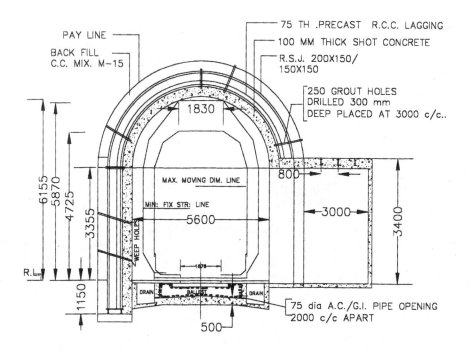

Fig. 2 : Typical section of tunnel with trolly refuge

It was advised that TBM may not succeed in such frequently and drastically changing strata with geological surprises even in every few meters. The experience of M/s. DSB, the French Contractors of Dul-Hasti Hydro-Electric Project, who brought a TBM of 8.3 m dia. with rotary cutters, manufactured by M/s. Robbins of USA, was not very promising. The progress achieved by TBM was much less as compared to the original projections. Use of TBM in tunnelling on JURL was also not preferred because of the following reasons:

 i)   High initial cost of the equipment.

 ii)   Lack of proven track record in varying strata and surprises in geological formation.

 iii)   Small lengths of tunnels.

 iv)   Long commissioning and decommissioning time of TBMs

 v)   The requirement of D shaped cross-section of tunnel for transport tunnels.

 vi)   Non-availability of indigenous technology.

Thus the conventional method of tunnelling by "drill and blast" has been used . The "full face" method *i.e.* tackling the full face at a time has generally been used except at few locations where "heading and benching" method has been adopted. The blasting pattern adopted is shown in Fig. 3. This has enabled the railways to carry out tunnelling successfully to the extent of 8.612 kms and tackle the problems as were encountered. However, hydraulic drill jumbo Tamroc 205 H Model has recently been inducted on the longest tunnel, which has reduced the drilling time considerably and improved the progress of tunnelling.

3.0 ALIGNMENT IN TUNNELS

3.1 The alignment of track inside the tunnels is straight, in curves up to 4 degree and even in 'S' curve. The track is normally in 1 in 100 rising grade which creates drainage problems on Udhampur end (higher end) face of the tunnel specially if it is followed by open deep cutting of long length. The alignment and levels are checked inside the tunnels even at every 5 m with the help of electronic distomats, 1" accuracy theodolite fitted with laser equipments and laser distomats. All permanent reference points are maintained and checked

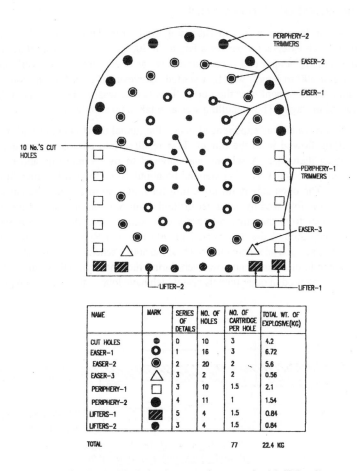

NAME	MARK	SERIES OF DETAILS	NO. OF HOLES	NO. OF CARTRIDGE PER HOLE	TOTAL WT. OF EXPLOSIVE(KG)
CUT HOLES	●	0	10	3	4.2
EASER-1	○	1	16	3	6.72
EASER-2	◉	2	20	2	5.6
EASER-3	△	3	2	2	0.56
PERIPHERY-1	□	3	10	1.5	2.1
PERIPHERY-2	●	4	11	1	1.54
LIFTERS-1	▨	5	4	1.5	0.84
LIFTERS-2	●	3	4	1.5	0.84
TOTAL			77		22.4 KG

Fig. 3 : Typical balsting pattern for tunnel with IM pull

from time to time. Some extra margins in the section of tunnel has been provided to cater for inaccuracies during working and due to poor strata. There is slight difference between centre line of track and centre line of tunnel to cater for curvature. Inspite of all possible care, there have been cases when the alignment, especially, in 'S' curve portion has been out by few centimeters, which for railway tunnels is not acceptable. It is due to the fact that the permanent reference points marked inside the tunnel get disturbed due to poor strata, muddy formation and loose falls.

4.0 SEQUENCE OF ACTIVITIES IN TUNNELLING

4.1 The sequence of activities in conventional method of tunnelling being adopted on this project areas under :

1. Surveying – alignment and level control, profile marking.
2. Drilling
3. Explosive loading and blasting
4. De-fuming – forced ventilation
5. Mucking
6. Support erection and placing of precast lagging.
7. Back fill concreting
8. Lining

5.0 DRAINAGE PROBLEM OF TUNNEL NO. 8

5.1 Tunnel No.8 is 2445 m long and is in 1 in 100 rising grade from Jammu end. Udhampur end portal is in 20 m deep rock cutting which extends further for a length of 300 m. The work was started simultaneously on both faces. The approach of Udhampur end is comparatively easier and progress of work on this face has been better. The open cutting in 300 m length was first started. However, it took time in completing the same. A nallah was also crossing the alignment near the portal. This nallah was made to cross the alignment over the cut and cover, constructed in stages. The tunnelling was also taken up simultaneously.

Heavy ponding of rain water was experienced inside the tunnel on Udhampur end in initial years as the entire run off from cutting flowed in to the tunnel. This resulted into complete stoppage of work not only during monsoon, but even thereafter. A scheme was devised to over come this and to solve the problem forever. The length of cut and cover was also extended by about 100m to guard against hill slope failures and to ease drainage problem. The open side drain in cutting for initial 80 m beyond the cut and cover face was made in reverse grade. It was followed by an under ground drain only on one side of the cutting, also having reverse grade. The rain water from the other side drain was connected to this underground drain through an RCC box culvert. The underground drain consisted of 900 mm dia RCC hume pipes with sufficient size manholes at every 40 mtrs spacing. The schematic sketch of the above scheme is shown at Fig. 4. This completely solved the drainage problem in the tunnel and now even during monsoon normal progress of tunnelling is being achieved. This scheme is being used in other tunnels also which have long cutting on the up grade side.

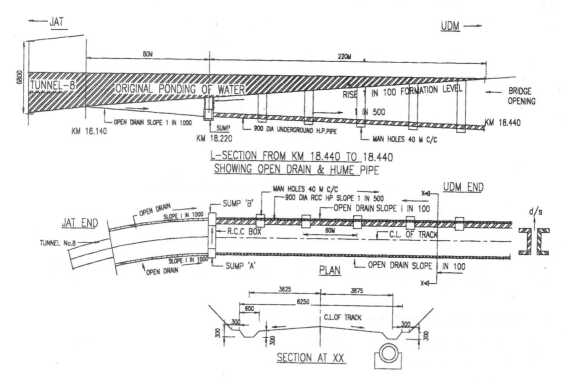

Fig. 4 : Jammu-Udhampur rail link drainage arrangement near tunnel no. 8

6.0 CHIMNEY FORMATION IN TUNNEL NO. 8

6.1 The work of tunnelling for Udhampur end progressed faster. However, a local shear zone was encountered at a distance of 314 m from Udhampur end. It consisted of plastic clay and pulverised materials.

256

During boring through this shear zone, heavy seepage took place along the shear zone bringing down the crushed materials into the tunnel, creating a chimney. This was further exacerbated due to presence of a spring nearby which got connected to the cavity through infiltration channel. This caused water flow and loose fall inside the tunnel resulting into enlargement of chimney formation. Initially the chimney was 8 mtr high and about 5 m in dia which finally enlarged to 16 mtr high and 9 m in dia. The entire tunnelling activity came to a halt and gave a severe jolt to the progress of work. Fig. 5 explains as to how this problem was tackled. The main problem was to do tunnelling through the uncompacted muck from the cavity and then to provide the adequate supports in the cavity portion as higher support pressure was expected in the collapsed zone. The work was re-started by adopting "heading and benching" method instead of "full face" method. The heading portion of the tunnel was taken up by "multi drift" method in three drifts namely left, centre and right drifts. Adequate fore-poling consisting of 25/32 mm dia tour steel rods and pressure grouting with cement grout through 4-5 m long pipes in the strata ahead was carried out to reduce seepage of water and to stabilise the strata. Excavation of drifts was taken up thereafter. Each drift was lined with the help of ISHBs (150 x 150) bent to the larger profile of the tunnel spaced at 40 cm c/c. These were supported with independent temporary vertical posts resting on the cross girder. Once the heading drivage was completed, conventional benching was carried out with precaution that all the operations of excavation, placement of vertical joists and lagging and concrete backfilling etc. were completed in a length of 40 cm in the entire cross section before the next length of the chimney was taken up. Thereafter, normal ribs to the designed profile were erected. The annular space between the two layer of ribs was filled with concrete (Fig. 6). The gap in between the rock and outer ribs was filled up with cement sand grout. Then the temporary vertical supports resting on the cross girders were cut. This provided solid concrete structure to bear the overburden load. The so tackled chimney formation is standing well with no sign of distress.

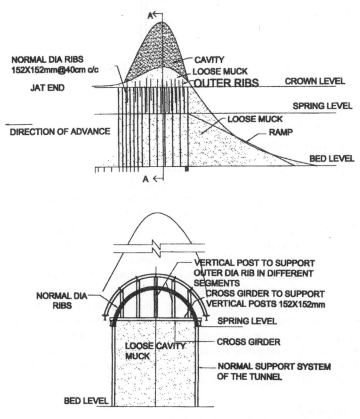

Fig. 5 : a. Longitudinal section showing cavity in tunnel no. 8
b. Erection of segmented outer dia ribs

257

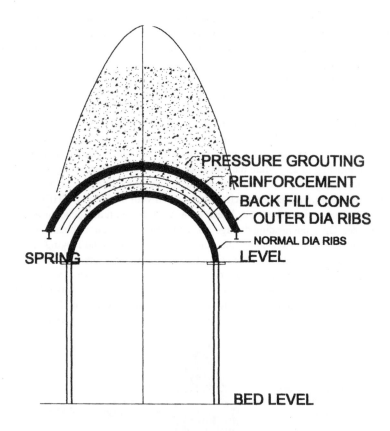

Fig. 6 : Section at AA in Fig. 5 (a)

7.0 CHANGE OF ALIGNMENT

7.1 Tunnel No. 10 –E, F is in close proximity of Dhar-Udhampur Road in some stretches. The alignment in this stretch passes across the hill with lateral ledge of about 20 mtrs only on the river side. When the work was progressing In reply to: this stretch, some movement of ground was noticed on the road side hill slope and the work had to be suspended. The detailed geo-technical investigations in this stretch were carried out by boring 5 bore holes on the alignment. The bore holes showed that the over burden in this reach of the alignment consisted of loose boulders/debris of about 10-12 mtrs thickness. This was considered unsafe to undertake tunnelling as over burden of loose boulders may not provide sufficient strength and may collapse during blasting. Luckily the hard rocky strata was sloping steeply towards river Tawi. The geological cross section across the most critical location is shown in Fig. 7. It was decided to change the alignment towards the up hill side by about 21 mtrs which ensured safe tunnelling. The strata along this revised alignment was of loose boulders/debris in the upper strata of about 10-12 mtrs thickness laid over the alternate bands of sand rock and shale. It provided at least 5 mtrs of good rock cover with further 10-12 mtrs. of boulders/debris over the crown of the tunnel. This also gave additional lateral stability to the tunnel. The tunnel was in 5^0 curve for some length. The change in alignment was adjusted with the help of a compound curve. With this change in the alignment tunnelling could be completed without much problems. The tunnel has since been made through.

8.0 SEEPAGE OF WATER

8.1 During tunnelling heavy seepage of water was encountered in some patches in tunnel No. 8 and 10-A. The hill top at these locations was surveyed. It was found that the hills have small nallahs flowing across the

258

alignment in these locations and there were some cracks in the bed of nallah. The diversion of nallahs was not feasible. It was decided to lay a thick layer of about 300mm thick mass concrete well vibrated in about 50 m length, 25m on either side of the tunnel alignment. Another source of water seepage was infilteration water which was controlled by contact and pressure grouting. This resulted in reducing the seepage considerably and enabled boring of tunnel with least problems.

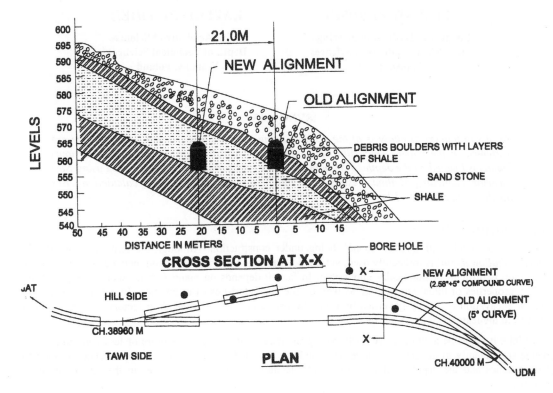

Fig. 7 : Geological cross-section across the most critical location

9.0 CONCLUSION

9.1 The young Himalayas have highly varying geological formations. There are drastic changes in geological formations even in few meters. No extensive geological investigations with conventional boring are sufficient. Geo physical methods using radar are also not fully reliable. Even otherwise, taking radar based equipments along the hill top, poses its own problem. Therefore, one has to be prepared to face surprises and problems while doing the tunnelling in Himalayas. Adequate resources have to be available at site to tackle such eventuality/problems through un-conventional methods. Any delay in tackling the problems as and when they arise, complicates the solutions and makes them costly. The observational investigations pay rich dividends especially in Young Himalayas. A close watch is essential to know even slight movements both inside the tunnel and in the close vicinity on the hill top which pre-warns the Engineer to take remedial measures.

10.0 ACKNOWLEDGEMENTS

10.1 The authors express their gratitude to Dr. J.L. Jethwa, Director, CMRI for his help in tackling the cavity in tunnel No.8 and Sh. Malbarna, Senior Geologist of GSI who lent his extensive support to make tunnelling a safe activity . The authors also feel indebted to the numerous workers who have worked hard and completed tunnelling in this difficult terrain and conditions.

INFLUENCE OF TUNNEL DEPTH ON ITS BEHAVIOUR DURING CONSTRUCTION

LESLAW ZABUSKI

Institute of Hydro-Engineering of
the Polish Academy of Sciences
Gdansk, Poland

KAZIMIERZ THIEL

Polish Academy of Sciences
Deptt. of Technical Sciences
Warsaw, Poland

SYNOPSIS

Results of numerical analysis of the shallow tunnel stability using Itasca's UDEC code are presented in the paper, which illustrate influence of construction depth on its behaviour. Tunnel is located on the depth 15-55 m in the flysch rock formation. It is excavated using New Austrian Tunnelling Method. Results show significant differences in three calculation cases (depth 15 m, 30 m and 50 m).

1.0 INTRODUCTION

The paper deals with shallow tunnel, being under construction. Although the term *shallow tunnel* (or *tunnel at shallow depth*) is frequently used, yet the determination of transition depth between shallow and deep excavation is difficult or practically impossible. In fact, it depends on many factors, such as, e.g. rock mass quality, properties of initial stress field, as well on the characteristics of excavation. In general, behaviour of shallow excavation is mainly governed by rock mass structure, whereas deciding factors in deep excavation result from the features of the stress field.

While it could be assumed, that rock mass properties at great depth are more or less constant, they can change significantly in the shallow range. If tunnel is excavated in this range, the dependence between rock characteristics and the depth should be taken into account, to avoid inaccuracies in the results of stability analysis.

The aim of the paper is to prove the influence of the depth on the stress distribution, deformations, failure phenomena in and around the tunnel and, in effect, on its stability. Thus, shallow zone - between 15 and 55 m - is divided into "sub-zones" with different geomechanical characteristics. Tunnels at three different depths are analysed, allowing for demonstration of the excavation behaviour depending on the depth.

2.0 SITE AND TUNNEL CHARACTERISATION

Water reservoir is realised in Carpathy Mountains in Poland. It is created by construction of earth dam. Two hydro-technical tunnels located at the depth up to 55 m, each of 6.5 m internal diameter are excavated inside the slope of the river valley. The tunnels service as a facilities for water intake and outflow in construction as well as in exploitation stage of dam and reservoir.

Rock mass is built of sedimentary quasi-homogeneous flysch formation. It is composed of strong and hard sandstone beds, with weak and soft clay shale interbeddings. The average content of sandstone is equal to ca 65%. Rock mass is intersected with two systems of joints, approximately perpendicular to bedding planes. Quaternary cover is 5-10 m thick. Effects of mass weathering depend on the depth. Zones in the nearest vicinity of terrain surface are intensively weathered, whereas the influence of weathering at the depth greater than 40-45 m is of minor importance. This "gradient of weathering" is taken into account in the analysis and is considered in the next chapter.

Tunnels were excavated according to the general principles of New Austrian Tunnelling Method (Müller, 1978). Construction procedure was divided into two phases; preliminary supported excavations are constructed in the first of them and permanent lining built of reinforced concrete was installed in the second. The first phase was divided into two stages – upper half (kalota) was executed on the whole length of the tunnel and lower part was excavated later. Preliminary support was composed of rockbolts, shotcrete and steel ribs (Fig. 1). Excavation works were accompanied by extensive monitoring of deformations. Roof settlement, convergence and delamination of the rock surrounding the tunnel were measured in particular.

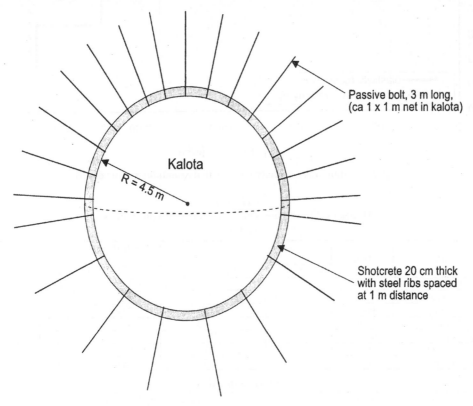

Fig. 1 : Tunnel excavation stages and elements of support

3.0 MECHANICAL AND NUMERICAL MODEL

Numerical analysis was carried out using UDEC code (Itasca 1991). Calculation domain was divided into fully deformable blocks.

As it was mentioned in the former chapter, rock mass properties depend on the weathering intensity and thus on the depth. Rock mass is strongly deteriorated and de-compressed in the near-surface layer, immediately under Quaternary cover. Random fissures here are very dense. Quality of rock increases with depth and density of discontinuities diminishes. Series of seismic measurements were carried out to determine the dependence between seismic wave velocity and depth. Next, assuming similarity between this relation and the relations between mechanical properties and depth, the later were established. It should be mentioned, that as a "calibrating" values served results from static in situ and laboratory tests. Example relationship, demonstrating the influence of the depth on the seismic wave velocity and – in consequence – on the elasticity modulus of the rock mass is shown in Fig. 2. Other parameters change in the similar way. The parameters used in calculations are assembled in Table 1. All parameters were determined on the base of laboratory and field tests.

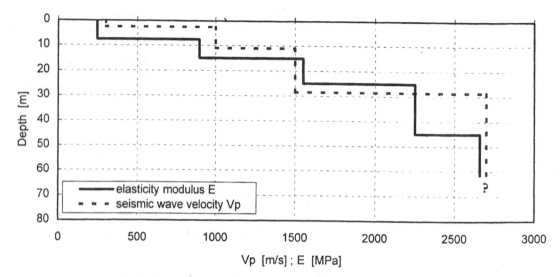

Fig. 2 : Seismic wave velocity and elasticity modulus versus depth

TABLE 1
Mechanical parameters of the rock and discontinuities
a) Rock

Parameter	Unit	Layer W(*), Depth H [m]				
		W.1: 0-7.5	W.2: 7.5-15	W.3: 15-25	W.4: 25 -45	W.5: > 45
Elasticity modulus E	MPa	243	896	1550	2250	2660
Poisson's coeff. ν		0.33	0.33	0.33	0.33	0.33
Unit weight γ	kN/m³	22	22	22	22	22
Cohesion c	kPa	15	55	95	135	175
Angle of friction ϕ	Degree	22.0	24.8	27.4	30.0	33.5
Tension strength σ_t	Kpa	7.5	27.5	42.5	67.5	87.5
Dilatation angle ψ**	Degree	11.4	13.0	14.6	16.1	18.35

b) Interbedding fissures

Parameter	Unit	Layer W, Depth H [m]				
		W.1: 0-7.5	W.2: 7.5-15	W.3: 15-25	W.4: 25 -45	W.5: > 45
Normal stiffness k_{nf}	MPa/m	110	410	700	1000	1290
Shear stiffness k_{sf}	MPa/m	110	410	700	1000	1290
Cohesion c_f	kPa	10.5	12.0	13.5	15.0	17.5
Angle of friction ϕ_f	degree	9.9	11.3	12.7	14.0	15.4
Tension strength σ_{tf}	kPa	0.0	0.0	0.0	0.0	0.0
Dilatation angle ψ_f	degree	5.0	5.7	6.4	7.1	7.8

c) Joints perpendicular to rock layers

Parameter	Unit	Layer W, Depth H [m]				
		W.1: 0-7.5	W.2: 7.5-15	W.3: 15-25	W.4: 25 -45	W.5: > 45
Normal stiffness k_{ni}	MPa/m	231	861	1470	2100	2720
Shear stiffness k_{si}	MPa/m	231	861	1470	2100	2720
Cohesion c_i	kPa	0	5	10	15	20
Angle of friction ϕ_i	degree	22.0	24.8	27.4	30.0	32.4
Tension strength σ_{ti}	kPa	0.0	0.0	0.0	0.0	0.0
Dilatation angle ψ_i	degree	11.4	13.0	14.6	16.1	17.6

(*) layers are shown in Fig. 3a; (**) values of dilatation angles are assumed

TABLE 2
Tunnel cases in calculations

Tunnel	Range X [m]	Range Z [m]	Depth of tunnel roof H [m]	Rock mass layer (Table 1 and Fig. 3a)
(A)	-35 - +35	0 - 40	15.5	3
(B)	-50 - +50	0 - 55	30.5	4
(C)	-60 - +60	0 - 75	50.5	5

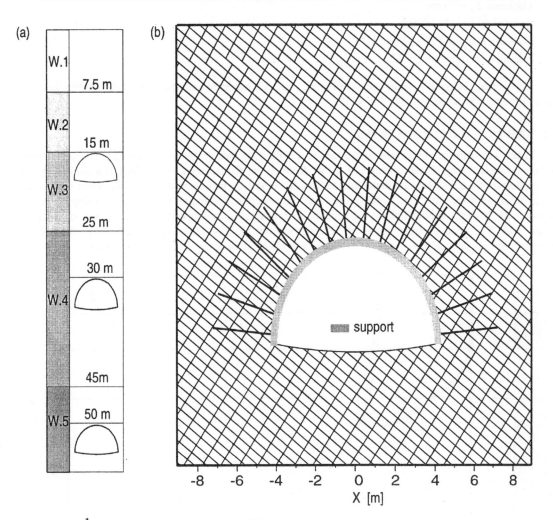

*Fig. 3 : Geometry of calculation domain; (a) rock mass layers and tunnel depth;
(b) rock blocks and excavation with support elements

    Rock mass in numerical model is modelled as a system of rock blocks. Spacing of interbedding fissures and joints is equal to 0.375 m and 0.75 respectively. Inclination of rock layers is equal to 35° with respect to positive direction of X axis. Rock and discontinuities in numerical model reveal elasto-plastic behaviour.

Characteristics of rockbolts (Fig.1)

Length of bolt L_k = 3.0 m

Diameter of bolt d = 22 mm

Elasticity modulus (steel) E_{st} = 210 000 MPa

Unit weight (steel) γ_{st} = 78 kN/m^3

Max. tension force P_r = 118 kN

Cohesion of contact between rock and bolt JCOH = 55 kN/m

Shear stiffness of contact between rock and bolt JKS = 10 000 MN/m/m

Shotcrete + steel ribs

Thickness d_{bn} = 20 cm

Unit weight γ_{bn} = 25 kN/m^3

Elasticity modulus E_{bn} = 23 000 MPa

Poisson's coefficient ν_{bn} = 0.2

Ideal contact between support and surrounding rock is considered and thus very high cohesion between these elements is assumed, equal to 1 000 GPa.

4.0 RESULTS OF CALCULATIONS

Figure 4 presents curves of vertical displacement in the zone between the tunnel (at three different depths) and terrain surface. Significant influence of tunnel depth on the displacement is observed. It is surprising, that the smallest displacement appears in the intermediate (B) tunnel. In fact, these results qualitatively confirm the results from measurements. Fig. 5 shows calculated settlement of the tunnel roof and – as comparison - its measured settlement. Although these settlements do not agree quantitatively, yet qualitative similarity is clearly seen. The reasons of quantitative differences can not be univocally explained. They probably come from both simplifications in mechanical model and technological inaccuracies of the construction.

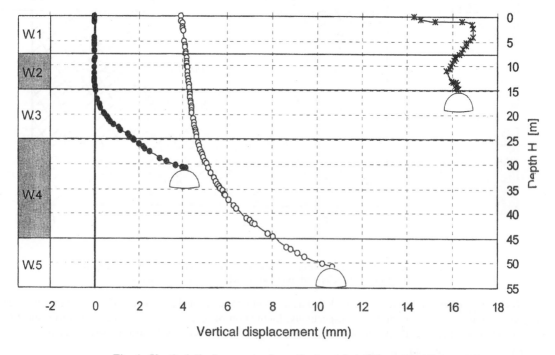

Fig. 4 : Vertical displacements above the tunnel at different depth

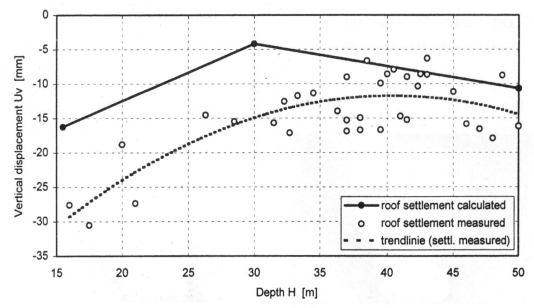

Fig. 5 : Measured and calculated settlement of tunnel roof versus tunnel depth

The explanation of smallest displacements in tunnel B seems to be obvious. Due to small thickness of layer above the shallow (A) tunnel, "closed" form of pressure arch is not created. The zone of decompressed and loosened rock mass reaches terrain surface. Therefore, all points between roof and terrain surface settle more or less identically. The settlement U_v of the roof and terrain is equal to 16.25 mm and 14.3 mm respectively. As it explains well known concept of *ground reaction curve* (*e.g.* Hoek, Brown, 1996), there exist two "kinds" of pressures acting on the support during excavation process, *i.e. deformation* and *static* pressure. The level of deformation pressure, which is connected with initial stress state is lowest in the most shallow tunnel (A), highest in the rock surrounding deepest tunnel (C) and intermediate in tunnel B. Three cases could be summarised as follows :

Tunnel A: low level of deformation pressure, high static pressure due to "connection" of pressure arch with terrain surface.

Tunnel B: intermediate level of deformation pressure, relatively low level of static pressure (pressure arch has closed form and is not connected with the terrain surface).

Tunnel C: highest level of deformation pressure, static pressure almost the same as in B case.

This summary allows to explain small settlement in B tunnel – above two kinds of pressures are relatively (*i.e.* in comparison to the cases A and C) low.

The difference of displacement fields is clearly shown in Fig. 6. Lines of equal vertical displacement in the A tunnel reach surface, whereas closed shapes of lines are visible in B case (displacement field in the case C is similar as in B). Shear bands in A tunnel are created along oblique lines from its invert to the terrain. Wedge limited by these lines, tunnel and terrain surface is composed of strongly loosened rock, which exerts static pressure on the tunnel support.

Fig. 7 shows the curves of equal vertical normal stress σ_y calculated for A and C tunnel. It illustrates differences in the zone between tunnel and terrain surface. Significant decompression and diminishing of the stresses occurs above shallow tunnel, whereas decompression and influence of excavation in deeper tunnel disappears below the terrain surface.

LEFT - TUNNEL A (roof depth 15 m)

RIGHT - TUNNEL B (roof depth 30 m)

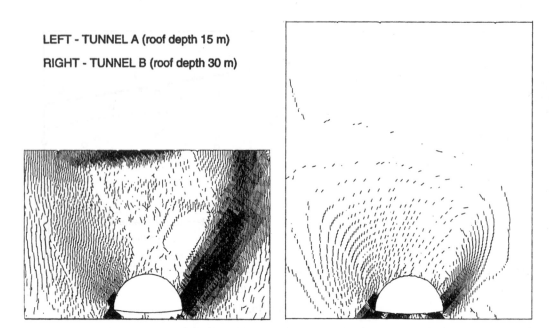

Fig. 6 : Lines of equal vertical displacements in the zone between tunnel roof and terrain surface

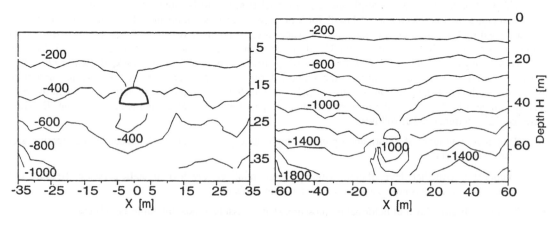

Fig. 7 : Lines of equal σ_y stresses – tunnel A (left), tunnel B (right)

Analysis of displacement field proves differences in considered tunnel cases. Relatively small displacements in intermediate tunnel (B) are the result of specific stress field. Figs. 8 and 9 present vertical and horizontal stress along the line between tunnel roof and terrain surface.

It can be seen, that vertical stresses in A and B cases are almost the same, whereas distinct difference in horizontal stress is obseived. It means, that arching phenomenon in B case – due to better rock mass quality - is more efficient and gives natural supporting effect. Average value of axial tension force carried by rockbolts in B case is therefore equal to only 75% of this force in case A. Although axial forces in shotcrete shell in B case ($N_{max} = 1157$ kN) are greater in comparison to A ($N_{max} = 803$ kN), yet their magnitude can be considered as low, when relation of tunnel depths in these to cases is taken into account. Differences of the stress level between B and C cases result on one side from the increments of initial stress and on other - from the improvement of rock mass quality with depth. Greater values of displacements and internal forces in the

266

support elements in C case result first of all from the elasto-plastic decompression of rock – not from increments of static loading.

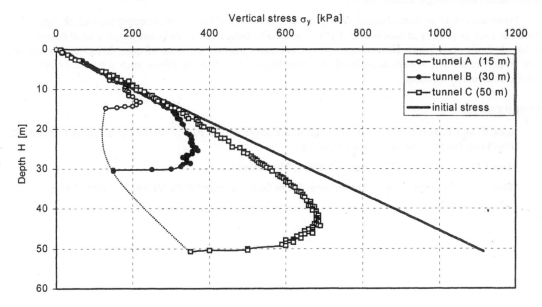

Fig. 8 : Vertical stress σ_y between tunnel roof and terrain surface versus depth

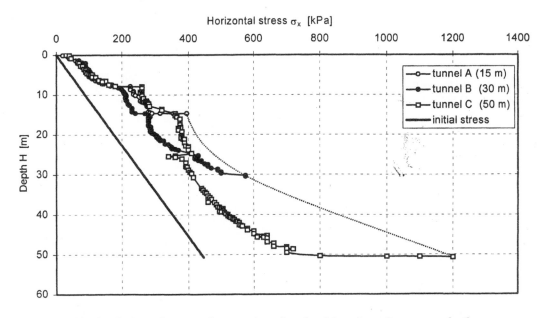

Fig. 9 : Horizontal stress σ_y between tunnel roof and terrain surface versus depth

5.0 CONCLUSIONS

Results of numerical calculations show complex stress and displacement fields in rock mass and around the tunnel, depending on the depth of excavation. There are two phenomena influencing stress/deformation distribution, *i.e.* increment of initial stress level and improvement of rock and joint properties with depth.

267

Although depth increment is the reason of augmentation of stress level and in consequence, pressure acting on the support, yet in the same time improvement of the properties of the medium are the main reason of smaller deformations and stronger arching effect.

Presented results and conclusions clearly indicate necessity of taking into account existing changes in the profile of rock properties in function of depth. It is usually done in cases of heterogeneous rock masses, built of few different rock kinds. However, as proves common practice, in cases of quasi-homogeneous masses, rock properties are not always considered as depending on the depth. In effect, such a practice could cause significant errors in solutions.

REFERENCES

Hoek E. and Brown E.T. (1996), Underground Excavations in Rock, E&FN Spon, London-Weinheim-New York-Tokyo-Melbourne-Madras, 527 p.

Itasca CG. Inc. (1991), UDEC User's Manual, Minneapolis.

Müller L. (1978), Der Felsbau, Dritterband: Tunnelbau, Ferdinand Enke Verlag, Stuttgart, 945 p.

SHOTCRETING AND MICRO TUNNELLING

USE OF POLYURETHANE GROUTS IN ROMERIKSPORTEN, NORWAY — A CASE STUDY

HELEN ANDERSSON   P. BORCHARDT
Geoteknisk Spets-Teknik AB, Gullbergs Strandgata 36 D, S-411 04 Göteborg, Sweden

SYNOPSIS

During 1998, a major post-grouting operation was performed in the railway tunnel Romerik sporten along Gardermobanen to the new international airport near Oslo. Polyurethane was used in combination with cement grout in order to meet the leakage demands needed to restore the groundwater level. A method called combination grouting makes it possible to adjust the cement grouting to the conditions in the rock mass by combining cement suspension and water reactive polyurethane in the same borehole. The positive sealing results were in good agreement with earlier experience of this grouting method. This paper also discusses the difficulties encountered when the cement suspension is flushed out of the fractures due to erosion of slow setting cement.

1.0 INTRODUCTION

1.1 Project Orientation

The tunnel Romeriksporten along the first high-speed railway in Norway, to the new international airport some 60 km from Oslo, is 14.5 km long (13.8 km in rock) and has an average cross-sectional area of 110 m^2. Fig. 1 shows the typical cross-section and the cross-section with a watertight concrete lining that was used in geologically difficult zones.

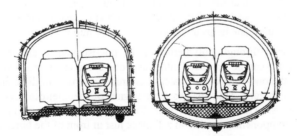

Fig. 1 : The railway tannel Romeriksporten : Typical cross-section with drained concrete lining (left) and cross-section with watertight concrete lining (right)

The rock mass in the Romeriksporten tunnel consists predominantly of gneiss of different types, but the tunnel also traverses the major fault between Oslofelter and the crystalline basement rock type in the stretch between Etterstad and Bryan. The rock cover varies between 20-250 m. with the minimum at the fault close to Bryn.

For the most part of the tunnel, conventional pre-grouting with cement suspension worked rather well. But the tunnel also traverses weak zones (e.g., Puttjern and Lutvann) with a large number of water-bearing fractures and pre-grouting with cement suspension in these zones was not adequate. It was also attempted to use the acrylamide grout Rhocagil to seal the large water leakages.

As a result of severe depletion of the groundwater level in the area, a small lake in the vicinity of the tunnel was found to be almost emptied of water in February 1997. Strict demands were then placed on the water leakage into the tunnel and a major post-grouting operation was performed during 1998.

1.2 Technical Background

During a period of around two years, approximately 365 tons of the acrylamide grout Rhocagil was used for pre-grouting at Romeriksporten. This comparatively limited use of Rhocagil was stopped along with all other chemical grouts, in connection with the pollution problems due to the extensive use of the acrylamide grout at the Hallandsas tunnel in the autumn of 1997. After that, the post-grouting was performed with accelerated cement grouts, instead of with chemical grouts. The sealing results were however far from satisfactory and chemical grouting was considered again. Very high demands were put on the materials to be used in the future work.

After a thorough examination of the influence on health and environment of the polyurethane product TACSS, two trial grouting rounds were performed in January 1998. During the trial post-grouting, possible pollution from the polyurethane grout (both to air and to water) was followed up. Since the pollution was well within the set health and environmental limits, use of this product was allowed for further post-grouting in order to meet the leakage demands needed to restore the groundwater level. By use of a combination of cement grout and polyurethane, the post-grouting performed between February and June 1998 reduced the leakage to 80% of that required.

In the following, the combination grouting with cement and polyurethane is presented, as well as polyurethane grouting in general. The paper concludes with the application of these methods for the post-grouting at Romeriksporten and discusses the difficulties encountered when the cement suspension is flushed out of the fractures, due to erosion of the slow setting cement, especially when the groundwater pressure around the tunnel is high.

2.0 POLYURETHANE GROUTING

2.1 Properties of Polyurethane Grouts

Polyurethane grout has been used for grouting of rock, as well as concrete and soils, for several decades. Mostly, polyurethane grouting has proven successful, when other grouts have failed. The most favourable property of polyurethane grouts is that they react with water and expand due to production of carbon dioxide gas. The expansion facilitates for sealing of fractures with flowing water, and the CO_2 gas pressure caught behind plugs of hardened polyurethane helps the grout to penetrate into very narrow fractures.

Polyurethane is produced through polymerisation of polyisocyanate and polyol, e.g., polyethers or polyesters. Polyurethane grouts can be produced either as a one-component or a two-component material. The main difference being that the one-component material is prepolymerised to different extent in the controlled environment of the polymer factory. This enables the designer of the chemical system to direct the order of build-up in the polymeric structure. The polymer formation for a two-component is carried out by simultaneously mixing the components together in a T-piece just in front of the borehole.

Prepolymers lowers the toxicity of the isocyanate since the reaction is partially completed, and other advantages that can be mentioned are : (i) less heat is generated during foaming, which slows the reaction time down thus enabling the grout to reach further into the rock mass, and (ii) a more efficient reaction is obtained. Karol (1990) claimed that the use of a prepolymer is to be preferred for grouting purposes, since simultaneous mixing of the ingredients for a two-component polyurethane requires closely controlled conditions, which is not very often found in rock tunnels under excavation.

2.2 Health and Environmental Aspects

Not even in the first phase of the post-grouting works at Romeriksporten, use of only microcement suspension gave a satisfactory sealing result, despite of the use of different additives to improve the properties of the cement. Therefore, the builder "Gardermobanen AS" applied with the controlling authority "Stations

Forurensingstilsyn (SFT)" for a trial grouting operation with the water reactive polyurethane TACSS. The purpose of the trial grouting was to investigate the environmental consequences from grouting with TACSS and also to study the sealing effects. In the beginning of 1998, two trial grouting rounds were granted after a thorough investigation of the polyurethane product. The trial grouting showed that the pollution was well within the set health and environmental hazard limits.

The thorough investigation of the polyurethane product TACSS (Aquateam, 1998), before using it for the future post-grouting works, regarded its health and environmental impacts. It comprised a comprehensive survey of the different chemical substances in the grout (DBP, MDI and HDMA) and an evaluation of the possible environmental hazards, as well as the risks associated with the handling of the product. Based on an assumption of the amount of polyurethane grout that would be used for the sealing works, calculations of the pollution to the recipients (*i.e.*, the river Alna and eventually Oslofjorden) were performed. Maximum contaminated discharge water from the tunnel and minimum water flow in the river are important parameters in such a calculation.

The evaluation of the environmental hazard focused on the phtalate DPB, since the risk for use of the isocyanate MDI was regarded very small. The examination concluded that it was acceptable, with regard to the environment, to use TACSS for post-grouting. However, it was recommended that a measurement program be set up for supervision of the amount of water leaving the tunnel, as well as the amount of DPB in the water. Analyses of the water (NIVA, 1998) showed that even with the conservative value of Predicted No Effect Concentration (PNEC) of 2.6 µg/1 for DBP, the area of influence will virtually never extend beyond 400 m from the discharge point of the river Alna into Oslofjorden. The distance of 400 m is recommended as an outer limit for the acceptable area of influence.

Based on the Norwegian standards for contamination in the working environment, the exposure of MDI, MDA and DBP was calculated for different situations : (i) when mixing TACSS, (ii) when transporting TACSS, (iii) when injecting TACSS, (iv) caused by leakage of TACSS components during grouting, and (v) when cleaning the equipment used for mixing and grouting. The analyses did not indicate that the work force involved in the grouting operations with TACSS polyurethane at Romeriksporten would be subjected to unacceptable health risks. Nevertheless, the Aquateam report recommended that the exposure be reduced as much as possible and that measurements of both air and water be performed regularly. Finally, the work force should be educated in the handling of TACSS polyurethane and relevant protection should be used with regard to skin contact and inhalation.

2.3 Application of Polyurethane Grouts

Even though the most widely used grout for sealing of rock masses is cement suspension, there are certain conditions where chemical grouts are used as a complement. One such condition is when the conducted pre-grouting is insufficient and grouting after blasting has to be performed in order to seal the remaining leakage. The main difference between pre-and post-grouting is the grouting pressure that can be used. In post-grouting, which is performed closer to the tunnel wall, the grouting pressure has to be held rather low so that the grout material is not flushed back into the tunnel before hardening. Thus, it is helpful to use a material, such as polyurethane, for which the grout penetration can be controlled by a set gel time.

This property is also favourable when sealing large water-bearing rock joints, in which the flowing water both dilutes and transports the cement grout away. Another difficult problem that is often faced in cement grouting, is when the fractures are too narrow for the cement particles to penetrate. In the described situations, the solution of the problem is grouting with water reactive polyurethane. Polyurethane grouts can be applied for sealing and stabilisation of dams and tunnels, for example. Moreover, the polyurethane material can also be used in combination with the cement suspension, and this application is simply called 'combination grouting'.

3.0 COMBINATION GROUTING

3.1 Description of the Combination Method

The so-called combination method makes it possible to adjust the cement grouting to the conditions in the rock mass by combining cement suspension and water reactive polyurethane in the same borehole. The polyurethane grout is transported with the cement grout to the most permeable area in the rock, where the reaction starts after a certain, set time. When reacting polyurethane creates a front against the tunnel surface, the inflow of cement suspension into the tunnel is stopped. However, the polyurethane reaction does not block the flow in the grout hole and the cement grouting can continue in order to fill the remaining fractures in the area around the hole. Since a one-component polyurethane has been prepolymerised in a factory, the reaction is not too sudden when the cement suspension and TACSS polyurethane is mixed. The prepolymerisation also renders the reaction of TACSS close to independent of the amount of water the grout is mixed with during the grouting.

As mentioned, the CO_2 gas pressure caught behind plugs of hardened polyurethane helps forcing the grout into narrow fractures. Fig. 2 illustrates the importance of the gas produced during the reaction between polyurethane and water, when a system of fractures is grouted.

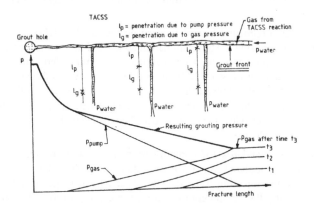

Fig. 2 : Importance of the gas production during the reaction between polyurethane and water when grouting a system of fractures (Bogdanoff, 1990)

In short, the philosophy of combination grouting can be concluded as : Instead of regarding grouting as a procedure where cement grout is simply forced into fractures from a borehole, a grouted barrier is created by use of polyurethane to limit the spread of cement. Within such a barrier, a cement suspension can then be used without risking for it to be transported away.

3.2 Reasons for Using the Combination Method

Post-grouting projects invariably involves sealing of water flowing or streaming into the tunnel. When only cement grouts are used for the post-grouting, there are always situations when the suspension in fractures and faults is flushed away by the streaming water or flows out of the fractures before it has cured. During the cement grouting in the first phase of the post-grouting in Romeriksporten, it was attempted to accelerate the cement suspension by use of different additives. This is however regarded a quite complicated process, and experience also shows limited sealing results from that type of procedure. The explanation is that there are fractures with varying width along the borehole. The cement grout penetrates faster into the fracture with the largest opening and/or the fracture with flowing water is filled first due to the influence of dilution (Fig. 3).

Although the flow into the largest fracture may be stopped when an accelerator is added to the cement suspension, much too often blockage is obtained in the borehole as well. Consequently, the spreading of suspension into the remaining finer fracture is blocked, and the rock mass remains permeable, as shown in Fig. 3. Therefore, new boreholes are required for the impending sealing work and this procedure may be repeated again. Thus, instead of using accelerated microcement suspensions, combination grouting with cement and polyurethane in the same borehole was found applicable for the post-grouting in Romeriksporten.

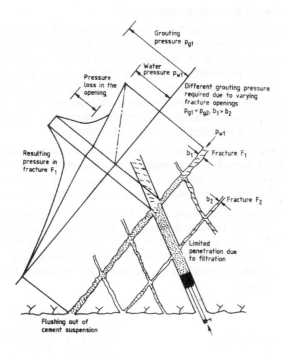

Fig. 3 : Different grouting pressure required depending on fracture width

4.0 PROJECT ROMERIKSPORTEN

4.1 Use of the Combination Method in Romeriksporten

Everyone of the 7 grouting teams in the tunnel were provided with equipment for immediate use of TACSS, as necessary, combined with cement grout. Combination grouting was used to seal large leakages in the roof and walls of the tunnel without causing blockage in the boreholes. The positive sealing results were in good agreement with earlier experience of sealing rock with combination grouting, facing conditions with largely varying degree of difficulty. A vast number of concentrated leakages (25-60 1/min) in the tunnel were further sealed effectively by use of polyurethane only.

In general, when rock with flowing water was grouted using the combination method a blockage was obtained at the tunnel surface (*i.e.*, downstream). Usually, the grouting was then terminated, but nevertheless it was observed that more fractures, closer to the grout hole (*i.e.*, upstream), were sealed when the still liquid TACSS worked against the water flow. This observation corresponded well with the findings in a Chalmers study on polyurethane grouts (Andersson, 1998).

A specific observation concerned a certain case with a large concentrated leakage in a difficult area. Since the grouting efforts with cement so far had only managed to move the water leakage back and forth, new holes with a different direction were drilled and cement grout was injected. Two distinct leakages were then observed, at a distance of around 5-8 m. When polyurethane was injected, a mixture of water and unreacted TACSS appeared from the largest of the two leakages and the pumping was terminated. After one hour, when the grouting was resumed, it was noticed that the polyurethane had worked its way towards the smaller leakages and sealed them, but that water and TACSS still appeared from the larger leakage. Continued grouting efforts made the large, concentrated leakage divide into several smaller, as the reaction of the polyurethane blocked the routes for the water in the rock mass. As the polyurethane reaction continued, the smaller leakages then decreased gradually and were completely sealed minutes later.

4.2 Grouting Problems in the Romeriksporten Tunnel

A group of experts, from consulting firms, contractors as well as representatives from a couple of suppliers of grout materials had been appointed after the problems encountered during the autumn of 1997. This group recommended a systematic grouting around the whole of the cross-section. Since the bottom of the tunnel was covered with an approximately 1m thick concrete slab, the grouting in that part of the tunnel was made difficult and the sealing work started in the roof and the walls. The work involved drilling and grouting of fans with a relative distance of 6 m. The length of the grout holes was 18 m, *i.e.*, the drilling was performed through the fan that already had been established during the pre-grouting. Therefore, large amounts of water were encountered and these had to be sealed with combination grouting.

Thereafter, the work continued with sealing of the bottom of the tunnel. Fans with a distance of 5 m were drilled with slightly inclined boreholes, to a depth of approximately 12 m. When the cement suspension escaped towards the rock surface, either below the concrete slab or towards the drainage ditches, combination grouting was performed. As for the rest, grouting with only a cement suspension of microcement was employed.

The recorded water leakage into the tunnel resulted in optimistic prognoses until the middle of April 1998. Then, it was discovered that a zone between the wall and the bottom of the tunnel had not been sealed. Thus, individual fans were drilled in this so-called "blind zone", which had not been included in the prescribed systematic grouting (Fig. 4). After completed grouting, it was observed that water was forced back through the actual boreholes, even though the packer with a backpressure valve remained in the hole. Obviously, the result of the sealing could be questioned.

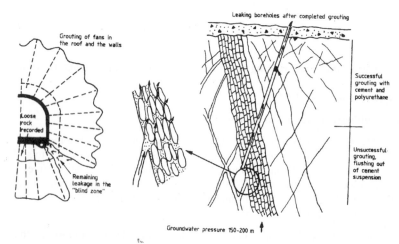

Fig. 4 : Design of the fan for systematic post-grouting in Romeriksporten (left), and grouting of the tunnel bottom, with leaking boreholes after completed grouting (right)

Despite intensive efforts and even with increased use of TACSS, the leakage measurements showed only minor decrease. The work continued until June 1998, when it was established that the demands set concerning the water leakage amounts were not fulfilled to 100% but only to approximately 80%. Even though the pore pressure was already on an acceptable level (except for in some local areas), it was decided to continue with the sealing. Thus, the planned opening of the Romeriksporten tunnel on October 8th 1998 was postponed.

The new group of experts, that was appointed for the sealing works after July 1st 1998, did not give any new recommendations. Still, systematic grouting should be performed. The sealing work along a stretch of 900 m was then started in the bottom of the tunnel and was performed in two stages. The fans consisted of slightly inclined 10 m long boreholes, and the grouting was done using microcement as first choice. The method called combination grouting was performed as necessary. The measured leakage decreased steadily until the middle of September 1998, but thereafter the measurements showed only small changes. At this point, it could be questioned whether it was at all possible to fulfil the leakage demands.

During the summer 1998, the measured groundwater pressures on the outside of the grout fan was around 130 m at Lutvann and 170 m at Puttjern. Soon after the start of renewed efforts for grouting, it was observed that after completed grouting the cement suspension was forced back out of the borehole. Flushing out of cement suspension, where the flow of "dirty" water (*i.e.*, mixed with cement) turns into clear water, can be explained as follows : When grouting highly fractured zones, where the rock mass quality is poor and the fractures are filled with clay, there is high probability of water bearing fractures that have not been sealed, although they are in contact with the borehole. Even fractures filled with cement grout can release water in narrow slots created when the pressure in the fracture decreases. Finally, the hydration of the cement is very slow at the low temperatures in the rock mass. In a laboratory in the tunnel, setting was not measured until after around 6 hours. When the suspension was flushed out due to lack of initial strength in the cement, the boreholes acted more or less as drainage holes (Fig. 4), especially with the influence from the high groundwater pressure outside the "grouted" fan.

An analysis, in which current parameters were transformed into stochastic variables, showed that the probability for the flushing out of the suspension can be limited to below 0.2% when the yield strength of the cement suspension amounts to at least 450 Pa. But for a stable suspension the yield strength is only between 10-20 Pa during the actual grouting, and therefore high demands are placed on a rapid strength development for the cement to stay in the rock mass. An analysis on the importance of the curing time showed that the probability of the flushing out of cement grout is approximately 10 times larger when there is no curing after 6 hours compared with after 2 hours. To withstand the groundwater pressure of 150-200 m, it was required that the cement suspension obtained strength of 7-15 KPa within 2 hours.

4.3 Results of Grouting Efforts

After the trial grouting rounds early 1998, grouting with polyurethane and cement was performed for sealing of different types of leaking zones. The post-grouting using a combination of cement grout and polyurethane, performed between February and June, reduced the leakage to 80% of that required. Fig. 5 illustrates the marked sealing effect obtained during the first period of the post-grouting in Romeriksporten with the water leakage into the tunnel versus bored meters of grout holes.

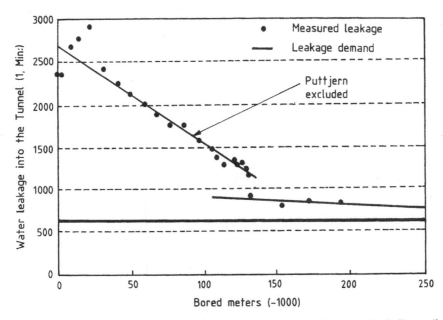

Fig. 5 : Measured water leakage into the tunnel versus bored meters of grout holes in Romeriksporten

277

Naturally, the remaining leakage was more difficult to come to terms with and needed much effort. Small leakages are more difficult to locate and, as the tunnel is getting tighter, the increasing water pressure also render the sealing more difficult. In fact, it was estimated that every litre of reduced water leakage was more than 3 times more expensive to obtain in the end of 1998 compared with in the beginning of the summer.

5.0 CONCLUSIONS

Combination grouting with cement and water reactive polyurethane, type TACSS, can successfully be used for sealing of a rock mass with good quality and water leakage through single fractures and channels, albeit high groundwater pressure and large water flow. In Romeriksporten, flushing out of cement grout was a central problem. When facing broken fracture zones with poor rock quality and clay-filled fractures, a cement suspension with both good penetration properties and sufficient initial strength increase is required, in order to prevent the flushing out of cement at high groundwater pressures. The strength development is an important aspect, which has not gained adequate attention. Instead, focus has been placed on the penetrability of the cement grout. It must however be regarded as important that the cement stays in place in the rock, as it is that it can penetrate it.

REFERENCES

Andersson, H., 1998. Chemical Rock Grouting, An Experimental Study on Polyurethane Foams. Ph.D Thesis, Department of Geotechnical Engineering, Chalmers University of Technology, Gothenburg.

Aquateam, 1998. Miljo-og helserisikovurdering for bruk av TACSSli tetting av lekkasjene i Romeriksporten, Aquateam rapport 97-152 (in Norwegian).

Bogdanoff, I., 1990. Fintatning av berg – Inträngning sförsok i smala spalter och sand med polyuretanprodukt och injekteringscement, Rapport B 90:1. Inst. for geoteknik, Chalmers TH, Göteborg (in Swedish).

Karol, R.H., 1990. Chemical Grouting, 2nd edition, Marcel Dekker, Inc., New York and Basel.

NIVA, 1998. Okologiske virkninger av utslipp av di-n-butylftalat (DBP) og akrylsyre fra Romeriksporten til Oslofforden, Norsk institutt for vannforskning, Rapport LNR 3899-98.

EARLY EXPERIENCES OF MICRO–TUNNELLING IN INDIA

J.D. BROOMFIELD

Director

Herrenknecht International Ltd., Wearfield, Southwick, Riverside, Sunderland, U.K.

SYNOPSIS

Every project presents its own set of challenges, which have to be overcome. The Mumbai project, despite experiencing some unexpected difficulties is nevertheless proceeding and every aspect of the TBM's and their management on site are being constantly appraised to minimise the risk of any further delays. The benefits of jacking over trench excavation remain a clear overall advantage in the busy suburban streets of Mumbai.

At the time of writing this paper, due to the difficulties in controlling storm water flooding shafts, the contractors has made the decision to halt all tunnelling works until the monsoon period is over. Up to this time a total of around 600 m have been jacked using the AVN800B and AVN1000 machines.

1.0 INTRODUCTION

The jacking of pipes and concrete structures below ground is not new to India. In Calcutta for instance shaft pipejacks under busy roads for sewerage transfer and complex *in-situ* box structure jacks for road and rail use date back to the 1980s.

Whereas the techniques in jacking equipment are believed to have come from abroad (particularly the UK) considerable innovation is known to have been adopted to enable its successful use, both from construction and economical aspects.

The softer alluvial materials found in Calcutta made these techniques particularly suitable for hand excavation, the traditional and by far still the prime means of working in India.

Open trench working has remained the mainstay for the installation of services from telephone cables through to large pipes for water and sewage (Fig. 1). Such trenches are difficult to maintain fenced. During rainstorms or when adjacent to leaking water mains and ditches are prone to becoming large septic tanks until backfilled. Hand excavated spoil dominates the roadside restricting access to adjacent properties. Sorting, selecting and packing of backfill is laborious and slow increasing the already extended road occupation time.

Fig. 1 : Sewer installation by conventional trenching

Explosives are used where rock is found in addition to labour intensive plug and feather techniques.

In a country where labour is cheap and abundant there is clearly a disincentive to the employment of highly sophisticated mechanical equipment to execute work which can be done, all but at a slower pace, by hand.

However, with the ever-increasing populations of the major cities with the accompanying need of an expanding infrastructure, microtunnelling technology is now being introduced to cope with the great public demand to control water pollution and improve the quality of urban waterways. There are literally thousands of kilometres of sewers that are required in highly populated urban areas as soon as possible. These must be constructed without disruption to public areas, railways and busy roadways.

Traffic disruption is a major consideration in the planning of underground work. Not only does it cause inconvenience and delayed business to the public; it also considerably adds to environmental pollution.

This has identified the great necessity for the introduction of environmentally friendly microtunnelling technology for the trenchless installation of underground pipelines. It is anticipated that the technology introduced will be quickly integrated into the Indian construction industry cohabiting with more traditional ways of working.

The Indian Society of Trenchless Technology are now actively promoting the use of this trenchless technology in India.

This paper describes the early experiences of a microtunnelling project in India.

2.0 MUMBAI SEWERAGE DISPOSAL PROJECT

With a population of circa 16 million, Mumbai (formerly Bombay) is the capital of Maharashtra State and located on the West Coast of India. It is actually an island connected to the mainland by road, rail and air services.

It has an annual rainfall of around 2500 mm most of which falls in the monsoon period between June and September. Mean daily temperatures range from 22°C in January up to 36°C in May / June.

Whilst having an underlying basalt bedrock with many outcrops, some of the urban areas are reclaimed. Groundwater table is generally high.

The city is administered by the Municipal Corporation of Greater Mumbai (MCGM) which is responsible for providing public sewerage facilities for the city and its suburbs; which cover an area of 430 km². There are seven catchment areas; Colaba, Worli, Bandra, Ghatkopar, Bhandup, Versova and Malad and 21 major pumping stations. 9 million people are currently served by the public sewer systems.

Under a recently completed Water Supply and Sewerage Project, funded by the World Bank, the corporation added 75 km of sewers, ranging from 250 - 1800 mm in diameter, in the western and eastern sewerage catchment areas.

However there were several sewer links mainly across railways and busy road junctions which could not be constructed under conventional open trench methods due to the unacceptable upheaval which would ensue.

With the foresight of public concern the Municipal Corporation of Greater Mumbai awarded a US$ 8 million contract to an international contractor. The contract also funded by the World Bank as part of the Mumbai Sewerage disposal project, proposes to construct the missing links by microtunnelling and pipejacking techniques and embarked on a special contract totalling 3.7 km of sewers from 350 - 1400 mm in diameter.

It was a requirement that the microtunnelling TBMs proposed be capable of excavating in a wide range of ground conditions from water charged clay strata with boulders to rock strata.

The project believed to be the first of its kind in India was put out to contract during 1999 with construction commencing in September 1999.

A company of consulting engineers were retained by the Municipal Corporation and the World Bank as Consultants to oversee the use of trenchless methods / microtunnelling for the installation of the missing sewer links without causing any disruption to the city's road networks and transportation system.

Through a thorough selection process, assessing general construction experience, experience in microtunnelling / pipejacking and expertise to handle the difficult ground conditions, the consulting engineers selected a suitable contractor.

The contractor selected was also responsible for the manufacture of the jacking pipes, a prerequisite for this contract was that the contractor had to have manufactured and supplied a minimum of 2,000 m of 800 mm diameter and above reinforced concrete jacking pipes in one year during the last five year period. No such pipes were previously available in India.

AVN800 and AVN1000 microtunnelling machines were selected and approved by the Engineer.

3.0 PROJECT SUMMARY

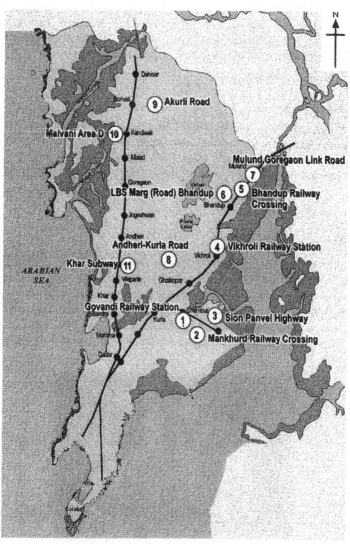

Fig. 2 : Mumbai sewerage disposal project

281

There are 11 sections of work spread across the northern suburbs of the city. The sections are situated at 10 different locations, (Fig. 2).

Section	Description
1.	Govandi Railway Station Crossing
2.	Mankhurd Railway Crossing and Sion Panvel Highway
3.	Sion Panvel Highway and Railway Crossing
4.	Vikhroli Railway Station Crossing
5.	Bhandup Railway Crossing
6.	LSB Marg Road – Bhandup
7.	Mulund Goregaon Link Road
8.	Andheri – Kurla Road
9.	Akurli Road
10.	Malvani Area (D)
11.	Khar Subway

Work to date has commenced on 4 sections; 2/3, 6 and 8, in 3 locations. Work at section 8 in the busy streets of Kalina is the most advanced with 9 shafts excavated and 3 tunnel drives completed, totalling 250 m. Section 6 in Bhandup has a further three drives totalling a similar length.

Three shafts have been sunk on sections 2/3.

4.0 GROUND CONDITIONS

At 6 and 8 bedrock is to be found within 4 m of the surface. It is fine grained, generally grey coloured basalt of variable strength from 60 - 220 Mpa. In certain locations it has been found to have been geologically altered by thermal action and presents itself in a black quartz laden denser form. The rapid variation of the rock in the ground presents onerous conditions through which to navigate small diameter microtunnelling machines.

At section 2/3 the ground conditions are more favourable being predominately clay deposites. Tunnelling has yet to commence in this area.

5.0 SITE ACCESS

The location of launch and reception shafts presents a continual challenge for site access.

All slow roadside drains clogged with various refuse and overflowing, (Fig. 3). Generally these are 400-600 mm wide and 600 mm deep. They are invariably constructed to rock and mortar and subject to collapse without warning which presents great problems to adjacent to working shafts. They are to be found on either side of virtually all roads and are ignored only at ones peril when setting up site and carrying out headworks excavation. Frequently they are constructed over other service facilities, particularly clusters of electric cables.

Fig. 3 : Existing roadside sewer

Traffic congestion presents a continuous problem, Fig. 4 shows one of the rail crossings where work has yet to commence. Traffic is queuing to cross after the train has passed. The launch shaft will be in the open area to the left of the photograph.

Fig. 4 : Rail crossing site access   Fig. 5 : AVN1000 launch shaft construction

6.0 SHAFT CONSTRUCTION

On sections 6 and 8 of the shafts have been sunk down to rock formation 3-4 m below road level using conventional steel walling frames and vertical timber poling boards (Fig. 5). In most locations the presence of service pipes and cables crossing within the pit have given rise to a degree of difficulty in constructing the headworks. Voids around existing services, leaking water mains and collapsing roadside drains have also had to be accommodated in addition to high water table levels, random strata / backfill overlying the bedrock and traffic congestion ever present in the busy streets. All the pits have been excavated by hand with aide and head carried pans, the traditional method here. The rock has been broken out by road breakers and plug and feather technique. A vibrating hammer is used for the piled pits with again hand excavation.

7.0 JACKING PIPES MANUFACTURE

A small casting yard has been set up adjacent to the airport runway at Kalina. Both 800 and 1000 mm diameter jacking pipes are cast (Fig. 6). Facilities allow the casting of 4 pipes per shift. Two shift working is in operation. The pipes are cast vertically, using moulds manufactured in Italy. The steel cages are manufactured locally and delivered to the yard ready for installation in the moulds. The 45 N concrete is delivered by truck mixer sufficient to cast 2 pipes at a time. External vibrators are built onto the moulds to ensure compaction is achieved. The steel bands forming the socket ends are also made locally and fixed to the cages at the casting yard. The pipes which are cast spigot end up are stripped within 12 hours and are covered with wet hessian until cured.

8.0 MANHOLE CONSTRUCTION

Manholes are constructed in local brick either circular or square plan. Work is all done by hand using traditional skills. The rate of construction of these manholes being circa 2.5 m across and having 450 mm thick walls is impressive with a 2.5 m height achieved in a single shift including mortar rendering inside and out, foot irons, backdrop pipes etc. Manholes are 6 to 12 m deep.

Fig. 6 : Jacking pipe manufacture

At the time of writing preparations are in hand for weathering the imminent monsoon. Exposed shafts are being strengthened and covered and brick bunds are being constructed around pits to prevent flooding.

9.0 GROUND PROBLEMS ENCOUNTERED TO DATE

Whilst the hardness of the basaltic rock has remained well within the capability of the roller cutters on both machines, some difficulties have been experienced with the AVN1000 in the deeper drives in section 8. A failure of one of the semi-gauge cutters, set obliquely to the rock face, resulted in the need to excavate a temporary rescue shaft.

Full analysis of rock samples taken from the rescue shaft have indicated unconfined compressive strength in the range of 190 to 210 Mpa. More important were the stress / strain curves calculated which indicate the cutability of the rock. In each of the tests carried out on 50 mm diameter rock samples, the elongation was observed to be in the range of 7-9 mm/m.

This quite simply means that a rock sample 50 mm diameter x 1,000 m long would have to be stretched between 7 to 9 mm before fracture would occur.

The curve in Fig. 7 clearly shows that there is no specific yield point or indication of brittle fracture. The rock encountered therefore is shown to be extremely elastic up to the point of fracture without any defined yield point.

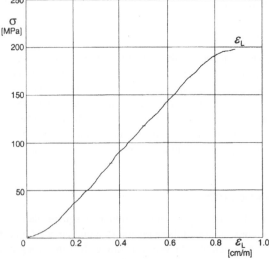

Fig. 7 : Stress/strain curve for rock samples taken from AVN1000 rescue shaft

The curve opposite also shows for this specific case an elongation of 9 mm/m was achieved before the sample fractured.

The rock characteristics encountered have in some cases resulted in reduced disc cutter life. In such instances high bearing loads have been induced as the cutter attempts to penetrate the elastic rock thus resulting in premature disc cutter failure.

New design cutters are now available providing increased permissible bearing loadings. These will certainly increase disc cutter life and thereby reduce the possibility of the failure of microtunnelling drives in such difficult ground conditions. To totally eliminate the risk of drive failure due to high cutter wear / damage, a larger diameter machine, and therefore jacking pipe, must utilised. The use of a larger machine permits man access to the excavated face for cutter replacement.

The strength and composition of the basalt has also been found to vary significantly over very short distances. On the most recently commenced 800 mm drive at a depth of only 6.5 m the grey basalt present in both the drive and reception pits (70 m apart) disappeared completely after 40 m of driving to be replaced firstly by a zone of completely weathered rock resembling sand and then by a ledge of very hard black coloured basalt. The former gave the machine the tendency to drive whereas the latter deflected the machine upwards and to one side. In this case although steering of the machine could be recovered, the 2.5 m long concrete pipes were unable to negotiate the misalignment resulting in high jacking pressures and damage to the pipeline.

These two problems have been difficult for both contractor and owner alike and have caused unwelcome delays to a tight programme. The problems have occurred in spite of all the safeguards imposed by the clients engineer in selection and approval of both the contractor and the microtunnelling machines themselves, the previous experience of the contractor and the proven history of the machines. Neither are the problems encountered unique. They are typical of the problems endemic with putting small diameter tunnelling machines into what can only be described as relatively high risk situations. No matter how much information is available at tender and no matter how much experience lays in the hands of contractor or machine manufacturer alike there can never be a 100% guarantee that the ground will not present a few surprises.

In tunnelling machines of 1200 mm internal diameter and above access to the tunnel face is can be accommodated for cutter maintenance. Power and steering capabilities also become improved. On smaller diameter machines face access through the machine is not physically possible and maintenance on the machine is extremely limited whilst it is in the ground.

For this reason serious consideration should always be given to utilising a larger diameter machine with the facility for face access.

Pipejacking though presents a different range of problems to those met in clays and sands. The following are some of the typical difficulties that can be encountered.

- Excessive cutter wear or damage through localised hard abrasive zones in the rock mass resulting in reduced rate of advance or total stoppage.
- Damaged overcutters, reducing size of bore and hence affecting steering capability of the TBM.
- Lenses of hard rock or boulders lying in one part of the tunnel face only can be difficult to penetrate with the machine tending to glance off them taking the least line of resistance. Correcting line/level under such circumstances is difficult and due consideration has to be paid to the alignment of the jacking pipes passing through the misalignment.
- Cavities in rock can be experienced where rock fragments fall into the annular space between the rock bore and the concrete jacking pipes. Such rock fragments can become wedged between the rock bore and the concrete jacking pipe, thereby creating high jacking loads due to the "wedge action" and high points loads to the outside of the jacking pipe, can be contributory to possible jacking pipe failure.
- Where rock excavation results in a high proportion of fine grain size particles, the particles can be carried out by slurry to the annular area between the jacking pipes and the rock bore. The collection of fine particles eventually results in drastically increased jacking forces.
- Over shoving the pipeline during excavation induces excessive stress on the disc cutters.

PRACTICAL APPLICATION OF STEEL FIBRE REINFORCED SHOTCRETE IN DESILTING CHAMBERS OF NATHPA JHAKRI HYDROELECTRIC PROJECT

SUBHASH MAHAJAN

Superintending Engineer

Nathpa Jhakri Power Corporation Ltd., Shimla, Himachal Pradesh, India

1.0 INTRODUCTION TO THE PROJECT

The Nathpa Jhakri Power Project being constructed by Nathpa Jhakri Power Corporation Limited (NJPC), is a run of river scheme located in the middle reaches of the river Satluj utilizing 488 m gross head available between Nathpa and Jhakri for power generation.

This is largest tunnel project of the country and is also the largest run of the river scheme.

The salient features of the project are as under :

1.	Cachment area	49,820 sq m
2.	Design discharge	405 cumec
3.	Type of dam	Straight concrete gravity
4.	Height of the dam	60.5 m
5.	Top of dam	154.0 m
6.	River sluices	5.0 nos.
7.	Desilting arrangement	4.0 underground chambers each of 525 m × 17.11 m × 27 m, having low velocity of 30 cm/s.
8.	Head race tunnel	27.3 km long 10.15 dia circular concrete lined under ground tunnel having water at velocity of 5.0 m/s.
9.	Power house	220 m × 20 m × 49 m cavern having 6 francis turbines each of installed capacity of 250 MW
10.	Tail race tunnel	10.15 m dia. 788 m long concrete lined tunnel.

2.0 THE DESILTING ARRANGEMENT

The Himalayan rivers carry heavy load of silt. As the project is a run of the river scheme the removal of silt is being done through desilting chambers. To remove the silt up to 0.2 mm particle size four no. Large under-ground chambers with following features are under construction as shown in Table 1.

Geologically the rock in which the chambers are under construction consists of following type of features :

1. Closely joined and weathered gneiss and augen gneiss
2. Amphibolite schistose in nature, slightly weathered along joints
3. Biotite schist
4. Moderately jointed quartz mica schist.

TABLE 1

1.	Type	Underground chambers
2.	Inlet structures	4.0 nos. 6.0 m dia horse shoe
3.	Intake level	EL 1469.0 m
4.	Size of each chamber	Length = 525.0 m Width (at centre) = 17.11 m Height = 27.5 m
5.	Flow velocity	0.30 m/sec
6.	Flushing tunnel	Length - 1580 m Dia = 5.0 m horse shoe Slope = 1 in 450
7.	Outlet tunnels	4.0 nos. concrete lined circular tunnel

3.0 ROCK SUPPORT SYSTEM – AVAILABLE OPTIONS

In tunnelling all the water conductor systems are generally concrete lined. The basic reason of concrete lining is :

(i)   a smooth surface,

(ii)   protecting original rock from the effect of weathering and

(iii)   prohibiting free flow of water from tunnel to the rock mass and vice - versa.

Traditionally, we have been providing the temporary support prior to lining concrete in the form of wiremesh reinforced plain shotcrete, either dry or with wet process.

The advent of SFRS and with its growing usage, NJPC used the opportunity to evaluate this form of rock support also in addition to the traditional method. Moreover, smooth surface not being the prerequisite for desilting chambers, opportunity to look around for other methods of rock support existed.

Steel fibre reinforced shotcrete lining was considered in comparison to the traditional concrete lining evaluation. Both the options were compared keeping in view the technical and commercial advantages and disadvantages. SFRS seemed more suitable for application due to the following advantages.

4.0 ECONOMICAL ADVANTAGE

On the cost front, SFRS cost around USD Mio. 15.00 as against USD Mio. 16.50 in concrete lining. The saving potential of around USD Mio. 1.5 clearly made it the no. one choice from the commercial angle as shown in Tables 2 and 3.

The cost saving mainly came from the reduced quantity of shotcrete compared to concrete option. It was significant to note that the quantity of concrete in overbreaks alone amounted to 43 per cent of total concrete required, which could be avoided totally in SFRS lining.

4.1 Technical Advantages

A   Higher flexural strength

B   Durable

C   Elasticity

D   Improved shrinkage behaviour as steel fibres form a narrow and three-dimensional steel fibre net in every cubic meter of mix. The homogeneously distributed steel fibre obstruct the formation of shrinkage stresses and decreases the tendency of cracking.

E   Higher impact resistance

4.2 Time Factor

As the single shell steel fibre reinforced shotcrete can be applied with the excavation operation simultaneously, it saved the time required for form work erection, dismantling and re-erection during concreting. The contractor showed a time saving of 8 to 10 months in SFRS application in comparison to traditional concrete lining.

4.3 Safety

The weight of the formwork for 27.5 m high and 17 m wide cavity would have been quite considerable, making the concrete operation very risky. SFRS, with it's simplistic logistics inside the cavity was found much safer.

TABLE 2
Cost for concrete lining

Sl.No.	Description of Items	Quantity	Rate, US$	Amount, US $
1.	100 mm th. shotcrete	1.25×121130 m^2	13.69	2073999
2.	Wiremesh in shotcrete	303 tons	1011.63	306523
3.	Admixture for shotcrete *i.e.* allowance for rebound	$18860 \times 450 \times .09 =$ 763830 kg	1.65	1260319
4.	Formwork curved	119850 m^2	18.02	2160087
5.	Formwork plane	1280 m^2	13.95	17860
6.	Bulk heads	4000 m^2	116.27	465116
7.	Concrete arch (35)	$39550 \times 0.35 =$ 13840 m^3	83.72	1158697
8.	Concrete walls	$81580 \times 0.3 =$ 24474 m^3	83.72	2048986
9.	Concrete to pay line	$121130 \times 0.15 =$ 18170 m^3	83.72	1521209
10.	Geological over break concrete	45500 m^3	73.49	3343720
11.	Rebars in concrete	1200 tons	639.53	767441
12.	Conc. admixtures	1389460 kg	1	1389460
		Total	**USD**	**16513417**

TABLE 3
Cost of SFRS lining

Sl. No.	Description of item	Quantity	Rate	Amount
1.	SFRS	28400 cum	432.55	12284420
2.	Micro Silica	766800 cum	1.65	1265220
3.	Superplasticizer	245745 cum	1.65	405479
4.	Accelerator	664957 cum	1.65	1097179
		Total	**USD**	**15052298**

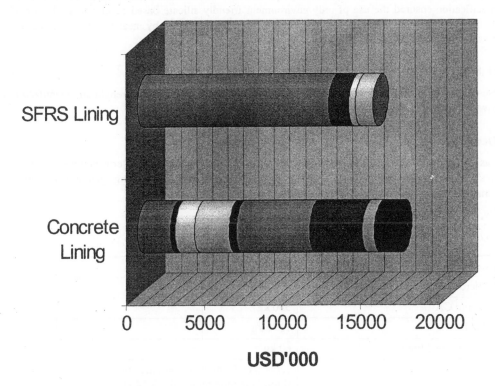

Fig. 1 : Cost comparison : Concrete vs SFRS Lining

5.0 CHARACTERISTICS OF SFRS

Being the first application of it's kind in India, the specification for SFRS was prepared with the objective of ensuring application meeting the technical requirements, restricting the use of improper material and increasing the durability of the shotcrete. A particular emphasis was laid on controlling the rebound. The contractors were asked to make available expert personnel for application with international experience and use only time tested and proven material.

The technical requirements were laid down as follows :

(a) Strength requirement
Compressive strength   3 days   > 10 MPa
(cylindrical strength)   7 days   > 18 MPa
                                   28 days > 30 MPa
                                   (Equivalent cube strength of 28 days > 35 MPa)

Flexural strength   28 days > 3.8 MPa
Toughness   28 days > 3.8 MPa

Material selection – The following criteria was fixed for the selection of admixtures and fibres.

5.1 Admixtures

A proven system to control the consistency, ensure good workability, low pumping pressure, adequate slump retention and low rebound was made mandatory. The contractors were asked to provide the technical documentation showing compatibility between superplasticiser and accelerators they proposed to use. The technical specification ensured the use of only environment friendly silicate based accelerators which could ensure rebound lower than 10 per cent overall and have lower fibre rebound by means of having thixotropic consistency when the shotcrete hit the substrate.

5.2   Steel Fibres

The use of steel fibre was restricted to only those with over 1000 map tensile strength and manufactured under approved quality assurance system.

5.3   Mix Design

NJPC asked the contractor to propose the mix design, including the choice of material complying with the specifications and the trial results at site. A series of mix designs, with the varying quantities of cement, microsilica and the accelerator were conducted at site.

Initially, the following mix design was put to use.

Quantity for 1 cum of SFRS

Cement	475 kg.
Micro- silica	7% BWOC ~ 33.25 kg
10 mm CA	550 kg
Sand	1170 kg
Dramix steel fibre	60 kg
Polyheed SG 30	1.7% BWOC max., depending upon slump/retention required ~ 8.64 kg
Water to cementitious Material ratio	0.41
Accelerator, MEYCO TCC 766	4%BWOC ~ 20.33 kg

The higher amount of microsilica tended to give very sticky mix, and was creating cleaning problems in the receiving hopper of the batching plant. Moreover, it was also taking away the workability fast. In order to provide more flexible mix design, in terms of increased pot life and less sticky mix at the batching plant, the mix design was revised as following :

Cement	482 kg
Micro-silica	27 kg
10 me CA	495 kg
Sand	1064 kg
Dramix steel fibre	60 kg.
Polyheed	1.7% BWOC ~ 8.64 kg
Accelerator MEYCO TCC 766	4.6% BWOC ~ kg
Water cement ratio	0.41
Slump	80 cm -160 cm

The selection of dmixture system mainly dependent on fulfilling the technical specifications. As the emphasis was on durable shotcrete with as low strength loss as possible, MBT's TCC system was found to be the best option. The two component system not only ensured successful spray at as less as 4 per cent accelerator dosage, it also enabled smooth pumping of fibrecrete at lower slumps, despite of several bends in the conveying system of the spraying equipment.

5.4 Rebound Calculations

Due to the higher input costs of SFRS ingredients, compared to traditional shotcrete, a particular emphasis was given to the rebound control. In order to control the rebound, a trained fleet of internationally experienced team was employed for application. In addition, the assistance from the admixture suppliers was taken to set up the correct application, avoiding all possible flaws generally encountered during fibre shotcrete.

The technical specification permitted a higher rebound upto 12 per cent compared to 8 per cent on the wall. In the beginning of the application, rebound was measured to be quite high. Also the Nozzleman however experienced, had to get familiar with the application system, equipment and the spraying pattern. Hence in early part of execution, the rebound was around 15 per cent the first three months. But later on the rebound was found to be as low as 6 per cent and was being controlled at the same level.

The rebound was measured by spraying measured quantity of material and collecting the rebound on the tarpaulin. The care was taken to ensure that the whole area where rebound material fell was covered.

The regular rebound test and control of the slump during spraying ensured that the rebound always remained within the acceptable limit.

5.5 Slump Range

Specification provided for the slump between 10 cm and 14 cm but this was not enough range as sometimes slump dropped heavily either due to admixture, detention of batched mix or temperature variation inside and outside the chambers. An allowance of ± 2 cms outside the specification range was given to take care of the fluctuation. Also, redosing was permitted whenever required.

6.0 CONCLUSION

The exercise at desilting chambers established the point in the country that SFRS does not only provide economical and technically more advantageous rock support system in Himalayan geology, but it is safer and less time consuming also.

At the same time, selection of the right ingredients and tight control over the application are key to success with SFRS.

USE OF ALKALI FREE ACCELERATORS IN
TALA HYDROELECTRIC PROJECT

P.K. TRIPATHY   **HIMANSHU KAPADIA**   **SHAILESH KUMAR**
L&T ECC Construction Group   Masters Builders Technology (MTB) India Pvt. Ltd.
Tala H.E. Project, Bhutan, India   Old Rajendre Nagar, New Delhi, India

SYNOPSIS

The Tala Hydroelectric project is an outstanding challenge in the worldwide tunnelling. The challenge is not only given by the dimension of the project but also by the unique combination of demanding engineering problems.

The paper describes in detail the following aspects: shotcrete requirements in the head race tunnel and the desilting chambers, due to aggressive ground water and poor rock condition and the concrete technology challenges offered by the crushed rock aggregates with high mica content. In order to cope with these circumstances, a trial programme was carried out for the evaluation of shotcrete with high performance levels and the results are summarized herewith.

1.0 INTRODUCTION

Tala Hydroelectric Project is situated in southeast Bhutan in the district of Tala and Chukha. The main underground works comprise of 22 km long head race tunnel and 3 desilting chambers, 250 x 14 x 18.5 m. The work is divided into five packages, C1 to C5 and awarded to three main highly experienced civil and tunnelling contractors. The ground condition is typical of the Himalayas, with good to fragmented rock and nil to excessive seepage at few places.

In addition, the high mica content and the potentially alkali reactive aggregates have further augmented the challenges. The excavation started in 1998 and since than, all the experience to date have shown a variety of tunnelling problems, especially in the desilting chambers and parts of the HRT at C2, C3 and C4. In particular, efforts are being made to cope the varied rock in desilting chambers, which is predominantly biotite-gneiss and schistose, foliated and having glossy surface.

2.0 HEAD RACE TUNNEL

At C3, 4km of head race tunnel is being constructed from one adit Geduchu. The HRT runs parallel to the river Tala and runs southwards upto Tabji. The HRT passes through a combination of rocks and also runs parallel to river at few places making tunnelling an interesting challenge. The rock in HRT is predominantly gneiss and schistose with planar quartzite and marble bands. The support system in the tunnel is rock-bolting combination with steel fiber shotcrete or plain shotcrete as temporary lining and a final permanent concrete lining.

3.0 PROBLEM AREAS

In parts of the adit, the HRT and the desilting chambers, the shotcrete has shown a very peculiar behaviour due to the slimey rock. A freshly sprayed concrete seems all right for a couple of days and than it starts de-bonding with the parent rock. At a few places the debonding with parent rock looks all right but the bond between foliations which are parallel to the alignment gives way.

The conclusions drawn from this behaviour are following :-

The existing specifications for shotcrete is not enough to produce the behaviour required in these geological conditions. The steel fiber shotcrete calls for the use of silicate based accelerators at maximum 5 per cent dosage. At these dosages, the shotcrete sets at slower rate, and combined with the effect of mica infested aggregates, the sprayable layer thickness in one operation is restricted to around 25 mm. At higher dosages, the strength loss of shotcrete is excessive and becomes impracticable.

Due to poor bond between the foliation, separation takes place between the cleavages due to the weight of shotcrete. Based on the above the conceptual solution being thought to provide a shotcrete which exhibits the following behaviour:

(a) Faster setting time so that the de-bond can be controlled and the shotcrete sets on the wall.

(b) Higher early strength gains so that the shotcrete portal stands on its own without transferring the load to the rock for a longer period.

(c) HRT:- While most of the rock in HRT is allowing enough standing time to carry on with the shotcrete works, at a few places of very fragmented rock the high water rush is probating smooth progress. For *e.g.*, At Padechu, C2, twice the tunnel has collapsed in a chimney formation and bringing the work to a halt the post damaged exercise and removing the muck and placing of ribs and back fill concrete has been extremely tedious and slowed progress. While this type of challenge is not typical, the rock in this region requires a suitable pre-injection system using micro-cement to avoid such re-occurrence. This type of proactive technology has not yet been practiced.

4.0 ALKALI FREE ACCELERATORS

Internationally the construction industry has been demanding safer accelerators with better performance. In addition, requirements for reliability and durability of shotcrete lined structures are increasing. Strength loss or leaching effects suspected to be caused by strong alkaline accelerators have forced the industry to provide answers and to develop products with better performances.

The answer to this problem in Europe was seen in the functioning of liquid alkali free and non-caustic products providing safe, high quality and cost effective sprayed concrete applications. The use of alkali free accelerators have seen the dramatic reduction in the use of caustic or silicate-based accelerators and increase in the use of alkali free accelerators.

Based on the effectiveness and the increasing usage of this system, MBT's Meyco SA160+ / Meyco SA 170 was chosen for trials. Comparative results of the trials are as following :

	Shotcrete 1	Shotcrete 2	Shotcrete 3
	(MeycoSA160+)	(Sodium Silicate)	(Sodium Silicate)
	kg/m^3	kg/m^3	kg/m^3
Cement	550	550	550
Micro Silica	27.5	27.5	27.5
Water	231	231	231
W (C+S)	0.4	0.4	0.4
Sand 0-5 mm	892	892	892
Aggregate 5-10 mm	552	552	552

Fibers	50	50	50
Slump	160	160	160
Admixtures			
Rheobuild 1100	. 1.5% w/w of binder	1.5% w/w of binder	1.5% w/w of binder
Accelerators			
MeycoSA 160+	5% w/w of binder	-	-
Sodium silicate	-	5% w/w of binder	8% w/w of binder
Setting time			
Initial	3 min.	> 10 min.	> 5 min.
Final	5.5 min.	>30 min.	> 15 min.

5.0 PERFORMANCE COMPARISON

The study was carried out to compare the traditional system with the alkali free accelerator system; the comparison points are listed below:

I TECHNICAL

Item	Alkali Free System	Silicate Based System
Setting	Fast setting characteristics provides high early strength development at controlled dosages (approx. 5%).	Not possible with silicate based system (unless used at very high dosages of 10-12%).
Strength Loss	Good final strengths and reduced loss of final strength, limited to the extent of 10% of the control mix.	Strength loss to the extent of 20-25% compared to base concrete.
Layer Thickness	Extremely large layer thickness (15-30 cms) in a single application over-head which leads to the substantial savings in time during construction.	Not possible with silicate system even at high dosages (maximum 25 to 35 mm in one layer) leads to wasting of time.
Alkali Aggregate Reaction	Low alkali content-hence suitable under alkali reactive aggregate situation.	High alkali content (7-9%) and hence undesirable.
Rebound	Less rebound compared to silicate system of approx. 3-4% at low dosages.	Rebound of app. 10-12% at dosage of approx. 7-8%.
Suitability In Poor Rock Condition	Highly suitable under poor rock geology due to immediate hardening/stiffening property of the alkali free accelerator.	Fairly suitable for the poor rock geology as the silicate base accelerator are having slow ardening/stiffening property to restrict the fall-outs of rock.

II COMMERCIAL

Item	Alkali Free System	Silicate Based System
	(Rs. per cum)	(Rs. per cum)
Accelerator cost	1400/-	900/-
Admixture cost	335/-	335/-
Total chemical cost	1735/-	1235/-
Rebound rate	4%	12%
Rebound cost	380/-	1150/-
Applied cost	**2115/-**	**2385/-**
Saving with AF	**270/- per cum.**	------

Note: - The above costs do not account for saving in time cost required for building up the desired thickness with silicate based system in multiple layers.

6.0 DOSING PUMP

It was observed that the gear pumps was not able maintain consistent dosage rate and also the life of the gear pump was suspected to be lower with alkali free accelerators.

From the available information the mono-pump (squeze pump) by Meyco Equipment and Cifa were found suitable.

7.0 CONCLUSION

The trials conducted at contract packages C1 and C3 have shown encouraging results, both technically as well as on the commercial front.

The possibility of spraying from bottom upto the crown at the thickness around 100 mm in single operation and rapid strength gain is now possible with the AF system. This may enable the shotcrete lining to take care of the rock support, without the danger of falling down or separating from the parent rock.

The combination of early strength gain and higher layer thickness, both at controlled dosages of accelerator, is a certain hope for the rock support in otherwise intricate Himalayan geology.

USE OF STEEL FIBERS AS REINFORCEMENT FOR UNDERGROUND CONCRETE STRUCTURES

MARC VANDEWALLE

N.V. Bekaert S.A., Belgium

1.0 The underground offers the space which mostly is not anymore available on surface.

Urban traffic is saturated and rapid mass transport is only possible with a good underground metro network.

Service facilities require a dense network of pipes to bring fresh water into the cities and a dense network of sewers to evacuate the waste water to the water treatment stations.

Communication, being the main characteristic of our present world community, needs an extensive network of cables and service stations.

The required space is only available under surface level.

Cities can be linked through air connections but saturated air space, overcrowded airports and lack of comfort in most of the aircrafts do not offer the business people the possibilities of a useful use of their travelling time.

The construction and the extension of the high speed railway network in Europe and Japan proves to be successful and will be connecting soon most of the European main cities even faster than air travel can do.

Also heavy traffic is being taken off the road mainly in mountaineous areas where winter conditions are prohibitive for fast transport during part of the year. The two most spectacular examples are the two long Alp Transit tunnels in Switzerland : Gothard (57 km) and Lötschberg (39 km). Also Lyon – Torino is expected to be realized soon and includes a 52 km long railway tunnel.

The use of natural resources such as hydropower and the excavation of minerals from the mines require the construction of impressive underground constructions.

The modern industrial principles do not only need a fast data communication system but also an efficient transport to get the goods on the right time on the right place. There is still a lot to be built and to upgrade in order to realize an efficient infrastructure.

2.0 The use of the underground used to be a very expensive solution but has proven in the last 25 years to become the most efficient way of solving problems.

Development of modern tunnelling techniques, construction materials and skilled mining people have highly influenced the attractiveness of the underground building technology.

One of the main breakthroughs was the change in mentality when designing a tunnel. Observational methods, such as N.A.T.M (New Austrian Tunnelling Method) and N.T.M (Norwegian Tunnelling Method), are strengthening the underground to become self supporting instead of supporting the rock mass above the tunnel opening.

This of course made it possible to build underground constructions in a much more economical way and much faster than what was done in the past. This, however made it necessary to develop new tunnelling equipment and adapted building materials. And of course trained people to build the tunnels. This human factor, still today, sometimes remain a critical issue.

3.0 Depending on underground conditions, rock quality, overburden tunnel diameter and length, ground water level, a tunnelling method has to be selected using or manual or mechanised excavations.

Manual excavation includes drill and blast and / or the use of a roadheader. Depending on the tunnel parameters the tunnel cross section will be excavated in one operation or in different steps. In the last case, more temporary linings will be required to stabilise the underground at any moment.

Mechanised tunnelling, however is cutting the full cross section in one operation. Impotant and spectacular progresses have been made in the field of Tunnel Boring Machines (T.B.M) being used in a wide variety of ground and rock conditions and for very large diameters.

Up to now the record is held by 4th Elbe Tunnel in Hamburg having a diameter of 14.56 m using a Herrenknecht mixed shield TBM. But larger tunnels are coming up soon and it can be expected that the race between East and the West for the largest tunnel diameter is going on for quite sometime.

The main parameters, however, driving the tunnelling method selection are speedy and safety, resulting mostly in the economically and most attractive solution.

4.0 Tunnelling methods using the manual excavation technique are still being used to build most of the underground constructions. This methods offers a high degree of flexibility both in easy and also in difficult ground conditions. This of course is a major advantage in a heterogeneous underground where ground conditions may change even within short distances.

Tunnelling experience and understanding of the ground conditions allow to apply the correct support system in each point depending on local parameters. Constant monitoring is an essential part of modern tunnelling and indicates if the underground has been stabilised after applying the selected support system.

Modern tunnelling mostly uses rock bolting, reinforced shotcrete and steel arches, or a combination of two or three of these techniques, in order to create inside the rock surrounding the excavated tunnel opening a new equilibrium forming a self supporting arch.

The cycle comprises blasting or roadheader excavation, mucking and stabilisation of the rock. It is important to stabilise the rock as soon as possible after it has been disturb by building the tunnel opening. Safety of the whole operation can be strongly enhanced by applying a reinforced shotcrete layer before sending mining people inside the newly built opening.

Shotcrete has become a standard technique and is used as a major tool to stabilise the rock in the early stage of the tunnel construction.

Shotcrete has a double effect : it glues the loose pieces of rock together forming a continuous outer shell and it develops strength in order to control and support the rock in it's early movements. Both effects contribute to create a new equilibrium and to help the rock to become again self supporting.

It is important that this shotcrete layer can be applied before someone of the tunnelling crew members has to enter the new opening as loose pieces of rock still can fall down. The use of a remote controlled robot allows to apply the shotcrete layer in safe conditions from outside the dangerous area of falling blocks and not exposed to the rebound of the shotcrete material.

Plain shotcrete like also concrete does, however, is a very brittle material which can hardly absorb any deformation. It needs to be reinforced in order to cope with tensile and flexural stresses. In fact the shotcrete layer has to be able to control the underground movement which actually has to result in a new equilibrium where the rock again is self supporting. Hence not the shotcrete strength, to *support* the soil, is the most important materials's characteristic but it's ductility in order to make the rock *self supporting*.

Shotcrete is brittle and has to be properly reinforced in order to become ductile and tough.

Steel still is the most common and most reliable and durable material to strengthen concrete. The efficiency and compatibility of the composite concrete and steel have been proven in reinforced and pre-stressed concrete.

Traditional wire mesh is difficult to fix to the irregular substrate of the blasted or excavated cross-section. Also this meshing operation takes a lot of time. Job data have shown that installing the mesh lasts 3 times more than shotcreting the same surface. The continuously changing position of the reinforcement within the shotcrete lining does not guarantee at all a uniform bearing capacity.

Even more than the lack of an efficient technical performance is the risk for accidents when installing the mesh on the freshly exposed rock surface. Mining people has to work in a very difficult and unprotected conditions.

Steel fibers have proven to be a more modern and technically better solution. Steel fibers are added as one more aggregate to the shotcrete mix, both wet and dry, and mixed together with the other shotcrete constituent materials. Steel fiber reinforced shotcrete is applied using standard equipment and when using a robot a reinforced lining will be in place before any miner has to enter the new opening, which has a major impact on the safety of the tunnel job.

Steel fiber reinforced shotcrete's main technical advantage, however, is it's high degree of ductility. Being able to absorb important movements without loosing it's bearing capacity it helps the underground to stabilise in a controlled way.

Ductility, however, is not a common characteristic for a usually brittle concrete. An appropriate test method to determine steel fiber reinforced shotcrete's ductility has been developed and proposed by SNCF / Alpes Essais (France). This test method has been approved by EFNARC and will also be included in the new European Standard on Sprayed Concrete, which is expected to be published in 2001.

Based on this test usually three steel fiber reinforced shotcrete classes are being defined :

500 Joules	for sound ground conditions
700 Joules	for medium ground conditions
1000 Joules	for difficult ground conditions

Steel fiber reinforced shotcrete is the standard for primary linings as it allows to build a ductile support in safe working conditions quite immediately after excavation.

5.0 Tunnels still mostly are built using the double shell concept : a primary stabilising lining of steel fiber reinforced shotcrete and a final lining mostly of plain cast concrete. In between a waterproof membrane is built in to avoid water leakages through the tunnel lining.

Steel fibers sometimes are used as a reinforcement for this final lining in order to have a better crack resistance of the concrete which has a positive influence on the durability of the concrete structure.

Also when future tunnelling in the neighbourhood of the tunnel is expected steel fibers enhance the ductile behaviour of the concrete lining and give it a better resistance against changing stress conditions.

Improved shotcrete technology, a wide range of admixtures keeping the shotcrete characteristics within the required workability limits and skilled and certified nozzlemen allow to produce now-a-days high quality and durable steel fiber reinforced shotcrete linings.

In the single shell method the full lining, including the primary and final layers, is built using only shotcrete. The total lining consists of different shotcrete layers which form a single shell. There is a need for a tough quality control of the total shotcrete lining as bond strength between the different layers is the key issue in order to guarantee the "one shell" behaviour.

As no waterproof membrane can be used in this single shell concept, as it should act as a bond breaker between the different layers, shotcrete durability and impermeability need special attention. It is strongly recommended to use silica fume and PFA in adequate dosages in order to improve the density of the applied shotcrete material.

298

Steel fibers are being used both in the first and the final shotcrete layer, be it for different purposes. Ductility is required in the first stabilising layer, while in the final layer crack control improves the durability of the lining.

The single shell method offers the advantage of being able to apply the final layer shortly after the first layer. This allows to shorten drastically the total construction time. In the double shell method very often the final cast lining only can be applied after the breakthrough as the mould obstruct the normal traffic in the tunnel.

6.0 More recently mechanised tunnelling largely has increased it's application field. Improved technology makes it possible to use a TBM (tunnel boring machine) in a wide variety of ground conditions and for large to very large diameters.

Although the investment of a TBM still is quite high, in some cases a tunnel length of 2.2 km already can justify economically the application of the mechanised tunnelling process.

This method of course offers the advantage of constructing a constant tunnel cross section, avoiding expensive overbreak which is the case in drill and blast. Particularly in urban areas the use of a TBM proves to be very interesting as it provides a continuous support, even at a small overburden. The tunnel opening also is lined immediately behind the TBM using precast concrete segments. This of course reduces the risk for settlement of buildings in the neighbourhood of the tunnel.

The segments are precast or at the job site or are brought in from a remote specialised precast concrete plant. The lining consists of concrete rings which are formed by putting several single pieces together. The ring is closed or by bolting the segments or by using a keystone which gives an expanded lining.

In most ground conditions the segments forming a closed ring only have to resist normal compressive forces in their final position. However, these segments are subjected to different loading conditions before they get into their final place. The precast segments have to resist bending moments and flexural stresses when being demoulded and transported to the storage facilities located outside the precast building. They have to resist tensile thermal stresses due to temperature changes at the storage area. The heaviest loading, however, takes place when the segments are being installed and have to resist the very high jack loads of the TBM when moving forward.

Cracking and spalling is the main problem of reinforced concrete segments.

The heavy jack loads are being applied at the outer unreinforced concrete skin of the segments. When spalling occurs the segment has to be repaired or even replaced for obvious reasons of durability concern. This however delays the tunnel construction progress and is a very expensive operation.

The very complex tensile stress pattern induced by the jack loads requires a quite complicated steel reinforcing cage, what makes it heavy and expensive.

Steel fibers are a technical valuable alternative for the traditional mesh and rebar reinforcement. The homogeneous fiber distribution makes it possible to absorb tensile stresses in any point and any direction of the concrete segment. Cracking and spalling resistance are considerably improved.

Various reference projects have proven that tunnels with a diameter up to 7-8 m can be lined with steel fiber reinforced concrete segments without the need for any other additional reinforcement. For higher diameters it is recommended to use a mixed reinforcement of mesh and steel fibers.

Some heavy fires that recently happened in some of the main European transport tunnels have focused the need for an improved fire resistance of the tunnel lining.

Extensive research, especially under U.K. initiative, has shown the positive result of a blend of steel fibers and polypropylene fibers.

7.0   There is a still growing use of the underground for infrastructure and utility projects.

Modern tunnelling methods are ruled by concerns on safety and speed.

Steel fibers which can be added as another component to the concrete and shotcrete mix allow to increase the tunnelling speed under safer working conditions and provide concrete and shotcrete with a high degree of ductility in order to allow a controlled movement of the underground in a process of becoming again self supporting.

Steel fibers do control concrete cracking much better and justify for durability reasons it's use in final cast or sprayed concrete layers.

Mechanised tunnelling becomes more popular and economically more attractive. Substantial savings are possible when using steel fiber reinforced concrete for segment manufacturing.

CONSTRUCTION OF INCLINED TUNNEL FOR HRT OF RANGANADI HYDROELECTRIC PROJECT AND A TYPICAL ADIT JUNCTION TO INCLINED TUNNEL — A CASE STUDY

R.D. VARANGAONKAR

Vice President (Projects)

Gammon India Limited, Gammon House, Veer Savarkar Marg, Prabhadevi Mumbai, India

SYNOPSIS

Ranganadi Hydroelectric project is in Arunachal Pradesh in North-east region. North-east region is situated in food-hills of Himalayas. Due to very heavy rainfall and rivers originating from Himalayas, rivers carry large quantities of water. North-east region have an estimated hydel power potential of approximately 30,000 MW. Ranganadi Hydroelectric project is having an installed capacity of 135 × 3 = 405 MW. The project is in very advanced stage of construction. The project have following main structures :

- *10 km long 6.8 m diameter(finished) horse shoe shape concrete lined tunnel*
- *30 degree inclined pressure tunnel 530 m long fully steel lined.*
- *A surge shaft 25 m diameter up to 25 m depth and then 16 m diameters up to 100 m depth. Surge shaft also have a lower expansion chamber of 6 m 'D' shape 50 m long.*
- *One diversion tunnel of 6 m diameter concrete lined 254 m long.*
- *Concrete gravity dam 68 m high and 355 m long*
- *One surface power house having 3 turbines of installed capacity of 135 × 3 = 405 MW*

Tunnel being a very long lone passing through very bad geology took maximum time for construction. Paper describes the difficulties in construction of main tunnel specially inclined tunnel under extremely adverse conditions.

1.0 INTRODUCTION

The layout plan of the Ranganadi Hydroelectric project is given in Fig. 1. The head race tunnel of the project takes off 400 m u/s of Ranganadi diversion dam. A 6.8 m (finished diameter) concrete lined tunnel is approximately 10 km long. Out of which 530 m tunnel is at 30 degree inclined steel lined pressure tunnel. Initially it was planned to bore tunnel from intake adit 1 and outlet *i.e.,* from four faces only. However actual tunnel construction was carried out from eight different working faces. Work adits 1, 2 and 3 were meeting the main tunnel along its horizontal path whereas adit 4 was the only adit meeting the inclined tunnel. Fig. 2 shows tunnel with different working faces along with surge shaft, lower expansion gallery and power house.

2.0 ALIGNMENT AND GEOLOGY

A section along the tunnel alignment is shown in Fig. 2 which shows tunnel passing through precambrian schist, precambrian gneiss, sand stone and carbonaceous shale with coal seams, Volcanic and metabasics belonging to Gondwana group soft triable sand stone with bands of clay/shale and streaks of coal belonging to Shivalik super group. Gondwana rocks are also folded, fractured and sheared. The shivalik comprises soft gray coloured medium grained pebbly sand stones with clay bands. Geological mapping of tunnel shows that tunnel is passing through mountain covered with thick vegetation and tunnel passing through the strata mentioned above.

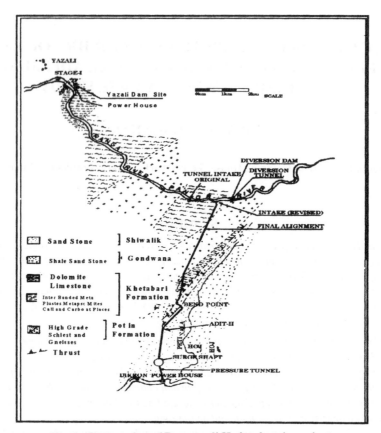

Fig. 1 : Layout plan of Ranganadi Hydroelectric project

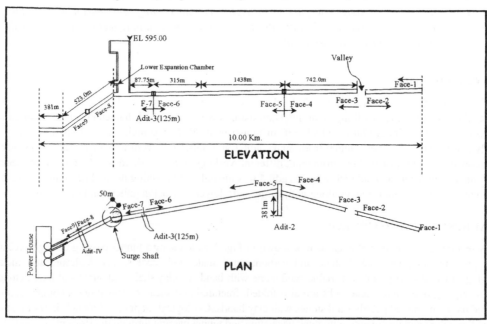

Fig. 2 : Longitudinal section showing adits, working faces plan and elevation

Tunnel boring was planned by using 2 boom drill jumbo and robo arm Aliva shotcrete machine by adopting NATM technique. RMR value found along the tunnel alignment is tabulated below :

R.D.	Average RMR value
0 to 1319	10 to 46
1319 to 5435	60 to 65
5435 to 6210	12
6210 to exit	30 to 42

Table shows the poor RMR value along the tunnel alignment except for the stretch between RD 1319 TO RD 5435. NATM therefore could not be adopted. It was felt that it is extremely difficult to excavate tunnel through the thrust zone and in view of this the tunnel alignment was revised thus avoiding the thrust zone. A plan of revised alignment is shown in Fig. 3.

3.0 INCLINED TUNNEL

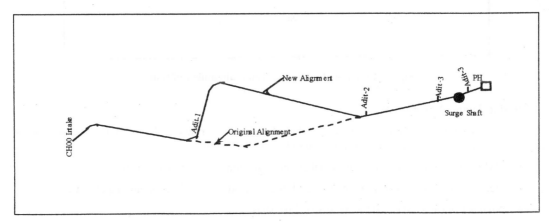

Fig. 3 : Change in tunnel avoiding thrust zone

In order to expedite tunnel boring it was decided to open Adit III located approximately 75 m u/s of surge shaft though this was contrary to initial planning to bore tunnel from only four faces.

Adit III was having a length of 142 m approximately. Boring of adit III posed many problems. Due to poor rock condition entire length of adit was supported by cold bend ISMB 250 steel ribs along with precast slabs and backfill concrete. While boring was at chainage 135 m supports at chainage 68 got deformed resulting in roof collapse and subsidence of hill 60 m above the crown. Remedial measures involved forepoling, backfill concrete and strengthening of present supports and thereafter grouting at low pressure. A second layer of rib was provided below the existing supports and then the muck was removed cautiously and strengthening the supports simultaneously. Remaining length of adit III was then bored through to reach junction point with main tunnel. Junction was the place where three portals were to be erected. One at tunnel from adit III and other two on u/s and d/s faces of main tunnel. RMR value of rock was 20-22. Therefore the junction was prepared by extensive supports of ISMB 250 at 0.4 m c/c and in addition ISMB 400 were also provided at both ends. Despite these extensive supports due to poor rock condition entire junction collapsed, approximately 800 to 1000m³ muck collapsed from tunnel crown. This situation forced a decision to realign tunnel by shifting it towards adit III by 12 m by providing smooth curve as shown in Fig. 4. Precautions were taken to construct RCC portal by embedding ISMB 400,both u/s and d/s faces. On completion of portals tunnel towards surge shaft was bored by heading only. Pull was restricted to 0.5 m. Supports were erected by using

ISMB 250AT 0.5 m c/c with precast lagging and backfill concrete. Benching was then taken up carefully by providing vertical girder to each rib. With such care tunnel was bored up to surge shaft from where inclined tunnel boring was to start at down gradient of 30 degree.

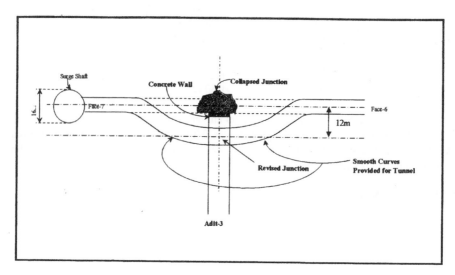

Fig. 4 : Collapse of juction - 3 and realignment of tunnel

3.1 Boring Tunnel Downwards

Boring inclined tunnel down from surge shaft face posed serious problem. Boring became almost impossible by normal boring method because of problems given below :

- RMR value of rock was very low and therefore stand time was less than one hour.
- Heavy seepage was observed with the result roof and sides were slipping causing cavities.
- No mechanical loader could work at 30 degree slope.

Works remain almost at stand still for over a month. Due to heavy seepage and very poor rock condition it was not possible to bore tunnel downwards. A high level meeting of technical experts was held and two proposals were discussed.

3.1.1 Proposal 1

Excavate vertical shaft of 6.8 m diameter below surge shaft to a depth of 100.35 m. Excavated adit 4 to meet inclined tunnel at chainage 9493.7 m. Bore the tunnel upwards at 10 degree instead of 30 degree and meet the further extended shaft at EL 396.812m. Excavate one pressure shaft at 192.539m from the horizontal stretch of tunnel from Exit1.

3.1.2 Proposal 2

- Create niches at spring level of tunnel on both sides.
- Drill 30 m long holes 5 degree inclined upwards, 50 mm diameter parallel to tunnel (this was with the intention to arrest seepage water and collect it in the niches for disposal).
- Drill three numbers of 6m long 50mm diameter advance holes at crown and at 30 degree on both side for pre-drainage.
- Advance only by taking blast for 0.5 to 1m, load mild charge.

- Immediately support by steel rib and precast lagging.
- Muck loading manually and hauled by rail trolley system installed on one side of tunnel.
- For next blast adopt forepoling for roof support.

Proposal 2 was approved and with all these precautions though tunnel boring was done successfully but due to manual loading and heavy seepage average boring speed was only 8 m/month in heading only.

To increase the speed of boring it was necessary to increase speed of muck loading and hauling it up to top for disposal. Being the boring downwards necessarily a backhoe type excavator was required. Ex 100 (shovel type) was used in horizontal stretch of tunnel was not useful. On conversion to backhoe it would not operate inside 6 m diameter heading of inclined tunnel. A newly launched EX60 excavator was deployed for loading muck in the rail wagons. In order to operate EX60 in 30 degree slope a false level platform was made on which EX60 was parked and load the muck on 1 m³ wagons 2 numbers at a time. These wagons were hauled up by positive drive electrically driven winches. These wagons were hauled on elevated platform from where they were unloaded directly by side tilting on dumpers, which was parked below the elevated platform. Dumper then used to haul and dispose off muck at dumping yard through adit III. EX60 was not removed from the tunnel and was protected by steel shield during blasting operation. After using EX60 for loading of muck the speed of mucking was increased. Cycle time was improved substantially and an average boring speed of 20 m/month was achieved.

3.2 Boring Inclined Tunnel Upwards

Boring from exit face for inclined tunnel upwards was also carried out. Boring from bottom upwards was carried out taking full tunnel section because the rock condition was better and there was negligible seepage at this face. Drilling upwards at 30 degree posed a problem since keeping drilling platform at 30 degree was not possible. Drilling jumbo could not work on the steep gradient or 30 degree. To create drilling platform semicircular plates were welded to the erected steel ribs at 2 m c/c. and pipes of 4"diameter cut to exact length were placed on it. Wooden plants 2"thick were then placed spanning pipe to pipe to create working platforms for drillers. A netted steel (jali) was suspended from crown to prevent falling of small stones on the drilling crew. After charging explosives the working platform was dismantled and blast was taken.

Removal of muck was again a problem :

- At 30 degree rock would not slide/fall by it self.
- No excavator was capable of working at 30 degree slope.
- Manual loading would take lot of time.
- None of the mobile hauling equipment was able to work on 30 degree slope.

In order to overcome the above problems, following method was adopted.

Excavator EX100 was selected for loading of muck. Since the excavator was not designed to climb 30 degree slope, it was decided to take the help of hydraulic cylinders of boom and bucket to support the movement upwards. A level platform was then made by using the available blast muck and EX100 was parked on the level platform so that it can be made operative. It would not have been possible; to remove excavator during blasting operation. In order to protect the excavator a steel shield was made in 8 segments, which was placed to protect excavator from damage. In order to haul the muck downwards a 600 mm gauge rail track was laid on one side of tunnel and two loaded wagons of 1 m³ each were descended at a time by using a winch which was installed at the bottom of the inclined tunnel by providing a small niche. Return rope from winch was taken through a pulley anchored at invert of tunnel by embedding a girder ISMS 250 and concreting it. Anchor was shifted after every 5 m advance. Tremendous load would come on the anchor and thus after boring 142 m from bottom boring from bottom was suspended for safety reasons.

4.0 ADIT 4

In order to expedite boring of inclined tunnel one more adit (named as adit 4) was proposed for boring. Length of adit 4 was 251 m. This was a typical adit. Unlike other usual adit this adit was intersecting inclined tunnel at 30 degree. Adit 4 was designed as 8 m wide x 7 m height. (These dimensions were adopted so that steel liner can also be erected from this face. Junction of adit 4 to inclined tunnel posed problems because adit 4 was horizontal tunnel meeting inclined tunnel at 30 degree which means one end of inclined tunnel to adit 4 will be higher than other by 4.4 m . From this adit inclined tunnel was proposed to be bored both upwards as well as downwards. Therefore a very well planned adit junction was constructed which is technically excellent example for tunnelling engineers.

4.1 Adit 4 and Inclined Tunnel Junction

Construction of adit 4 tunnel of 251 m was decided for multipurpose functions :

1. Boring of inclined tunnel upwards 85 m.

2. Boring of inclined tunnel downwards 73 m.

3. Erection of steel liner for downward tunnel.

4. Concrete behind steel liner.

Adit 4 was meeting inclined tunnel at ch. 9493.7 m. Keeping all above four functions in mind a well planned junction was constructed which was one of the difficult construction under prevailing geological conditions. A plan and alignment of adit 4 is shown in Fig. 5 and a section indicating the typical junction is also shown in same figure (Fig. 5). It will be seen that at junction height of structure comes to 11.85 m. *i.e.* crown of adit 4 gradually rising from EL 354.65 to EL 360.00 and invert level moderately lowering down to EL 348.15 from an EL348.95 m. Therefore at ch.239 of adit 4 muck was filled up and a slope was made for the excavation of junction up to crown level of 360.2m. Adit 4 was extended beyond the intersection point to accommodate the winches, proposed to be installed for handling muck wagons in both directions *i.e.* upward and downwards. While the support to adit tunnel for crown was given by steel ribs, it was necessary that at perpendicular direction these supports are rested on a heavy beam (ISMS 600) spanning across the width of inclined tunnel. This heavy beam was then rested on vertical girders erected on both sides of inclined tunnel width. Thereafter both the vertical as well as horizontal beams were embedded to form R.C.C. portal. R.C.C. portals were constructed on both sides of adit *i.e.* one portal to take care of higher side of inclined tunnel and second one at lower side of inclined tunnel. Adit roof above the inclined tunnel was supported by placing the cross girders spanning both the portals. Vertical girders over these horizontal girders were erected and circular steel ribs were supported by welding these to steel ribs. In order to support the cross girders at same elevation vertical girders over the lower portal were erected so that cross girder is at same level. Junction of typical intersection between inclined tunnel and adit 4 was thus structurally supported and made safe.

Functionally adit junction was now required to be prepared for :

(i) Boring tunnel upwards

(ii) Boring tunnel downwards.

Adit 4 junction required preparation by making some adjustment on levels in the adit to suit the differential level of 4.4 m between u/s and d/s face of adit 4. It was planned to excavate tunnel upwards by using EX100 excavator for loading and rail wagons for hauling the muck down. Rail tracks of 600 mm gauge were laid in a specially designed manner which was capable of multiple function :

(i) It could unload muck directly in dumper parked by the side of horizontal stretch of rail track. A specially designed winch was used to tilt the bucket from rail trolley.

(ii) Alternately rail trolley was brought to lower platform. In this case bucket was lifted by a EOT crane and unloaded in the dumper parked in adit 4 tunnel.

306

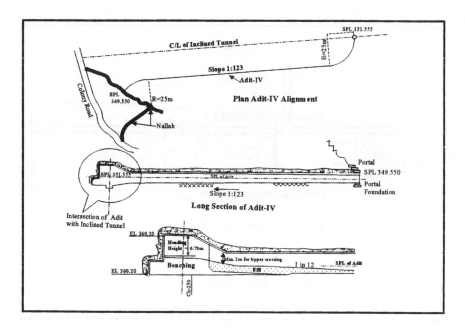

Fig. 5 : L section of adit and junction

(iii) If for any reasons wagon was not unloaded, trolley was brought on turn table and loaded wagon and trolley was parked in the niche portion.

(iv) Rail back was extended for the d/s side of inclined tunnel (for excavation of heading only).

For operating all above functions following preparation of infrastructure was done in adit 4 junction.

1. Lay the rail track in adit then take slope for one platform on u/s side (for tilting bucket).

2. Install turn table at adit portion and lay cross track towards niche.

3. Lay rail track on d/s side tunnel.

4. Install 3 numbers positive drive winches-

   (a) One winch for hauling muck on u/s side of tunnel.

   (b) One winch for hauling muck on d/s side of tunnel.

   (c) One winch for tilting the bucket for bucket for unloading directly on dumper.

5. Install EOT crane of 5MT capacity spanning across inclined tunnel.

6. Prepare parking place for dumper (by the side of tilting bucket platform).

7. Install safety steel gates on u/s and d/s portion of rail track.

8. In order to arrest seepage water construct a RCC wall at u/s side and embed 1"diameter pipe so that dewatering is done by gravity.

Figures 6 and 7 showing the fall details of above installations.

On installation of such elaborate arrangement full face tunnel boring to u/s side to a length of 85 m was completed. On d/s side only heading was first done to a length of 73 m. On completion of these activities last activity left out was benching towards d/s side and removal of horizontal portion from the junction so that inclined tunnel u/s and d/s of adit 4 is joined in one line. Excavation of junction portion was started from u/s portal column towards d/s face. Rail track laid for heading was removed and re-laid at invert level of inclined tunnel and benching of d/s side tunnel was completed. While rail track was laid at invert level on inclined

307

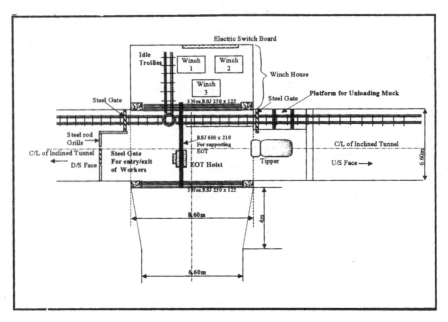

Fig. 6 : Plan of installation at adit-I and inclined tunnel junction

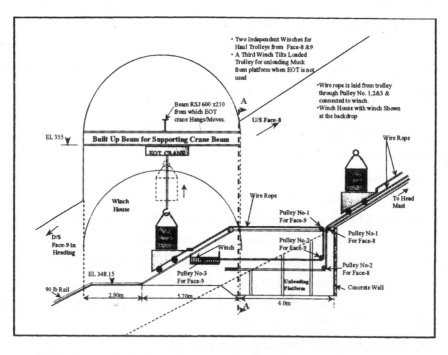

Fig. 7 : Section through junction showing rail trolley and EOT arrangements

tunnel a horizontal platform was made in balance portion of tunnel by girders and planks so that dumper can be parked on it to enable EOT crane to unload muck bucket in it. Muck loading on d/s face was done by using EX60 excavator. A schematic arrangement of these activities is shown in Fig. 8.

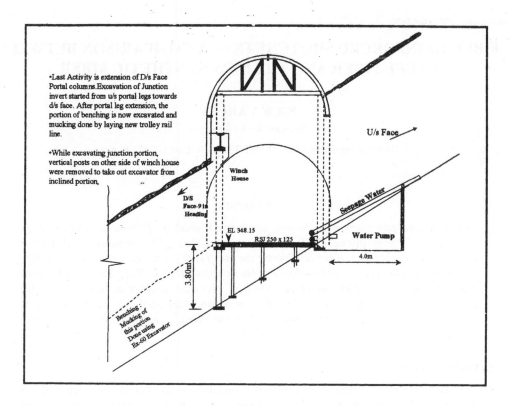

Fig. 8

5.0 CONCLUSION

- Proper and detailed geological investigation must be done before the start of the project. This will avoid change in construction plans delays arising out of it.

- Inclined tunnel should either be made steep enough so that after blast the muck rolls down by itself to the bottom of tunnel. Alternately the slope should be mild enough so that excavators can work on such slope and hauling equipments can negotiate the slope.

- Survey of such complex junction has got to be done with utmost accuracy so that alignment is proper on daylighting.

FIBER REINFORCED SHOTCRETE — A COMPARISON BETWEEN STEEL FIBER AND THE NEW SYNTHETIC FIBER

NICK VARLEY

Business Sector Manager

Underground Construction, MBT Middle East L.L.C., U.A.E.

SYNOPSIS

Until now, steel fibers have been the only alternative to mesh reinforcement. Lately the High Performance Polypropylene (HPP) fibers have reached a performance level that is comparable to steel fibers. When considering the use of such fibers, the normal response from new potential users is to query how does this fiber compare to steel fiber? Steel fiber has been used for many years and has been most successful. This short paper is an attempt to provide some answers to that question, explain the test methodology used and make reference to information from recent tests carried out by university and other reputable testing laboratories.

1.0 INTRODUCTION

The advantages of fiber reinforcement in shotcrete have been demonstrated in a number of projects and applications around the world. When using state of the art technology, the technical performance of fiber reinforcement is generally equally good or better than traditional mesh reinforcement. Additionally, it is giving a number of other advantages :

(a) Overall productivity when applying on drill and blast rock surfaces is often more than doubled.

(b) Substantially improved safety, shotcrete with reinforcement can be placed by remote controlled manipulator. (nobody venturing below partly supported or unsupported ground to install the mesh).

(c) No areas of poor compaction behind overlap areas of 3 to 4 layers of mesh, causing very poor concrete quality and high risk of subsequent mesh corrosion and concrete cover spalling.

(d) The intended thickness and overall quantity of shotcrete can be achieved quite accurately and the problem of excess quantity to cover the mesh on rough substrates is avoided.

(e) One layer of mesh will be placed at varying depth in the shotcrete layer and cannot be placed in the cross section to target tension zones. The fibers will be present in the whole cross section, irrespective of where the tension will occur.

(f) Logistic advantage of avoiding handling and storage of reinforcement mesh under ground.

To be able to achieve the above advantages the wet mix application method has to be used. The reason for this is the lack of control and the large amount of fiber rebound in the dry process. Typically the steel fiber rebound in the dry process is above 50 per cent and using synthetic fibers it is likely to be even higher. In comparison, steel fiber rebound is typically 10 to 15 per cent in the wet mix process (Fig. 1).

2.0 DESCRIPTION OF NEW S-152 HPP FIBER

Developed at the University of British Columbia, the S 152 HPP concrete reinforcement is a polymer based material specifically designed to extend the performance of concrete under stress. Mixed into the concrete using conventional mixing techniques, these high strength fibers allow for relatively high fiber volumes of 1 to 2 percent (9 to 18 kg/m^3 of concrete). In certain situations, higher fiber proportions can yield structural performance levels that extend beyond the level achieved using secondary reinforcement.

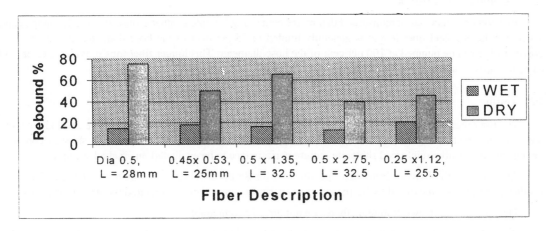

Fig. 1 : Typical results for steel fiber rebound in shotcrete (D.Wood *et. al.*)

The HPP fibers are specifically shaped to resist matrix pullout, enhancing the concrete performance even after it has developed stress cracks. Tests have shown that HPP fibers improve the load deflection performance of stressed concrete. This performance is desired in many tunnelling applications.

Packaged in 0.9 kg water-degradable bags, which are easily mixed with the fresh concrete, this product allows for customised dosage rates to give the finished product the desired end-use properties.

Property	Value
Specific gravity	0.91
Nominal filament diameter	0.9 mm
Fiber lengths available	35; 40; 50 mm
Deformation spacing	3/10 mm
Elastic modulus	3500 MPa
Ultimate elongation	15%
Water absorption	Zero
Melting temperature	1750 C
Ignition temperature	3600 C
Acid/alkali resistance	High
Thermal conductivity	0.2 W/mK at 20°C
Electric conductivity	Very low

2.1 Site Batching and Mixing

The S-152 HPP fiber is added to the shotcrete mix during the mixing process. For a large-scale application, automated dispensing is preferred, the vibratory dispenser used for steel fiber is not possible due to the low density of the synthetic fiber. Using practical experience from site it has been discovered that these synthetic fibers may be added in a controlled manner by using a modified auger dispenser similar to that used for silica fume. Addition rates of between 3 and 9 kg/m³ are easily achievable. The dosing of the fiber becomes an integral part of the mixing process and thus an increase in mixing time is not generally required.

2.2 Pumping and Spraying

Experience to date has demonstrated that mixes containing flexible synthetic fibres are easily pumped and sprayed. Whereas steel fiber length is generally limited to 75 per cent of the hose diameter, the more pliant polymer fiber can be pumped at 100 per cent of the hose diameter. This longer fiber thus provides greater load carrying capacity across cracks with deformations in excess of 4 mm.

2.3 Testing

The fiber reinforcement imparts significant ductility, crack resistance, impact resistance and toughness to the concrete lining. Toughness is best described as the load deflection performance of a stressed concrete. Specifications for fiber reinforced shotcrete require performance to be evaluated on cracked specimens.

According to EFNARC – European Specification for Sprayed Concrete, toughness is specified by either residual strength class measured by beam testing or energy absorption class measured by plate or panel testing.

The results of these two testing methods are not directly comparable.

Beam Testing - There are many different test methods but the common three are the American test to ASTM C 1018 which derives a toughness index as a performance indicator, the Japanese Standard JSCE-SF4 which measures flexural strength at maximum loading *i.e.* "Modulus of Rupture" and thus derives an Equivalent Flexural Strength as a toughness measurement. Lastly, the EFNARC third point beam test which is commonly specified. This is a well-established procedure using specimens of 75 x 125 x 550 cut from larger sprayed specimens. The test uses third-point loading apparatus, measurement is over a span of 450 mm. Residual strength is estimated from the load / deflection curve using flexural stress values between 0.5 and 1, 2, or 4 mm depending on the specified deformation class as shown in following Table.

Residual strength class definition points – EFNARC specification for sprayed concrete

Deformation class	Beam deflection (mm)	Residual strength in MPa for strength class			
		1			
	0.5	1.5	2.5	3.5	4.5
Low	2	1.3	2.3	3.3	4.3
Normal	3	1.0	2.0	3.0	4.0
High	4	0.5	1.5	2.5	3.5

Plate/Slab/Panel Testing - There are three main testing methods, the French Slab method developed by the French Railways in conjunction with Alpes Essais Laboratory of Grenoble. The test slab is 600 mm x 600 mm x 100 mm, supported on all four edges and has a centre point load of 100 mm x 100 mm. The EFNARC plate test which has been adopted from the French Slab method and the Round Determinate Panel developed by Bernard. The round panel is 800 mm in diameter, 75 mm thick and is supported on three radial points located at 750 mm diameter.

These tests measure load and deformation and the property described as 'toughness' is derived from the energy absorbed in units (Joules) calculated by integrating the area under the load – displacement curve.

Concrete exhibiting a higher energy absorption will be able to carry more load in a cracked state. For specified values according to EFNARC test methods as shown in following Table.

Energy absorption class – EFNARC specification for sprayed concrete

Toughness Classification	Energy absorption in joule (J) for deflection up to 25mm
A	500
B	700
C	1000

3.0 FURTHER SUPPORTING EVIDENCE

The following references illustrate the considerable investigation and research that has gone into the evaluation of the synthetic fibre.

3.1 Contribution of Bernard (1999)

The School of Civic Engineering and Environment at the University of Western Sydney, Nepean, Kingswood NSW Australia has issued the Engineering Report No.: CE9 with the following title :

'Correlations in the Performance of Fibre Reinforced Shotcrete Beams and Panels'

The report contains 18 pages in total and presents the results of testing of 204 beams and 204 panels. Using one concrete mix design for the spraying into plywood moulds, only the type and quantity of fibers were varied. The tests carried out cover EFNARC third point beam test, centrally loaded beam test the EFNARC panel test and the Round Determinate Panel test. Three specimens were used for each test method for each mix.

The basic concrete mix used for the tests (with variable fiber and quantity) was :

Ingredient	Quantity (kg/m^3)
Coarse aggregate (7/10 mm)	500
Coarse sand	775
Fine sand	415
Cement (GP)	360
Silica fume	40
Water	200
Fibers	Varied
Slump	70-80 mm
(Note -Slump adjusted with superplasticiser)	

From the 34 mixes tested, the following have been selected for comparison here :

Mix No.	Mix design	Reinforcement	kg/m^3
3	1	Drainix ZP 30 fibers	25
4	1	Dramix ZP 30 fibers	25
5	1	Novotex 0730 fibers	50
8	2	50 mm HPP fibers	9
9	2	50 mm HPP fibers	13

Here it must be noted that mix numbers 1 through 5 had a slightly different concrete mix design. The main difference being 20 kg more cement and 40 kg fly ash added. When comparing results for mix no. 3 to 5 this has no influence. When comparing results of 3 to 5 with 8 and 9, the difference in concrete mix design can have an influence. It is likely that this difference goes to the advantage of results measured for mixes 3 to 5. The selected mean of three specimen results are reported as follows :

Mix No.	UCS (MPa)	EFNARC panel energy (J)	Round panels energy (J)
3. 25 kg/m³ Dramix ZP30		765	256
4. 50 kg/m³ Dramix ZP30		1026	439
5. 50 kg/m³ Novotex 0730		1096	439
8. 9 kg/m³ HPP 50 mm	59.8	1171	455
9. 13.5 kg/m³ HPP 50 mm	58.5	1450	513

3.2 Contribution of Banthia and Yan

As part of a larger test programme conducted by AGRA Earth and Environmental Ltd., six wet-mix shotcrete panels were shipped to the University of British Columbia, Vancouver, where they were tested using the standard EFNARC test procedure. All panels contained as reinforcement the 50 mm long S 152 high performance polypropylene fiber at a dosage rate of 1 per cent. This report presents data for four panels; two panels tested at an age of 7 days and the other two tested at an age of 28 days. The remaining two panels will be tested for long term toughness assessment at an age of 96 days.

3.2.1 Test Procedure and Results

The procedure adopted for the tests in outlined by Bernard (1991) (§ 10.4). In brief, during a test, the 600 mm x 600 mm x 100 mm shotcrete slab was simply supported on its four edges and a centre point load was applied through a contact surface of 100 mm x 100 mm (Fig. 1). The rough side of the panel was at the bottom during the test, i.e. the load was applied opposite to the shooting direction. A 1.78 MN capacity stiff machine was used for the test where the load was applied at a cross-arm displacement rate of 1.5 mm/min. The plate was stored in water before test and kept moist during the test. The load deformation curve was obtained, and from the load deformation curve a second curve was drawn1 of energy absorbed as a function of panel deformation.

Notice that the two 7-day old slabs absorbed energies, respectively, of 1012 J and 1047 J at a load point displacement of 25 mm. At the age of 28 days, on the other hand, the two slabs absorbed slightly higher energies, respectively, of 1122 J and 1177 J. When the ultimate loads are considered, the 7-day old slabs supported an average ultimate load of 58.81 kN and those with an age of 28 days supported an average ultimate load of 63.86 kN. The load displacement responses are generally linear to about 50 per cent of the ultimate load after which the curves became distinctly non-linear with a substantial amount of strain-hardening. The descending part of the curve depicted stable softening with large amounts of energy absorbed. An observation of the fracture surfaces revealed that most fibers had pulled out with imprints from the pull-out process.

3.2.2 Observations

It was found that wet-mix shotcrete panels reinforced with 50 mm long S 152 deformed polypropylene fiber at a dosage rate of 1 per cent when tested according to the EFNARC procedure absorbed at both ages of 7 and 28 days energies in excess of 1000 J at a load point displacement of 25 min.

The remaining two panels will be tested for long term toughness assessment at an age of 96 days.

3.3 Contribution of EFNARC (1996)

EFNARC - (European Specification for Sprayed Concrete, European Federation of Producers and Applicators of Specialist Products for Structures, Hampshire UK, 1996) document contains 5 pages of illustrations; 1 Table and Fig. and 4 pages with load deflection and energy absorption curves.

3.4 Contribution of Morgan

In a memo Dr. Rusty Morgan presents the following evaluations to Rick Smith of Synthetic Industries, regarding HPP fibers compared to steel fibers used as reinforcement in shotcrete :

"There are various advantages and disadvantages with either steel or synthetic fibre reinforced shotcretes for underground support in mines. With respect to the plastic shotcrete, properly designed high volume synthetic fibers are typically more *user-friendly* in pumping, shooting and finishing compared to steel fibres. Synthetic fibres result in lower pump pressures and less wear and tear on equipment, hoses and nozzles than steel fibres. Where -finished surfaces are required, the synthetic fibre shotcretes typically are easier to cut, trim and finish than steel fibre reinforced shotcretes. All the preceding comments apply to wet-mix shotcretes. We have not worked with high volume synthetic fibres in dry-mix shotcretes.

With respect to the hardened shotcrete, steel fibre reinforced shotcrete typically is better at providing higher post-first crack residual load carrying capacity than high volume synthetic fibre reinforced shotcretes at low deformations when crack widths are narrow. By corrolary, steel fibre reinforced shotcrete [SFRS] with a sufficient fibre addition rate [say 50 to 60 kg/m^3] is generally preferred in permanent mine works where little ground movement is expected and near watertight conditions (with narrow crack widths) are needed, *e.g.* underground hoist rooms, crusher stations, electrical and pump rooms, etc.

By contrast, high volume synthetic fibre reinforced shotcretes [SnFRC], [1.0 to 1.5 per cent vol.) typically display better residual load carrying capacity at larger deformations compared to SFRS. Refer to the paper *on Comparative Evaluation of System Ductility of Mesh and Fibre Reinforced Shotcretes* 1 presented in Brazil at the Shotcrete for underground support VII conference. In this respect SnFRS behaves more like mesh reinforcement. As such its use is preferred in applications where larger ground deformations and hence wider cracks developing in the shotcrete are to be expected. Examples might include squeezing ground, collapsing slopes, deforming pillars and shotcrete subjected to rock burst type conditions. SnFRS is being used quite widely in the deep gold mines in South Africa for this purpose.

Finally, another advantage of SnFRS over SFRS is that any protruding fibres will not cause cuts, scrapes and skin lacerations that can be caused by steel fibres. This has been a contention issue with miners in certain mines.

3.5 Contribution of Geopractica (1999)

The spraying of test panels was carried out with low strength concrete, giving compressive strength from 20 to 26 MPa. The accelerator used was on silicate basis, insufficient supply of compressed air to the nozzle created poor compaction, which is reflected in the shotcrete density of 2.16 to 2.25 (should have been around 2.3). Consequently, the failure energy tested on EFNARC panels is generally on the low side.

Harex steel fibers of 1100 MPa steel quality was tested in dosages from 20 to 40 kg/m^3 and HPP fibers 30 and 50 mm from 5 to 10 kg/m^3 of mixed shotcrete. The most interesting results are the 40 kg dosage of Harex, compared to 7.5 and 10 kg of 50 mm HPP. The results (average of 3 panels for each result) are :

40 kg/m^3 of Harex steel fibers	329 Joules
7.5 kg/m^3 of 50 mm EPP fibers	467 Joules
10 kg/m^3 of 50 mm HPP fibers	650 Joules

REFERENCES

Bemard, E.S. July 1999, "Correlations in the Performance of Fiber Reinforced Shotcrete Beams and Panels", The School of Civic Engineering and Environment at the University of Western Sydney, Nepean, Kingswood NSW Australia, Engineering Report No. CE9.

Banthia N. and Yan, C., "Toughness of Fiber Reinforced Shotcrete Panels (EFNARC) with S 152 Deformed Polypropylene Macro-Fiber, Department of Civil Engineering, The University of British Columbia, Vancouver, BC, Canada, VGT IZ4.

Morgan, R., "Memo to Rick Smith of Synthetic Industries, AGRA Earth and Environmental Ltd.

Geopractica C.C., 1999, Roodepoort, PO Box 227 Maraisburg, South Africa, "Moab Khotsong, SFRS Wet Shotcrete Application - Spray Trials Conducted on 17 December, Jobnwnber: 99 1 00"

Garshol Knt F. – July 2000, Comparing Steel Fiber with new Synthetic Fiber, www.ugc.mbt.com.

Kimball W. and Galinat M.A. – 1999, A Synthetic Spray Solution – Concrete Engineering International June/July.

SPECIAL TUNNELLING TECHNIQUES

CONTROLLED BLASTING FOR REMOVAL OF CONCRETE PLUGS IN DRAFT TUBE TUNNELS AT SARDAR SAROVAR PROJECT

G.R. ADHIKARI   A.I. THERESRAJ   R. BALACHANDER   R.N. GUPTA

National Institute of Rock Mechanics, Champion Reef Post, Kolar Gold Fields, India

B.B. VASAVA

Deputy Executive Engineer

Sardar Sarovar Project, Kevadia Colony, India

SYNOPSIS

The Sardar Sarovar (Narmada) Hydroelectric Project is under construction in the state of Gujarat. The underground powerhouse complex includes six pressure shafts, powerhouse cavern with six turbine units, six draft tube tunnels, exit tunnels and tailrace tunnels. An unprecedented flood entered the powerhouse cavern through draft tube tunnels on 7th September 1994 and caused damage to underground structures. After this incident these draft tubes were plugged in order to prevent the entry of the flood. The specification of concrete plugs and reinforcement in the draft tube tunnel are described.

After heavy gates were erected at the exit end of the draft tube tunnels, the concrete plugs had to be removed. With the varied reinforcement/concrete support and change in shape the excavation of draft tube tunnel by drilling and blasting was very complicated and challenging. Moreover there was concern for damage to the concrete lining and the ribs erected. Proper care while charging the perimeter holes had to be taken to avoid the damage.

Controlled blasting was designed and executed for removal of these plugs in the same manner as for tunnel blasting. The concrete plugs have been successfully removed. The paper highlights the interesting observations in respect of drilling and blasting in the different grades of concrete and reinforcement.

1.0 INTRODUCTION

One of the largest river valley project, the Sardar Sarovar (Narmada) Hydroelectric Project is under construction in the state of Gujarat. The project envisages the construction of a concrete dam of 128 m high and 1210 m long across the river Narmada. The underground powerhouse complex comprises six pressure shafts, powerhouse cavern with six turbine units with an installed capacity of 6 x 200 MW, six draft tube tunnels, collection pool and exit tunnels.

The tailrace system of the project comprises six draft tube tunnels connecting the machine hall to an open collection pool. The finished and excavated diameters of these draft tube tunnels are 10.0 m and 11.1 m respectively. But they vary from 30.0 m chainage to 0.00 m chainage. The chainage are marked from the downstream wall of the machine hall towards the collection pool. The length of the draft tube tunnels varies from 150 m to 175 m. The minimum rock ledge between any two successive draft tube tunnels is 9.0 m on the downstream wall of the machine hall. It gradually increases to 13.0 m at 30.0 m chainage, which is constant for the remaining length up to the collection pool. These draft tube tunnels were excavated by drilling and blasting. The layout of the draft tube tunnels is shown in Fig. 1.

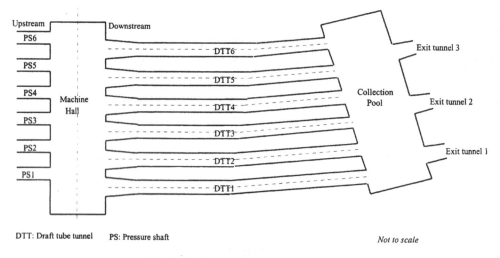

DTT: Draft tube tunnel   PS: Pressure shaft   *Not to scale*

Fig. 1 : The layout of the draft tube tunnels

On 7th September 1994, an unprecedented flood entered the machine hall through the exit tunnels, collection pool and draft tube tunnels. At that time the gates of the draft tube tunnels were not installed. After this incident, in order to prevent flood entry from the downstream side of the dam into the machine hall during monsoon, these draft tubes were plugged. After installation of the heavy gates for the draft tube tunnels, the concrete plugs were to be removed.

2.0 Details of the Concrete Plugs

The rock excavation including benching in draft tube tunnels 1 and 2 was completed. The 16 m long plug was provided from chainage 148.65 m to 164.0 m in draft tube tunnel 1 and from chainage 143.87 m to 159.87 m in draft tube tunnel 2. The design of the concrete plug was same in both the tunnels.

The draft tube tunnel 1 at the collection pool side is D-shaped with a height of 10.0 m and the width varies from 8.5 m at 164.0 m chainage to 10.0 m at 148.65 m chainage. The ribs were installed at an interval of 0.5 m. For the first one meter from the collection pool side the concrete was of M20 MSA20 for the full face. Steel rods of 20 mm diameter were also provided at 0.3 m spacing in horizontal and vertical directions. In addition to this, 25 mm diameter and 4.0 m long anchors were grouted at the crown and sides of the tunnel. From 163.0 m chainage, the M20 MSA20 concrete was provided only for the top portion of 7.0 m which gradually decreased to 2.4 m at 148.65 m chainage. The lower portion was concreted with M10 MSA150 though initially it was planned to use M10 MSA150 concrete only. After reaching a certain height, it was difficult in providing the plug manually with this concrete. In order to complete the work before the monsoon, the top portion of the plug was provided mechanically. As the cement mixture pump would not handle MSA150, the top portion was plugged with M20 MSA20.

The draft tube tunnel 1 was not lined with concrete, but ribs were erected. In draft tube tunnel 2 the lining was completed for 11 m and the unlined portion was for only 5 m from the collection pool. The longitudinal section of the plug in draft tube tunnel 1 and 2 is shown in Fig. 2.

In draft tube tunnel 3 and 4, the rock was excavated only for the heading portion. The lengths of concrete plugs were 4.0 m from chainage 21.0 m to 25.0 m. It had reinforcement consisting of 36 mm diameter at the spacing of 0.25 m for vertical rods and 16 mm diameter at the spacing of 0.25 m for horizontal rods. The longitudinal section of the plug in draft tube tunnel 3 and 4 is shown in Fig. 3.

In draft tube tunnel 5, the length of the rock heading to be excavated was 6 m and that of benching was 25 m. The length of concrete plug was 6.0 m from chainage 6.0 m to 12.0 m with grade of M20

MSA20. The longitudinal section of the plug in draft tube tunnel 5 is shown in Fig. 4. In draft tube tunnel 6, the position of benching was same as in draft tube tunnel 5 but the heading to be excavated was 15 m. Considering the longer length of the unexcavated portion of the heading in draft tube tunnel 6, a plain concrete plug was provided only for a length of 2.2 m from chainage 17.25 m to 15.0 m. Even though the rock excavation was yet to be completed, the plugs in these tunnels were provided to restrict the seepage of water through the joints and fissures.

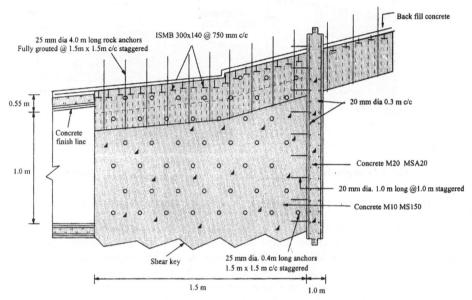

Fig. 2 : Longitudinal section of the concrete plug in draft tube tunnels 1 and 2

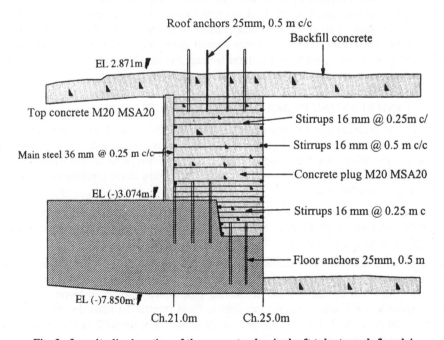

Fig. 3 : Longitudinal section of the concrete plug in draft tube tunnels 3 and 4

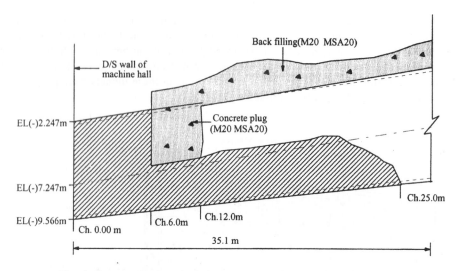

Fig. 4 : Longitudinal section of the concrete plug in draft tube tunnel 5

3.0 THE METHOD ADOPTED FOR REMOVAL OF THE CONCRETE PLUGS

The concrete plug was removed by drilling and blasting in the same manner as for tunnel excavation. The holes were of 34 - 38 mm in diameter and drilled with jack hammers. Wedge cut pattern was followed in all the draft tube tunnels. The perimeter holes were drilled at spacing of 0.3 m. Before drilling, the position of the holes were marked on the face. For drilling at higher levels a scaffold was assembled near the face and wooden plànks were placèd to form stages. The holes were charged with Superdyne, the only available slurry explosive at the site. Each cartridge of explosive was of 25 mm diameter, 200 mm long, weighing 0.125 kg. Even though it was recommended to charge alternate perimeter holes with a detonating cord, the holes were charged with Superdyne using bamboo spacers since the detonating cord was not available at the site. The holes were initiated with a combination of short and long delay electric detonators to reduce the maximum charge per delay. Because of the varied reinforcement / concrete support and change in shape the excavation of draft tube tunnel, drilling and blasting was complicated and challenging. Further, blasting had to be conducted without causing any damage to the concrete lining and to the ribs already erected, not to mention the rock ledge.

4.0 DRILLING AND BLASTING IN DRAFT TUBE TUNNEL 1 AND 2

To start with the excavation of the concrete plugs in draft tube tunnel 1 and 2 was taken by full face method for 10 m diameter. Due to the long cycle time per blast it was decided to go in for heading and benching method. If the holes were drilled dry, there was a severe dust problem and when drilled wet there was frequent rod jamming. The holes got jammed even after drilling due to the presence of pebbles in the M10 MSA150 concrete portion of the plug.

In 1999, when monsoon was over and after completion of dewatering from the collection pool, it was decided to start the excavation work from the collection pool side. The advantages of the excavation from this side were natural ventilation, quick dust dissipation and the availability of day light.

A heading of 5.0 m diameter was excavated first so as to allow the excavator to go and operate near the face. The blast design for heading is given in Fig. 5. Some of the holes could not be drilled to the required depth due to the obstruction by anchors and dewatering pipe which resulted in lesser pull. The steel rod reinforcement exposed after blasting required arc cutting. Even though the blast design remained same for these two tunnels, the perimeter holes were charged lightly in draft tube tunnel 2 as compared to that of draft tube tunnel 1 to avoid damage to the concrete lining.

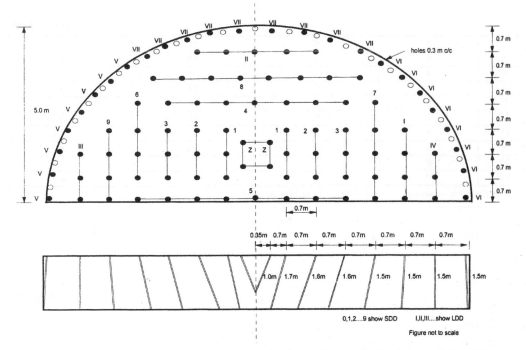

Fig. 5 : Blast design for concrete plug excavation by heading in draft tube tunnels 1 and 2

Although M20 concrete portion was stronger than M10, blast results were relatively better in M20 compared to M10 concrete. Drilling of holes was also easier. As the draft tube tunnel advanced towards the machine hall the blast results were poor with many blown out shots due to the increase in M10 concrete area and due to the presence of voids in this concrete. Because of this, the top M20 portion was advanced little more than the lower M10 concrete resulting in an irregular and sloping face.

As the cross-section of the draft tube tunnels varied for every meter, cut holes being the same, other holes were adjusted from blast to blast. As it was difficult to drill 2.0 m depth in the concrete plug because of the choking of holes, the depth was restricted to 1.5 m. The charging pattern is given in the Table 1. The total charge per blast was 60.75 kg. Figs. 6 and 7 shows the number of holes drilled and total charge used in draft tube tunnel 1 and draft tube tunnel 2. The number of holes varied from 114 to 165 in draft tube tunnel 1 and from 111 to 162 in draft tube tunnel 2. Depending on the face condition, the total charge per round varied from 41.0 kg to 70.0 kg in draft tube tunnel 1 and from 50.0 kg to 59.75 kg in draft tube tunnel.

TABLE 1
Charging pattern for heading in draft tube tunnels 1 and 2

Delay No.	No. of holes	No. of cartridges per hole	No. of cartridges per delay	Charge weight per delay (kg)
0	4	4	16	2.0
1	6	6	36	4.5
2	6	6	36	4.5
3	6	6	36	4.5
4	7	6	42	5.25
5	7	6	42	5.25
6	5	6	30	3.75

(Table 1 Contd...)

(Table 1 Contd...)

7	5	6	30	3.75
8	8	6	48	6.0
9	4	6	24	3.0
I	4	6	24	3.0
II	5	6	30	3.75
III	3	6	18	2.25
IV	4	6	24	3.0
V	8	2	16	2.0
VI	8	2	16	2.0
VII	9	2	18	2.25

0, 1, 2....9 Short delay detonators   I, II, III...... Long delay detonator

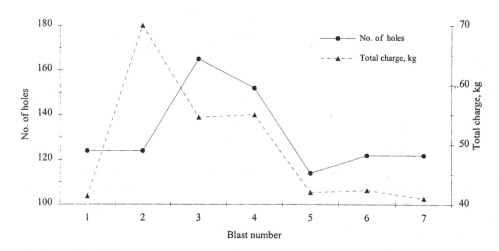

Fig. 6 : Number of holes drilled and total charge used for draft tube tunnel 1

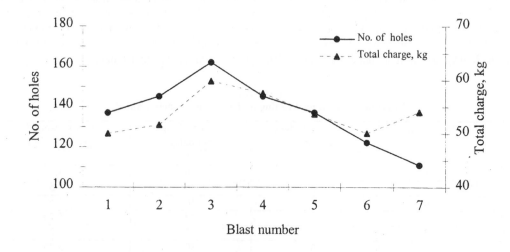

Fig. 7 : Number of holes drilled and total charge used for draft tube tunnel 2

It was desired that the ribs erected should be exposed after blasting for lining the tunnel with concrete and straight charges equivalent to that of production holes were followed to expose the ribs and to shatter the reinforcement in between the ribs. However the concrete between the ribs that did not come out with face blasting was removed by trimming blasts which required drilling of short holes and charging them lightly. Rock breakers were also used to remove the portion of the concrete entrapped between the ribs.

Due to the risk of damage to the lining, the perimeter holes in the lined tunnel were drilled at 20 mm away from the perimeter. The uncharged perimeter holes in the lined tunnel did not help much. Fortunately, the contact between the concrete lining in the tunnel and plug behaved like a weak plane and helped to get a smooth profile.

Breakage, as anticipated, was difficult in the reinforced concrete (one meter from the collection pool side) compared to the plain concrete with same specifications.

5.0 DRILLING AND BLASTING IN DRAFT TUBE TUNNELS 3 AND 4

The blast design standardised for the semicircular heading in draft tube tunnel 1 and 2 (Fig. 5) was tried first. However the breakage was very poor. The reinforced concrete was much stronger and difficult to blast. Hence the blast design was modified. The plug was completely removed by five blasts in each tunnel. Fig. 8 shows the number of holes and total charge used in removing the plug in draft tube tunnel 3 and Fig. 9 for draft tube tunnel 4. The number of holes varied from 156 to 191 in draft tube tunnel 3 and from 97 to 211 in draft tube tunnel 4. Depending on the face condition, the total charge per round varied from 97.5 kg to 111.8 kg in draft tube tunnel 3 and from 52.4 kg to 124.4 kg in draft tube tunnel 4. After the blast the exposed reinforcement rods were cut before starting the next cycle of the blast. The blast results were good and the plug of 4.0 m was completely removed by 5 blasts in each tunnel.

6.0 DRILLING AND BLASTING IN DRAFT TUBE TUNNEL 5

The width and the height of the tunnel was 14.0 m and 5.0 m. Blasting was conducted at 12.25 m chainage to remove the concrete plug. Holes were drilled to a depth of 1.5 - 1.8 m. The number of charged holes in this round was 131 and uncharged holes were about 30. It was found that the pull was very poor because most of the holes were either blown out or explosives energy wasted due to the presence of voids in the concrete. These voids in the concrete plug were not usually visible on the face before the blast.

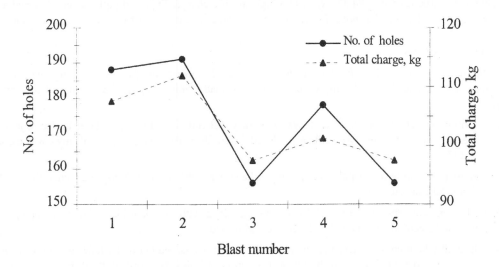

Fig. 8 : Number of holes drilled and total charge used in draft tube tunnel 3

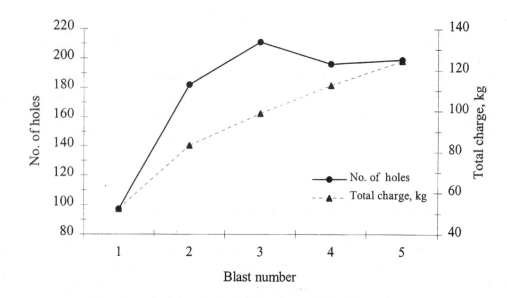

Fig. 9 : Number of holes drilled and total charge used in draft tube tunnel 4

As the tunnel advanced, the width increased needing slight modifications in the blast design. Holes were drilled to a depth of 1.5 m. The number of charged holes were 129 and uncharged holes were about 30. There was a large cavity in the concrete plug at the crown of the tunnel. The total explosive consumption was about 56.125 kg. In perimeter holes, two cartridges of Superdyne were distributed along the length of the hole using PVC pipes. The blast gave a pull of 1.25 m with a specific charge of 0.79 kg/m^3.

7.0 DRILLING AND BLASTING IN DRAFT TUBE TUNNEL 6

In this tunnel, as the plug was of plain concrete, holes were drilled to a depth of 2.2 m equal to the length of the plug. The total number of charged holes were 130 and uncharged holes along the perimeter were about 30. The total charge used was 84.375 kg with a high specific charge. Expected pull was not obtained and it was found that the charging was not sufficient to break it. Hence one more blast was taken to remove the plug.

8.0 CONCLUSIONS

The concrete plugs in all draft tube tunnels were removed successfully. The concrete plugs were made of two grades of concrete. The following observations were made while drilling and blasting of the concrete plugs.

1. Drilling in M10 MSA150 concrete was extremely difficult. Wet drilling of the holes led to jamming the holes/loss of holes. Dry drilling produced severe dust. However, drilling of holes was easier in M20 MSA20.

2. Blast results in terms of pull, volume of breakage was better in M20 MSA20 concrete than in M10 MSA150 concrete.

3. The concrete between the ribs did not come out along with the main blast. Several stripping blasts were to be carried out and rock breakers were also used.

4. In draft tube tunnel 2 where the tunnel was lined, the uncharged perimeter holes did not help much. Due to the risk of damage to the lining, the perimeter holes were drilled at 20 mm away from the

perimeter. However, the contact between the concrete lining in the tunnel and plug behaved like a weak plane and helped to get a smooth profile.

5.   Concrete plugs of 16 m long in draft tube tunnel 1 and 2 probably had a very high factor of safety which may not be required for similar situations elsewhere. As the cost of concreting with M20 MSA20 is only about 25 per cent compared to M10 MSA150, the full section of draft tube tunnel 1 and 2 would have been made of M20 MSA20.

6.   Breakage, as anticipated, was difficult in the reinforced concrete compared to the plain concrete.

ACKNOWLEDGEMENTS

We are thankful to the engineers of N P Power House, Civil Construction Circle, Sardar Sarovar Narmada Nigam Ltd for their kind cooperation and untiring efforts in the execution of this work. We are also thankful to the site engineers of the Jayaprakash Associates, Kevadia for their support.

UNDER WATER LAKE PIERCING IN
KOYNA HYDRO ELECTRIC PROJECT STAGE-IV

S.N. HUDDAR　　　　**V.M. SOMAN**　　　　　　**V.M. KULKARNI**

Chief Engineer　　　　*Executive Engineer*　　　　　*Executive Engineer*

Koyna Project　　　　　　　　　　　　　　Koyna Design Circle

Sinchan Bhavan, Pune　　　　　　　　　　　Kothrud, Pune

Maharashtra, India　　　　　　　　　　　Maharashtra, India

SYNOPSIS

Under water lake tapping is a Norwegian technique developed in that country mainly to tap the inland lakes located high up in the mountains below their normal levels for electricity generation and drinking water supply. This technique is even used in sub-sea tunnels for oil and gas activities. In this technique, a shaft is sunk on the fringe of the lake / reservoir to a pre-determined depth from the bottom of which a tunnel is excavated underneath the lake to reach the lake bottom leaving a break-through rock plug which is finally blasted to connect the lake with the pre conceived water conductor system. The blast is designed in such a way that vibrations produced in the adjoining rock mass and the resultant hydro dynamic pressure build up in the system are kept at minimum acceptable levels, thus protecting the adjoining structure well.

This technique has been used for the first time in India as well as in Asian in region on Koyna Project. This paper is a case study describing the various aspects from investigations to actual execution of this technique on said project. With the successful lake tapping experiment in this country, there is a technology transfer; thus opening a new avenue for adopting this technique on several other projects involving improved utilization of water from the existing reservoirs.

1.0 INTRODUCTION

The Koyna Hydro Electric Project (KHEP) is located in the State of Maharashtra at the foothill of the Sahyadri hill range in the peninsular part of the country. The project has unique geographical as well as geological features viz. abundant rainfall (+5000 mm), the presence of Sahyadri hill range with westerly natural fall of 500 m, presence of good quality compact basalt making it one of the ideal site for an underground hydro-power project. The hydropower development on this project has taken place in stages over the past 40-45 years.

The project was initiated in late fifties with the construction of a 103 m high rubble concrete dam across Koyna river with a gross storage of 2797.40 Mm^3 (98.78 TMC). The dam has multiple use with 1918.40 Mm^3 of water for power generation through west-ward diversion and balance for Irrigation d/s on east side along the banks of the Koyna river through a small dam foot power house (2 nos. x 20 MW). The Stage-I & II was completed in 1966 which consists of a 3.748 km long underground head race tunnel taking off from the Koyna reservoir and leading to an underground power house through four pressure shafts utilizing the natural head of 475 m with the installed capacity of 560 MW (4 nos. 65 MW and 4 nos. x 75 MW). The Stage-III development was completed in 1976. The water released through tailrace of Stage-I & II is diverted in the adjoining valley leading to Kolkewadi dam with a dam foot underground powerhouse, having installed capacity of 320 MW nos. x 80 MW).

The Stage-I & II is developed as a base load station with 60 percent load factor. However, the Stage-III is developed as a peaking station with 24 percent load factor. The Koyna Project laid the foundation for the rapid industrial development in the State; thus termed as lifeline of the Maharashtra then. However, over the years since then, the power development in the State is primarily though thermal power stations, it's share being 80-85 percent today. But the thermal power stations are unable to meet the increasing power demand particularly in the peak hours, the Koyna Stage-IV project was taken up in 1990, which has been recently completed. It is a replica of Stage-I & II[1]. The system consists of a 4.25 km long headrace tunnel, 4 Nos. under ground pressure shafts with 496 m head leading to an under ground powerhouse with the additional installed capacity of 1000 MW (4 Nos. x 250 MW). It has several unique features viz. high head Francis turbines, gas insulated switch gear located in side the power house, typical intake structure with twin lake tapping etc. After completion of Stage-IV all the four stages put together will act as peaking stations with the combined load factor of 18 percent but the total water use for Electricity generation remaining unchanged.

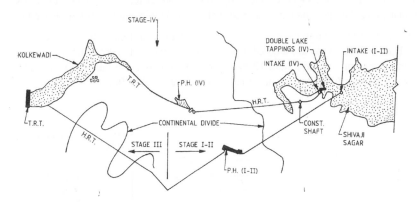

Fig. 1 : Koyna H.E. project (Stage I-II-III-IV) schematic layout

2.0 KHEP STAE-IV - INTAKE STRUCTURE:

In case of KHEP-IV (Photo 1) since the source of water is from an existing reservoir, the location of the intake structure was a real challenge. Due to perennial use of the reservoir for power generation, the natural depletion of the reservoir is not easily possible and even the forced depletion is possible some time by end May, just before the on set of monsoon in that area. As such locating the conventional over ground intake tower with approach bridge from the fringe of the reservoir like Stage-I &II was not possible due to very limited or no period available for construction.

Photo 1 : Intake structure of K.H.E.P. (Stage-IV)

329

Thus, in case of Stage-IV, an intake structure has been identified on the fringe of the reservoir above Full reservoir level. The general layout arrangement for intake structure consists of four shafts, each shaft is 68 m deep and has been constructed to house service gate, trash rack and two stop log gates from d/s to u/s towards lake side respectively. All the shafts converge into a big cavern at the bottom. From the bottom of the two stop log gate shafts, two intake tunnels 6.5 m dia concrete lined 240 m and 180 m long each traverse towards lake side underneath the lake. At the end of each intake tunnel a muck pit has been provided from the other end of which an inclined lake tap tunnel of the same size rises at 45 degrees to reach the lake bottom. Break through rock plug of 7.5-m dia and 6 m thick was left at the lake bottom for each inclined tunnel.

The roll of all these components is functionally inter related hence all these components are designed considering their interdependence. For KHEP Stage-IV design discharge is 260 Cumecs, as such Instead of providing one large lake tap two lake taps of 6.5 m diameter each are planned. These are located 35 m apart. The dimensions are decided by restricting the velocity to 4 m/s (i.e. velocity permissible in concrete structures). The gate shafts are located on the fringe of the lake at Koyna Reduced Level (K.R.L.) 658.00m. The invert of intake tunnel at gate shaft and inclined shaft are 590.00 m and 591.00 m respectively. Considering

the nearness to the lake the minimum rock cover of 20 m (three times diameter)is kept over the crown of the intake tunnel. The inclined shafts are aligned at 45 degrees to meet the lake bottom. Further based on Model studies to avoid air entrapment, minimum 12 m water cushion above the lake tap is provided. The minimum draw down level is fixed at 630.28 m; against the lake tap floor @ 618.00m.

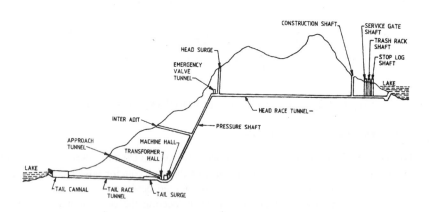

Fig. 2 : Schematic layout of K.H.E.P. (Stage-IV)

3.0 UNDERWATER LAKE TAPPING

BACKGROUND : Initially, As per contract provisions, it was envisaged to tap the lake from the top by conventional dry piercing with the lake depleted for the last two years in succession. However, the submerged piercing was very much under consideration while deciding the over-all layout of the intake structure. The contract provisions also envisaged the removal of over burden in the lake tap area by dredging and under water blasting, however, the first experience in this regard was not found to be encouraging. The lake was forcefully depleted in the last week of May 1995 and the removal of balance over burden, the protection works around the lake tap area, the constructions of staining walls around lake tap etc. was undertaken on war footing. At that time, additional geological investigation like bore hole in the lake tap area, permeability test of the adjoining rock mass, rock stabilisation around the lake tap by driving rock bolts etc. was done. Some Norwegian consultants also informally visited the site and based on the rock conditions in the lake tap area viz. hard and unweathered basalt with low joint frequency, they indicated favourable conditions both for dry as well as submerged piercing. Further keeping both options open the balance over ground works were completed in the next working season during May-June 1996 by depleting the lake once again.

In the meantime, the contractors M/s. Patel Engineering Company, Mumbai were officially instructed to hire the services of M/s. Norconsult for planning and designing of submerged lake piercing for KHEP-IV. They submitted their first report in August 1995. A team of engineers from Government as well as contractor side visited Norway in August 1996 to witness the actual lake tapping and had deep and thorough discussion with the Norwegian consultants for under-water piercing to be adopted for KHEP-IV. M/s. Norconsult submitted their detailed report in August 1997 describing the detailed methodology, functioning of the designed lake tap, pressure build up on stop log gates, mass transportation after blasting etc. for KHEP-IV.

The work was in progress still keeping both the options open. By March 1998, the progress of both civil and electrical work was such that the first machine was getting ready by December 1998. However, the power generation would have been differed till Oct. 1999, if the dry piercing as envisaged in the contract was to be adopted which was possible only in May / June 1999. As such in September 1998 a final decision was taken in consultation with the panel of experts and the World Bank to go ahead with the submerged piercing as per Norwegian method to accomplish the early generation through first and subsequent units. The contractor was finally allowed to hire the services of M/s.Norconsult to supervise the whole exercise right from procurement of material till the final blast. The final blast was successfully and ceremoniously performed on 13-3-1999. The engineering details of this historic event are as follows.

4.0 INVESTIGATION, SELECTION OF SITE

Careful investigation is required prior to selection of the site (Johansen and Mathiesan). The selection of site, for lake tap involves number of aspects viz., the geology of the area, location of gate shaft, the length of the intake tunnel, availability of minimum water cushion etc. The first phase will always be a geological survey. Arial photography gives information of rock structure, cracks and fissures. Seismic refraction studies are more preferable. Visual inspection can be carried out by divers or remotely controlled submarines. Divers are useful in shallow waters. Remotely controlled submarines are recommended for depth of 50 m and above. Since the Koyna lake was depleted twice ,the visual inspection as well as exploratory boring around lake taps was possible. This was an added advantage because such eventuality is rare for normal under-water piercing.

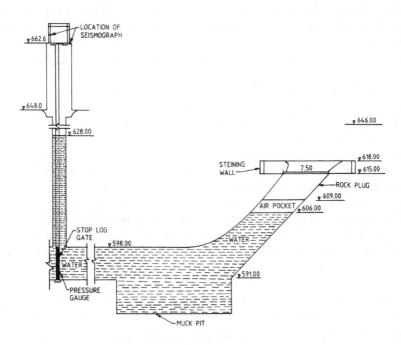

Fig. 3 : Layout of laketap

In K.H.E.P. Stage-IV, as already stated the compact basalt with tight joints was observed. The rock is classified with Bartan's "Q" and Bieniawaski's "RMR" system . On this system 'Q' and R.M.R. values are 14 and 69 respectively. Initially Under water seismic reflection and refraction survey was carried out by Central Water and Power Research Station, Pune near the proposed lake tap location to ascertain lake bed levels, probable rock levels and it's quality. The compression wave velocity of bed rock is 4800 m/s. This indicated the rock of very good quality. The overburden was more.

5.0 LAKE TAPPING-METHODS AND CHOICE MADE

Lake tapping is a technique for connecting the water conductor system to the lake either natural or man made. The lake can be pierced by both dry and wet (submerged) methods. In dry piercing the lake is depleted forcibly and the rock plug is blasted from top (open blasting). In submerged piercing the rock plug is blasted from bottom without depleting the lake. A submerged piercing can be executed by two different methods· viz. closed method and the open method.

In closed method, the control gate is on the upstream side of the shaft and thus the tunnel is not connected to atmosphere. The tunnel is dry or partly filled with water. In closed system shooting of the plug can take place shortly after loading. However the hydrodynamic conditions are uncertain. Unacceptable pressure rise at the gate can be generated if the distance between the plug and gate is short. The plug fragments and sediment above it are likely to be carried uncontrolled into the tunnel and can cause damage to the gate. In closed method the tunnel friction plays very important roll. This method is better suited for long tunnel lengths and low reservoir heads (Sharma 1988).

In open method, the control gate is on down streamside of the shaft, and there is a direct connection between the tunnel face at the plug and the atmosphere, through the gate shaft. The tunnel system is partly filled with water, keeping a sufficient pressurised air pocket underneath the plug. In this system, hydrodynamic conditions are clearly set out and the pressure rise against the gate can be calculated to a fair degree of accuracy. However, the task of water filling is complicated and there is a considerable time lag between explosive loading and shooting. In this system, the velocity of water in the tunnel after blast is low and the debris is easily trapped in the muck pit. In open method of lake piercing volume of compressed air pocket has a key roll which depends on quantity of explosives. Thus proper estimation of explosives and the sufficiency of air pocket volume are required to be ascertained at layout stage only. Otherwise the dynamic pressures on the gate after plug blasting are excessive and the gate design becomes very complicated.

The design of the K.H.E.P. Stage-IV intake structure is as per open method of lake piercing.

6.0 FINAL ROCK PLUG THICKNESS

Experience has shown that a moderate overburden of 1 to 3 m is favorable as the soil seals the joints and fractures in the surface rock mass preventing excessive leakages. In Koyna about 2 m thick overburden over the top surface of rock was therefore retained. Under normal rock conditions with overhead water depth ranging between 10-80 m the rock plug thickness is usually taken as 1.2 times the shorter size of the plug cross section (rectangular) or equal to the diameter of the plug (circular). The thickness of the plug is influenced by the quality and the permeability of the rock mass.

For KHEP STAGE-IV, based on the overall consideration of the lake tap area a circular rock plug with a thickness of 6 m was found to be suitable. Depending upon the total volume of the plug rock mass, a muck pit almost 2 to 2.5 times the said volume is provided to facilitate easy trapping. The shape of the muck pit is such that the further movement of the fallen debris along the flow in to the intake tunnel is prevented. Thus, a rectangular muck pit with vertical sides has been provided.

7.0 DESIGN AND AIR VOLUME REQUIREMENT BELOW THE ROCK PLUG

In the open method of submerged piercing for successful final blast, the design requirements are as under.

- An adequate air pocket below the rock plug to prevent transmittal of shock waves to adjoining rock mass after the final blast.

- Sufficient air volume below the plug to receive the gas from the explosion without any excessive pressure rise.
- The water level in the gate shaft after water filling to create sufficient air pressure below the plug to squeeze the air through rock plug in to the lake .
- The pressures on the control gate within acceptable limits.
- The upsurge in the gate shaft to be contained within the height of the gate shaft.

In case of Koyna, the water filling envisaged was such that the tunnel system would be submerged above the tunnel crown level to avoid any direct passage from inclined tunnel to stop log gate shaft resulting otherwise into unacceptable hydraulic condition after the blast. At the time of final blast, the reservoir level was at 646.00 m, 28 m head above lake floor at 618.00 m. At this head of water, the water filling in the tunnel system was planned at 628.00 m. with the precompression of trapped air pocket below the plug from initial 1150 cum to approximately 338 cum and corresponding water level under the plug at 605.20 m. The resultant air pocket pressure was 3.4 bar. There was marginal difference in the air pocket parameters between the two lake taps. The anticipated upsurge in case of both the lake taps was expected to be within the shaft at KRL 658.00 m or so.

8.0 DEVELOPMENT OF DYNAMIC PRESSURE AND DESIGN OF GATE

The explosives in the plug on detonation create gas in the air pocket. This gas pressure without any air pocket, would propagate more or less undamped and most likely to cause great damage to the gate. The gates are required to take this shock loads (Huddar *et al.*, 2000) The explosion gas expands into the air pocket and the pressure in the pocket increases rapidly. The pressure propagates through the water towards the gate. The rise of the pressure at the gate is twice the amplitude of the pressure wave.

Design pressure on the gate $Pd = Ps + 2 \times (Pg - Pi)$

where

Pd = Design pressure on the gate.

Ps = Static head on the gate prior to plug blast.

Pi = Air pocket pressure prior to plug blast.

Pg = Air pocket pressure prior to plug blast.

The ratio between quantity of explosives and quantity of entrapped air is decisive for the resulting explosion gas pressure. The quantity of explosives is fixed beforehand, while the quantity of entrapped air is adjusted to control the pressure rise.

In case of Koyna two alternatives with reservoir levels at 635.00m and 640.00m were studied. Compression of air pocket to 1.56 bars and 2 bars (air pocket volume of 770 cum and 600 cum) were assessed. These alternatives give expected pressure build up on the stoplog gate of 46 and 69 m of water column respectively. In normal condition, maximum static head on the stoplog gate is equal to 73m of water. The reservoir level at the time of plug blast is uncertain hence the gates are designed to resist dynamic pressure of 130 m of water column. For static conditions gate is designed with permissible bending stresses as 0.55 Yp and permissible shear stresses as 0.4 Yp. As the dynamic pressure at the time of blast is momentary, increase in permissible stresses to an extent of 33% are allowed.

The stop log gate in the form of 7 needles of size 5.82 m (L)x 1 m (H) was designed. The gates are with downstream sealing arrangements and are designed to be operated under balanced conditions. The water filling valves are provided in the top elements of stop logs.

The gates should be leak proof, otherwise water filling and maintaining the desired water levels in the tunnels becomes difficult task. Hence all precautions even from design stage are required to make the gates watertight. The single leaf type gate can also be an alternative in this context; but which will require higher hoisting capacity to handle it.

9.0 EXCAVATION TOWARDS THE FINAL ROCK PLUG (MUCK PIT AND INCLINED TUNNEL)

Normal excavation procedures are usually followed up to a point 30 to 100 m from the final plug location depending on the quality of rock mass. In case of Koyna, the excavation and concrete lining for the circular intake tunnels up to d/s face of the muck pit was initially completed. The excavation beyond this point up to face of the rock plug was divided in to three main excavation sequences viz. The lake tap tunnel (horizontal portion above muck pit). The piercing inclined tunnel and the muck pit. For all the excavation the blast holes 41 mm dia were used. The lake tap tunnel was excavated with full face by conventional drilling and blasting with 2.5 to 3 m long rounds. The muck pit was excavated using bench blasting. The piercing shaft was excavated using under cut pilot and successive roof slashing; but the rounds were gradually reduced from 2 m to 1 m as well as face was trimmed from being perpendicular to the shaft axis to horizontal at the end. The spot rock bolting during excavation and systematic rock bolting after excavation was done for all the three components. The lake tap tunnel and the muck pit was supported with 50 cm thick M20 RCC lining, so also the piercing tunnel for first 9 m only. The lining of the piercing tunnel was stopped well below the bottom of the final rock plug to provide sufficient space for drilling the contour holes of the final blast. It must be kept in mind that sufficient space below the plug was provided for drilling as well as charging the contour holes. It is significant to note that there was no seepage through the rock mass even though the excavation was going on underneath the lake.

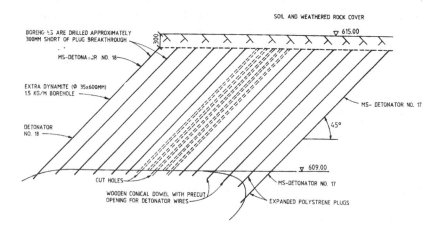

Fig. 4 : Details of practical charging performance

10.0 DRILLING OF FINAL PLUG

Based on the experience in the Koyna rock, 139 Nos. of production holes, 48 mm dia. and 16 Nos of cut holes 89 mm dia. were drilled, limited to 0.3 m from the top surface of the rock. These holes were drilled at 45 degrees inclination parallel to the axis of the piercing tunnel (Fig. 4)

The excavated face of the plug was at K.R.L. 609.5 m. The wooden platform was erected at K.R.L. 607.3 m. This platform was specially designed to take live load of drilling equipment, 20 persons working on it and also for the vibrations during drilling. The unique feature of this platform was that it was totally insulated, so as to safe guard against any electrical current. The platform was elliptical in shape measuring about 10.60m and 7.5m along major and minor axis respectively. The platform structure was formed of 75mm thick timber planks which were supported on 100 mm G. I. pipes spanning along minor axis and spaced at about 1.5 m interval. The G. I. pipes were supported by 22 mm diameter insulated steel rock anchors, which were anchored 2.5 m in rock and protruding 1m outside. Braced wooden props from bottom further supported the platform.

M/S Atlas Capco had specially designed the jig on which the BBC 120F drifters were supported in such a way that drilling at 45 degrees was possible with convenience and required accuracy. The frame was mounted with air motor operated chain feeds with both longitudinal and lateral movements to conveniently access each drill hole location. The BMS screw feed assembly was permanently fixed on main frame at 45 degrees inclination.

In each lake tap four representative exploratory drill holes were taken to ascertain the top profile of the rock. The rock profile was found to be fairly uniform at K.R.L. 616.81m. The plug thickness at each drill location was estimated from this data and inventory for each bore hole was prepared. The planned drill length was 0.3 m less than the estimated plug thickness at each drill location. Rock mass condition was inspected before commencement of drilling. In both the lake taps basalt was heavily fractured with random joints, however the rock mass was more or less dry. Hence initially no rock supports were provided. Only the manual scaling of the rock mass was done prior to the exploratory through-hole drilling.

Prior to the commencement of the final drilling rock fall occurred in one of the lake tap probably due to introduction of drilling water. Hence rock support in the form of 3m long 14 rock bolts was given. In the other lake tap the rock condition was still worst. Hence as a precautionary measure rock bolting (67 rock bolts) and 25 mm thick coat of sealing shotcrete was done. The alignment of these rock bolts was strictly in plane parallel to the blast hole direction and in between these blast holes. There after the actual drilling was started and completed as per the designed drilling pattern. The drilling time required for each lake tap was about 5 days. For large diameter hole the pilot and reamer head bits were used. The entire drilling activity was managed with local team of drillers, assisted by foreign drillers and under the expert supervision and guidance of the Norconsult. The technicians of the Atlas Capco trained the drillers. The common difficulties experienced during drilling were interception of already driven rock bolts, jamming of the drill rods, keeping accuracy in drilling due to lack of rigidity in drilling equipment in accurate collaring, the box platform was untidy.

The packers with innovative design and manufactured locally proved to be effective in sealing the leakage through holes. The water leaking from holes drilled through was never a major problem, however as a precautionary measure grouting agency was kept ready at site. Polyurethane (hardening plastic type Chemical grout)with the capability of expanding at least 10 times its initial volume was attempted. However, as the same was found to be toxic causing respiratory trouble, in the first attempt, the same was discontinued. All the drilling activities were closely monitored from control room with closed circuit T.V. sets. The average length of each drill was @ 7.8 m and the total drilling length for each plug was @ 1100 m.

11.0 EXPLOSIVES

A special type of explosives, detonators and exploders were specified for the under water blast. The important specifications of the same are as under.

Sr. No.	Particulars	Specifications
1.	Explosives(3500 Kg)	Type : Extra Dynamite. Slurry type polythene cartridge sticks.
2.	Detonators(650 pieces)	Millisecond delay detonators (1 to 18 No with protective sheathing.)
3.	Shotfiring Cable 2 pieces of 300m and 350m length.	• $2 \times 2 \ mm^2$ • Resistance 2.1 Ohams per 100 m length. • Insulation > 5000V. (Minimum 3000V)
4.	Connecting Wire 350m	• 0.7 mm diameter. Resistance 4.5 Ohams per 100m
5.	Exploder 2 Piece	• Cl 160 VA Type. • Resistance 4.5 Ohams per 100m length.

The quantity of explosives required depends on plug volume, type of rock, sediment overburden and water pressure on the plug etc. In case of Koyna specific charge of 4.2kg/cum was designed. The specific charge for under water blasting is 50 to 100% higher than the charge for normal tunnelling. Detonator requirement

is two per drill hole. The explosives, detonators, exploder and other accessories were specially imported from Norway/Sweden and arrived at site in due time. A pretest of extra dynamite and detonators stored at 40 m water depth was carried at site.

12.0 CHARGING AND CONNECTIONS

Prior to charging, the length of each hole was measured and also the holes were checked for possible cave in or collapse by inserting plastic stick. All the precautions were taken to avoid any accidental leakage of current. All the electrical connections upstream of the stop log were strictly removed. A lightening conductor was erected on the top of service gate hoist structure. Further to avoid any leakage of static electricity from atmosphere due to storm/lightening an early alarming system capable of giving indications of any storm approaching at 160 km radial distance was installed. All the explosive loading work was done with battery operated minors lamps.

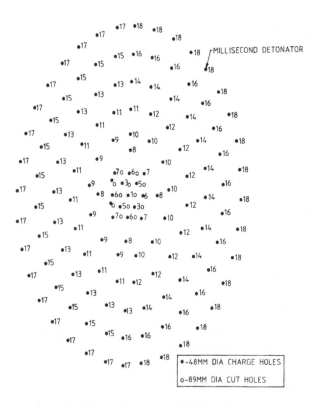

Fig. 5 : Charging and ignition plan

The drill holes were loaded with explosive sticks. The plastic charging rod was used for pushing the explosives cartridges in to the hole without any hollow in between. In all the charged holes two number of millisecond detonators with same delay number and normal sensitivity were used. One detonator in the top most cartridge and another in the middle of the outer part of the hole was placed. The charging plan giving the details of appropriate series of detonators to be used in each drill hole was given by Norconsult. As per the charging plan delay numbers of the detonators were increasing from center to outer periphery (Fig. 5). The same was strictly followed. Before placing the detonators in position each detonator was checked for appropriate resistivity with Ohammeter. Last uncharged length of about 0.3 m of each hole was stemmed with expanded polystyrene plugs. Thereafter each hole was packed with wooden conical dowel with precuts through which the lead wires of detonators were taken out. The loading work in both the tunnels was done simultaneously. The loading of the explosives took 4 days.

The connections were made to form two parallel circuits The bottom detonators of all the holes were connected in one circuit and the top in another circuit. During making connections, after joining of every 10 detonators the total resistivity of completed circuit was checked to detect accidental damage of any detonator during loading. The two circuits were ultimately connected to the shotfiring cable in series and the shotfiring cable was taken out from stop-log gate shaft.

The local workers also did all the charging work along with foreign experts. The workers and the experts experienced headache as well as vomiting due to fumes of the explosives. Even with the medicines it was not possible to work for more than half-hour continuously. Hence the workers were required to be replaced by about every half an hour.

13.0 INSTALLATION OF MONITORING DEVICES

After completion of loading, following monitoring devices were installed.

- Pressure gauges on second lower element of stop log to measure the dynamic pressure that will build up during plug blasting.
- Two video cameras, specially imported from Scotland, which were capable of recording even under 300 m of water head were installed on lining offset of the piercing tunnel. The batteries of these cameras were kept on top, which were made on only after the charging.
- Trigger elements with overflow buckets were fixed at predetermined water filling levels to monitor the water filling level under the plug.
- One transducer was installed in each lake tap tunnel at K.R.L. 608.00m to monitor the air pressure underneath the plug. The connection of this transducer was given to the computer on the top.
- The seismographs were installed on stop log shaft (at about 90m from the blast site) to register the blast induced vibrations.

14.0 WATER FILLING

There was a slight deviation between design assumption and actual conditions at the time of lake tap especially regarding excavated volume in the piercing tunnel, quantity of explosives and reservoir level thus requiring revised water filling estimates.

Hence the water levels underneath the plug and in shaft corresponding to designed air pocket were decided as 605.20 m and 628.0m respectively using following graphs based on the geometry of the lake tap tunnel.

- Water elevation underneath the plug V/S Air Pocket volume.
- Water Elevation under the plug V/S Water elevation in the shaft.

Prior to commencement of water filling all the foreign material including platform was removed from upstream of the stop-log. Stop-logs were lowered in position. The muck pit was already filled with leakage water through drill holes in the plug. For water filling six 40 H.P. and two 180 H.P. pumps were installed on the fringe of the lake. The time required for filling each tunnel was about 22 Hours. It was completed about 5 hours prior to scheduled blast time. During water filling the leakages through the stop logs was a major problem, particularly in lake tap No. 2. The same was however, overcommed by controlling the leakage and increasing the pumping capacity.

337

15.0 FINAL BLAST

The resistance of the shot firing cables was once again checked and the shot firing cables were connected to the exploders. The electrical current setting calculated was 13 A (6.5 A per circuit). Both the lake tap plugs were detonated simultaneously. The lake tap was successful. The large amount of detonation energy was directed towards the lake forming two mushroom shaped water fountains of about 6m height and creating only moderate dynamic pressure build-up on the stop log gates.

16.0 SAFETY PLAN

The detailed safety plan was prepared and implemented. Major safety precautions are listed below.

- **Access and Exit to the work area**
  At the time of drilling and there onwards the access was restricted to minimum possible actual working and supervisory staff. The identity cards were issued to these personnel who were kept on the control board at the time of entry. This is to ascertain the evacuation of all the personnel entered in, at the time of emergency.

- **Communication**
  Radio communication with Motorola Model GP 300 was provided between the foreman at work area, cage operator and control room. Seiren was also installed to alert the workman as well as monitoring personnel.

- **Evacuation**
  To enable immediate evacuation during emergency one cage was always kept on the bottom of the shaft. The training of evacuation within minimum time was given. Cage movement was controlled by radio communication /bell/hand signals. In emergency ambulance van was kept ready at site round the clock.

- **Ventilation**
  The ventilation system capable of reverse flow was installed in both the tunnels. Action levels of air monitoring were as under.

Oxygen	19.5 – 22 %
Methane and Hydrogen Sulphide	< 10% LEL (Lower Explosive Level)
Carbon Monoxide	< 50 PPM

- **Dewatering**
  Sufficient pumps with standby arrangement were kept ready. The muck pits were always kept half filled. So as to achieve sufficient water cushion in case of failure of platform and also in case the leakage starts it should provide sufficient water absorbing capacity till the workmen reaches towards cage.

- **Grouting**
  To attend heavy leakages through the drill holes, if any a grouting team with men, material and machinery was kept on vigil round the clock at work site.

- **Work Area Illumination ·**
  During drilling halogen lamps were provided. These lamps need to be shut off when the action level of combustible gas is noted. Before bringing the explosive materials in the tunnel all the electrical installations were strictly removed and the minor's lamp with hard hat were provided to the workmen for illumination.

- **Explosive Handling**
  The lightening conductor was erected near the work area. A lightening detector was also installed to provide adequate advance warning of the approaching thunderstorm. On receipt of such warning the detonators should be removed from work area to non-conductive enclosure and further loading of the explosives should be stopped. Smoking was strictly prohibited in tunnels and even on outside in a radius of 40m. All the electric installations and heat producing items were removed from tunnel before the explosives were taken in. Only one box of explosives was taken inside the tunnel at a time.

- **Loading**

  Each drill hole was tested for its length prior to loading. Air at plug face was tested for combustible gas before commencement of loading activity. The loading was done strictly as per the charging plan of the consultant. The resistance of each detonator was tested before loading. The delay series of detonators was watched carefully before loading every hole. Tamping was done with plastic rods. The portable radios were turned off. No other simultaneous activity was allowed in the tunnel at the time of loading. Loading was strictly prohibited during drilling. All the circuit connections were maintained watertight. A 10kg fire extinguisher was kept on the platform. Only the concerned persons were allowed entry in to the work area.

17.0 OBSERVATIONS AFTER THE PLUG BLAST

- The pressure gauges installed on the stop log gates recorded the pressure of 9.5 bars and 6.5 bars as against the anticipated pressure of 8.6 bars.
- The seismograph recorded the peak particle velocities 3.73mm/s. and 4.32 mm/s. These are well within the limit of 50 mm/s.
- The piercing tunnel was inspected by deploying the services of the divers and no damage was noticed. The entire muck has collected in the muck pits provided for and mostly of appropriate size.
- The stop log gates were inspected from downstream side and no damage was noticed to the gates. However, the leakage through the sealings of the stop logs was increased. This may be due to pulsating dynamic loads of the blast.
- After few days the cameras installed were removed by the divers. They were safe. The recording revealed no damage to the concrete lining of the lake tap shaft as well as the smooth peripheral blasting.

18.0 CONCLUSIONS

The open method of under water lake piercing is more preferable as dynamic pressure can be estimated to fair degree of accuracy and the debris is smoothly collected in the muck pit. The high specific charge gives high shock loads on the gate. Reducing the specific charge on the contrary increases the risk of failure of the blast. Hence it is better to be conservative in deciding the specific charge. In open method of under water piercing high degree of quality control is required while manufacturing the control gate. The water tightness of the gate is very important. If leakage starts it is very difficult to maintain the desired water levels. If rock bolting is required for plug then extreme precision is necessary. If the rock bolting is not parallel to the alignment of charge holes, it is extremely difficult to drill the charge holes. The rigidity of the drilling equipment is very essential to keep drilling accuracy. The high degree of precision is required at each and every step especially in drilling the holes to correct alignment, charging, deciding adequate air pocket volume and maintaining it. The innovative packers manufactured locally proved to be extremely successful in stopping the leakage from through holes.

This technique of under water lake piercing can be used for using the artificial and existing man made lakes economically for hydropower generation.

ACKNOWLEDGEMENTS

The authors are grateful to Government of Maharashtra for according permission to present this paper. The authors keeps on record their feelings of gratitude towards engineers of the department, contractors Patel Engineering Co. Mumbai and NORCONSULT who worked with devotion for the project.

REFERENCES

Johansen, J. and Mathiesen, C.F. "Modern Trends in Tunnelling and Blast Design".

Sharma, H.R. "Underwater Tunnel Piercing". International Symposium on Tunnelling for Water Resources and Power Projects, CBIP 1988, New Delhi.

Norconsult. "Koyna Hydro Electric Project, India, Lake Taps, Detailed Engineering Design Report".

Huddar, S.N., Kulkarni, V.M. and Dravid, R.G. "Development of Dynamic Pressure During Under Water Lake Piercing of Koyna Hydro Electric Project – Third International R&D Conference, C.B.I.P., Jabalpur, 2000.

FASTER AND ECONOMIC TUNNELLING BY CONVENTIONAL DRILLING – BLASTING

S.Ṛ. KATE
Dy. General Manager

**IBP Co. Ltd., Business Group-Chemicals
NOIDA, U.P., India**

VINOD KUMAR
Vice President

**Continental Foundation
New Delhi, India**

SYNOPSIS

Drilling and explosives are the basic input requirements for commencing any construction and mining activity. In tunnelling too, these two aspects are not only basic but are most important and significant due to added constraints on account of geology, strata condition, space, light and ventilation. The drilling operation is most tedious, time consuming and extremely high cost item. Similarly, explosive is also a high cost component item and is a core item influencing the progress of tunnelling on account of pull, over/under breaks, fumes, toxicity of fumes and toxicity of explosives in handling and transport apart from its impact on drilling component and involved wide range safety aspects. The techno-economics of tunnelling thus, consists and is influenced by two main items – one - drilling and second - explosives which depends on their individual costs, efficiency and application. Amongst the two, drilling is a high cost constituent and has higher influence on time and productivity. Methods of reducing drilling, practical field application and techno-economics, have been discussed in the paper.

1.0 INTRODUCTON

Drilling and explosives is a basic need of any excavation work. Drilling is the first activity that has to be started to commence the excavation or mining activity. Drilling assumes great importance since it is most tedious, time consuming and very high cost component in the overall progress as well cost structure. It is necessary to take into account the real influence of drilling aspect while assessing the total performance and cost. It is thus, obvious that faster drilling rate and less drilling meterage/number of holes holds the key for success both for performance and cost, which has to be accomplished with the use of compatible dia and energy explosives resulting into desired blast performance.

2.0 BLAST PERFORMANCE

In case of blasting in tunnels, the blast performance is evaluated mainly on the pull in m and pull as percentage of drill hole length/depths : fragmentation : over and under breaks; secondary blasting requirements for boulder, under breaks and floor level blasting.

The blast performance is termed as normal and adequate when -

- The pull achieved ranges between 75 and 80% of drill hole lengths. The pull in the range of 80-90% and above 90% are termed as very good and excellent blast performances.
- The fragmentation is said to be normal when the loading equipment can load the blasted muck smoothly.
- The boulders are said to be ones whose size is more than 2/3 size of the bucket (of loading equipment).
- The secondary blasting is within range of 5 to 10% of the main blast.
- The under-cuts in sides, face, roof and floor and boulder blasting forms the part of secondary blasting and should be kept within the above range.

- Throw, muck profile and muck tightness influences the efficiency of loading equipment and thus, should be such that the length of boom is in a position to deal with the muck pile. Muck pile height should not be more than the boom length. Thus, the throw should be adequate which give required muck profile, height and compactness to allow efficient handling of the muck.
- Overbreaks should be as minimum as possible, visibility of drill marks and smooth profile on sides and roof are to be achieved to call it a very good blast.
- The blast performance depends upon four main factors, *i.e.*, drilling pattern; accuracy in drilling; explosive; quality/strength and quantity and charging and firing pattern.

3.0 REDUCTION IN DRILLING

Reduction in drilling obviously, provide distinct advantages like lower cost, time saving, higher productivity and thus overall cost reduction. Reduction in drilling can be achieved by application of three methods -

1. Use of increasing diameter holes.
2. Use of higher diameter explosive cartridges.
3. Use of high energy strength explosives.

The above aspects are described in detail in following paragraphs. Field application of these methods and results obtained are given subsequently.

3.1 Explosives Energy and Reduction in Drilling

Explosive is a chemical energy which is released instantaneously on detonation in the form of gases at a very high temperature and pressure. When this explosive is placed in the "Hole" and on detonation, the gases which are released under high temperature and pressure exerts pressure to break the hole surrounding, *i.e.*, breaking of the rockmass. Thus, the volume and type of fragmentation of rock broken shall depend upon the quality and strength of explosives *i.e.*, total energy content in the hole and its distribution in the hole. Lower energy shall break lower volume or poor blast performance, viz., higher energy shall remove higher volume or better blast performance whereas, quantity of explosive, *i.e.*, explosive column height and % occupancy of explosive, *i.e.*, distribution of energy in hole plays major role in the matter of impact on fragmentation.

There are different explosive products having different strength/grades and are represented by energy per kg of explosives (amount of energy in kilo calories/kg) just like different grades of coals which has different thermal energy/calorific value per tone or different size and grade of bombs whose destruction power differs accordingly.

When all blast design parameters are held constant except for the explosive type, the disturbance or crush velocity through the stemming column can be used as a reliable heave index, for relative explosive performance and in predicting crater results (Chiappetta, 1993). Fig. 1 illustrates the results of four different explosive formulations superimposed on a graph of displacement versus time. Three of the explosive formulations consisted of the same emulsion/anfo blend, but with different quantities of aluminium of 0, 3 and 9%. Since higher the Al.% - higher is the energy. The fourth explosives consisted of standard anfo and was used as the control explosive. Since most mine operators work on a volume basis, the explosive length was kept constant at 6.0 feet (1.83 m). Each hole was detonated separately and in total confinement to eliminate any end boundary effects or disturbances from adjacent detonating boreholes.

As expected, the explosive containing 9% aluminium resulted in the best overall performance, followed by the 6% Al., 3% Al., 0%Al. and anfo. The disturbance length above the top of the explosive column into the stemming correlated very well to the surface measurements in terms of the crater radius, heave volume, material displacement for the purposes of obtaining some statistical significance and respectability. High speed photography, seismographs and manual measurements of the surface effects were also used to substantiate and correlate results.

It was concluded in this particular environment that the minimum and maximum crater heave heights and crater radii could be predicted for the different explosives based solely on VOD results. Performance in terms of crater height and crush distance into the stemming and radius are presented in Table 1 and Fig. 2. Since

341

substantial improvements were evident with the addition of metallic elements to the explosive formulation *i.e.*, the ENERGY CONTENT, they are highly recommended.

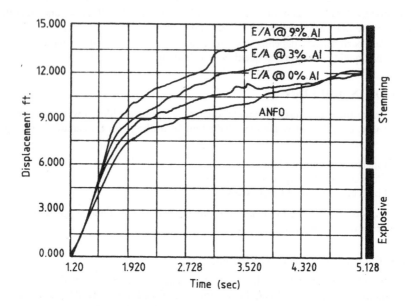

Fig. 1 : Aluminized emulsion / Anfo blends versus Anfo

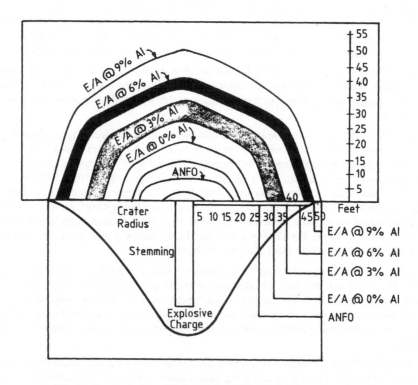

Fig. 2 : Crater test results in terms of Crater height and radius

TABLE 1
Effect of explosive energy

Explosive	Crater heave height (m)	Crater radius (m)
Anfo	(1.5-3.0)	(7.5-9.0)
E/Al. @ 00%	(3-7.5)	(9-10.5)
E/Al. @ 03%	(6-10.5)	(10.5-12)
E/Al. @ 06%	(9-12)	(12-13.5)
E/Al. @ 09%	(12-16.5)	(13.5-15)

Al. = Aluminium

The performance of explosive thus, basically depends upon the strength/energy content, rate of reaction, *i.e.*, velocity of detonation and degree of confinement. As illustrated with the increase in strength-energy content of explosive/kg the performance in terms of either volume per hole or if volume per hole is kept constant then the degree of fragmentation and throw goes on increasing. Thus, energy content of explosive can be used most advantageously to improve the blast performance fragmentation or to reduce the number of holes per round (because of increase in volume per hole). Since reduction in number of holes gives direct cost and time saving, this impact needs to be taken into account while selecting the explosives.

3.2 Explosive Cartridge Diameter and Reduction in Drilling

Another way of reducing the drilling is application of appropriate diameter explosives. The size of tunnel determines the size of drill hole diameter which can be adopted advantageously. It has been studied and concluded that for the smaller cross-sections blasting with cartridge diameter < 25 mm should be adequate and economic. In the cross-section area between 10 to 20 sq m cartridge diameter of 30 to 35 mm is adequate. Above 20 sq m the range of application for the 40 mm cartridge diameter starts and above 40 sq m cross-section generally 50 mm cartridge diameter can be used.

The use of larger diameter hole (and thus higher diameter cartridges) accommodates higher quantity of explosive per metre length of hole. Thus the energy content of the hole per metre goes up. This allows more volume to be broken per hole and thus it is possible to reduce the number of holes per round of blast. However, in another study in large open cast mines, it has been established that for a given fragmentation the energy factor is almost constant. This implies that when we are reducing the number of holes in a blast round we should derive the energy factor with higher quantity of explosive charged due to application of higher dia. holes. To achieve the same or higher/improved fragmentation it would thus be necessary to use higher energy explosives basically to meet the need of matching or increasing the energy factor required to achieve the objective namely Same or Better fragmentation. In general, mining and tunnelling industry a larger number of tests using 40 and 50 mm cartridge was carried out and compared with 25 and 30 mm cartridge diameters, Table 2.

Excavation cross-section : 27 sq m
Rock type : Argollaceous shale
Advance per round : 0.5 m

TABLE 2
Parameters of test rounds using 25 mm, 30 mm, and 40 mm cartridge diameters in a stone drift

	Cartridge diameter		
	25 mm	30 mm	40 mm
Borehole number	105	81	55
Borehole length, m	367.5	283.5	192.5
Boreholes/sq m	3.39	2.61	1.77
Borehole length/cum	3.39	2.61	1.77
Explosive consumption (kg ammonia-gelatine)	121.0	71.75	75.20
Spec. explosive consumption (kg/cu.cm)	1.12	0.66	0.69
Spec. explosive consumption (kg/m borehole length)	0.33	0.28	0.39

As a rule of thumb, each additional millimetre cartridge diameter provides for a reduction of borehole number by some 3%. This rule of thumb has turned out to be almost accurate in many blasting tests and practical applications. Percentage wise the reduction of number of hole has been established Table 3.

TABLE 3
Cartridge dia and reduction in drill hole numbers

Cartridge dia	% Reduction of number of holes
25 mm	0, *i.e.*, base
30 mm	20-28
35 mm	30-35
40 mm	40-45
45 mm	45-50
50 mm	50-52

With the larger diameter cartridges, the examination showed that the borehole spacing and thus, the burden can be increased in linear with the cartridge diameter. For example in a large surface shaft shows the linear increase in the burden in relation to the cartridge diameter at the same time reducing the borehole number.

The obtained findings for borehole reduction resulting from the use of larger cartridge diameters gave rise to the expectation that a particularly great economic success would be achieved for large cross-sections. These considerations led to detailed comparative tests in shafts using 25 mm, 30 mm, 40 mm and 50 mm cartridge diameters. Exemplary results are presented in form of a test series, Table 4.

TABLE 4
Parameters of test rounds using different cartridge diameters in shaft

Cartridge dia	25 mm	30 mm	40 mm	50 mm
Advance per round, m	2.0	2.0	2.0	2.0
Excavation cu cm	138.62	138.62	138.62	138.62
Borehole number	137	97	59	44
Borehole length bl.	274.0	194.0	118.0	88.0
Borehole bl./sq m	2.0	1.4	0.9	0.6
Borehole length bl./sq m	2.0	1.4	0.9	0.6
Explosive consumption kg	77.5	110.0	111.25	106.5
Spee. explosive consumption kg/cu cm	0.75	0.79	0.80	0.72
Spee. explosive consumption kg/bl.	0.377	0.567	0.94	1.22
Detonator consumption pcs/cu m	1.34	0.71	0.43	0.32

4.0 NEW SLURRY/EMULSION EXPLOSIVE

Blasting in tunnels/drifts is the most difficult and important operation which has direct bearing on progress and cost of tunnelling. Since decades all over the world, NG explosive was considered to be the most suitable because of its strength and the ability to give pull despite NG's many serious drawbacks. Constant efforts were being made to find an alternate explosive which can give pull and good fragmentation and at the same time eliminate the serious drawbacks associated with NG explosives. New generation slurry/emulsion explosives have already replaced NGs in the developed countries and is on its way out in other countries too. In India, IBP has developed new slurry/emulsion explosives which have been proved to be an excellent performer. Construction industry now can avail and enjoy all the advantages of slurry/emulsion explosives without sacrificing the basic need of pull and fragmentation.

344

Traditionally, NG explosives are preferred because of their efficiency, performance and ability to suit the hard rock conditions. Because of meeting these basic needs, other serious drawbacks like more vibrations, shattering effects, longer throw, more fly rocks more proneness to accidental blasting due to higher sensitivity, pungent smell, head-ache during transport, use and handling; cough, burning, irritation to eyes and throat by the large quantity of post-blast fumes associated with NGs have been overlooked.

It will be pertinent to note here that almost all explosive accidents occurred are that from NG explosives. It is because of this and other drawbacks affecting human beings associated with explosives, that NG technology has been discarded by the developed countries and is on its way out from other countries also.

In contrast, the slurry/emulsion explosives demonstrate low strength, basically due to its inherent water content, low VOD and low density. However, the low density helps in obtaining better explosive distribution and more column length of hole, therefore, it gives controlled throw and more importantly less damages to rock mass or widening of existing cracks and lower over breaks, etc. Thus, slurry/emulsion explosives are more suitable for working through geologically disturbed and earthquake prone area.

The other slurry/emulsion advantages related to safety and physiological effects are absolute and established. Slurry/emulsion are very safe and are detonated only by required intensity of shock imparted by the detonators. These explosives are free from accidental blasting due to fire or friction, etc., moreover, the pre and post blast fumes in these case are very less, are free from any smell and do not give cough, burning or irritation to eyes and throat, etc. Thus, the slurry/emulsion are very safe and free from adverse physiological effects and attributes to the benefits of the "workmen."

IBP Co. Limited, a leading manufacturer of slurry/emulsion explosives in their stride to achieve the break-through for NG equivalent performance/efficiency explosive, particularly related to pull and fragmentation, have developed and introduced a product called Indotunneler (K) for the use in construction field, drifting, tunnelling through even the hardest granite like rock conditions and have been proved to be better in performance in the tunnelling operations undertaken by the Konkan Railway Corporation through Western Ghats of Maharashtra, Goa, Karnataka and for tunnelling operations in Nathpa Jhakri Hydroelectric Project in Himachal Pradesh. The details of advantages and savings are enumerated below.

4.1 Tunnelling in Konkan Railway Corporation Projects

KRCL was engaged in laying down a new railway line between Bombay–Mangalore which passes through western coastal area of Maharashtra, Goa, Karnataka states and has to encounter very difficult terrain of western ghats. The total length of 75 number of tunnels under construction was about 78.5 km. The longest tunnel being 6.5 km in length. The cross-section area of tunnels was 35 sq m.

The tunnelling operations were done by drilling and blasting method with hydraulic jumbo drill machines from Atlas Copco. The diameter of blast holes drilled by jumbo drill is 45-48 mm and hole depth varies from 3.5 to 4.0 m. The number of holes vary from 80-90 holes per round in 35 sq m area. NG based 80-90% strength of 32 mm dia cartridge explosives were used. Use of NG 90% was more common. The pull varied from 2.75 to 3.48 m per round, average being around 3 m normally, 130-150 kg of NG explosives along with 130 to 150 kg ammonium nitrate alone was used for blasting, except in one tunnel where about 250 kg of NG explosive alone was under use.

4.2 USE of New Slurry/Emulsion Explosives

The product Indotunneler (K) was introduced and tried in various tunnels under the jurisdiction of Chief Engineers of Mahad, Ratnagiri (South), Ratnagiri (North), Goa and Karwar. Of the various locations, tunnelling in Karwar section was through extremely hard granite rock where the drilling and blasting was the most difficult and slow process where average 2.7 m pull per round was obtained and drill bit life was less than 400 m compared to 600 to 700 m in other sections. Hard, difficult and highly geologically disturbed rock conditions were encountered in this area. Tunnelling in Koyana and other parts of Western Maharashtra, Karnataka were carried through earthquake prone and geologically sensitive area. Following advantages have been achieved with Indotunneler (K).

4.3 Reduced Drilling

Against the requirement of 80-90 holes per round with NG 80-90% strength explosive and ANFO only 70-75 holes per round required with Indotunneler (K). Only 60-65 holes were required per round with full slurry against 85 holes per round with full NG 90% explosives.

Thus, about 20-25 numbers of drill holes per round blast could be reduced. As a result savings in time and costs of drilling was obtained.

4.4 Less Explosive Consumption

Against (K) the consumption of 130-150 kg of NG 80-90% explosives with equal amount of ANFO, only 115-125 kg of Indotunneler (K) having higher energy/strength than NG 90% explosive with ANFO was required thus, 10 to 15% savings in explosives was obtained. In full NG and full slurry blasts, the explosive consumption was 200-215 kg against 250 kg of NG explosive.

4.5 Higher Pull

With both the reduced number of holes and explosive quantity the pull was more than that obtained with NG 80-90% explosives. Against an average pulls of 2.7 to 3.4 m obtained with NG explosive, an average pull of more than 3.5 m was obtained with Indotunneler (K).

In case of tunnels, the average pull of 3.25 m was obtained during trials against previous average pull of 2.76 m with NG explosives. However, in the subsequent months, a pull of more than 3-5 m , i.e., 87.4% on an average, (3.7 to 3.8 m pull being more common), was obtained. Thus, the pull obtained varied between 85 to 90% of hole lengths. This brought down the cost of tunnelling at this location to Rs. 64.44 lakhs from 94.14 lakhs per km giving a performance boost in this area. A clear edge and preference for slurry explosive use and reduced drilling.

4.6 Physiological Effects and Safety

There has been no physiological effect either in handling of the explosives or from post-blast fumes, i.e., no headache, irritation to eyes or throat, cough, burning sensation as is usually associated with NG explosives and their post blast fumes.

4.7 Reduction in Fumes and Defuming Time

A full slurry/emulsion explosive blast with 200 to 215 kg of Indotunneler (K) against 250 kg of NG 90% explosive has demonstrated exceptionally small quantity of fumes which were cleared within 15-20 minutes and the next cycle could be started within 20-25 minutes of the blast against 1.30 to 2.0 hours normal defuming time obtained with full NG explosive blasts. The visibility was excellent when entered in tunnel for checking/examination even 5-10 minutes after the blasts.

4.8 Reduction in Overall Time

The use of new explosive has given substantial time saving in terms of defuming, charging, drilling, etc., details are as given in Table 5.

4.9 Cost Saving

A comparative cost saving details in one of the granite rock tunnel is given in Table 6.

5.0 TUNNELLING IN NATHPA JHAKRI HYDROELECTRIC PROJECT

Nathpa Jhakri Hydroelectric Project in Himachal Pradesh is a gigantic project which involves substantial tunnelling work. Because of the strata conditions, size and lengths of each tunnel, the tunnelling work is extremely difficult and hazardous.

Due to serious drawbacks of NG explosives as explained above and major fatal accidents with NG explosives, one of the projects showed keen interest in trying the new generation explosive. The trial results of tunnel blasting are summarized below.

5.1 Techno – Economics

The techno-economics assessment details have been computed on the blasts conducted and ignoring the abnormal blast results which occurred due to various site conditions including drilling inaccuracies, non availability of required accessories, etc.

TABLE 5
Advantages of new explosives

Particulars		Practice with NG explosive (A)	Practice with slurries (B)	Variance (B-A)
Time/blast				
Drilling time	Minutes	150	120	-30
Charging time	Minutes	120	90	-30
Defuming time	Minutes	300	30	-270
Mucking time	Minutes	150	160	10
Survey time	Minutes	60	60	0
Misc. Time	Minutes	60	60	0
Total time/blast		840	520	-320
Time/km				
Drilling time	Minutes	46875	346820.8	-12192.92
Charging time	Minutes	37500	26011.56	-11488.44
Defuming time	Minutes	93750	8670.52	-85079.48
Mucking time	Minutes	46875	46242.77	-639.23
Survey time	Minutes	18750	17341.04	-1408.96
Misc. Time	Minutes	28125	26011.56	-2113.44
Total time/km	Minutes	281875	159859.54	-112915.46
Total time/km	Hours	4531.25	2649.33	-188.92
Total time/km	Days	188.8	110.39	-78.41

TABLE 6
Comparative performance of slurries in granite / km

Particulars		Practice with NG explosive (A)	Practice with slurries (B)	Variance B-A
Size of tunnel	sq m	35	25	
Rock type	Name	Granite	Granite	
Dia hole	mm	45	45	
Requirements/km				
Drilling meterage	m	108242.61	79807.23	-28435
Explosive quantity	kg	77760.5	62168.67	-15591.82
Detonators SDD	Nos	3732.5	3180.72	-551.78
Detonators LDD	Nos	23328.15	1760.24	-6557.06
Cost/km				
Drilling meterage	Rs.	5412130.64	3990361.45	-1421769.19
Explosive Quantity	Rs.	3110419.91	2486746.99	-623672.92
Detonators SDD	Rs.	37325.04	31807.23	-5517.81
Detonators LDD	Rs.	37325.04	31807.23	-5517.81
Total cost/km	Rs.	8597200.62	6540722.89	-2056477.73
Total cost/km	Rs.	85972.2	6540.72	-2056.48

5.2 Upstream Tunnel A

The size of the tunnel is 73 sq m. Against the performance with NG explosives and tunnel average of 145 holes/round, 450 kg explosives/round and pull of 2.83 m /round in 4 m length drill holes of 45-48 mm dia., (*i.e.*, 71% of drilled hole length) the average results with Indotunneler (K) explosives have been 132 holes per round, 340 kg explosives per round and 3.25 m pull per round 4 m length drill holes (*i.e.*, 81% of drilled holes lengths).

The approximate savings in Rs/km of tunnelling at CFJV is mentioned below.

Savings	Per km/Tunnelling
Drilling	42.506 m
Explosive	54.412 kg
No. of blast holes	46 rounds
Time savings for drilling, explosive charging and less blast rounds	To be assessed and taken into a/c
The cost savings	**Rs. per km/Tunnelling**
Drilling (assumed @ Rs. 98.00/m)	41,65,000
Explosive	10,11,000
Detonators	1,16,000
Total cost saving	52,94,000

5.3 Downstream Tunnel A

The tunnel size – 73 sq m

Performance with NG explosive and against the tunnel average of 145 holes/round, 475 kg explosives per round and pull of 3.12 m per round in 4 m length drill holes of 45-48 mm dia., (*i.e.*, 75% of drilled hole lengths) the average with Indotunneler (K) explosives have been 125 holes per round, 420 kg explosives per round and 3.25 m pull per round in 4 m length drill holes (*i.e.*, 85.75% of drilled holes lenghts).

Savings	Per km/Tunnelling
Drilling	31,590 m
Explosives	22,878 kg
No. of blast rounds	13 rounds
Time savings for drilling, explosive charging and less blast rounds	To be assessed and taken into a/c.
The cost savings	**Rs. Per km/Tunnelling**
Drilling (assumed @ Rs. 98.00/m)	31,25,220
Explosive	7,37,000
Detonators	87,70,000
Total cost saving per km of tunnelling	24,75,900

5.4 Downstream Tunnel B

The tunnel size – 73 sq m

Against the tunnel average of 160 holes/round, 535 kg explosives/round and pull of 3.00 m /round in 4 m length drill holes of 45-48 mm dia., (*i.e.*, 75% of drilled hole length) the average results with Indotunneler (K) explosives have been 125 holes per round, 470 kg explosives per round and 3.43 m pull per round 4 m length drill holes (*i.e.*, 85-75% of drilled hole length).

348

Savings	Per km/Tunnelling
Drilling	40,741 m
Explosives	41,310 kg
No. of blast rounds	42 rounds
Time savings for drilling, explosive charging and less blast rounds	To be assessed and taken into a/c.

The cost savings	Rs. Per km/Tunnelling
Drilling (assumed @ Rs. 98.00/m)	32,92,610
Explosive	16,100
Detonators	1,12,000
Total cost saving per km of tunnelling	40,88,000

In spite of large difference in landed cost of explosives between conventional NG – 80% explosives and new generation slurry/emulsion Indotunneler (K) explosive, the cost savings in explosive itself is about 7 to 10 lakh per km of tunnelling. The cost of traditional NG 60-80% explosives and its equivalent 60-80% slurry/emulsion explosives is almost same.

6.0 WORLD TRENDS

The construction industry has always given more importance to the "Progress/Pull" obtained from explosive use to judge explosive performance and ignored all other drawbacks and indirect advantages of explosives which has definite influence in overall performance and cost including safety, accidents, loss of human lives etc. etc. The world wide "maximum average pull" obtained in tunnelling with Ng/slurry/emulsion are in the range of 90 to 95% and the average pull obtained by new generation slurry/emulsion particularly Indotunneler (K) is also in the same 90-95% range.

The results obtained in Norway as reported in the 22nd Annual Conference of Society of Explosives Engineers, USA are reproduced below (Table 7) which are said to be the best as average in tunnelling so far.

TABLE 7
Blasting performance in Norwegian tunnels

Tunnel	Length of tunnel km	Size sq m	No. of holes per round	Dia. of drill bit mm	Length of holes m	Pull m	Per cent
Aurland tunnel	24.5	58	98	45	5.45	4.8	88
Matre tunnel		82.5	127 (119 + cut holes 4 + 4)	45	5.45	5.07	92.9

The possibility of manufacturing slurry/emulsion explosives in any required diameter cartridge has added advantage of reducing the number of hole and total meterage of drilling per round which shall result into substantial cost saving. Using the highest possible diameter holes drilling, along with use of matching explosive cartridge diameter, explosive energy (strength) in a product an appropriate drilling pattern is the way to look at "Faster and economic tunnelling with conventional drilling and blasting method".

7.0 CONCLUSIONS

1. Use of higher diameter drill holes helps to reduce number of holes per round of blast, *i.e.*, reduced drilling.

2. The application of appropriate higher diameter explosive helps in reducing the number of holes, drilling meterage, time and related costs.

3. The use of appropriate diameter explosives, drilling pattern and compatible energy explosives reduces the quantity of explosives to be used per km of tunnelling. The establishment of final set of drilling pattern and explosive quantity has to be accomplished after observing practical results and requirements.

4. The savings on account of drilling meterage, number of holes, pull, *i.e.*, number of rounds per km is not only significant but noteworthy and has severe cost impact.

5. Traditionally, Ng explosives of 60, 80 and 90% strength in 25/32 mm are produced. The new generation slurry/emulsion explosives of equivalent strength (*i.e.*, 60/80%) as well as more stronger than 90% Indotunneler (K) are produced in any desired diameter cartridges. The energy content of the explosives provides the comparison of explosive strength*.

6. The "time saving" and "fume characteristics" of slurry/emulsion explosives is another most important aspect which heeds to be considered most appropriately and taken into account since it has substantial overall cost impact.

7. The application of the methods explained in the article are the sure and tested ways to achieve Faster and Economic tunnelling which ultimately gives higher productivity and overall cost reduction.

*The facilities for such testing are available in R & D centres of explosive manufacturers. IBP R & D facilities which are recognized by DSIR can be utilized by consumers to know the quantity of the explosive products.

BIBLIOGRAPHY

-   Frank R. Chiappetta, Continuous Velocity of Detonation Measurements in Full Scale Blast.

-   Walter Heinz Wild Germany, "The Optimal Borehole and Cartridge Diameters".

-   Las Haakon Seim and Bjoern R. Petterson Norway, "World's Longest Highway Tests Site Sensitized Emulsion".

-   MO Sarthi : Delay Blasting, An Inexpensive Tool for Reduced Total Mining Costs.

-   Ghosh, A.K. and Others, Blasting Economics and Overview.

-   Manoj Raoot, IBP : His field work and supervision of blasting in tunnels.

VARIOUS ASPECTS OF VENTILATION SYSTEM FOR UNDERGROUND WORKS — EXPERIENCES AND USEFUL SUGGESTIONS

M.M. MADAN

Chief Engineer

National Hydroelectric Power Corporation Ltd., Sikkim, India

SYNOPSIS

Ventilation in long tunnels forms the most important requirement not only for the working of machines but also for the sustenance of the personnel. Out of the various requirement of sustenance of the personnel, possibly some relaxation can be accommodated for a short period in facilities like food, water light, and transport, etc., but any shortcoming in the ventilation system can be disastrous and can be produce long term demoralizing effect on workmen and others. Ventilation management is the most important aspect of underground works during construction phase. The progress of construction depends largely on a good ventilation management and efficient ventilation system. The accurate design of a system involves many intricate formulas and calculations based on assumptions. It has been practically observed that even after designing the ventilation system and installation of the system as per the design, the system fails and full efficiency is not obtained. In the opinion of the author during construction of underground works, it is not essential to follow the intricate design procedures but some calculations must be done in order to establish the necessary requirements for the ventilation management. After setting the parameters, the system shall be monitored regularly and based on velocity measurements the requirements of the fans must be established. This system has been followed in some of the projects and found very effective. Author has tried to explain difference in various systems of ventilation which can be selected as need based and ventilation can be managed with continuos follow up by measurement of air flow to the best advantage, useful suggestions have also been given in concluding part.

1.0 IMPORTANCE OF VENTILATION

Foul and poisonous gases are generated by blasts in tunnel, by exhaust of diesel operated machines, breathing out foul gases by working crew and besides, a tunnel may cut through strata being trapped foul poisonous gases which would get released on excavation. So, unless there is a foolproof arrangement for speedy and evacuation of such gases, these can result in serious mishaps. The heat released by the rock, exhaust of machines, electrical power transformers and cables etc., raises temperature inside the tunnel to noticeable extent. All these factors demand the standards of design and performance of the ventilation system. In fact, ventilation is the essential factor in controlling the efficiency, performance and progress of an underground project.

Developments in science and technology have given sufficient tools to make tunnel work as comfortable as required but there are always limitations of resources specially in developing country like India. Very often, financial constraint becomes a major deciding factor in selecting design of the ventilation system.

To adopt a simple ventilation system, which would serve the purpose without causing serious hardship to the working of men and machinery and does not put unbearable financial strain on the project has been explained in this paper.

2.0 ELEMENTS OF THE SYSTEM

An auxiliary ventilation system is a closed system in which one or more fans located outbye end of the heading drivage are connected to a long line of ducting extended upto inbye end of heading drivage. If all the fans employed are clustered together at the outbye end of the duct line, this arrangement is called a simple auxiliary ventilation system. In series layout, additional fan/fans, apart from those at the outbye end, are spaced along the duct line at intervals, essential elements of the installations are the fan/fans, ducting and their respective positions forming the desired layout.

2.1 Fans

Fans employed are auxiliary ventilation system are invariably electrically driven small fans, the motor being rated for power consumption of about 7.5 to about 40kW. Centrifugal fans are less commonly used. Although bifurcated in – line centrifugal fans have been deployed abroad for high pressure application, they are not being used in India where majority of installations employ in-line axial flow fans in following designs :

1. Single-stage or two-stage (mounted on same shaft) fans.
2. Two-stage contra rotating fans and
3. Bifurcated axial flow fans.

2.2 Ducting

The ducting employed may be rigid, flexible or semi-flexible. These ducting can be made in any diameter but commonly used diameters are 600, 700, 750, 800, 900, 1000, 1200 mm.

3.0 AUXILIARY VENTILATION SYSTEMS

3.1 Simple System

The air supplied for ventilation at face travels through the ducting in forcing system and hence is discharged to the face at considerable velocity. In case of exhausting system the same air reaches the face through the roadway section and thus is at a very low velocity.

3.2 Multiple–fan Systems

In very long drivages, multiple fan units are required to overcome the high resistance of ducting. Clustering these fans at the outbye end of the drivage. Proves to be uneconomical as leakage along the duct length are considerably increased and actual gains in terms of increase in air quantity at the face may not be significant. To obviate this waste, fans are spaced along the duct-line. This method although economical in terms of power-consumption, results in uncontrolled re-circulation of air in a self-contained closed circuit, which in gassy seam may lead to dangerous firedamp buildup.

4.0 SOURCE OF GENERATION OF OBNOXIOUS GASES

The requirement of Ventilation to defume obnoxious gases generating out of the following activities in underground works and to provide fresh air for the working crew inside tunnel and clean atmosphere.

Activities which generates obnoxious gases :

1. Blasting: Fumes generated due to the use of explosives.
2. Drilling: Dust generated due to the drilling activity.
3. Gases generated due to working of diesel equipment.
4. Fresh air for humans working is inside the tunnel.
5. Heat generated due to electrical appliances/equipment.

5.0 REQUIREMENTS

The total quantity of air required ventilating the tunnel and the work face that must be sufficient to ensure an adequate environment Compatible with health and safety requirements of tunnel personnel. For ventilation planning the major pollutant, sources considered are Diesel engines and blasting operations. These two sources mostly produce CO_2, CO and NO (Oxides of Nitrogen). Under normal working condition where persons are required to work the Gas, concentrations are set out in Mines regulations. The relevant values are—

CO_2 not exceeding 0.5% by Vol. (5000 ppm)

CO not exceeding 0.005% by Vol. (50 ppm)

NO_2 / NO not exceeding 0.0025% by Vol. (25 ppm).

The oxygen percentage in Air volume should not be less than 19% and the dust concentration level in the Air should be kept in the range of 3-5 Mg per cu m of Air. The concentration of explosive Gas (mostly CH_4) if any, should not be allowed to exceed 0.5%. A minimum air velocity inside the tunnel shall be kept as 0.5m/sec. Normally Air velocity in duct should be 6-12m' sec. Velocities in excess of 15m/sec may be avoided and for long tunnels in excess of 8m/sec.

6.0 CONSTRAINTS

The following constraints have to be considered for designing and managing a successful ventilation system. There is no doubt that larger the dia of the Duct the less friction it would create but the Main constraints would be the size of tunnel which will decide the dia of the Duct otherwise the equipment would not move freely and ducts would be damaged regularly. The Main constraints may be described as follows :

(i) Size of tunnel

(ii) Type of equipment used
  – Diesel operated
  – Electric operated
  – Pneumatic
  – Hydraulic
  – Battery operated

(iii) Temperature difference between outside and inside tunnel.

(iv) Humidity of the Area.

(v) Flow of Natural Air.

(vi) Proximity of other openings.

(vii) Type of Explosives used.

7.0 SOME THUMB RULES

- Suction system is effective while using Battery operated or Electrically operated equipment like rail mounted battery Locos and wagons, Miners, Electro Hydraulic Drill Jumbos, TBM's etc.

- Forcing system is effective when Diesel operated equipment are used. Although the Airway remains dusty but the use of Airway after mucking is Minimum. Therefore, the working area at the heading remains clean of smoke therefore better atmosphere of work.

- Reversing type fans have some operational problems, if the fan's operation is controlled at one point then the system may be successful, but if different starters are installed at different points then the chances are that the fans may run in opposite direction thereby causing air lock in the tunnel and may damage the whole system. Also when the system is used in the Exhaust mode then the four air, carbon etc. is deposited in the Airway and as soon as the fans are reversed then the deposits will blow out at the face causing more unhygienic conditions.

- Damaged fan must be removed at the earliest so that it does not cause hindrance in air velocity in duct.
- In forcing system whole area remain clean including duct line and temperature at face is controlled because of blowing of fresh air at the face.
- Air lock is created because the fresh air does not go beyond certain reach in tunnel and many times no fresh air reaches working face. In leaking condition, the short circuit circulation of air is very common.
- Temperature keeps on rising at the heading.

7.1 Size of the Tunnel

The size of the duct is limited by the size of the tunnel and equipment size, however, following diameter of the duct with respect to various sizes of tunnel may be considered for planning of duct diameter.

Size of tunnel	4 m	5 m	7 m	> 7 m
Dia of Duct	600 mm	900 mm	1000 mm	> 1200 mm

- Cost of fan and ducting increases due to increase in size of Ducting and vice-versa.
- Flexible ducts are more economical considering handling and installation costs.

8.0 SELECTION OF SYSTEM

8.1 Exhaust System (Pull) vs Forcing System (Push)

This difference in velocity at the face has two fold effect. High velocity in forcing system increases the cooling power of air and causes through mixing of any firedamp produced at the face with air. The second effect of high velocity at the face is that it raises the dust produced at the face and makes it airborne. This airborne dust is unlikely to settle in course of its return passage through the heading and complete heading is very dusty when mucking operating is in progress.

The other advantages of the forcing systems are that intake air travelling inside the ducting does not have the chance to pick up heat and moisture which in exhausting system is inherent as Intake air travels through the drivage; and that in forcing system unsupported flexible duct can be used whereas in exhausting system supported flexible ducts have to be used which are about twice costly and offer about thrice high resistance to air flow compared to unsupported ones.

In forcing system, a tendency among workmen has been observed to puncture the duct at places of their work, away from face, to get a high-pressure jet of leaking air to feel comfortable in hot conditions. In exhausting system, such intentional tampering is useless. However, unintentional leakages are easily detected in forcing system and hence can be repaired immediately. Detection of leakage in exhausting system is not so easy and may require the help of smoke tube etc.

In forcing system, the fan is in Intake side of feeder airway and handles clean air. The exhausting fan, on the other hand, deals with vitiated return air, which is generally dust-laden gas-air mixture. With this in view, bifurcated fans have been designed in which motor of the fan is not exposed to corroding effects of return air. However, these fans are costly.

8.2 Exhaust System

The main feature of the system is :

(i) Contaminated air and fumes exhausted through duct :
   The dust, fumes etc., produced by the machines or otherwise is pulled out from the main working area and fresh air is replenished through the main airway or tunnel to the manpower working at the face.

(ii)   The tunnel atmosphere is full of fresh air :

As the fresh air is replenished through the tunnel or main airway, therefore, the whole length is filled with fresh air. This system is workable when battery operated equipment or electrically operater equipment like Locomotives, Rail Mounted Wagons, Road Headers, Miners etc., are used because foul exhausted air is not produced enroute. While using diesel operated equipment, such as Tippers, Dumpers, Loader etc., which move, though the entire length gets filled with exhaust gases and the same gases travels through the entire length of tunnel and goes up to the face and then these are exhausted through the ducts. This system creates breathing problems for manpower as well as for equipment. Because the speed at which fresh air replenishes is very slow. The fumes get concentrated at face where actual works is to be carried out. The dust, smoke, carbon, etc., fills the total duct line and fan motors are covered with thick carbon deposits which increases maintenance and replacement costs.

(iii)   The system has a restricted reach and turbulence is created only in the immediate proximity of the duct end (1 to 1.5 m).

(iv)   Leakage of air through the joints reduces the effectiveness of ventilation at the working front.

(v)   Because of slow speed of air the working area remains hot and the fresh air travels through the entire reach of tunnel the temperature of air also rises and the air may not remains as fresh as required at face.

(vi)   Because of temperature difference, natural air current direction, humidity inside the face, air lock develops at certain places in the length of tunnel, which results in less air availability at the face.

(vii)   Breakdown of intermittent fan causes reduced suction capacity. Thereby reducing fresh air availability.

8.3 Forcing System

The main feature of the system is :

(i)   Tunnel gases or fumes are not brought to the working face where people operate. Air moves out at a relatively at high velocity from the end of the ducting and sweeps at distance of 10 to 15 m from the duct and cleaning the gas and the dust form the face.

(ii)   It ensures fresh air blast at the working face and creates turbulence.

(iii)   Leakage through ducting helps in diluting polluted air in the tunnel.

(iv)   The whole length of the tunnel or Airway is filled up with contaminated air. So working at any other point of the tunnel except face may prove unhealthy.

(v)   However, in construction of tunnels not much work is done at any other part except the face, therefore this system does not provide any negative effect.

(vi)   Workmen have to travel through the contaminated atmosphere in the tunnel to reach the working spot, but due to large area the dilution of gases takes place therefore workmen will not face much difficulty in travelling through contaminated atmosphere.

8.4 Forcing Main Plus Exhaust at the Face

This system is satisfactory where the working face is cleared of fumes very quickly. It ensures improvement over the forcing system ventilation.

8.5 Exhaust Main Plus Forcing at the Face

This system combines the merits of both exhaust and forcing systems. The forcing fan eliminates the restricted reach and restricted turbulence of the exhaust system and work spot gets fresh air. Contaminated air is removed via the exhaust duct and the tunnel is replenished with fresh air throughout its length.

8.6 Flexible vs Rigid

Flexible: "Flexible ducting" of non-reinforced type can only be used when the ventilation system is forcing. In exhaust, system spirally reinforced ducting shall be used. The pitch of spiral and thickness of spiral wire shall be decided depending on suction pressure and proximity of fan location. Ducting installed near the fan must be of close pitch so that it can resist the negative pressures developed by the fan.

Steel Ducting: "The Steel ducting" is preferred many times because of following :

  (i) Low resistance to airflow.

  (ii) Recommended for suction type of ducting.

  (iii) Can be fabricated locally.

  (iv) Resistance to damage.

  (v) Withstand more Ventilating Pressure. However, practically it has been observed that it has more disadvantages than advantages (Refer comparative statement at Annexure – 1).

9.0 DESIGNING A VENTILATION SYSTEM

Detailed design steps are very cumbersome and time consuming. Most of the parameters are inter-related. Trial and error has to be employed by assuming certain parameters. However, even after doing a perfect design the system fails in field and it is found that effective ventilation does not take place. Therefore, it is essential to design a system based on simple assumptions and calculations. The system chosen shall be monitored in field with simple instruments like Air Flow meters. Whenever it is observed that velocity of Air is reduced at the face then additional fan must be installed immediately. Side by side, it must be ensured that the proper maintenance of system is done regularly and managed in such a way to have maximum effectiveness of the system so that efficiency of the work increases. The formulae used for designing have not been discussed in this article. For designing a system, refer various books and papers on the subject.

10.0 EXPERIENCES

In Rangit Project, the Head Race Tunnel upstream package was taken over from private contractor for construction by department. One of weak point observed was poor ventilation management of tunnel and other underground works. The private contractor was using 600 mm dia ducts in a combination of Steel/Flexible/GI sheet ducts. All types of ducts were in use. Only 10 HP Fans at a spacing of 150 m to 200 m were used. The system adopted was exhaust (suction) system. The joints were highly leaking. In order to expedite the progress it was essential to design and manage a good ventilation system. The system was designed by project engineers and implemented strictly as per the requirement. Instead of exhaust system, forcing system was used. 20 HP Fans at a spacing of 100 m/150 m were installed. Complete duct line was changed and only flexible reinforced type duct line of 900 mm dia with leak proof banded coupling were installed. The progress of tunnelling which was on an average 10 m per month during the period of contractor could be increased to average 50 m/month with maximum 60 m in one month from one face.

11.0 CONCLUSION AND SUGGESTIONS

With the experience of many projects, the following can be concluded and the suggestions given can be followed for the efficient management of ventilation system.

11.1 Forcing vs Exhaust System

  (a) In exhaust system the explosive fumes and exhaust from all the diesel equipment operating both within the tunnel and at the mouth of the portal travels inside, thereby further compounding the pollution at the face of heading and all along the length of the tunnel. Even inspite of waiting for more than 2 hours after blasting, the fumes makes the mucking operation very difficult and the face workers have to come out for fresh air. By adopting forcing system, the ventilation system would not be over loaded with the exhaust of diesel engines.

(b) The ducting line is an exhaust system is never close to the face and as such the face workers are exposed to the static atmosphere full of noxious fumes. Since it is difficult to keep the ducting line extended close to the faces, a forcing system would overcome the draw back.

(c) In forcing system, fresh and brisk air would be available at face for workers soon after the blasting operation. The mucking operations could be commenced within 30 minutes after blasting, hereby accelerating the pace of advance considerably.

(d) Since the success of tunnelling operations depends largely on good standard of ventilation, locating points of leakage in the ducting line is very vital and this can be easily detected in a long ducting line so that due care is taken to plug the leakage and take remedial measures to stop such occurrences in future.

11.2 Joints of Flexible Ducts

The joints should be made properly with the help of only flexible couplings of proper design and construction placing all the flexible ducts in front of each other and joined together with the help of 3 or 4 pieces of wires puncturing through the ducting cloth is not recommended and should be avoided otherwise leakage's cannot be controlled. Besides damaging the flexible ducts, the joints become heavily leaking, as flexible couplings are not utilized. It has been established that leakage rate in such ducting can be as high as 33 per cent for 100 m length of ducting.

11.3 Installation of Flexible Ducts

Proper installation of flexible ducts is very essential for controlling leakage of air and effective utilization of the full potential of the auxiliary fans. The over head wire-rope suspension assemblies with proper turn-buckle arrangements, in lengths of 25m/30m should be used. Longer lengths are difficult to handle and keep the ducting in a straight line, as it becomes difficult to negotiate deviations in the drivages/headings. The flexible ducting should be suspended only from the overhead wire rope suspension assemblies and not with the help of vertical wires attached between the roof of heading and the suspension-hooks provided in the ducting with the sole idea of maintaining a gradually sloping ducting line. Sings or wires should not be used for suspension of flexible ducting. While joining the flexible ducting, due care should be taken to ensure that the PVC coupling band of the flexible coupling fully traps the end rings the ducting all along it's hold by any subsequent inadvertent jerk or vibration after installation. Care should be taken that undue weight of the duct should not be allowed on a single suspension hook. All suspension hooks provided at every 0.5 to 0.6 m apart should be used for suspension of the ducting. If due to any reason, the hooks are lost these should be immediately fitted. While fitting new hooks, care should be taken to avoid any puncture into the ducting cloth.

11.4 Reducers

Since the Reducers provided between the auxiliary fans and ducting are the biggest culprits for leakage of air, it is essential to improve the Design/Construction of the reducers. The diameter of flat end of the reducers where flexible ducting will be connected should be such that the difference between the flat portion of the reducer and internal dia of the flexible duct should not be more than 15mm. It is observed that the reducers are connected To Whom It May Concern: the fan without providing any gasket packing between the flanges of reducer and fan. It is essential to provide an effective gasket-packing to make the joint fully leak proof, especially in view of the fact that the ventilating pressure of the fan is maximum at the inlet and outlet of the fan. Once a proper joint is made, the effort in doing so will go a long way in effective utilization of the fans, thereby, improving the standard of ventilation.

11.5 Operation of Fans in Series

Ideally, the fan in series should be interlocked so that when one of the fans stops, the remaining fans, in series, should also remain stopped. Since it is cumber-some to ensure such inter-locking, steps should be taken to ensure that all fans are operating. Mere observation of vibration of the steel frame may give a wrong impression that the fan is running. It is better to put an Ampere-meter on the electrical starter.

11.6 Repairing of Flexible Ducts

In case of damage to the ducts, two types of repairing kits namely hot and cold repairing kits shall be required. Two numbers of such repairing kits consisting of welding-iron (220V x 180W) with special welding tip, soft metal brushes, a repair of heavy duty scissors, spare PVC coated ducting cloth shall always be kept ready. Persons shall be trained to repair the ducts. The repair involving relatively larger cuts can be repaired outside the portal. However, small cuts / punctures can be repaired right at site with the help of cold-repairing kit consisting of a special adhesive. While replacing the damaged ducts in the already installed ducting line, it may be necessary to use small distance pieces of flexible ducting complete with suspension hooks and flexible coupling.

11.7 Spacing of Fans

It is necessary to refer to the pressure vs quantity characteristics of the fans supplied by the manufacturer by observing the total ventilating pressure of the fan. While in operation, it is easy to know the exact spacing so that fan continues to deliver the desired quantity of Air. It is possible to increase the spacing of fans by measurement of Air Flow to face and pressure measurement. It is, therefore, necessary to fix a nipple on the reducer on the forcing side of the fan for measuring the water gauge.

11.8 Selection of Auxiliary Fans

The rate of flow of air should be about 25 cu m/second to provide air velocity to 1m per second for fast evacuation of noxious fumes necessary to speed up the tunnelling operations. It is, therefore, necessary to ensure that the ducting line is leakless, straight and fully stretched so that the resistance of the ducting line is not unduly increased with consequent reduction of flow of Air.

11.9 Water Accumulation inside the Ducts

When the ducting line is not uniformly sloping, water accumulation inside the ducting line becomes unavoidable. This exerts unduly heavy pressure on the flexible ducts. The ducts will be damaged. A nipple with valve can be fitted at the bottom most part of the ducting to drain water whenever necessary.

11.10 Periodic Inspection

After installation of the new ducting line, it is very essential to tighten the flexible coupling after every seven days for atleast a month and later the periodicity could be increased further. The ducting line should be inspected and remedial steps taken to ensure proper suspension of ducting and it is leak proofness from all joints of ducting, reducers and fans.

REFERENCE

Madan, M.M.: "Ventilation management – design and operation of ventilation system for underground works – Improvement based on continuous follow up of measurements" , Indian Journal of Power and River Valley Development, Vol. XLVIII, Nos. 1 & 2, January – February, 1998, pp. 8-24.

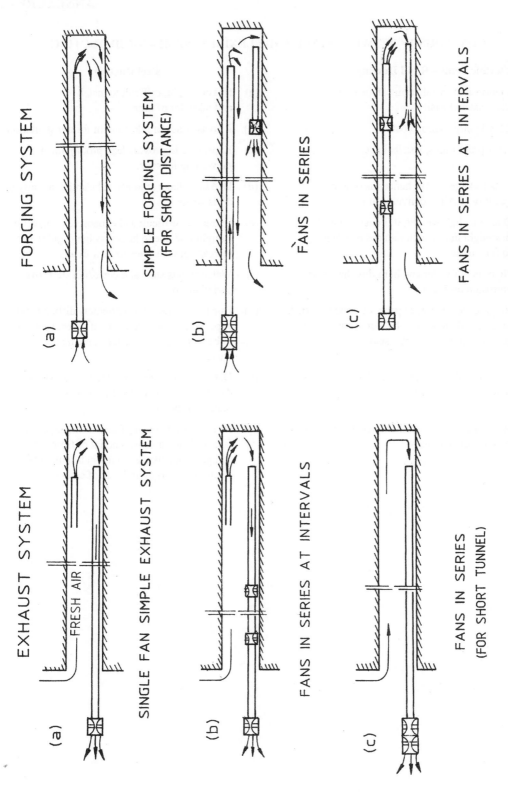

EXHAUST SYSTEM

FORCING SYSTEM

(a)

FRESH AIR

(a)

SINGLE FAN SIMPLE EXHAUST SYSTEM

SIMPLE FORCING SYSTEM
(FOR SHORT DISTANCE)

(b)

FANS IN SERIES AT INTERVALS

(b)

FANS IN SERIES

(c)

FANS IN SERIES
(FOR SHORT TUNNEL)

(c)

FANS IN SERIES AT INTERVALS

Fig. 1 : Schemetic diagrams of the different systems of ventilation

359

COMPARISON BETWEEN STEEL AND FLEXIBLE / SEMI – RIGID DUCTING

Flexible Semi – Rigid Ducting	Steel Ducting
(a) Lightweight installation can be done very fast and easily.	(a) Heavyweight, difficult to install and takes lot of time.
(b) Less Transportation cost.	(b) It costs considerable amount for transportation.
(c) Leak proof joints, less leakage.	(c) Expensive to provide leak proof joints for preventing leakage.
(d) Less leakage of Air, hence maximum quantity of Air is delivered at face.	(d) Due to leakage at joints, substantial quantity of air is wasted.
(e) Due to its lightweight, it is easy to install in straight line, which reduces resistance to flow of air.	(e) Extremely difficult and expensive to install in straight line due to heavy weight/rigidity hence increase in resistance to flow of air.
(f) No effect of water and humidity, hence no corrosion problem.	(f) Water effects steel and corrosion starts over a period of time.
(g) Requires less storage space, i.e., one twentieth of its length hence can be stored easily and economically in covered space.	(g) Due to bulkiness it is extremely difficult and expensive to store in covered space, hence it is to be stored in open space at the mercy of weather.
(h) Due to its flexible nature of shape deformation or dents does not arise.	(h) Dents and deformations due to any external damage create resistance to flow of air and reduce of the fan.
(i) As total air is delivered at Face, there is no wastage of electricity.	(i) Needs more number of fans or higher capacity fans to deliever same amount of fresh air to compensate leakage, which in turn increases electricity charges and cost of equipment.

ALP TRANSIT GOTTHARD BASE TUNNEL, SEDRUN ACCESS AND VENTILATION SHAFT

ROBERT MEIER

Chief Resident Engineer

Electrowatt Engineering Ltd., Sedrun, Switzerland

SYNOPSIS

After the Swiss population accepted the proposed concept for financing public transportation in a referendum at the end of 1998, construction of the base tunnel under the Gotthard Massif has started in ernest.

At 57 km long, the twin tube Gotthard base tunnel is the longest tunnel in the Alp Transit scheme. The tunnel is divided into 5 lots, the twin northern lots at Erstfeld and Amsteg, and the three southern lots at Sedrun, Faido and Bodio.

The rock of the Tavetsch Massif in the Sedrun lot is expected provide the most difficult tunnelling conditions in the Gotthard base tunnel and lies on the critical path for the tunnel completion. The access shaft at Sedrun was tendered as a separate construction contract to allow an early start in this zone. The project includes site installations for the future main tunnel drives for the Sedrun lot with four working faces for the base of the shaft.

The shaft was sunk to a depth of 800 m by drill and blast methods in the crystalline gneiss of the southern Tavetsch Intermediate Massif between 1998 and the beginning of 2000, using a 20 m high working platform that incorporates a special shaft drill Jumbo.

1.0 GOTTHARD BASE TUNNEL WITHIN THE EUROPEAN RAILWAY NET

The 57 km long Gotthard Base Tunnel is the longer of two base tunnels to be built as part of Switzerland's Alp Transit Railway project. The project has been in planning in various forms for over 50 years.

Although the existing Gotthard mountain railway, which is all of 116 years old, and the 85-year old mountain railway via the Lötschberg may offer impressive panoramas during the trip and represent outstanding engineering accomplishments, nowadays they are far too slow in competition with road and air. To reach the portals of the existing tunnel, the railway has to climb from around 500 m above sea level to nearly 1100 m. The result is that rail traffic is constantly losing its market segment to the road network. Switzerland lies on the most important north-south European transit route between Germany and Italy and over one fifth of the transalpine goods traffic is via the Swiss alpine routes. Switzerland does not want the rapidly expanding traffic to use the Swiss road systems, which are already at full capacity.

As a result, the engineering vision of a level railway below the Alps was taken up once again towards the end of the 1980s. The idea was to establish a modern high-performance railway trunk-line with a large clearance cross-section so that all kinds of goods can be transported through the tunnels and with generous curved radii so that high travelling speeds are possible.

The Swiss government decided on the development of two axes through the Alps to the north of Italy. A clear majority in two Swiss referendums over the scheme, accepted the project in its existing form. The main transit axis is via the Gotthard. It leads from the north via Zurich into the south of Switzerland and on to Milan in Italy. The second axis leads from Berne, via Lötschberg and Simplon to Milan.

2.0 CONCEPTS OF GOTTHARD BASE TUNNEL

2.1 Routing

The routing of the Gotthard Base Tunnel and the entire Gotthard axis had to be accomplished in such a way as to allow high-speed passenger trains to travel at up to 250 km/h and freight trains at up to 140 km/h.

As a consequence, maximum gradients of 1 % are permitted, the portals must be located roughly 600 to 800 m lower than the portals of the existing Gotthard Railway Tunnel.

The geological conditions had to be taken into account, especially the attempt to avoid difficult geological formations or to pass through zones with minimal expansion. Further considerations were the overburden as well as ensuring that safety gaps were adhered to regarding reservoirs. The outcome was a slightly z-shaped horizontal routing of the tunnel.

With the northern portal near Erstfeld and the southern one near Bodio, the tunnel is just under 57 km in length.

2.2 Tunnel System

Following intensive investigations into possible tunnel system variants with regard to aspects such as construction operations, capacity, maintenance of facilities, safety and last but certainly not least, the construction and operational costs as well as the risks pertaining to construction costs and duration, it was decided to select a system comprising two single-track tunnels running parallel to one another.

The tubes will be linked to each other at regular intervals by means of cross-cuts.

At two cross-over points within the tubes, trains will be able to transfer from one to the other. Each cross over point features also an emergency station with rescue installations.

2.3 Intermediary Points of Attack

In order to arrive at an acceptable construction period for the 57 km long tunnel as well as to cater for its ventilation when finally operational, intermediate points of attack have to be created.

These intermediate points have been set up within the topographical possibilities in such a way that on the one hand, they divide up the tunnel into roughly equal sections and on the other, permit zones, which are tricky in technical terms, to be tackled at the earliest possible stage.

Construction programme and cost investigations with a large number of concepts relating to the points of attack led to the following intermediate points of attack being selected for the Gotthard Base Tunnel in the preliminary project. (Fig. 1)

- Erstfeld: Northern portal
- Amsteg: A horizontal, approx. 1.2 km long access tunnel
- Sedrun: A blind shaft with a depth of 800 m and a clear diameter of 8.0 m opened up through an approx. 1 km long horizontal access tunnel. An additional inclined shaft connection from the head of the shaft to the surface serves as ventilation when operational.
- Faido: An inclined tunnel with a gradient of approx. 12 %, which is roughly 2.7 km in length and overcomes a 330 m difference in height.
- Bodio: Southern portal

2.4 Geological and Constructional Aspects

Over 53 km, the tunnel geology is marked by three major gneiss complexes (Aar-Massif, Gotthard-Massif and Penninic Gneiss-zone), which can largely be regarded as favourable formations for construction purposes. In the zones with high overburden, these formations are to be found in an almost perpendicular position and will be cross-cut at a right angle during tunnelling.

The Tavetsch Intermediate Massif and several younger sedimentary zones, which are located in between, represent the technically difficult section of the tunnel. (Fig. 2)

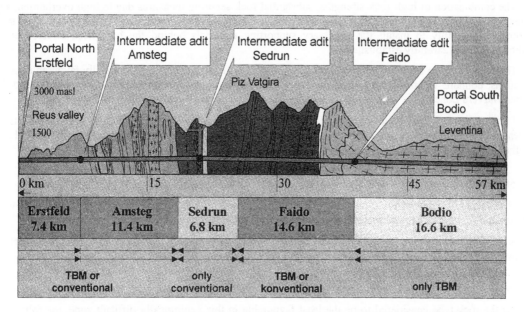

Fig. 1 : Intermediate points of attack, construction lots, portals, adits and shafts

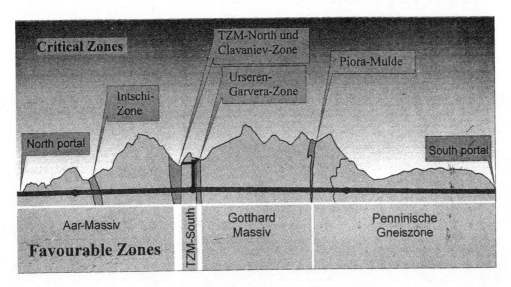

Fig. 2 : Gotthard base tunnel, geological zones

The overburden will be very high over major sections of the tunnel. This can be seen from the following figure. It amounts in fact to more than 1,000 m for roughly 35 km to more than 1,500 m over 20 km and even over 2,000 m over approx. 5 km. The maximum overburden amounts to 2,300 m

2.5 Tunnelling Techniques

The driving of more than 42 km of the two tunnel tubes will proceed using a TBM. Technically favourable geological formations will be driven using normal methods. All TBM drives, however, are expected to encounter somewhat problematic fault zones or areas with poor rock conditions.

The combination of high rock strengths, substantial rock securing measures due to high overburden, and the necessity of penetrating complex sections with a TBM present unique conditions for the operation of such machines. The ability to install the necessary supporting materials behind the bore head under specific conditions must be carefully established.

The tunnel section from the bottom of the shaft in Sedrun, in addition to the technically favourable gneiss formations, contains technically difficult squeezing formations of the Tavetsch Intermediate Massif and the Urseren Garvera Zone.

Two construction approaches have been chosen for these difficult formations which must be headed using conventional methods. One is the application of the driving method with divided face and crown removal and the other is the full-face extraction method.

2.6 Standard Profile Design

The standard profile of the tunnel tubes is currently based on a two-shell support system secured with shotcrete and reinforced with wire mesh, rock-bolts and, if necessary, steel arches and a membrane seal.

In the TBM driven sections the excavated diameters range between 8.8 and 9.5 m. A typical cross-section is shown in Fig. 6.

2.7 Particular Design Aspects in Squeezing Rock of Tavetsch Intermediate Massif

For the rock type considered to be the least favourable in this geologically difficult zone, the following properties were assumed :

- Elastic (*i.e.* Young's) modulus $E = 2$ MPa
- Angle of internal friction $\emptyset = 23°$
- Cohesion (drained condition) $c = 250$ kPa

The main design elements for the Middle Tavetsch Massif and the Clavaniev Zone, in the order of their importance, are the following (Fig. 3) :

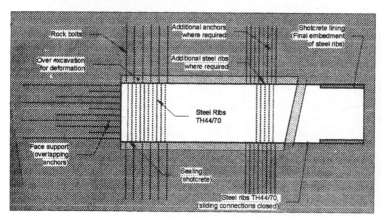 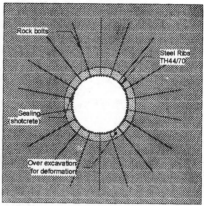

Fig. 3 : Over-excavation and proposed support measures in the Tavetsch Intermediate Massif

- Circular tunnel cross-section
- Full face excavation
- Uniform systematic anchoring of the face
- Over-excavation to accommodate the convergence
- Steel arch linings closed to a ring with sliding connections (Toussaint-Heintzmann)
- Uniform radial anchoring around the cross-section
- Closed ring of shotcrete lining behind the face.

364

Detailed investigations showed that a top heading excavation could not be carried out in the most unfavourable rock zones because it could lead to uncontrollable construction situations. Therefore, only full-face excavation was considered further.

Lances in the roof region and the immediate support of the face guarantee safe working conditions. The length of the anchors in the face are at least 6 m, this is achieved by overlapping the original 12-18 m long anchors. The anchors are glass fibre anchors to allow easy destruction during the course of later excavation.

The steel arch linings form, at uniform convergence, two rings, one lying within the other. Both a considerable lining resistance and a high level of safety against lateral buckling exist.

The shotcrete lining is to be applied after the closure of the steel arch linings and the full exploitation of the estimated convergence, respectively, to prevent a further reduction in cross-section of the tunnel opening.

The final lining with a thickness of maximum 1.20 m of unreinforced cast-in-place concrete follows the tunnel excavation at a distance of about 300 m. Its dimensioning is based on the assumption that the temporary support in the course of the long operating life of the tunnel (roughly 100 years) may completely loose its structural function due to corrosion. Besides the high rock pressures the final lining has to withstand a water pressure corresponding to a height of about 100 m.

The variations of the various support measures to the different geological conditions are as follows:

• Tunnel excavation radius		5.09 - 6.54 m
• Over excavation		0.30 - 0.70 m
• Area of full section		81 - 134 m^2
• Length of round		1.00 m
• Steel arches	Spacing	1.00 - 0.33 m
•	Weight per m'	2.5 - 9.4 tons
• Shotcrete behind face		0.35 - 0.50 m
• Rock bolting	Length	8.00 - 12.00 m
	Ultimate load	320 kN
	per tunnel m'	96 - 288 m
• Anchoring of face	Length	12.00 - 18.00 m
	Ultimate load	320 kN
	per tunnel m'	80 - 210 m
• Final lining		0.30 - 1.20 m

The most remarkable feature is the size of the full section of 134 m^2, which might result in the most difficult geological conditions. The increase of the excavation radius up to 6.5 m is necessary because of the large over-excavation for achieving the convergence, with a thick shotcrete lining and a greater thickness of the inner lining. In Fig. 4, the concept of the TH steel profiles allowing radial deformations of up to 2 x 75 cm is shown.

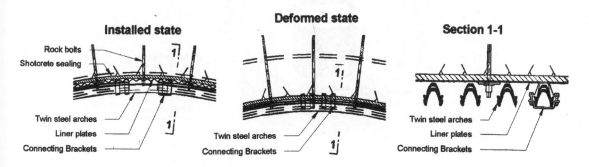

Fig. 4 : TH-steel profiles to allow limited deformation

3.0 INTERMEDIATE ACCESS OF SEDRUN (ACCESS TUNNEL, SHAFT TOP CAVERNS, VENTILATION TUNNEL, SHAFT)

The intermediate point of attack at Sedrun consists of field installations, access tunnel lot and the 800 m deep vertical shaft. At the foot of the shaft, start of the heading operations will begin to the north (about 2 km) and to the south (about 4 km). (Fig. 5)

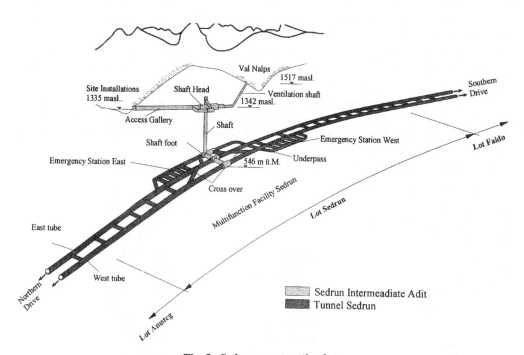

Fig. 5 : Sedrun construction lot

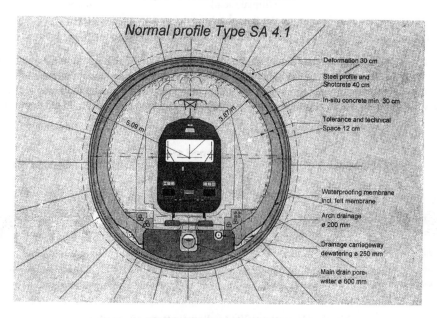

Fig. 6 : Typical cross-section

366

During construction, the shaft head caverns will accommodate the necessary installations for construction and operation of the Sedrun shaft and be used for material transport. During tunnel operation, the shaft head cavern will house the ventilation control centre.

3.1 Design Criteria for the Shaft

The 800 m deep vertical shaft is the only access to the four tunnel headings starting from the bottom of the shaft. So the layout of the shaft had to consider the following points :

- Transportation of materials in the tunnel: like machines, concrete, other lining support material.
- Transport of muck material from the headings through the shaft and through the access tunnel to the surface.
- Fresh air for the tunnel ventilation (fresh air in / waste air out).
- Additional pipes for water supply, compressed air, electrical supplies, communications and so on.
- The safety of the workers in all possible conditions has to be guaranteed.

In the very beginning of the shaft-layout some decisions had to be made about the kind of logistic concept from the headings through the shaft to the surface. It had to be decided between truck, conveyor or locomotive transport in the tunnel tubes and in the access tunnel. This leads to the hoisting equipment in the shaft: skips or cages for wagons or a combination of both.

3.2 Evaluation of required Shaft Capacity for Four Headings

To make this decision it was necessary to know the capacity of transports which have to be fulfilled by or through the shaft. Which is the necessary transport capacity for muck material, for concrete, anchors or steel arches and how many people have to be brought in and out every day?

The four main headings have to be logistically supplied by the shaft. Each of these headings has - depending on geological conditions - different advance rates and also different maximum advance rates. It was fixed that the capacity of the shaft should not limit the advance rates of the tunnel headings, so the shaft capacity depends on the maximum capacity for the four headings reaching maximum advance rates.

But which is the maximum that is possible in 4 more or less independent tunnel headings? The possibility that all of these 4 headings are going with a maximum advance rate at the same time is quite small.

Using probability methods it was possible to get the following results for the necessary capacity:

- 6'000 tonnes of muck material per day
- Handling of 50 wagons with material.
- 0.4 h a day for special transports
- 1.6 h a day are necessary for the main man riding of the shifts
- 1.0 h a day for additional man riding
- 2.0 h a day are necessary for maintenance

3.3 Comparison of Different Hoisting Systems

For maintenance, man riding and special transports about 5 hours a day are necessary. So 19 hours a day are left for hoisting of muck material and transportation of wagons.

Based on these figures, different types of hoisting equipment where compared. In addition to the necessary capacity when making the comparision, it was also important to have a look at the point of safety. In Switzerland there are no regulations for construction and using deep shafts like the shaft Sedrun. So regulations from the German coal mining industry were used. In these regulations it is fixed that in each shaft with man riding two different and independent hoisting equipments for people have to be installed.

From these conditions the following hoisting concepts were compared :

- Hoisting of muck material by skip or double skip hoisting machine, man riding and material transportation with a big cage and a counterweight.

- One big cage with a counterweight and a second independent safety man riding cage.
- Double cage hoisting and a second independent man riding safety hoist.

In the comparison the logistic concept from the heading to the shaft and from the shaft to the surface has to be involved. The comparison showed that all three possibilities are feasible.

3.4 Layout of the Definitive Hoisting Installations, Shaft Cross-section and Shaft Diameter

The solution with one big cage with a counterweight and a second independent man riding hoist led to the smallest shaft dimensions.

Therefore this system with the following main data for the big cage hoist was choosen :

- Double deck cage   for wagons with 11 m^3 capacity
- Cage dimensions   2600 mm x 5840 mm
- Type of winder   Koepe 4 rope winder
- Installed capacity   4200 kW
- Maximum speed at transport   16 m/s
- Speed at man riding   12 m/s
- Payload incl. wagons   50.8 t

The independent man riding hoist has a capacity of 18 persons.

The contract for this lot has been awarded to the German company Siemag Transplan.

For this hoisting equipment and the additional installations as mentioned an inner shaft diameter of 7.9 m was necessary.

3.5 Design of the Shaft

As there is no access to the bottom of the shaft, a pre drill is impossible. So the shaft has to be constructed by full-face excavation with drilling and blasting from the top.

The shaft is built in technically favourable gneiss formations. The overburden increases from about 550 m at the top of the shaft to 1350 at the bottom.

These rock conditions and the necessary inner diameter of 7.9 m led to a rock support system consisting of :

- Anchors, reinforcing mesh and shotcrete as well as steel arches for possible bad rock conditions as a first rock support applicated at the shaft bottom.
- In a maximum distance of 20 m to the bottom the concrete support ring with a thickness of 25 cm has to be constructed

With the rock support thickness between 30 and 45 cm and possible deformations of 5 to 10 cm the excavating diameter ranges between 8.6 and 9.0 m.

4.0 TOP DOWN CONSTRUCTION WORKS FOR THE SHAFT

4.1 Main Shaft Sinking Installations

The main shaft sinking installations comprise (Fig. 7) :

- Headgear, bankdoors, and loading silo
- Kibble winch
- Safety cage and winch
- Working platform and winch
- Shaft jumbo
- Compressed air plant
- Shotcrete equipment

368

The headgear is installed in the 50 m high shaft tower above the bank level, the two bankdoors are operated pneumatically, the loading silo for the excavated rock has a capacity of 40 tons or 25 m³.

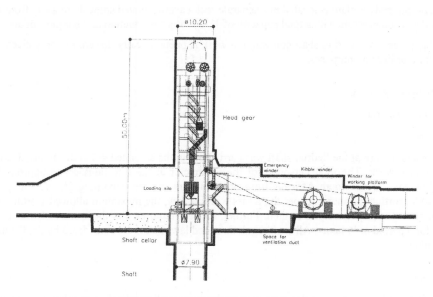

Fig. 7 : Longitudinal section shaft top

The kibble winch can operate two kibbles of 2 m diameter and a volume of 6 m³ each independently. The installed capacity of the kibble winch is 2.1 MW and the kibbles can be lifted with a maximum speed of 11 m/sec.

The safety cage has a diameter 1 m and is capable of transporting the whole personnel of a complete shift in one lift out of the shaft. The travel speed of the emergency winch is 0.8 m/sec, it's driven by an electric motor powered by an emergency diesel in case of a power failure.

The working platform consists of 5 different floors and has a total height of 16.5 m. The platform allows for the passage of the two kibbles, the safety cage and 2 ventilation pipes of 1400 mm diameter (Fig. 8). The cactus grab used to lash the excavated rock is suspended to the underside of the working platform. The stage winch with an installed capacity of 224 kW moves the stage at a maximum speed of 0.13 m/sec or 7.8 m/min. The total weight of the working platform is 85 tons.

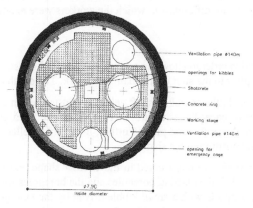

Fig. 8 : Layout working platform

369

The shaft Jumbo is an Atlas Copco pneumatic shaft sinking rig, equipped with 6 booms and rock drills diameter 51 mm. The rig can be folded up and closed to a diameter of 2 m in order to pass the openings in the working stage for transportation through the shaft. In the working condition, the 6 booms of the Jumbo are spread like the spokes of an umbrella.

The compressed air plant is located in a separate rock cavern situated some 40 m away from the shaft. The plant includes 6 compressors with a total capacity of 4'500 m^3 air production at 7 bar per minute.

All equipment used during shaft construction work is pneumatically driven, the only electricity provided in the shaft is for lighting purposes.

4.2 Shaft Sinking Works

4.2.1 General Aspects

Working Time

Construction work at the Sedrun shaft was carried out in the so-called 4/3 shift modus, 3 shifts at 8 hours a day, 7 days a week. The total personnel of the Contractor at the Site numbers to approx. 150 staff and workers. Thereof, 1/3 are South Africans from Shaft Sinkers and the remaining 2/3 are Europeans from the four Swiss Consortium partners. As per the Swiss working law, the maximum allowed working time per day, per month and year respectively is limited. Further, the law requires that the workers leave the shaft during their shift for a period of ½ hours. All such regulations and laws had to be considered by the Contractor in their work planning.

Safety Aspects

In Switzerland, as mentioned there are no actual mining codes and safety regulations available which are applicable for a shaft construction work from top. Therefore, the SUVA, the Swiss accident insurance company which is the responsible institution for the safety and health of workers, is following the German Mining Codes and technical requirements for shaft sinking equipment. For the implementation of safety measures during the construction works, it is basically the Contractor who is responsible that governmental and local regulations are fulfilled. However, the Client shares the responsibility and makes sure, through its Consultant, that all required safety measures are carefully followed. In addition, the SUVA is carrying out periodic site visits with respect to control of safety and health conditions at the site.

Local Conditions at Site

The installation area of the Sedrun shaft site is located in the Swiss Alps at an altitude of 1350 m.a.s.l. In winter, masses of snow make the management of the installation area difficult and, depending on the local danger of avalanches, the access to the underground work may be closed and work interrupted. The air temperatures may drop in winter time to minus 25° C, underground, in the shaft, the temperature is in the range of 20° C, a considerable temperature difference to which the workers were exposed.

4.2.2 Pre-sink

The pre-sink was required in order to allow the assembly of the working stage and to provide for a minimum distance from the bank level prior to the start of the main sink work. The depth of the pre-sink is 78 m.

The first 8 m of the pre-sink, the so-called shaft cellar, is constructed in a rectangular shape, 13 by 12 m wide. The shaft cellar serves to carry the structural steel work of the head gear and provides space for the ventilation pipes coming from the ventilation plant. Further, in the shaft cellar, all supply and utility pipes and cables are routed to the correct position in the shaft.

From - 8 to - 78 m, the shaft was constructed in the circular shape, excavation diameter 8.6 m. Manual drilling equipment was used to drill the blasting boreholes and the boreholes for the anchors. Lashing was done with a small shovel and all transports with a small kibble driven by the emergency cage winch. Shotcrete and anchors were applied after each round.

During the construction of the pre-sink, the erection of the steel structure of the head gear was carried out in parallel. For safety reasons, an intermediate cover was used between the two working areas.

After the pre-sink was completed, the working stage was assembled in the shaft and the electromechanical equipment such as stage and kibble winches, transformers, bank doors, silos and other equipment was installed.

4.2.3 *Main Sink*

Drilling and Blasting Procedure

With beginning of the main sink work (Fig. 9), the 6 boom shaft Jumbo was used to drill the blasting boreholes. The length of a round varied between 2.0 and 3.9 m and averaged at approx. 3.0 m over the whole shaft depth. The number of boreholes drilled for each round was 170, the penetration rate was about 1 m per minute which resulted in a total boring time of approximately 3½ hours including installation and de-installation of the Jumbo.

Drill water is pumped into the kibble and transported to the shaft top where it is pumped out and released into the drainage system.

Plastic explosive Gamsit and Tovex was used, the specific load being between 3.5 and 4.0 kg/m^3. The total amount of explosives therefore was between 620 and 700 kg for one round of 3 m length (shaft area 58.1 m^2, volume for a 3 m round 175 m^3). This relatively high specific load was due to the hard rock which tends to break in big blocks if less explosive is used. Big blocks are difficult to be loaded by the cactus grab into the kibble, they may be blocked in the kibble and thus create a problem while emptying the kibble into the silo on top of the shaft. Further, the silo may be damaged if it's filled with too big rocks.

The ignition system used is a non-electrical system using the Dynashoc detonator tubes system developed by Dynamit Nobel of Germany. Only the initial detonation is carried out by one electric detonator. This method is a very safe procedure and guarantees absolute safety during the loading process. Only at the very end of the loading process, the electric detonator is connected to the rest of the completed loaded system. This system however has the disadvantage that the functioning of the whole system can not be verified with the electric Ohmmeter, therefore partial misfires can happen if Dynashoc tubes are damaged during or after the loading process.

For the execution of the blast, all personnel had to leave the shaft and all equipment to be removed from the shaft bottom. The working stage was raised by some 100 m in order to prevent damages to the platform.

Lashing Procedure

Half an hour after the blast, the workers were allowed to re-entry the shaft and after the working stage has been brought into the working position, the lashing of the muck was started. The muck is loaded into the kibble by the cactus grab. A minimum of 3 kibbles are used, one is located at the shaft bottom and is being loaded by the grab, the two others are moved up and down the shaft in a contra-rotating operating modus. The muck is emptied via a chute to the silo at the shaft top from where it is transported by means of trucks out of the 1 km long access tunnel onto the intermediate storage area for further use.

Approximately 1 m of muck is in a first step left in the shaft in order to allow the execution of the anchor works and the application of shotcrete. This last 1 meter of muck is removed directly before the execution of the blow over, the cleaning of the shaft bottom prior the start of the drilling procedure of the next round.

During the whole lashing process, the grab operator must take care not to damage the completed anchors and shotcrete along the shaft walls. The whole lashing process takes approximately 6 hours.

Rock Supporting System

The immediate rock supporting system consist of anchors and shotcrete. Approximately 12 numbers of 3 m long Swellex anchors were set per 1 m of shaft length which means 1 anchor per 2.25 m^2. Reinforcing mesh 150/150 spacing and bar diameter 6 mm were placed on the rock prior to the application of shotcrete. The thickness of the shotcrete is theoretically 5 to 10 cm, caused by over profile in the rock excavation, more shotcrete was partially applied in order to fill up cavities. Anchors and meshes were set depending on the actual rock conditions considering the dipping of the rock layers and the existence of joints.

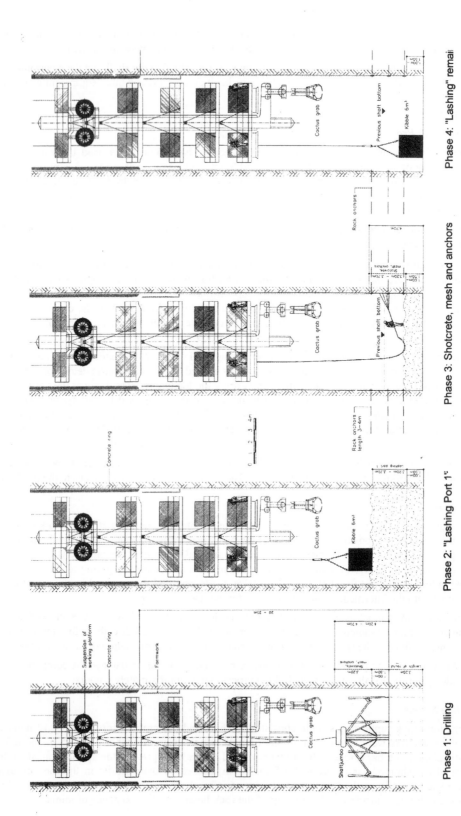

Phase 1: Drilling

Phase 2: "Lashing Port 1"

Phase 3: Shotcrete, mesh and anchors

Phase 4: "Lashing" remai

Fig. 9 : Work sequence, shaft construction

The process of applying the rock supporting system required approx. 5 hours.

The quality control required testing of 1% of the anchors, therefore 5 anchors were tested for every 500 anchors which is a test every 40 shaft meters. The ultimate load of the Swellex anchors is 11 tons, the applied test load was max. 10 tons.

Quality control of the shotcrete required core drills for every 150 m^3 applied shotcrete. The tests on the core drills included compressive strength (40 N/mm^2 28 days), and water tightness (50 mm max. penetration).

Blow Over

The blow over, as mentioned before, includes the cleaning of the shaft bottom prior the start of the drilling procedure of the next round. Further, visible blasting boreholes are cleaned out to make sure, that no remaining explosive is located in such boreholes. The blowover is performed with a 3" compressed air tube, air pressure 7 bars. The pipe is held and operated manually by 4 to 5 miners. The duration of the blow over is approximately 2 ½ hours.

Concrete Support Ring

All above mentioned work happens at the shaft bottom, the time between 2 rounds was in minimum 18 working hours and averaged at approx. 22 to 24 hours. In parallel to the work executed at the shaft bottom, the concrete supporting ring had to be constructed. The concrete supporting ring has a minimum thickness of 25 cm, at areas with over profile, the thickness increased to maximum 120 cm, the average thickness was 60 cm.

The concrete ring is constructed from the working stage which generally was positioned between 18 and 25 m above the shaft bottom. The shaft concrete is divided into 6 m long rings. The concrete for each ring was placed in three phases. Phase 1 consists of the 1 m high curb at the bottom, Phase 2 is the 2 m high barrel and phase 3, the filler at the top is 2.7 m high. The remaining 30 cm form an open joint. In every fourth joint, a dewatering channel is formed in the top of the concrete to collect the water in the shaft and lead it to a drainage pipe.

The formwork of start ring and barrel was fixed by means of chains to the filler of the upper concrete ring. The concrete of the start ring required an early strength of 5 N/mm^2 after 8 hours, this strength was required before the chains are untied and all the load of the concrete was transferred to the rock via friction.

The concrete was manufactured in the concrete batching plant located at the outside of the access tunnel and brought to the shaft top by means of concrete trucks. From there, the concrete was filled into a drop pipe of diameter 150 mm. At the end of this pipe, the concrete passed an energy destroyer and was distributed to two flexible tubes.

The quality requirements of the concrete are 5 N/mm^2 compressive strength after 8 hours for the start ring and 40 N/mm^2 after 28 days for the whole concrete ring. Further, the water tightness was specified to allow for a maximum penetration of 50 mm. As per the shotcrete, quality tests had to be performed for every 150 m^3 applied concrete.

Utilities

The working stage is the main distribution station for the utilities comprising compressed air, water, electric power and communication cables. Further, the concrete drop pipes end at the working station. All these utilities as well as the ventilation pipe are installed and extended from the working stage. Power and communication cables had to be dismantled when the working stage was lifted for blasting and re-installed when the stage was brought back to its working position.

Cover Drill

Originally, the concept of dealing with the groundwater was to construct pump niches beside the shaft wall every 90 shaft meters, to collect the ground water in these niches and to pump it to the shaft top. The total expected water quantity was estimated to be some 30 lt/sec.

The Contractor made a proposal to seal the shaft using the so-called cover drill method. Every 36 m of shaft length, 8 numbers of 42 m long boreholes were drilled with the shaft Jumbo at an angle of 10 degrees to the vertical and tangent to the shaft walls. When water was encountered in the boreholes, cement injections were applied with pressures of up to maximum 120 bar. The cement was pressed into fissures and joints and forming a sealed rock cylinder of approx. 17 m diameter. The maximum water flow in the boreholes was approximately 6 lt/sec before the application of the cement injections and could be reduced considerably by the cementation. If no water was encountered, the holes were gravity filled with cement.

The concept of cover drill proofed to be very successful, the total water quantity which accumulates along the total shaft depth of 780 m amounts to less than 0.5 lt/sec. The number of pump niches was reduced from the original ten to two, located at - 300 and at - 600 m depth.

4.3 Ventilation Concept

Under normal operating conditions, the fresh air is blown by a 440 kW capacity ventilator via the ventilation pipe of 1400 mm diameter to the shaft bottom. The used air mounts in the shaft section and leaves the underground via the access tunnel. After a blast, the system was reversed for a duration of approximately ½ hour and the used air sucked out through the ventilation pipe allowing a quick removal of the gases from the shaft. The maximum air volume transported through the ventilation pipe was 40 m^3/sec.

4.4 Control Measurements in the Shaft

Deformation measurements were carried out at various shaft depths. For that, 4 bolts were set in the concrete wall rectangular to each other and the intermediate distance between the 4 bolts surveyed. After one and two further rounds, the bolts were surveyed again. Deformations could be observed mainly between the first two measurements, deformations decreased considerably later on. The maximum measured deformation between two opposite survey points was 25 mm or 0.3% of the shaft diameter. With increasing shaft depth, an increase of the deformation per time interval could be observed. Further, the deformations are dependent on the shaft depth.

Further, the rock temperature was measured in the shaft. This was done by drilling a 4 m long borehole into the shaftwall and the borehole then filled with fresh water. After 6, 12 and 48 hours the water temperature was measured. The temperature gradient is 2.9°C per 100 m resulting in a temperature at the shaft bottom of 33° C, which corresponds to the predicted rock temperature at this depth.

4.5 Recycling of Excavation Material

The excavated rock material was classified into 3 classes

- Class A: Material to be re-used for concrete aggregates
- Class B: Material to be re-used for road substructures
- Class C: Material not usable and deposited

Criteria for the classification of the excavation material are:

- Petrographic criteria (lithological designation, hardness defined by the point load index, brittleness index)
- Valuation of the aggregates (less than 5% non-usable fraction, free mica content less than 35%)
- Alkali aggregate reaction

The latter is a phenomenon which can result in considerable damage to the concrete due to the expansion created by the reaction between reactive aggregates and the porewater in the concrete.

Out of the total rock volume of approximately 50'000 m^3 solid, nearly 60% was classified to be class A material and was processed in a crushing plant and produced to concrete aggregates. The non usable material is deposited, together with the processes sludge in a nearby pit.

4.6 Construction Programme

The construction time of the 78 m deep pre-sink was 10 weeks, the time to complete the electro-mechanical installation was completed 4 weeks after the end of the pre-sink. The main sink started by the end of January 1999 and the top of the caverns at the shaft bottom at a depth of - 785 m was reached on February the 22nd of this year.

In the 13 months of construction time, 49 working days were used for the execution of the 20 cover drill stages, 24 days for the construction of the two pump niches and a water reservoir at level - 690 m and at 39 days, the work was interrupted to official leave.

The overall progress per day for the sinking of the main shaft was 1.80 m per day (393 days). Considering the actual working days only of shaft sinking (281 days), the rate averaged to 2.50 m per day.

After completing the first phase of the cavern excavation at the shaft bottom (4'100 m^3), the final 50 m of the shaft was excavated. This work includes the building of the shaft sump, a pump niche with high pressure pumps, two rock passes and the concrete work for the new hoisting equipment to be installed prior the start of the actual tunnel excavation work.

Currently work is proceeding with the construction of the caverns and working areas at the base of the shaft, from where the 4 drives – twin parallel tunnels to the north and to the south – will be excavated.

The contracts have been tendered for the major tunnel lots. The tenders are expected at the end of 2000/beginning of 2001 with the main sections being awarded in summer 2001, to start in 2002. The planned construction program is as shown in Fig. 10.

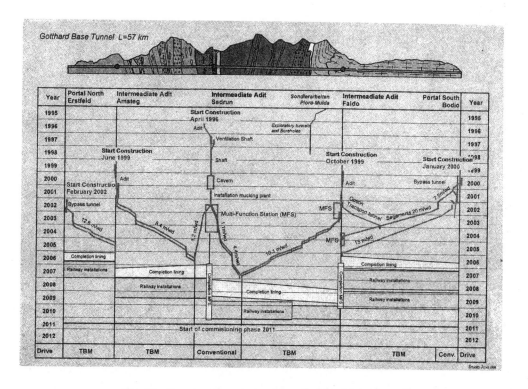

Fig. 10 : Construction programme, Gotthard base tunnel

REFERENCES

Blank K.E., Fellner D.: TBM-Vortriebe beim Gotthard-Basistunne,l Felsbau Heft 5, Okt. 1999.

Blank K.E.: Ueberlegungen zum Einsatz von Spritzbeton für lange, tiefliegende Tunnel in Konkurrenz zum Ortbeton. Spritzbeton-Technologie. 5. Internationale Fachtagung, Innsbruck - Igls, 18. Januar 1996.

Ehrbar H., Pfenninger I.: Umsetzung der Geologie in technische Massnahmen im Tavetscher Zwischenmassiv Nord, Symposium GEAT 1999, ETH Zürich, Zürich 1999.

Eppinger G.: Der Gotthard-Basistunnel: Abschnitt Sedrun, Bauingenieur Bd. 74, Nr. 6 - June 1999.

Gehriger, W.: Das Vorprojekt des Gotthard Basistunnels,Tagungsbericht Int. Symp. „Basistunnel durch die Alpen" (ed. R. Fechtig, K.Kovári), ETH Zurich, 1994.

Kovári K., Amberg F., Ehrbar H., Squeezing Rock in the Gotthard, World Tunnelling, June 2000.

Flury S., Ehrbar H., Henke A., Rehbock-Sander M., The Gotthard Base Tunnel: Construction Work progressing on 4 Part Sections, Tunnel, 4/2000.

Kellenberger J., Zbinden P.: Die neuen Eisenbahntunnel am Gotthard und ihre Besonderheiten, Felsbau Heft 5, Okt. 1999.

Rehbock-Sander M. and Meier R. Sedrun access and ventilation shaft for the new Saint Gotthard railway base tunnel. Project criteria, design, construction works from the top, ITA Congress, Durban 2000.

Schneider Dr. T.R. AG, Büro für Technische Geologie: Gotthard-Basistunnel, Teilabschnitte Erstfeld, Amsteg, Sedrun, Faido, Bodio / Bauprojekt / Spezialberichte Geologie, Geotechnik und Hydrogeologie, 1999, Uerikon/Sargans.

TUNNELLING AT PONG DAM – A CASE STUDY

R.R. OBEROI
Chairman

Bhakra Beas Management Board
Chandigarh, India

G.D. GUPTA
Chief Engineer

Beas Dam, Bhakra Beas Management Board
Talwara Township, India

SYNOPSIS

The art and practice of tunnelling dating back to ancient times has undergone tremendous changes mainly on account of advancement in mechanisation and automation. Nevertheless, the tunnelling is still a tuff and arduous job in water resources development projects and is likely to remain so for a long time. The construction of tunnels at Pong dam, the highest earthen dam in the country on the river Beas, was very complex in nature but was executed and managed efficiently. The experience confirms that if there is one thing certain about the tunnelling it is its uncertainty especially in the Himalayan region where its young rock formation springs surprises. With all the ifs and buts duly taken care of, there were some isolated incidents to the tunnels probably due to the invisible and invincible forces of the nature. The paper highlights two tunnel incidents and how those were tackled successfully by the project engineers.

1.0 INTRODUCTION

The Pong dam is a multipurpose river valley project, on the river Beas, constructed departmentally exclusively through indigenous expertise. The dam is located in District Kangra (Himachal Pradesh) about 480 km north-north west of Delhi. The project consists of an earth core-cum-gravel shell dam with a maximum height of 132.6 m, 13.7 m wide at top and about 610 m at the base. There are five tunnels of 9.14 m finished diameter with aggregate length of over 5017 m originally used for diversion of the river, two of these have been converted into irrigation outlet tunnels (T_1 and T_2) and the other three into penstock tunnels (P_1, P_2 and P_3) feeding Pong power plant having six units of 60 MW each. A chute spillway with a discharging capacity of 12375 cumecs has been provided on the left bank of the dam. Fig. 1 shows the general layout of the dam and appurtenant works. Pong reservoir, the largest in the northern region with water spread of about 260 sq km, 42 km long and 16 km wide (maximum), has gross storage capacity of 8570 Mm^2 out of which 7290 Mm^3 constitute the live storage which excels even that of the world famous Bhakra dam (7191 Mm^3). There is a minimal submergence of 30 ha of forest area out of the total submerged reservoir area of 28,271 ha. The work on the project was started in 1961 and involved 30 Mm^3 of rock excavation, 35 Mm^3 of fill materials and 1.13 Mm^3 of reinforced cement concrete. The storage in the reservoir commenced in the year 1974 and various units of the powerhouse were commissioned during 1978 and 1983. The project was constructed at a total cost of 3252 millions of rupees.

2.0 TUNNELLING AT PONG DAM

2.1 Geology

The Pong dam site is situated in the Himalayan foothills in the upper Shivalik formations consisting of sand rock, clay shale/silt stone. They have been subjected to folding. The clay shale bands are generally massive, well compacted and included silty and sandy portion with occasional calcareous modules which show air slackening (*i.e.*, break into fragments when exposed to air). The sand rock bands are generally coarse grained, massive, friable to firm and are seemingly pervious. All the three strata-sand rock, clay shale and silt stone are very poor with respect to self-supporting properties. Moreover, they do not take any grout.

At the intake area of tunnels, the dip of the rocks was 6-8 degrees due north to northeasterly direction, *i.e.*, externally. Along the outlet tunnels, there was an asymmetrical anticlinal fold and the dip of the rocks was 35-40 degrees southwest near the downstream portal. In the penstock tunnels also, the anticlinal fold was seen but it was very open and almost a flat topped anticline.

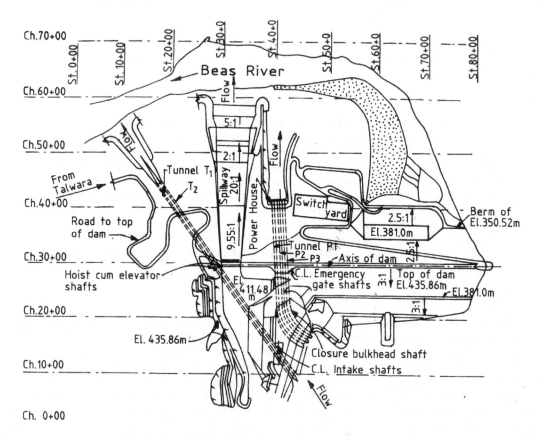

Fig. 1 : Layout plan of Beas dam and its appurtenant works

2.2 Salient Features of Tunnels

All the five tunnels are located on the left abutment of the dam. T_1 and T_2 tunnels are straight in alignment whereas P_1, P_2 and P_3 are also straight except for small reach at the upstream end where it had to be curved due to geological considerations. Separate stilling basin for each tunnel was provided except for P_1 and P_2 for which a combined stilling basin was provided. To ensure adequate rock cover between the excavated tunnels, P_1 and P_2 were spaced at about 38 m while P_2 and P_3 at about 41 m for provision of a divide wall in between the stilling basin. Fig. 2 shows the profiles of outlet and penstock tunnels.

The intakes of the tunnels during diversion stage were located at the river bed elevation. For their utilisation as irrigation outlets and power penstocks in the ultimate operational stage, permanent intake structures with trashrack have been provided at the dead storage elevation of 384 m. Minimum thickness of lining, inclusive of steel rib supports, is 76 cm. During the diversion stage, the flows emerging from the tunnels had high velocity in the range of 18-21 m/sec. The contact between the concrete lining and the rock was grouted at 2.11 kg/cm² through sets of three 40 mm dia, 1.52 m deep holes located above the spring level spaced at 3.05 m along the tunnel length. The holes in alternating sets were suitably staggered.

378

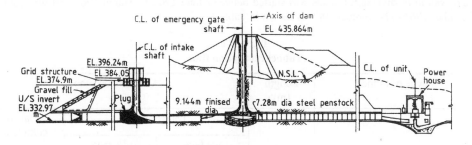

Fig. 2(a) : Profile along centre line of tunnel P_2

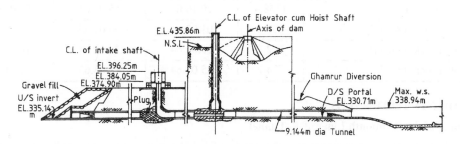

Fig. 2(b) : Profile along centre line of tunnel T_2

After the diversion stage three of the five diversion tunnels were converted into penstock tunnels with the provision of 3.05 m x 6.40 m rectangular constriction for the emergency gates, in the second stage concrete, at about the middle of each tunnel. The emergency gates in the tunnels can be lowered and raised through vertical shafts connecting tunnels, through three independent control structures located in the open at the top of the dam. The gate shafts, each of which is 92 m high and 6 m dia are partly in rock and partly in the core of the dam.

The two remaining diversion tunnels were converted into irrigation outlet tunnels, with the provision of second stage works at about the middle of each tunnel. In each tunnel two rectangular constrictions of 2.13 m × 3.05 m were provided to accommodate in each conduit a set of emergency and regulating slide gates which can be operated through vertical shafts about 92 m high.

Hydraulic model tests were done at the different research stations in the country to check the layout of transition upstream and downstream of the constriction. Photo elastic studies were also done at CWPRS, Poona for structural design of the second stage concrete transitions and rectangular section.

2.3 Excavation of Tunnels

Each of the five tunnels was excavated to an average diameter of 10.67 m. In special reaches, *i.e.*, at the junction of intake, emergency gates and hoist-cum-elevator shafts, the diameter of the tunnel was increased even up to 15.85 m.

Full-face method of excavation was adopted for all the tunnels. Tunnelling excavation through such poor strata as encountered at this site is normally done by pilot heading method but it was the opinion of the project engineers that the tunnel be driven in two stages, *i.e.*, with pilot heading technique. This method was time consuming and ultimately it was decided to adopt the full-face method. This was the first tunnelling job done in the country with the full-face technique for such a large diameter tunnel and though having poor strata. The tunnel excavation including steel supports and lagging was done at a rate (Rs. 59 per cu m) minimum in the country in the sixties.

The excavation of diversion tunnels was started between June 1963 to March 1965 and was completed during December 1964 to November 1965 as shown in Table 1.

TABLE 1
Tunnel excavation time

Sl. No	Type of tunnel	Length (m)	As diversion tunnel		
			Date of start		Date of completion
			u/s face	d/s face	
1.	Outlet tunnel T_1	1363.4	10.10.63	15.06.63	31.12.64
2.	Outlet tunnel T_2	1338.4	10.10.63	15.06.63	23.01.65
3.	Penstock tunnel P_1	808.0	01.02.64	05.01.65	05.11.65
4.	Penstock tunnel P_2	772.7	25.01.64	26.12.64	10.08.65
5.	Penstock tunnel P_3	734.2	25.01.64	08.03.65	17.11.65

3.0 INCIDENTS TO TUNNELS

3.1 Ripping of the Bellmouth Liner of T_2 Tunnel

After depletion of the reservoir below intake bench at E1. 374.8 m, emergency gate of T_2 tunnel was lifted and the upstream conduit was inspected in January 1975. The inspection revealed that in the case of right conduit, 20 mm thick steel liner at bellmouth entrance had separated from the base concrete in a length of about 0.9 to 1.5 m and in case of left conduit, there was an indication of start of separation of steel liner from the base concrete at the bellmouth. The ripped off steel liner location is shown in Fig. 3.

3.1.1 *Causes of Ripping*

The reason for the damage appeared to be that high-pressure water found its way somewhere behind the liner at places where there was no bond between the concrete and the liner. It exerted differential pressure on the liner during text running because of conversion of pressure head into velocity head inside the liner, whereas the pressure of the water which had entered behind the liners could not get released through drainage holes of 10 mm dia provided for the purpose. The entry of water appeared to be from the upstream face of the plug housing the conduit liner.

3.1.2 *Repair/Remedial Measures*

The damage to the bellmouth liner was discussed in the meeting of the Beas Board of consultants held in March 1975. The treatment based on such discussions and recommendations was carried out as shown in Fig. 4 and comprised the following :

(a) The ripped off 20 mm liner was cut in pieces and removed. Fabrication of the damaged portion of the bellmouth was done in small sections in the project workshop at Talwara Township. It was also revealed that at some locations concrete was also washed and accordingly further chipping of the damaged concrete was done. Concreting of the bellmouth between the liner plate and old concrete was done side by side as its erection in small sections.

(b) (600 mm high) progressed. Thus, it was restored to its original profile. Surface of liner in contact with concrete was cleaned with wire brushes and the old concrete surface and the lift line were epoxy painted. In all the works done after epoxy repair, it was ensured that the liner was not heated to a high temperature so as to avoid any damage to the epoxy. The liner joint was suitably welded and grounded flush.

380

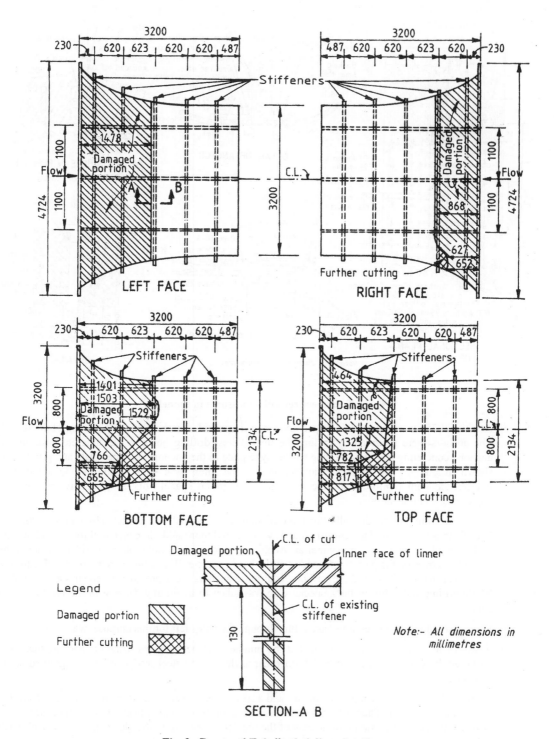

Fig. 3 : Damaged T₂ bellmouth liner details

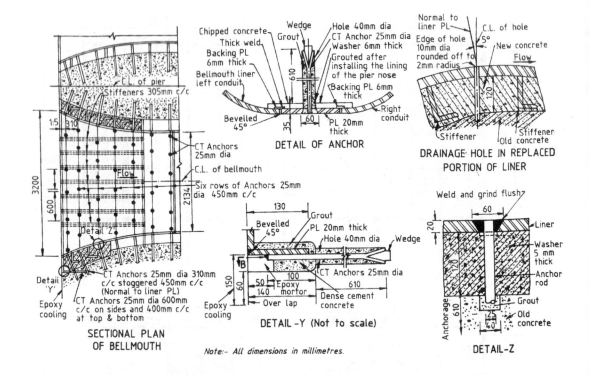

Fig. 4 : Treatment/remedial measures in tunnel T_2

(c) Additional 25 mm diameter 'V' shaped anchors were provided in the bellmouth portion of the tunnel. The anchors were provided by cutting the liner plates, drilling holes in the concrete, inserting of anchors, concreting and then welding washer type plate to the anchors and liner plate and grinding as shown in Fig. 4. The exact location/spacing of the liners was slightly adjusted to avoid their interference with reinforcement, etc., keeping a minimum spacing of 325 mm between any two adjacent rows of anchors.

(d) Upstream bellmouth divide wall of both the conduits in the tunnel was protected by joining bellmouth liner flanges by 24 mm thick cut-off plate. It was secured thoroughly in concrete through anchors and high strength concrete and epoxy mortar. Anchors provided in the holes already drilled anchored the liner plate covering the pier of left conduit to the base concrete. Where the separation extended to area other than that covered by these anchors, additional anchors were provided to cover such area also. While drilling holes for anchors, special care was taken to avoid any damage to the reinforcement bars.

(e) Epoxy grouting of the liner, after concreting through the holes left for this purpose, was done.

(f) A seal plate, connecting the two liners at the middle nose between two conduits was provided and anchored through 25 mm diameter deformed bar anchors, with steel wedge at bottom, grouted in the concrete.

(g) Drainage holes about 10 mm in diameter and 150 mm deep were drilled in liners starting from the bellmouth in a length of about 14 m to release external pressure, if any. Such drainage holes were also provided in the liner portion downstream of the regulation gate up to the glacis lined with steel. Each hole covered an area of about 310 mm x 740 mm. The holes were drilled inclined by 5° to the normal (towards downstream) in the liner plate to minimise cavitational effect.

3.2 Bulging of Liner of P$_2$ Tunnel

The transition liner in the penstock tunnel P$_2$ was installed and last lift of 2nd stage concreting was placed on May 7, 1975. The grouting between 2nd stage concreting and 1st stage concreting was done at 3.4 to 4.8 kg/cm^2 on 11-12th August, 1975 and concrete grouting between the steel liner and the 2nd stage concreting was done on 14-16th August, 1975 at 1.4 kg/ cm^2. The grouting was done with struts in position. No grouting was proposed for vertical faces and hence was not done. The concrete in the penstock portion just downstream of transition was placed in January 1976.

Flushing of the choked drainage pipe (meant for drainage of bonnet chamber of emergency gate) was done a number of times in December 1976. January 1977 and February 1977. The maximum water pressure applied for the opening of these pipes was 4.8 kg/cm^2 and sometimes air pressure was also applied up to 5.4 kg/cm^2. The reservoir level on 28th March 1977, *i.e.*, at the time of steel liner failure was $\pm$ 395 m, which corresponded to a static pressure of $\pm$ 5.6 kg/cm^2 at tunnel invert elevation. The maximum reservoir level during 1976-77 was about 420.6 m.

In April 1977 it was noticed that steel liner on right side in the transition downstream of emergency gate shaft in P$_2$ penstock had bulged. The bulging was noticed in some area of about 59 sq m and maximum bulge was about 320 mm near spring line and decreasing gradually to zero at the overt and invert. The affected area was triangular in shape extending over the entire vertical transition length.

A hole was drilled in the bulge steel liner and it was found that water rushed out of it under pressure. A piece of about 600 x 400 mm was cut from the liner to observe backside condition. The concrete was found sheared at the face of reinforcement at about $\pm$ 170 mm and this concrete which provided cover to the reinforcement was found sticking to the back face of the liner.

3.2.1 *Probable Causes of Bulging*

The liner in the vertical transition portion could not withstand the external water pressure developed behind it due to surcharge of the strata on filling up of the reservoir. It is an established fact that the rectangular transition shape is adverse compared to a circular liner against external water pressure. The bulging of the penstock liner downstream of emergency gate in the vertical transition portion was observed when the installation of the penstock steel liner in the tunnel was in progress. Fig. 5 shows the bulged portion of the liner and the treatment carried out.

3.2.2 *Repair/Remedial Measures*

A special study team recommended the following additional measures for safety of steel liner installed in P$_1$ and P$_2$ penstock tunnels in September 1976, *i.e.*, before the bulging of steel liner in P$_2$ Penstock tunnel.

(i)   20 mm 'V' notch shall be chipped and properly treated for the application of M-seal. The notch shall then be filled with M-seal wet setting type compound duly caulked and made flush with concrete and steel liner. The M-seal compound shall be prepared by mixing fine inert powder like silica powder in suitable epoxy resin.

(ii) 10 mm dia drainage holes spaced at 1.5 m centre to centre shall be provided in between stiffeners on vertical sides of steel liner. Similar holes shall also be provided at the centre of the panels on the roof and floor liners. The aforesaid drainage holes shall be provided only in 4.6 m length from upstream edge of liner.

(iii) 10 mm dia drainage holes spaced at 1.5 m centre to centre on vertical sides and similar holes in the centre of each panel at roof and floor liners shall be provided in the transition reach. These drainage holes shall be provided in portion starting 3 m downstream of centreline emergency gate slot and 2.4 m upstream of junction of penstock header.

(iv) 4 No. drainage pipes installed around the transition liner shall not be punctured in the portion downstream of emergency gate groove up to a point about 3 m downstream of the start of penstock headers.

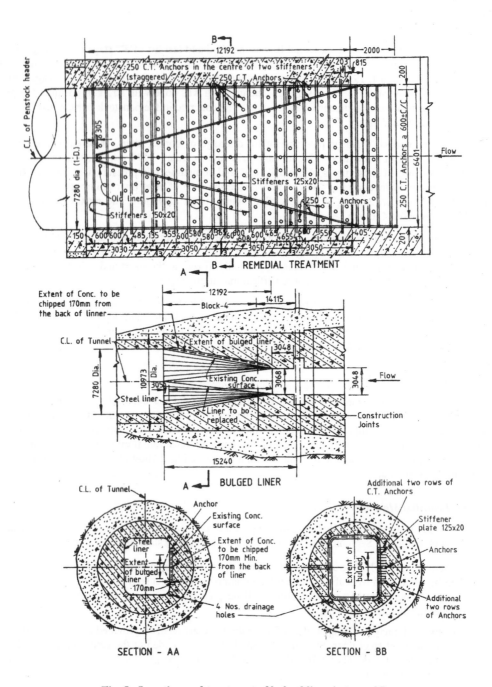

Fig. 5 : Location and treatment of bulged liner in tunnel P₂

(v)   Provision of water seals (wash-basin type) with removable joints shall be made at the out-fall of drainage pipes to prevent entry of any atmospheric air.

The additional measures for the safety of the liner were implemented in P_1 penstock tunnel. Before these could be implemented in P_2 penstock tunnel the bulging of liner downstream of emergency gate in the vertical

384

transition portion occurred. The same special study team suggested the following measures in May 1977 and was implemented in the field to rectify the bulged liner.

(i) 25 mm dia C.T. anchors (wedge shaped) were provided in between the stiffeners at about 0.61 m × 0.61 m spacing. The size of stiffeners was about 20 mm × 125 mm. Also two rows of anchors were provided beyond the replaced liner on all the four sides, where possible.

(ii) The chipped jetted/cleaned surface of old concrete was kept wet for at least 24 hours prior to the placement of new concrete behind the replaced liner. The new concrete was of 210 kg/cm² strength at 28 days.

(iii) No grouting was done except around the top lift of new concrete (viz. top contact of new and old concrete).

(iv) 10 mm dia drainage holes spaced at about 0.75 m centre to centre were provided in between the stiffeners up to old concrete. Also two rows of 10 mm dia drainage holes were provided beyond the replaced liner on the four sides, where possible. The pattern of drainage holes adapted and coincided with observed leakages and seepages.

(v) The four number drainage pipes installed around transition liner and penstock liner were not punctured up to a point about 5.5 m downstream of the start of penstock header.

The team also suggested some measures for penstock tunnel P₃ in which steel liner was yet to be installed and were implemented in the field as under.

(i) Provision of steel liner in transition was not considered necessary in penstock tunnel P₃.

(ii) Suitable cutoff and 'V' notch filled with epoxy material was provided at the start of penstock header.

4.0 CONCLUSION

The case study is about both the incidents and the remedial/corrective measures comprising providing additional anchorage for strengthening the steel lining against external pressure and also at the same time to release the external water pressure developed behind the steel lining, due to charging of the rocky strata on filling up of the reservoir, by providing proper drainage holes penetrating in the steel liner as well as in the concrete lining in varying depths. No damage of any sort was noticed in the tunnels during the subsequent closure, inspection and maintenance conducted in the years 1985, 1988, 1994 and 1999. The tunnels have withstood the vagaries of the nature during all these years of their continuous operation and this establishes the perfect effectiveness of the methodology by which these incidents were tackled by the project engineers. The incidents should not be sized off as these teach extremely valuable lessons to the engineers engaged on the planning, design, construction and operation in achieving the cost effectiveness and timely implementation of the river valley projects.

REFERENCE

Criteria for Operation, Maintenance and Observations of Pong dam, Volume II.

EXCAVATION AND SUPPORTING OF TRANSFORMER HALL IN POOR ROCK CONDITIONS (NATHPA JHAKRI HYDROELECTRIC PROJECT 1500 MW) — A CASE STUDY

V.K. SHARMA
General Manager (Project)

KRANTI GUPTA
Addl. Superintending Engineer

PARVEEN PURI
Asstt. Executive Engineer

Nathpa Jhakri Power Corporation Ltd., Jhakri, Himachal Pradesh, India

1.0 INTRODUCTION

The 1500 MW Nathpa Jhakri Hydroelectric project located in Shimla and Kinnaur districts of Himachal Pradesh is run-of-river scheme. The project consists of four underground desilting chambers each of size 525 x 15.00 x 27.5, 10.15 m dia and 27.3 km. head race tunnel terminating into 21 m dia and 301 m deep surge shaft. Three pressure shafts each of 4.9 m dia off-taking from surge shaft to feed six generating units of 250 MW each housed in an underground cavity of size 222 x 20 x 49 m. The generating power is to be stepped up in an underground transformer hall of size 196 x 18 x 27 m. The upper floor of it shall have GIS switch yard. This project has some unique features like one of the deepest surge shafts in the world, one of the longest power tunnels and one of the biggest cavities to house 6 units of 250 MW each (Fig. 1).

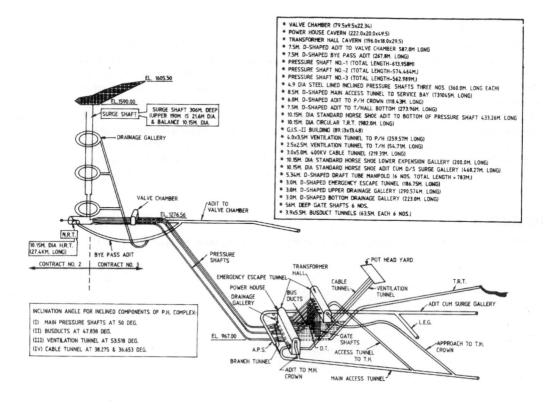

Fig. 1 : Nathpa Jhakri H.E. project (1500 M.W)

The underground cavities of power house (222 x 20 x 49 m) and transformer hall (196 x 18 x 27 m) were originally planned to be excavated and supported by providing rock reinforcement in shape of varied length patterned rock bolts. The power house could be excavated as per envisaged methodology of central heading, side slashing and subsequent stage down benching and concurrent rock reinforcement followed by welded wire mesh and shotcrete. But in case of transformer hall the envisaged methodology and design had to be amended due to geological occurrence different than that envisaged as discussed in the succeeding paragraphs.

2.0 GEOLOGY OF TRANSFORMER HALL

The rock encountered in transformer hall cavern was more or less same in its general characteristics as that found in power house cavern *i.e.*, quartz mica schist, but most of the area was dominated by lenses of biotite schist with quartz bounding having unfavourable sets of joints mostly having a shear of 5 to 10 cm. The situation further worsened due to saturated conditions beyond RD 100 with continuously seepage of water. A prominent shear of 50 cm. to 150 cm. width was encountered sided on both sides with shattered and fractured rock, dipping towards north and cutting u/s wall at RD 14, d/s wall at RD 32 m.

3.0 METHODOLOGY FOR HEADING EXCAVATION

The access tunnels to excavate the transformer hall cavern were 5 m D-shaped adit to its crown at invert EL 1062.00 m main and 7 m D-shaped at its service bay level *i.e.*, ± 1042.00 m. The envisaged construction methodology for heading was excavating central gullet of size 7 m x 7 m with concurrent supporting and subsequent side slashing (Fig. 2).

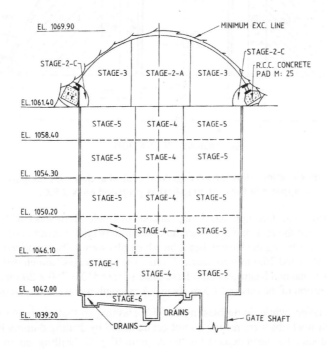

Fig. 2 : Typical section of transformer hall showing the excavation stage

Accordingly, the work was taken-up with excavation of central gullet of 7 m x 7 m and supporting with 25 mm dia rock bolts 5 m and 7 m and 100 mm thick shotcrete. After excavation of 15 m initial reach a shear zone was encountered which resulted in loose fall and subsequent stoppage of work. It was decided to reduce the section to 5 m x 5 m and advance further with temporary steel supports. Accordingly, the shear zone was crossed with this smaller heading upto RD 40 m. Beyond RD 40 m central gullet of 7 m x 7 m with immediate support of expansion shell type rock bolts was provided upto RD 120 m without any significant problem.

As the excavation of central gullet (7 m x 7 m) reached the RD 120 m, a big rock fall occurred with chimney and the supports provided in the central gullet in the form of rock bolts started loosening tension in the already supported reach adjacent to rock fall which tracked back upto RD 100.00 m and further progressively thereafter upto RD 80.00 m and beyond. Keeping in view the poor geology and loosening of already provided supports it was decided to provide the rib supports in the entire reach of transformer hall. As phenomenon of dilation of rock was travelling backward it was decided that central gullet already excavated be supported with steel ribs matching the final profile supported on temporary steel columns to be removed later (Fig. 3).

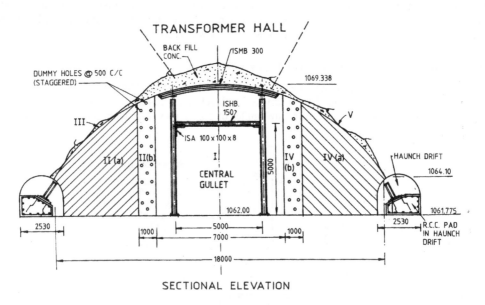

Fig. 3 : Selection elevation (showing arrangement of steel arches with temporary vertical supports in the central gullet) as proposed on N.J.P.C.

Immediate supporting was done upto RD 50 m, concrete of Grade M-10 was backfilled behind the ribs to give proper contact between the rock and the concrete. In the meanwhile excavation of side drifts, on both the sides of transformer hall for laying the haunch beam on which ribs were to be supported, started. For this cuts of 4 m width at RD 0.00 m and 36.0 m were made. From these approaches haunch drifts from RD 0 to 71 m were excavated and RCC haunch beam was laid. Base plate of size 410 x 250 x 20 mm was fixed to support the steel ribs, straight portion of the ribs and 1.5 m about 1.5 m length was welded with the base plate.

Utmost care was taken, while doing blasting for side slashing so that minimum impact is caused and steel columns on which steel ribs were resting do not get damaged by drilling dummy holes of 3.5 m length. The blasting pattern adopted has been depicted in the diagram (Fig. 4). Drilling length was kept 2.5 m so as to give a pull of 2 m, enabling the installation of ribs during the stand up time. After the excavation rib segment was erected and jointed with the ribs already provided on the steel columns in the central gullet on one side the straight portion of the rib already fixed on the base plate on other side. Vertical steel props were provided between the ribs and rock and packed by welding the steel plates between the rock and the top of props. A lag was being kept between the two sides for excavation, to minimize the damage to the vertical steel columns. The same cycle was repeated on the other side and then the over break behind the

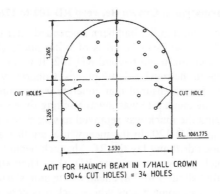

CUT HOLES

CUT HOLE

EL. 1061.775

ADIT FOR HAUNCH BEAM IN T/HALL CROWN
(30+4 CUT HOLES) = 34 HOLES

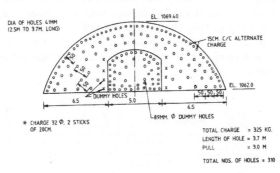

DIA OF HOLES 41MM
(2.5M TO 3.7M. LONG)

EL. 1069.40

15CM. C/C ALTERNATE
CHARGE

EL. 1062.0

DUMMY HOLES

89MM. Ø DUMMY HOLES

* CHARGE 32 Ø, 2 STICKS
OF 20CM.

TOTAL CHARGE   = 325 KG.
LENGTH OF HOLE = 3.7 M
PULL   = 3.0 M

TOTAL NOS. OF HOLES = 310

Fig. 4 : Drilling pattern for T.H. crown

ribs was backfilled with M-10 concrete after providing RCC lagging on the inner face of ribs, with the help of concrete pump.

Steel ribs of section ISMB 300 x 140 or built up section of ISMB 30 x 140 with steel plate of 160 x 12 mm at spacing of 0.5 m c/c were provided. This methodology was adopted from RD 50 m to 102 m and between RD 29 m to 33 m.

4.0 DISCUSSION ON SPECIAL REACHES

4.1 Tackling of Shear Zone Portion RD 0 to 40 m

First of all the portion between RD 29 m to 33 m where shear zone appeared just at the centre of the crown was widened to support the ribs matching the finished section and supported on steel columns.

While excavating the side haunch drifts in this reach shear zone material was scooped out upto the accessible depth and then backfilled with M-15 concrete upto the line and grade of haunch beams. The RCC haunch beams was subsequently laid. Sleeves of 45 m internal dia were left in the concrete to provide the 25 m dia directional rock bolts, for stitching the shear zone after concrete is set. However, due to constraints of the space it was not possible to drill with the boomer. Therefore, drilling was taken-up with the jack hammers, but the drilling rods were getting jammed after drilling. So it was not possible to install these rock bolts. It was, therefore, decided to pend the installation of directional rock bolts till the roof widening, when drilling will be done with boomer after grouting. Side slashing of both the sides was subsequently done and full rib section was provided (300 x 140, with cover plates of 160 x 12 on both the faces) between RD 0 to 29 m and RD 29 to 50 m. In every cycle, drilling, blasting of 2.5 m length with effective of pull 2 m was taken to minimise the overbreak and so as to provide the rib support with in the stand up time of rock. Immediately after excavation, ribs were installed with propping concrete lagging placed between the ribs and thereafter backfilling was done.

4.2 Tackling of Rockfall/Cavity Formation at Crown Between RD 102 to 126

The work beyond RD 102 in the central gullet had been suspended after the rockfall which caused a chimney formation in the crown. There was heavy ingress of water and rock-fall and huge muck was lying between RD 102 to 126. It was unsafe to go beyond RD 102. It was decided to fill the excavated portion beyond RD 102 with concrete in the muck pile by pumps. Concrete was poured above the muck already lying between RD 102-126. It was ensured that concrete was pumped in steps using the steel stoppers. The poured concrete above the muck pile was given a setting time so as to fill the space created due to chimney formation. After setting the muck was removed in stages and supporting the area created after muck removal with steel sets upto RD 120. In the meanwhile, another approach to side drifts was made at RD 106 to excavate the haunch drift and lay the RCC haunch beam, was done in advance to the central gullet excavation. The haunch pad was cast upto RD 136. Thereafter, widening of roof was started taking the blast of central gullet about 5 to 7 m ahead of the side slashing. Full section of the rib was provided with in the stand up time. At some places where rock was very weak, to increase the stand up time, shotcrete was also applied, Again at RD 153 m, side cuts were made to approach the side drifts and lay the haunch beam from RD 153 to 196. The excavation of this patch from RD 153 to 196 was the toughest as the length of drifts in this case was maximum *i.e.*, 39 m. The temperature was very high and ventilation could not help at the dead end. Fig. 5 showing the side cuts made for approaching the haunch drifts and laying RCC haunch beam.

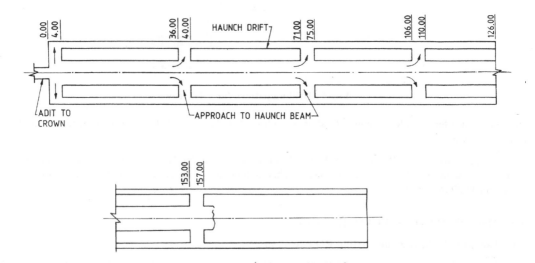

Fig. 5 : Plan showing excavation sequence of T.H. crown

After erection of the ribs contact grouting between the rock and backfill concrete was done at a pressure of 1 to 1.5 kg/sq m with grout of water cement ratio ranging between 0:5:1 to 1:1.

5.0 METHODOLOGY FOR BENCHING EXCAVATION

Excavation for benching between EL 1042 to EL 1062 m was done in 5 stages as indicated in Fig. 2. In fact, longitudinal gallery of the same size as that of access to transformer hall *i.e.*, 7.5 m D shaped was made through even before start of heading excavation along the transformer hall with objective of taking up excavation of bus ducts and gate shafts concurrently with the heading excavation of transformer hall. The mucking operations were carried out partly through 5 m adit to the crown of transformer hall and mainly from access to transformer hall bottom by making a ramp in the process of excavation. The general supporting which was done concurrent to excavation is shown in Table 1.

TABLE 1
Wall support details of transformer hall

Sl. No.	Wall	Rock bolt dia	Rock bolt length	Rock bolt spacing	Shotcrete
1.	Noth wall (Adit site)	32 mm	5 to 7 m	@ 6000 x 1500 c/c staggered and varying.	100 mm thick with welded wiremesh.
2.	South wall	32 mm	6 to 7.5 m	@ 6000 x 1500 c/c staggered and varying.	100 mm thick with welded wiremesh.
3.	East wall (U/s wall)	32 mm	6 to 9.5 m	@ 6000 x 1500 c/c staggered and varying.	100 mm thick with welded wiremesh.
4.	West wall (D/s wall)	32 mm	6 to 9.5 m	@ 6000 x 1500 c/c staggered and varying.	100 mm thick with welded wiremesh.
5.	Shear zone area	32 mm	7.5 m to 9.5 m	@ 3000 x 1500 c/c staggered and varying.	100 mm thick shotcrete with welded wiremesh.

5.1 Benching in Shear Zone Portion

Shear zone portion *i.e.*, RD 0-40 m in transformer hall was required to be supported with directional rock bolts of 32 dia. length 7.5 and 9.5 m perpendicular to the foliation so that they cross the shear band of 1.5 m width and stitch it to the other side rock. Benching in stages, 2 m at a time, was done and the gauzy material from shear zone beneath the haunch beam was scooped out and M-15 concrete was filled after placing the mesh of reinforcement at the lower most portion.

The drilling for directional rock bolts was done but the drilling rods were getting jammed in the gauzy material. The rock bolts, therefore, could be provided with great difficulty, were not taking tension. The area of shear zone area was grouted at a pressure of 1 to 1.5 kg/sq m Drilling was restarted after allowing setting time to the grout. The grouting and the rock bolts could be installed successfully.

Utmost care was taken while taking benching blasts, Proper prespillting was done. Very small blasts in a small stretches were taken limiting it only 2 m bench was taken at a time. As there was ingress of water on the right side wall, proper drainage holes were provided to give the free passage to the seepage water.

6.0 CONCLUSIONS

The total time taken for excavation of heading and benching against anticipated total period of 12 months is as below :

Heading excavation　　=　2 years 11 months and 26 days
　　　　　　　　　　　　for 33, 700 cum of rock excavation.

Benching Excavation　=　13 months
　　　　　　　　　　　　for 77, 700 cum of rock excavation.

It has been experienced that the sequence of excavation is totally dependent on the type of support system. The modality for excavation gets drastically changed with change over from conventional steel support system to rock reinforcement system in case of big caverns where central benching and side slashing is involved. It is very intricate and time consuming to change over from sequence of excavation with rock reinforcement system to conventional steel support system after the start of excavation with the former. It has also been experienced that it is also not possible practically to club both the options together in a single cavity as it is difficult to match the profile and execute the system. This is primarily difficult because of requirement of haunches for the seating of steel ribs. There is no separate access for the excavation and concreting of the haunches as the same has to be constructed prior to complete widening. It is, therefore, concluded that if there is any likely-hood of such week planes where steel ribs cannot be ruled out, it is preferable to execute the operations by planning excavation sequence with steel rib support system, nonetheless adjusting the size and sections of ribs as per design requirement. This would help in projecting reasonably fair schedule besides taking care of the surprises considerably. The overall cost when related with time would also be compatible.

INNOVATIVE CONSTRUCTION METHODOLOGIES

AN INNOVATIVE CONSTRUCTIVE METHOD FOR THE RETAINING STRUCTURES OF THE EXCAVATION OF THE UNDERGROUND PARKING DR. ÉNEAS DE CARVALHO AGUIAR, NI SÃO PAULO, BRAZIL

E.M. MAFFEI, CARLOS
Full Professor

H.H.S. GONÇALVES
Professor

M.C. GUAZZELLI

Escola Politécnica da Universidade de São Paulo, Maffei Engenharia

SYNOPSIS

This work presents a heady constructive method used for the retaining structures of the underground parking Dr. Eneas de Carvalho Aguiar, built in Sao Paulo. The project of the parking, of 15.5 m wide for 630 m long and 4 undergrounds, foresees the execution of the work in two phases. The first phase that includes 315 m of work, it was concluded in the end of 1998 and the beginning of the second phase of the work does not still have forecast.

The retaining structures of the excavation was executed through screw piles of 80 cm diameter, with 23 m of length, in the two faces of the excavation, spaced at every 2.50 m. Between the piles a fine layer of shotcrete of 2 cm of thickness was applied. The slabs of structure of the parking had been set up with precast beams of prestressed-concrete and the bottom slab with cast in-in-place reinforced concrete. The excavation was accomplished in two phases, the first one was 5 m deep and the walls worked as cantilever; after the installation of the upper slab working as struts, it was made the final excavation.

The final excavation till 11.60 m deep was made presenting as only support the piles spaced every 2.5 m, and the upper slab. To confirm the good operation of such light support, results of appropriate instrumentation were used, that checked the viability of the execution of the final excavation without the use of the berms or of any other additional support.

The screw piles are constituted, generally speaking, an economic solution for foundations, as is not common to use them as retaining wall due to the constructive difficulty of reinforcing this type of pile, tests were executed previous to the beginning of the work to confirm the executive viability of the piles with 19 m of reinforcement.

1.0 INTRODUCTION

São Paulo, capital of Mercosul, is a dynamic metropolis that does not stop of growing and of presenting challenges. Everyday 500 new vehicles enter in circulation to dispute the few spaces that the city offers. In face of this same problem, capitals all over the world found underground solutions showing the way to be followed by the city of São Paulo. Thus, critical areas indentified by the City Hall were placed in competition for the construction of underground parkings over the concession regime with resources originating from of the private initiative. It fit to the Consortium Dr. Enéas de Carvalho Aguiar to assist to the area of greatest concentration of hospitals of América Latina. In the parking underground clinics there were incorporated unpublished constructive techniques in the country, trying to interfere with to the minimum with the people's life that would transit in this area during the construction (Photos 1 and 2).

The great challenge in a work of this type is to have the skill to execute it with the best possible quality and minimum cost. So that this was possible, there were studied several alternatives of support project besides having had an investment in testing and appropriate instrumentation.

The initial project foresaw the execution of screw of 70 cm of diameter at each 2.5 m with 16 m long. The structure of the parking foresaw the execution of two central columns, whose foundations would be formed by piles of 1.40 m of diameter and 14 to 18 m long, between axes 18 to 26 and 55 to 66, respectively. Between the other axes concrete tubing piles would make the foundation of the central columns with diameter of the fuste equal to 1.40 m and diameter of the base of 3.65 m. The height of the enlargement was of 1.95 m and the distance between the bottom slab and the beginning of the enlargement should be larger than 1.5 m. However, to promote the insertion of the columns in the foundation structures it would be necessary the execution of a "chalice", whose constructive process was shown very difficult.

So it was decided to modify the project, the structure should be executed without intermediary columns and precast beams of prestressed-concrete that won the empty space of 15.5 m should be used. There were studied new support alternatives, being adopted screw piles of 80 cm of diameter, that were used so much as retaining structures of the excavation as well as for foundation of the structure of the parking. The piles were arranged at every 2.5 m in the two faces of the excavation, and they had the length increased for 23 m and the steel reinforcement introduced in the piles presented 19 m. The structure of the parking was conceived with the upper slab and the intermediary levels beam, in precast beams of prestressed-concrete and the bottom slab with cast-in-place reinforced concrete. The bottom slab works as complement of the foundation of the structure, because the piles just have load capacity compatible in relation to temporary phase, that is to say, without being considered the overloads in the floors of the underground.

The work began with the remotion of the interferences and execution of the piles. The placement of the upper precast beams was accomplished with the aid of a portico of 30 tons of capacity. The constructive method was just constituted basically in two excavation phases. The first excavation phase presented 5 m of depth and the walls worked as cantilever; after the installation of the upper slab that worked as struts, it was executed the final excavation. Due to the excellent characteristics of the soil a fine layer of 2 cm of shotcrete among the piles was just applied.

Temporary berms had been foreseen during the phase of final excavation, whose removal would depend on the instrumentation results that were constituted in measures of tension of the steel of the reinforcement of the pile and horizontal displacements of the piles. The obtained results checked the viability of the execution of the excavation without the use of the berms.

After the excavation and execution of the bottom slab, the installation of the intermediary levels slabs was proceeded and the piles restrained.

2.0 GEOLOGICAL-GEOTECHNICAL PROFILE

The geological-geotechnical profile of the area, shown in the Fig. 1 presents a layer of approximately 1 to 1.5 m of embankment of silty clay, followed by a layer of 10 to 12 m of silty clay with brown and red fine sand (porous) that presents index SPT variable among 10 to 20. Below this layer it comes a layer of 6 to 10 m of silty clay with a lot of variegated fine sand, SPT among 8 to 16, over a layer of silty-clayed sand of varied granulation. The level of the water was found at 17 m deep, therefore below the final excavation.

3.0 METHODOLOGY OF CALCULATION OF THE RETAINING STRUCTURES

The active pressure was calculated through the method of Rankine and the obtained values were smaller than the minimum values that are considered for clay. Therefore the pressure values were calculated by the diagram of minimum active pressure proposed in the norm "Calculation of the works executed by the method of the trench", of the company of the Metropolitan of São Paulo.

The underground of the area is very favourable to the excavation, because it is constituted in its first 12 m of a porous clay with cohesion estimated in 40 kN/m^2, that would allow vertical cut of 8.40 m. It is important to remind that the water level was meet below the bottom level of excavation. There were adopted for calculation of the pressures the following param for the several soils layers :

Embankment	$c=0$	$\phi=25°$	$\gamma=17kN/m^3$
Porous silty clay	$c=40kN/m^2$	$\phi=20°$	$\gamma=19kN/m^3$
Silty clay with a lot of fine sand	$c=20kN/m^2$	$\phi=27°$	$\gamma=17kN/m^3$

The diagram of minimum pressure is presented in Fig. 2.

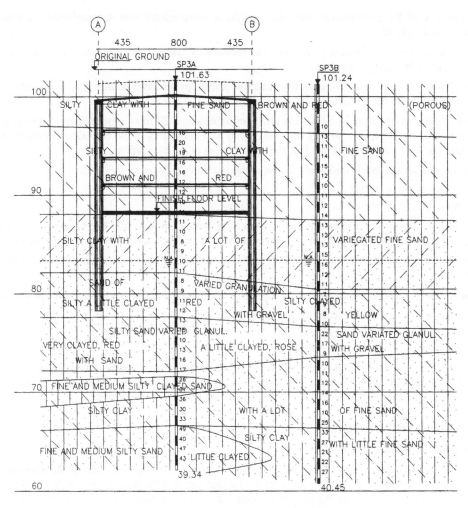

Fig. 1 : Geological-geotechnical profile

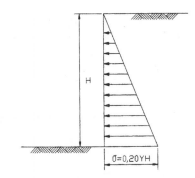

Fig. 2 : Diagram of minimum pressure

To determine the efforts and the displacements due to the excavation the evolutive and no-evolutive models were used.

397

In function of the geological-geotechnical profile, a compartment was accomplished; in which were determined three types of piles :

Type "light"-axes 0 to 17-trench constituted of porous silty clay red, level of water of 3 m above the end level of the embedded length, or below it.

Type "average"-axes 35 to 53-trench constituted of a homogenous layer of silty clay very sandy. The level of water was above 3 m the end level of the embedded length.

Type "heavy"-axes 18 at 34 54 to 68-trench constituted of 3.70 m of embankment, presenting till the bottom of the excavation a layer of silty clay very sandy and in the area of the embedded length the silty red porous clay. The level of water was met 1 m above the bottom of the excavation.

The piles were calculated considering the following materials :

- Concrete fck = 20 MPa.
- Steel fyk = 500 MPa.

The main reinforcement was constituted of special bars with 19 m of length. The razing level was the final level of the support of the beams and the polluted concrete was removed and the surface regularised with grout.

The piles were calculated to present 127 tons of vertical load capacity to permanent load and 18 tons for accidental load.

The maximum displacements of the piles foreseen through the evolutive model were the following ones :

Type "light"	-	$\delta=10.81$ mm
Type "average"	-	$\delta=16.00$ mm
Type "heavy"	-	$\delta=20.03$ mm

To guarantee the good acting of the work and to check the adopted param, temporary berms were left during the final phase of excavation and appropriate instrumentation it was done. Before the remotion of these berms the instrumentation results were compared with the one foreseen by the project.

4.0 CONSTRUCTIVE METHOD

4.1 General Constructive Method

The outlines of the constructive method are presented below. After the remotion of the interferences the screw piles were executed in the whole contour of the parking and metallic piles were installed with pre-hole in the extremity of the parking where the ramp was located that served as service road during the works and exit of the parking after the conclusion of the work (Fig. 3).

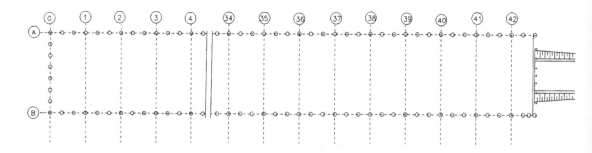

Fig. 3

In other to turn it possible the excavation till the installation level of the upper slab, with the walls presenting 5 m of cantilever, it was used an support understanding part of the piles and wood board among them. This was possible introducing a rib in the fresh concrete through vibration, settling forms for the concreting of a halfmoon (Photos 3 and 4).

Among the half-moons, the excavation was executed, being placed in the same time the boards that would be involved with geotextile blanket where there was risk of carriage of material (Photo 5).

After the concreting of the crowning beam the excavation was made till the level of the upper slab. The crowning beam sustained the portico, weighing 15 tons that transported the precast elements during the whole work, turning the process faster (Photos 6 and 7). After the partial demolition of the remaining reinforcement of the pile, until the razing level, coincident with the installation level of the upper slab, there were then installed the beams and the tie-rods that were executed in the work niches (Figs. 4a and 4b).

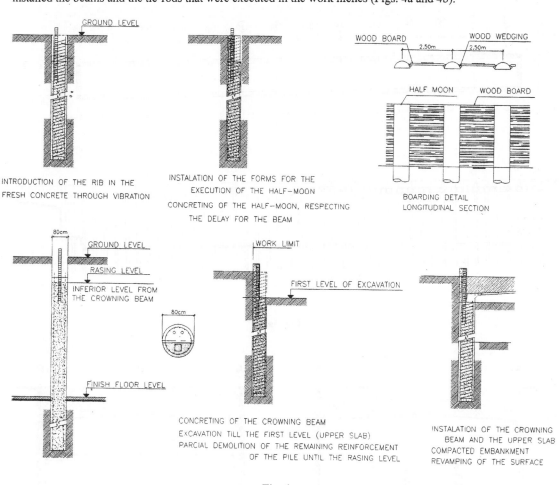

INTRODUCTION OF THE RIB IN THE FRESH CONCRETE THROUGH VIBRATION

INSTALATION OF THE FORMS FOR THE EXECUTION OF THE HALF−MOON
CONCRETING OF THE HALF−MOON, RESPECTING THE DELAY FOR THE BEAM

BOARDING DETAIL
LONGITUDINAL SECTION

CONCRETING OF THE CROWNING BEAM
EXCAVATION TILL THE FIRST LEVEL (UPPER SLAB)
PARCIAL DEMOLITION OF THE REMAINING REINFORCEMENT OF THE PILE UNTIL THE RASING LEVEL

INSTALATION OF THE CROWNING BEAM AND THE UPPER SLAB COMPACTED EMBANKMENT REVAMPING OF THE SURFACE

Fig. 4a

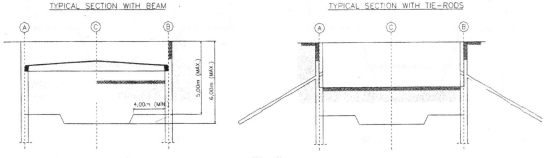

TYPICAL SECTION WITH BEAM

TYPICAL SECTION WITH TIE−RODS

Fig. 4b

Subsequently it was executed the upper slab, the compacted embankment and the revamping of the surface.

The excavation was made under the upper slab with 8.6 m of free empty space, continuing until the final level, berms were maintained with 4 m of wide and 3 m high in the whole turn of the parking. The berms were protected with shotcrete or mortar lining (Fig. 5a and 5b). Among the piles, the excavation was made in arch form and the soil mass protected with 2 cm of shotcrete (Photos 8 and 9).

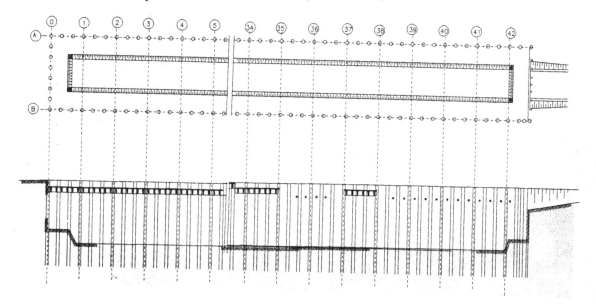

Fig. 5a

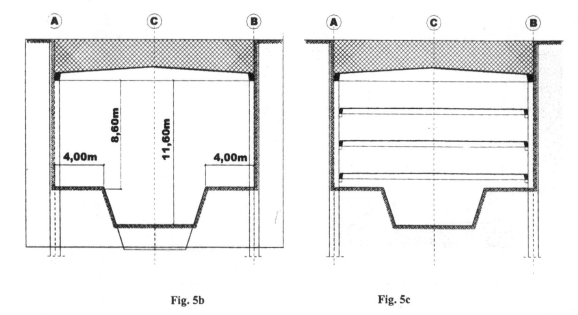

Fig. 5b   Fig. 5c

Next, the beams of the intermediary levels were installed and piles restrained (Fig. 5c). Subsequently the berms were removed and the bottom slab executed (Fig. 5d; photos 10, 11 and 12).

The remotion of the berms before the installation of the intermediary levels beams was only authorised after the confirmation of the validity of the project hypotheses through the results of the instrumentation as presented in the item 5.3 (Photos 13 and 14).

After the execution of the bottom slab, the inside structure of the parking was completed; it means the walls, the intermediary columns and the layers over the precast beams. Soon, the upper slab and the embankment were concluded in the areas used as work niche.

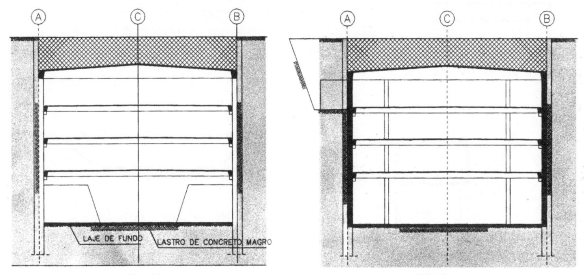

Fig. 5d   Fig. 5e

In the first underground there was the need to enlarge two samll areas of the parking for entrance and exit stalls of the vehicles. The excavations were made in slopes; the piles located in the areas of the stalls were cut (Fig. 5e). The spacing between the piles was decreased close this area with the purpose of supporting the concentration of loads in these locals (Photos 15 and 16). The construction method used in these areas is schematically presented in Fig. 6.

4.2 Constructive Method of the Screw Piles

The great challenge for the use of the screw piles, with 19 m of reinforcement, was the capacity of introducing the reinforcement, since the concrete should have peculiar characteristics to make possible going down 19 m of the frame. Two tests were accomplished and it was possible to point out some indispensable conditions for the success of the pile execution :

-    use of a crane compatible with the height of the frame to guarantee its gone down in the plumb line;
-    precise planning for the arrival of the concrete, concreting and immediate introduction of the frame.

The excavation of the piles was mechanised to the point level foreseen, 23 m below the level of the surface, without aid of bentonite slurry, and then the piles were concreted close to the surface level. The introduction of the main frame of 19 m long in the fresh concrete was executed through vibration (Photos 17 and 18).

4.3 Instrumentation used to Determine the Behaviour of the Excavation and the Possibility of the Remotion of the Berms

It was proposed an instrumentation programme, because in function of these results it could not have the need of the use of berms. The instrumentation was constituted in the installation of inclinom and anchorage points for the measure of horizontal displacements of the piles and verification of the tensions through extensometric test in the bar of steel closer to the inside face, in the section of maximum calculated bending moment.

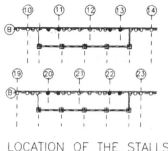

LOCATION OF THE STALLS
IN RELATION TO THE AXES

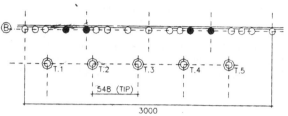

LOCATION OF THE AIR—COMPRESSED
CONCRETE TUBING

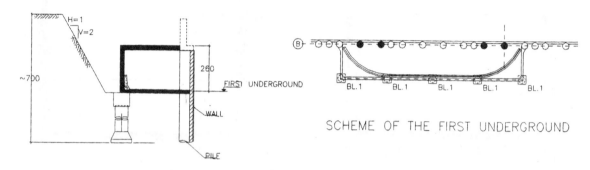

SCHEME OF THE FIRST UNDERGROUND

Fig. 6

The inclinometers were installed in the axes 8, 26 and 39 one-meter far from the longitudinal axes of the piles. The results of the inclinom were harmed by the proximity of the same ones to a cable of the Electricity Company. The maximum convergence observed in the several axes, among the piles, did not surpass 5 mm.

The results of the instrumentation supplied for the requested bar a tension of the order of 100MPa. As the tension of work of the frame is of the order of 300MPa, it was made a retroanalysis that allowed refining the param and through new calculations allowed the feasibility of excavating without berms.

5.0 CONCLUSIONS

The work presented an unprecedented constructive method used in the retaining structures of the underground parking Dr. Eneas de Carvalho Aguiar.

The opening of a trench of 15.50 m wide, 315 m of extension and 11.6 m of free empty space between the level installation of the upper slab and the final level of the excavation, presenting as only retaining structures screw piles at every 2.5 m restrained by the upper slab, it was accomplished with great success.

The comparison among the foreseen results and the observed one, checked the efficiency of the used support, therefore so much the efforts as the displacements of the retaining structures were smaller than calculated one. This constructive method was shown an economic and fast solution.

Photo 1 : Aerial view of the parking area
during the works

Photo 2 : Aerial view of the parking
after conclusion

Photo 3 : Detail of the half-moon

Photo 4 : First phase of excavation - detail of
wood boards between the half-moons

Photo 5 : Detail of the crowing beam

Photo 6 : View of the portico used to transport
the precast elements

Photo 7 : View of the berms of the first phase of excavation

Photo 8 : Internal view of the parking, excavation till the final levels and the lateral berms

Photo 9 : View of the extremity of the parking, execution of the shotcrete

Photo 10 : Execution of the supports for the intermediary levels beams

Photo 11 : View of the work in the phase shown in figure 5c

Photo 12 : View of one of the areas used as work niche, piles supported by tie-rods

404

Photo 13 : Overall view of the excavation with
the berms removed

Photo 14 : Removal of the berms with
machines

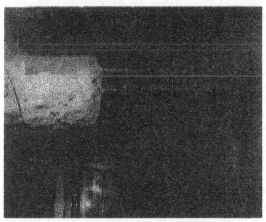

Photo 15 : Pictures of one of the piles cut
in the stall area

Photo 16 : Details of the decrease of space
between piles in the stall area

Photo 17 : Details of the execution of
the screw piles

Photo 18 : Overview of the jobsite, with some
piles frame already prepared

EXPERIENCE ON BHIMA SINA FEEDER TUNNEL

V.V. GAIKWAD

Chief Engineer (S.P.)

**Irrigation Department, Pune
Maharashtra, India**

CHANDRASHEKHAR HANGEKAR

Executive Engineer

**Sina Madha Project Division, Bhimanagar
Solapur, Maharashtra, India**

SYNOPSIS

In the state of Maharashtra, Sina river is the main left bank tributary of the Bhima river and flows from west to east through Ahmednagar and Solapur districts. Both the districts have large drought prone areas, they have rainfall less than 700 mm. The feeder tunnel takes off from village Kandar near Ujjani reservoir and joins the Sina river at village Kave through 26.255 km long link canal. It includes open cuts 1.56 km and 4.44 km on upstream and downstream respectively, remaining 20.255 km being a tunnel portion. This tunnel has a rectangular shape of 7.0 m × 3.0 m and arch portion of 7.0 m × 3.5 m in cross-section. Bed gradient is 1:3000.

1.0 INTRODUCTION

Maharashtra state is divided into five river basins namely Krishna, Godavari, Tapi, Narmada and basins of west flowing rivers in Konkan region. Distributions of rainfall in these five basins are varying from 9000 mm in Konkan region to 600 mm in Central Maharashtra. However, culturable land for agricultural purpose is mainly available in Krishna, Godavari and Tapi basins. This geographical situation demands for transfer of water from one valley to another to increase the standard of living by maximising the crop yield in this part of the Maharashtra state. Index map of the project is shown in Fig. 1

Bhima river is one of the main left bank tributaries of Krishna river. It originates at Bhimashankar in Sahyadri ranges and flows from west to east till it joins river Krishna in Karnataka state. Sina river is the main left bank tributary of the Bhima river and flows form west to east through Ahmednagar and Solapur districts. Both the districts have large drought prone areas, they have rainfall less than 700 mm. There is an assured rainfall in Sahyadari ranges, hence there is an assured storage of water in Ujjani reservoir in Bhima basin. So it was conceived to divert the surplus water from Ujjani reservoir to Sina river by constructing Bhima Sina feeder tunnel. It was proposed to construct 10 weirs of Kolhapur type across Sina river for storing this water which will be utilised to irrigate the 48000 ha fertile land on both the banks of Sina river by private lifts. From Ujjani reservoir 211 million cu m (7.47 thousand million cubic feet) water will be diverted to Sina river for this purpose.

The feeder tunnel takes off from village Kandar near Ujjani reservoir and joins the Sina river at village Kave through 26.255 km long link canal. The total length of canal is 26.255 km includes upstream 1.56 km and downstream 4.44 km open cut and remaining 20.255 km tunnel.

It is a free flow tunnel. Maximum discharge passing through this tunnel is 26.00 cu m/s. The tunnel is 'D' Shaped.

2.0 DETAILS OF TUNNEL

(a) Rectangular portion 7 m × 3.0 m
(b) Arch Portion 7 m × 3.5 m
(c) Bed gradient 1:3000

Cross-section of the tunnel is shown in Fig. 2.

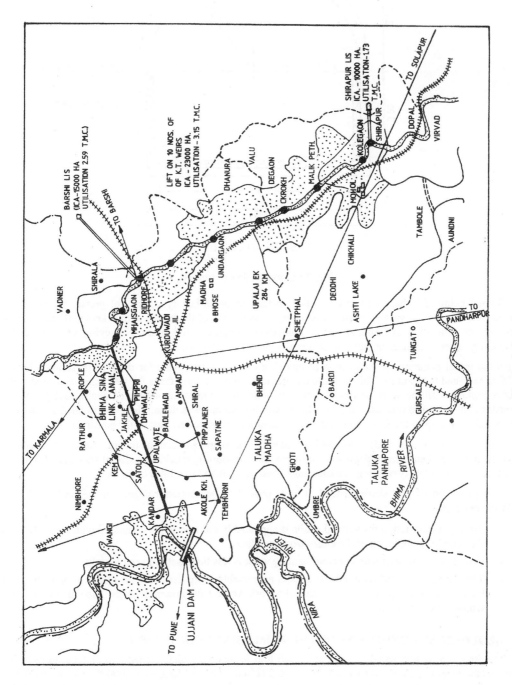

Fig. 1 : Index plan

407

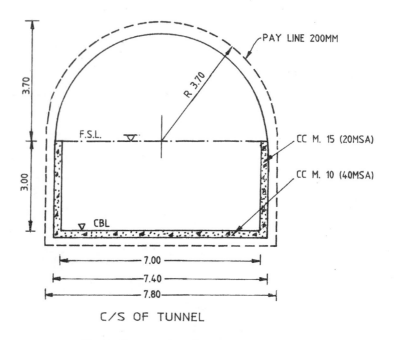

Fig. 2 : Cross-section of tunnel

Tunnel being 'D' shaped in cross section instead of horseshoe type, we get greater width at invert. Consequently it provides greater construction ease. Cross-section of the tunnel is thus kept sufficiently wide to carry out the tunnelling operations. Heavy tunnelling equipment and muck conveying vehicles, concrete transporting mixers and agitator cars can pass easily through the tunnel.

As per the initial planning for excavation of tunnel, a provision was made for three adits having a length of 3.725km each and two portals. This would have provided eight working faces for the tunnel excavation. But in order to complete all the work of tunnel including concreting within the stipulated contractual period of 42 calendar months, revised provision for five shafts with two portals was done. This has provided 12 faces in all and proved to be helpful for faster completion of excavation and mucking part of the construction.

Shape and size of the shaft is decided considering the shape and size of tunnel, quantity of muck to be lifted and the type of equipment deployed for lifting and transporting muck. The size of shafts is 11.0 m x 6.40 m so as to accommodate two lifting systems and a passenger lift and two lines of air duct.

This provision of five shafts instead of adits proved to be economical and can be revealed from the following features.

- Time of construction for vertical shafts is appreciably reduced when compared with the adits. Hence, the time for completion of the entire work is reduced.
- Cost of vertical shaft with concrete staining work is less than that required for adit.
- The quantity of land required for adits is appreciably more than that for shafts. Moreover, farmers being reluctant to part with their lands the work would have lingered longer. Also the cost of project would have increased.

3.0 GEOLOGICAL CONDITIONS OF THE BHIMA SINA LINK TUNNEL

The Bhima Sina link canal lies in the deccan trap formation, which is made up of thick piles of basalt flows. As the length of Bhima Sina feeder tunnel is 20250m with a gradient of 1:3000 the elevation difference between the crown at the upstream portal and the crown at the down stream portal is 6.73 m as shown in the Fig. 3.

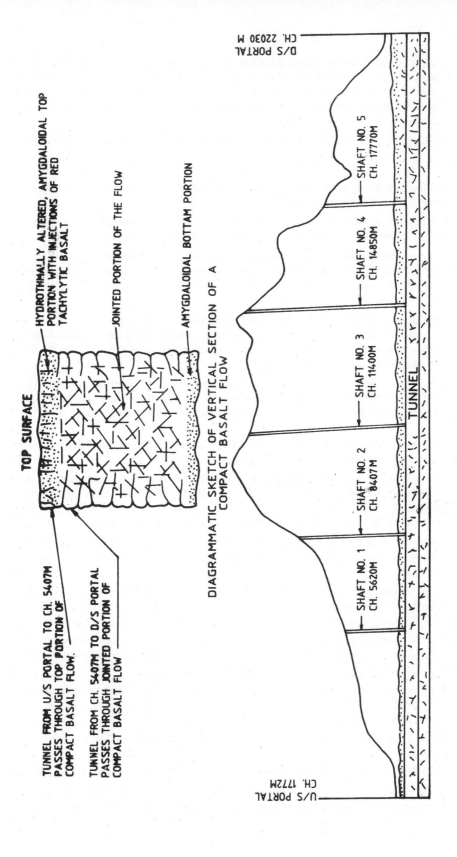

TOP SURFACE

HYDROTHMALLY ALTERED, AMYGDALOIDAL TOP PORTION WITH INJECTIONS OF RED TACHYLYTIC BASALT

JOINTED PORTION OF THE FLOW

AMYGDALOIDAL BOTTAM PORTION

TUNNEL FROM U/S PORTAL TO CH. 5407M PASSES THROUGH TOP PORTION OF COMPACT BASALT FLOW.

TUNNEL FROM CH. 5407M TO D/S PORTAL PASSES THROUGH JOINTED PORTION OF COMPACT BASALT FLOW

DIAGRAMMATIC SKETCH OF VERTICAL SECTION OF A COMPACT BASALT FLOW

U/S PORTAL CH. 1772M

SHAFT NO. 1 CH. 5620M

SHAFT NO. 2 CH. 8407M

SHAFT NO. 3 CH. 11400M

SHAFT NO. 4 CH. 14850M

SHAFT NO. 5 CH. 17770M

D/S PORTAL CH. 22030 M

TUNNEL

Fig. 3 : Geological conditions along the alignment of the tunnel

409

The compact basalt flows are always thick and extensive. Therefore entire tunnel from its upstream portal to downstream portal passes through only one compact basalt flow. However, there is a wide difference in the field characters of the flow, as far as its top and bottom portions are considered. As tunnelling conditions depend upon field characters, problems in the tunnelling work depend upon the fact as to through which part of the flow it passes.

As mentioned above, compact basalt flows are thick and extensive, having their top surfaces almost horizontal.

3.1 Tunnelling Conditions in the Top Portion of Compact Basalt Flow from Upstream Portal at Chainage 1780 m to Chainage 5407 m

Immediately below the top surface of compact basalt flow, for some thickness, rock is vesicular and amygdaloidal. It is also purple or reddish in colour due to hydrothermal alternation.

Due to a net work of tachylytic veins and injections, rock is broken up into angular fragments of different shapes and sizes. Rock appeared like a volcanic breccia in which fragments were held together in red tachylytic lava matrix. Due to the presence of tachylytic injections, problems like overbreaks and rockfalls were observed in the tunnel.

The peculiarity of the red tachylytic basalt is well known. It, on exposure to atmosphere, deteriorates very fast and gets converted into powder like material. Therefore, if rock having ramifications of tachylytic basalt, it creates problems of rockfalls from the crown. Due to deterioration of tachylytic rock, the fragments of rock became loose and subsequently they collapsed.

The condition is met with in Bhima Sina link tunnel between upstream portal at Chainage 1780 to Chainage 3459 m and onward upto Chainage 4689 m. To overcom this, immediate layers of shotcrete 50 to 70 mm thick were applied to avoid exposure. This solved the problems satisfactorily.

3.2 Tunnelling Conditions in Middle and Lower Parts of the Compact Basalt flow *i.e.* from Chainage 5407 m to Down stream Portal Chainage 22030 m

The middle and lower portions of the compact basalt flow were free from vesicles and amygdales and occurred in true sense as compact rock. Joints, which were the contraction cracks developed during cooling and solidification of lava were present in these portions. Due to presence of joints, rock in these portions occurred in dissected condition. This resulted in heavy loosefall along the axis of the tunnel.

Generally excavated surface of the tunnel in compact basalt is uneven having deep, angular depressions and wedge shaped protrusions with sharp edges. It is due to the presence of intersecting joints from where rock falls have occurred.

Some rock blocks fell down during blasting or immediately after the excavation. However, some rock blocks remained stable for a longer time. But after that, they too, collapsed down. The bridging capacity of the rock at the crown and the tunnelling ease is dependent upon the pattern and spacing of joints.

When joints were inconsistent and no interconnections were there, no rock falls were seen to occur from the crown. But when the joints were closely spaced and were interconnected rock falls were noticed to occur.

Rockfalls were particularly more when wedge shaped rock block, in the form of inverted keystone, was met with at the crown. As such deep depressions in the form of chimney occurred at the crown. This situation was observed between shaft no. 3 and 4. To overcome it immediate remedial measures like double chain linking and rock bolting were taken.

Geological conditions along the alignment of the tunnel are presented in Fig. 3.

4.0 TREATMENT GIVEN TO THE CROWN PORTION OF BHIMA SINA FEEDER TUNNEL

In order to keep the tunnel section stable and safe throughout its length, some kind of treatments by way of shotcreting/rock bolting, plain cement concrete lining, provision of permanent support lining were given.

4.1 Between Chainage 1780 to Chainage 5407 m

From Chainage 1780 m to Chainage 5407 m the tunnel work lies in the upper portion of the compact basalt flow in which heterogeneous rock is occurring with injections of red tachylytic basalt. Rock falls were seen from the crown on deterioration of injections of red tachylytic basalt. Therefore as a temporary measure, rock bolts with chain linking were provided to avoid accidents by rock falls. To avoid deterioration of tachylytic injections on exposure to atmosphere, the rock in the crown portion was concealed by providing shotcreting having thicknesses varying from 75 to 100 mm.

In the region where cover is thin, made up of weathered rock and large sized rock falls have occurred, permanent supports are erected for a length of about 1.5 per cent of total length.

4.1.1 *Occurrence of Rock Falls during Blasting*

Though joints are present in the rock in crown portion, rock falls have not taken place. Therefore, no rock bolts are provided in that portion. But it was observed at some locations that after every blast small rock blocks had fallen down. Falling away of rock blocks during blasting indicated that joints in that portion may get opened out in due course of time. In these reaches four meters long rock bolts have been provided to avoid rock falls during construction and in future.

Compact basalt is highly resistant to weathering; therefore, it does not require cover to avoid exposure to atmosphere. At most of the places of the tunnel, water percolates from the crown due to very much closely spaced jointed nature of the rock. To avoid rock blocks getting loose due to percolation of water and also to stop percolation of water, plain shotcreting or shotcreting with chain linking has been provided.

Due to closely spaced jointed compact basalt extensive rock falls have occurred resulting in the formation of chimneys. In such areas rock bolts and chain links do not serve the purpose. Hence, in such stretches permanent supports are erected.

4.2 Between Chainage 5407 m to Chainage 22030 m

From Chainage 5407 m to Chainage 22030 m the tunnel lies in the jointed portion of the compact basalt flow. Due to presence of joints, rock blocks occur in loose condition. Therefore, rock falls occur from the crown. It was also observed that vertical joints are predominantly occurring as two vertical joints intersect with each other. In such portions deep notches were seen to cause overbreaks.

To avoid rock falls from the crown portion, rock bolts are provided to anchor the rock blocks in the overlying rock.

Wherever, joints are broadly spaced and are inconsistent rock bolts is not required. On the contrary, in the region where joints are closely spaced and are interconnected large number of long rock bolts were required having close spacing.

5.0 SHAFT EXCAVATION

Size of shaft adopted in 11.00 × 6.40 m so as to accommodate two lifting systems and a passenger lift and two air duct lines. For mucking activity of proper shaft, CK-90 excavators were deployed and at each cycle. Same was being lowered with the help of 40 tonnes capacity diesel crane. Due to this the critical activity of shaft excavation *i.e.,* mucking was brought under control. By adopting this procedure shafts were completed upto invert of tunnel.

6.0 TUNNEL EXCAVATION

The excavation of Bhima Sina link tunnel is done by full face excavation except for a length of 1700 m between Chainage 1780 m to Chainage 3480 m. Due to occurrence of weak strata, heading and benching was adopted for tunnel excavation.

The multideck drilling jumbos were provided at each face for drilling operations. With nine Jack hammers, of Atlas Copco make were deployed for drillling 106 No. of holes. In weaker strata, controlled perimeter drilling and blasting method was deployed. Alternate periphery holes were charged to minimise the overbreaks. 375 m drilling was carried out per cycle, thus 26,20,000 m drilling was done on this project manually. The contractor adopted the conventional drilling and blasting method for excavation of tunnel and achieved the national record of 213 m of tunnel excavation through shaft on single face.

The various drilling and blasting patterns are adopted during the course of excavation, depending upon geological conditions.

6.1 Details of Tunnel Excavation

- Cross-sectional area of tunnel excavation   50.421 sq m
- Average length of drilling   3.50
- Average length of pull   2.90
- % of recovery with respect to drilling   83%
- Average no. of holes required per blast   106 no.
- Gelatin per blast   70 kg
- ANFO per blast   90 kg
- Average quantity of excavation per pull   146 cum
- Total length of drilling per pull   375 m
- Index of
  - (a) Drilling   2.56 m/cum
  - (b) Explosive   1.1 kg/cum
  - (c) No. of holes/C.S.A.   2 nos./sqm
  - (d) Average time cycle   14 to 16 hrs.

6.2 Details of Time Cycle

1. Survey : 0.30 hrs.
2. Jumbo erection : 0.30 hrs
3. Drilling arrangement and drilling : 4.00 hrs
4. Loading and blasting : 1.00 hrs
5. Degassing : 1.30 hrs
6. Loose removal : 1.00 hrs
7. Mucking : 5.00 hrs
8. Bottom cleaning : 0.30 hrs

Total hours : 14.00 hrs

6.3 Mucking

For transportation of muck, the contractor had deployed EX-200 excavator and seven tippers for each face inside the tunnel and four tippers on top of shaft for conveying it to muck yard area. Transport of muck was done by 5.5 cum capacity side opening buckets mounted on tippers. Muck was loaded at the excavation face into buckets mounted on tipper chassis by means of EX-200 excavator. Tippers conveyed the muck through tunnel and were positioned under the vertical shafts. Only bucket is lifted by the winch, muck is dropped in the tipper standing on the ground after operating the side opening of bucket manually and then transported to dumping area. Empty buckets are lowered on the chassis of tippers positioned on tunnel invert near the shaft. For this purpose 2 no. of 15 tonne capacity EOT cranes are installed on each shaft. For these capacities of EOT cranes, uninterrupted electric power supply was required. To achieve this, captive electric generator sets of 500 kVA capacity at each shaft site were installed. Main aspect in Bhima Sina link tunnel was, that 8,53,000 cum of muck was lifted through shafts.

7.0 VENTILATION OF BHIMA SINA LINK TUNNEL

For adequate progress of the tunnel excavation proper ventilation is more important. Reversible method was adopted in this project. Blowers of 566.25 cu m/min (20,000 cubic feet/minute) capacity were deployed at interval of 300 m and duct line of 1.1 m diameter was placed in the tunnel. Separate duct lines were provided for each face of the tunnel through shafts. Over and above, ventilation shafts at interval of 500 m having diameter of 2 m were provided where the tunnel overburden was less than 40 m. Total 19 no. of ventilation shafts were provided. Two ventilation shafts having depth of 50 m each were executed by "**Drop Raising Method**".

CONSTRUCTION OF DIVERSION TUNNEL FOR DOYANG
HYDROELECTRIC PROJECT —A CASE STUDY

R.K.KHALI
Dy. Project Manager

R.D. VARANGAONKAR
Vice President (Projects)

Gammon India Limited, Gammon House, Veer Savarkar Marg, Prabhadevi, Mumbai, India

SYNOPSIS

Construction of large diameter tunnels in foliated and moderately jointed rock mass for diverting river is very difficult and takes longer time for execution. Problems due to bad approach roads, roof collapsing, improper geological data, frequent changes in construction drawings, wrong estimation of quantities, hostile local conditions and serious law and order situation. All these problems occurred during construction of 12 m diameter and 654 m long diversion tunnel for Doyang Hydroelectric Project in the state of Nagaland. This paper describe various problems encountered during construction and innovative methods, spot solutions adopted for completing the project successfully.

1.0 INTRODUCTION

Tunnelling in bad rock strata is a serious engineering problem. This problem gets further complicated when the size of tunnel is large. Doyang Diversion Tunnel happens to be one of the most important structure of the project as the whole river was diverted through this in order to facilitate the construction of rockfill dam. Tunnel also comprised construction of 6 m diameter 52 m long depletion tunnel inclined at 45° and vertical gate shaft 85 m deep semi-elliptical in shape having major axis 16 m and minor axis 12.2 m. Boring of tunnel was done through massive hill on left flank. Fig. 1 shows the longitudinal section of diversion tunnel with details of other connected structures. The contract was for a period of 12 months, however, despite all modern facilities and very hard non-stop efforts, it took more than 72 months to complete and hand over tunnel to the clients, the North Eastern Electric Power Corporation in the year 1996 for diverting the river.

2.0 SALIENT FEATURES

The diversion tunnel covers a length of 654 m with a slope approx. 1 in 654 m with horizontal bend at chainage 123.72 and transition at the bottom of gate shaft at chainage 332.88 m and one shear zone at chainage 574 m from inlet face with gooseneck depletion tunnel 6 m in diameter and 52 m in length starting from chainage 123.72 and meeting with diversion tunnel at chainage 176.44 m. One gate shaft 16 m x 12.6 m in shape and 85 m deep at chainage 332.88 m. Tunnel was bored from two faces *i.e.*, inlet and outlet by using conventional heading and benching method.

3.0 GEOLOGY

Rock geology was very poor. Soft sand stone and hill wash material encountered during boring operation Advancing of tunnel was done with controlled blasting pull to avoid chimney formation. As a result cycle time of boring increased. Sand stone used to turn in to slush when it came in contact with water causing serious problems for dewatering and plying the mucking vehicles inside the tunnel.

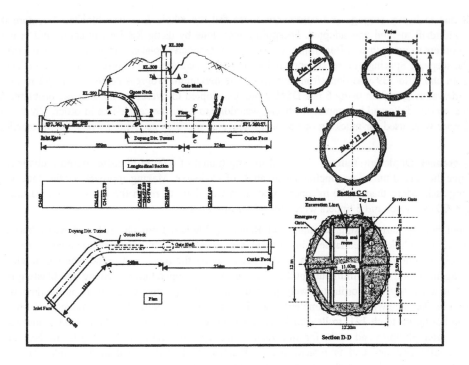

Fig. 1 : General layout of Doyang diversion tunnel

4.0 METHODOLOGY

4.1 Boring

Tunnel was to be bored through sediment of sand stone from two faces *i.e.* inlet and outlet. Open excavation was done at both faces to reach the tunnel face. Approximate quantity executed was 44,800 m^3 (29,800 m^3 at inlet face and 15000 m^3 at outlet face). One CK-90 excavator, one Tata Hitachi 083 LC excavator, 12 dumpers and one D-80 dozer were deployed for smooth execution of excavation work.

4.1.1 Inlet Face – At this face portal was erected from Ch.-1.5 m to 1.5 m and shotcreting was done to stabilise the hill above the face using Aliva Shortcrete machine. Boring was done by conventional drilling and blasting using Heading Benching method. Wedge cut drilling and blasting pattern used for a maximum pull of 1.8 m. Special care was taken for maintaining the C/L of tunnel by way of drawing ordinates after each blast. Steel rib supporting systems with precast lagging and back filling was adopted to stabilise the roof. Position of meeting point of depletion tunnel and horizontal bend clearly marked at site to avoid any possible deviation in the alignment of tunnel. Heavy seepage was observed, proper drains were made and suitable dewatering pumps were deployed to dewater the seepage water. Despite our this arrangement seepage water created serious problems for dewatering operation and also in plying mucking vehicles because of typical sand stone. Soft sand stone in the tunnel used to create slush/pulp when coming in contact with water with the result drains and sumps used to get choked. Dewatering sump were made at each 75 m to overcome the water-logging problem inside the tunnel. The maximum boring progress achieved was 55 m from this face in one month. The total length bored from this face was 326 m.

4.1.2 Outlet Face – After completing the open excavation of approx. 29,800 m^3, portal was made at ch. 654 and shotcreting was done by using Aliva Shotcrete machine to stabilise the loose strata above portal. Drilling pattern for 1.8 m pull was used with Heading and Benching Method. Tunnel roof was supported by steel ribs at 50 cm c/c and precast lagging and backfill concreting afterwards. When progress was at ch. 574 m shear zone was encountered and huge cavity about 7 m was formed above crown on left side. Work came to standstill due to heavy seepage and continuous loose fall the whole face was blocked for more than two months. Multidrift method was adopted. Forepoling was done by using RS Joist in series and providing the roof for workers to work inside. Backfilling was done using concrete placer. Grouting was also carried out to control the seepage. Spacing of steel supports reduced to 30 cm c/c and pull also reduced to 1.0 m. Total 315 m boring from ch654 to 339 was done from this face. The maximum progress achieved was 52 m in a month. Effective dewatering system using 6 nos DSM-3 pumps adopted to overcome the seepage problem. Accurate centre-line survey, followed by graphical offsets, was done to keep the centre-line in order. CK-90 excavator and 6 dumper were used for mucking operations. Excavator was also used extensively for rib erection at this face.

4.1.3 Gooseneck Depletion Tunnel – This 52 m long tunnel of 6.00 m diameter posed a challenge during boring since the axis of tunnel was inclined at an angle of 45 degree. Making approach for this face was itself a challenging task. More than 3,500 m³ open excavation was done to reach the ch. 121 *i.e.*, starting point of tunnel. CK-90 excavator and 6 tippers were utilised for this work.

Tunnel boring started using Heading and Benching Method from ch. 121 and when boring was at chainage 133, the whole face collapsed smashing all the rib beneath. Multidrift method adopted for stabilising the face. Ribs erection was done at open space after cleaning the muck and portal was constructed. Boring operation again started with a minimum pull of 1.00 m. At every operation of drilling and blasting the survey was done meticulously. The ordinates were plotted graphically to ensure that there is no deviation in the intersection of chainage with 12 m diameter tunnel. Rib spacing was reduced for 0.5 m c/c/ to 0.3 m c/c. After completing boring up to ch 141, trolley track arrangement was adopted for mucking. One 5t electrical winch fixed on elevated platform and tracks for movement of trolley were laid up to heading for movement of muck trolleys specially designed and fabricated at site. Blasted muck was loaded in the trolleys manually and unloaded directly in the dumper with the help of chute (Fig. 2). Steel supports were used to support the poor rock with RCC lagging and backfill concrete speed to boring achieved 8 m/month. The diversion tunnel work was in progress, hence the muck was not allowed to fall by gravity therefore pilot hole and widening method was ruled out.

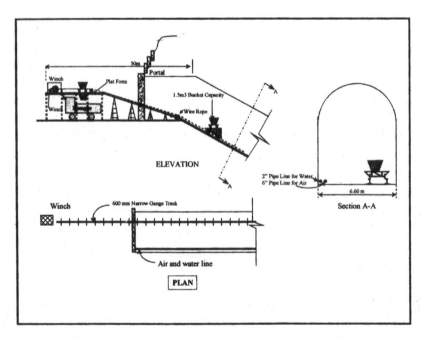

Fig. 2 : **Arrangement for mucking of gooseneck depletion tunnel**

4.1.4 *Gateshaft* – Boring operation for gateshaft took place from top at ch. 308 m. Total depth of gate shaft was 85 m *i.e.,* 50 m below ground level and 35 m above ground level. Conventional drilling and blasting with full face method adopted for this 16 x 12.2 m shape elliptical gate shaft. Mucking was done by using 10T guy derrick crane, operated with the help of two numbers electrical double drum winches. Muck was loaded in the bucket manually and unloaded in the dumper at EL 308 and then transported to dumping yard. Rock mass was supported by elliptical shape, steel ribs with precast lagging and backfill concreting. The maximum progress achieved was 10 m in one month.

There was no approach below the gate shaft as it was opening inside the tunnel. Thus use of raise climber was not possible and only way was to haul muck up by EOT/crane etc.

4.2 Lining

Lining in tunnel is technically an important activity and generally constitutes 30 to 40% of total cost of tunnel. Special care was taken for lining of this 12 m diameter 654 m long tunnel. The requirement on account of curvature thin section and difficulty in placing identified. Two 12 m diameter and 6 m long shuttering gantry were designed and fabricated in such a way so that dumper could pass below this. Same was erected inside tunnel *i.e.,* one at inlet face near ch.326 and one at outlet near ch.338. Weight of gantry was 25MT. Kerb concreting completed at both the faces to facilitate the movement of shuttering gantries.

4.2.1 *Inlet Face* – PCC overt lining started using the gantry erected near ch.326. Concrete pump was placed near gantry and concrete is transported through transit mixer and agitator cars and directly unloaded in to the hopper of concrete pump for placing in location.

Total 60 m length of PCC overt lining completed from this face and suddenly there was change in construction drawings. PCC lining earlier proposed was replaced by RCC lining. Due to this we have to mobilise extra efforts, shuttering reinforcement gangs, procurement of reinforcement steel, fabricate separate gantry for fixing reinforcement at 12 m height and other connected works.

It was decided to keep the PCC overt lining earlier done as it is and balance to be converted in RCC. Overt lining with RCC started again by transporting the concrete to concrete pump for placing. This has a disadvantage that as the time of transportation increases, the quality of concrete get affected through over mixing and certain amount of initial set could take place before actual placement. It has been observed that blockage in pipeline was frequent when concrete pump was placed nearer to gantry due to the backflow of concrete under gravity which choked the entire pipeline. The cleaning of pipeline itself is a tedious and time consuming operation. To overcome this concrete pump was placed near inlet face and three numbers 10/7 mixers placed on elevated platform and concrete was directly unloaded on concrete pump hopper with the help of chute pipe and longest length of concrete pumping done was 330 m horizontal plus 12 m vertical successfully without much breakdowns (Fig. 3). Shutter vibrators were placed on the gantry shutter for compaction. Plasticizers were also used to maintain the slump and workability. The cycle time for reinforcement fixing, gantry travel and alignment and pouring of 6 m length was brought down to 40 hours by several rounds of optimisation efforts. The geometry of tunnel is shown in Fig. 1. It can be seen from geometry that depletion tunnel (gooseneck) of 6 m dia intersects the 12 m dia tunnel at ch. 172.23 at the crown. Due to non-finalisation of designs, we had to skip 36 m (ch. 156 to 192 m) of overt lining of 12 m dia tunnel and continue overt lining from ch. 156 to 00.

The gantry arrangements made for overt lining was weighing 40MT. Initially it was planned to drag the gantry back to a distance of 156 m and start overt lining below gooseneck. The constraints apprehended were as under :

(i)   Negotiating the curvature of radius at ch.121.00 m

(ii) The friction between completed lining and gantry shutters.

417

(iii) Just for retrieving the gantry we have to maintain the kerb and thereby affecting invert concrete programme.

(iv) Time Loss.

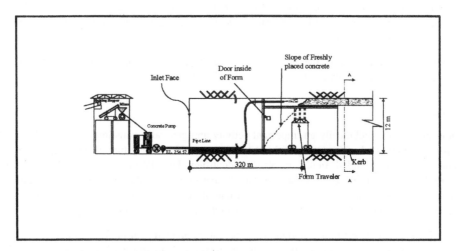

Fig. 3 : Arrangement for overt concrete from inlet face

Having above factors in mind a suspended type shuttering idea was conceived at site utilising the structural steel supports which were erected during heading of tunnel (Fig. 4). Due to this arrangement site could gain in following activity :

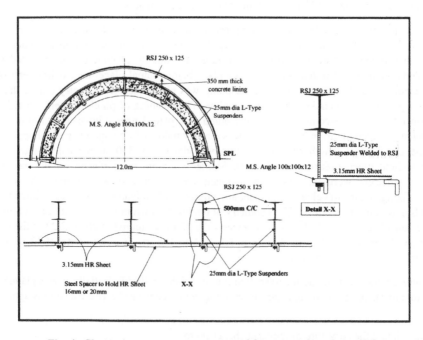

Fig. 4 : Shuttering arrangement in tunnel for concreting above SPL

1. The gantry which was reached at ch.00 was utilised for concreting of inlet portal transition and slab with zero staging arrangements.

4.2.2 *Outlet Face* – P.C.C. lining started at this face, approx. 12 m PCC length executed at this face. For this concrete production was done with the help of 30 m³ capacity Elba batching plant and concrete then transported to pump location by transit mixers and transit cars. Then RCC started as per new construction drawings. Concrete pump was placed near outlet face and one pan mixer type 15 m³/hr B.P. was erected on top ledge above the crown and concrete was directly unloaded on concrete pump hopper (Fig. 5) with the help of chute pipe and a record 332 m (320 m horizontal and 12 m vertical) pumping has been done successfully without much breakdowns.

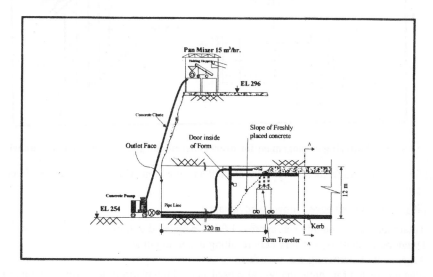

Fig. 5 : Arrangements for overt concrete from outlet face

4.2.3 *Gooseneck* – Having completed the boring of tunnel the methodology for placing lining concrete and shutters was to be finalised. Initially it was decided that a gantry will be erected for placing tunnel lining concrete. As per the design of the gantry we apprehended that the operation of pulling gantry after every pour will pose extra efforts mainly due to friction and angle of inclination. Finally with due thought of the various constraints which we would have to come across was studied in depth and then it was decided that operation of lining shall be as under :-

   (i)   Completion of invert concrete up to springing level.

   (ii)   Having completed the invert we decided to erect the tubular staging throughout the tunnel and have a moving shutter, having segmental length of 3 m.

   (iii)   The moving shutter details :

As per Fig. 6, number of trusses were fabricated and 3.15 mm sheet was welded over it. The truss were erected on 4 number guide channels which were always kept in parallel to axis of the tunnel. The adjustment for the channels were done by using revel pins which were acting as jack.

After every pour of concrete, just by adjusting the revel pins the total shutter including guide channels were released from the concrete surface.

After release of the shutters the total unit of 3 m segments were pulled forward with two number chain blocks of 1.5 MT capacity.

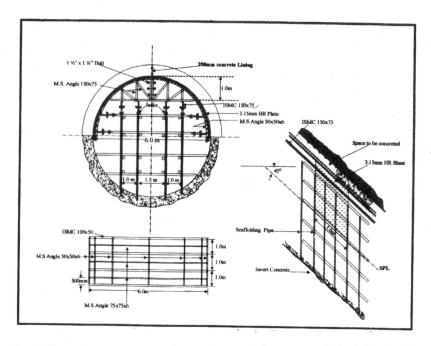

Fig. 6 : Shuttering arrangement for overt concrete for goose neck depletion tunnel

The using revel pins adjustment, alignment of shutters were done. Concrete was placed in position by 250 BP concrete pump.

The arrangement so used gave following advantages :

(i)   Total fabrication quantity was approx. 1.0 MT compared to initial proposed gantry of 7.0 MT.

(ii)   Dragging of shutters was very easy resulting in gaining time.

(iii)   Alignment of shuttering was very fast.

(iv)   Staging erected for shuttering was also used as working platform for fixing of reinforcement.

On Both sides of gooseneck crown concrete of diversion tunnel was left out as the junction details of gooseneck with diversion tunnel were not finalised which was done by innovative method of suspended shuttering, since shuttering gantry had advanced to outlet and not possible to travel back.

4.2.4 *Gateshaft – Pumping concrete in shaft below ground level* – The concrete lining below ground level has been done using 250 HD pump. Two number hanging working platforms were used for shuttering and reinforcement fixing work and positioning the pipeline as per requirement. One 30 m^3 batching plant was erected about 500 m d/s of gate shaft concrete from batching plant was transported by 4 nos of transit mixers and poured directly in concrete pump.

Earlier concrete was pumped to the required place using 90 degree bends at top and bottom as and when required. Due to long vertical distance and continuous pumping frequent damages at 90 degree bend portion have taken place. To overcome this 'S' type bend was provided on top portion to reduce the damage and concreting has been done successfully (Fig. 7). The maximum and minimum output achieved at Doyang was 21 m^3/hr and 10 m^3/hr. with average being 15 m^3/hr. Importance of gate shaft was to construct the gate grooves very accurately. A steel shuttering set with soldiers was specially fabricated for such accuracy.

4.2.5 *Pumping concrete in shaft above ground level* – The concrete pump was placed at ground level and staging pipes were fixed for reinforcement fixing and positioning the concrete pipeline as per requirement of pouring. 'S' bend was provided at ground level to minimise the damage. Needle vibrators were used for consolidation and smooth surface. As breakdowns were less cycle time also reduced to as low as twelve hours and concreting completed successfully (Fig. 8).

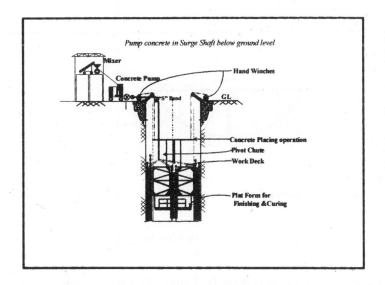

Fig. 7 : Doyang diversion tunnel

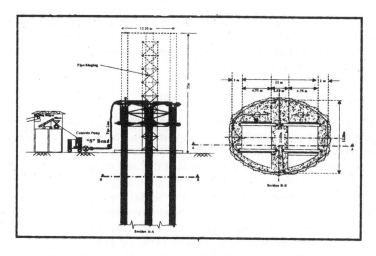

Fig. 8 : Pump concrete in surge shaft above ground level

4.3 Tunnel Invert Lining

After completing the overt lining dismantling of drill jumbo and shuttering gantry was done. Breaking of kerb started throughout the tunnel length. After completing the kerb breaking, cleaning of invert was done and splicing of reinforcement bars advanced @ 30 m/day/face. Specially designed invert template used for invert concreting and (Fig. 9) poring was done by directly pumping the concrete from face. 30 m to 50 m invert lining progress achieved per cycle. Finishing was done immediately after removing the invert template. Total invert concreting of tunnel completed successfully in one and a half months' time.

5.0 VENTILATION

Ventilation system was adopted by using semi-rigid ventilation ducts of 1 m in diameter and 40 HP reversible fans at an interval of 100 m for fresh air inside tunnel and taking out blasting and other toxic fumes from tunnel. System worked efficiently and found to be excellent.

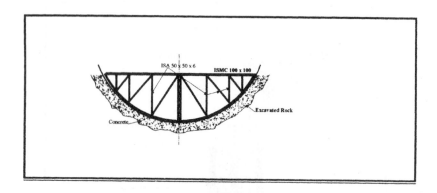

Fig. 9 : Invert concrete template

6.0 GROUTING

Grouting operation started after completing the lining of the tunnel. Drilling for grouting was done first and injection of grout took place later. Colomono grout pumps were used for grouting and pressure was kept at 2 kg/cm².

Grouting at crown of tunnel, at 12 m height was again a problem. For this activity two nos of 4 m wide light gantries were fabricated which were mobile on wheels. One gantry advanced with drilling activity and second gantry was used for grouting. Both the gantries had provisions of platforms at different elevations to tackle drill and grout holes at different levels.

7.0 TRAVELLING GANTRY MOUNTED ON DUMPER

After completing the overt lining of 12 m dia diversion tunnel and grouting, plugging of grout holes and finishing was required to be done. Normally, a travelling gantry is fabricated for such work. We find that steel gantry fabrication is very costly and requires about 15 days time for erection and also a special arrangement for pulling it at each location. To avoid this, it was decided that one dumper will be utilised by mounting a scaffolding arrangement on the dumper body. Dumper can move to the location along with the staging which was designed to reach upto the crown of the tunnel (12 m height). Having made such arrangement, complete 600 m tunnel finishing was done by moveable staging thereby avoiding fabrication of gantry and arrangement of pulling gantry to the location. This staging arrangement was found to be very mobile and economical. Proper ladders were provided for the workers to climb up to the top of the staging and to reach crown of the tunnel. Proper platforms were made to reach periphery of the tunnel to enable masons to reach to the concrete surface.

All the innovations described above were conceived, designed and developed at field level successful implementation of which resulted in time saving of several months in addition to cost saving of several lakhs of rupees.

8.0 CONCLUSION

Problems in tunnelling can occur any time and they cannot be predicted always. Construction of large diameter tunnel requires adequate planning and innovative spot solutions to prevent large scale damage to the tunnel and protect life and property. Innovative solutions described have been unique and may be adopted where similar situation arise.

REMINING OF POWER TUNNEL FACE-3 IN CHAMERA PROJECT HIMACHAL PRADESH – A CASE STUDY

B.C.K. MISHRA
Senior Manager (Civil)

A.K. MISHRA
Chief Engineer (Civil)

**National Hydroelectric Power Corporation Ltd.,
Kurichu Project, Bhutan**

SYNOPSIS

9.50 m dia, 6.48 km long power tunnel is situated mainly in phyllites and lime stones of Dhundhiara formation of lesser Himalayan range. The excavation of the tunnel has been carried out from seven faces and in two stages viz. heading and benching.

Though Chamera power tunnel encountered innumerable geological problems, the major mind boggling one was encountered on Face-3 which starts down stream of junction of access adit-I with main tunnel. The paper deals with analysis of the problem its solutions and further remedial measures.

1.0 INTRODUCTION

Chamera, a 540 MW hydropower project, a pioneer for accelerated development of hydro Power in northern region, consists of a 125 m high concrete arch gravity dam across Ravi, 6.4 km long 9.5 m dia horse shoe shape power tunnel on right bank of Ravi which will carry water to an underground power house through a 157 m deep 8.5 m dia vertical pressure shaft and 2.41 km long 9.5 m dia tail race tunnel to discharge water back to Ravi on its left bank. A layout of the tunnel showing various access adits and faces is shown in Fig. 1.

With the total installed capacity of 540 MW, the project has 3 units of 180 MW each for generating 1664 MU to provide peaking power to northern grid through a double circuit 400 kV transmission line.

The project, located in district chamba of HP has been constructed by NHPC under Canadian collaboration and has started smooth generation of power in March 1994.

2.0 GEOLOGY OF POWER TUNNEL FACE-3

The rock unit encountered on power tunnel face-3 down stream of adit-I junction in its initial 120 m reach was quartzitic phyllites with limestone bands. The general strike of rock unit varies from N 15 deg.– 60 deg, W to S 15 deg – 60 deg E with moderate to steep dip towards north. Beyond RD 120 and upto 600m the rock unit encountered is carbonaceous phyllite with thin bands of lime stone intersected by thinly foliated graphitic schist of variable thickness. The carbonaceous phyllite is closely foliated and runs sub-parallel to the tunnel axis. Presence of sheared zones was a characteristic feature of the rockmass. A few stretches in 450 m reach of the tunnel were observed to be moist with dripping stretches in between. Strength wise rock could be classified as fair to poor rock. A typical section of the tunnel showing the orientation of the foliations is shown in Fig. 2 and geological section of the power tunnel is shown in Fig. 3.

3.0 HISTORY OF EXCAVATION AND ROCK SUPPROT

Power tunnel access adit-I, 148 m long, was started in the month of September 1986 and completed in the month of January 1987. Junction of adit-I with power tunnel could be supported with extensive rockbolting

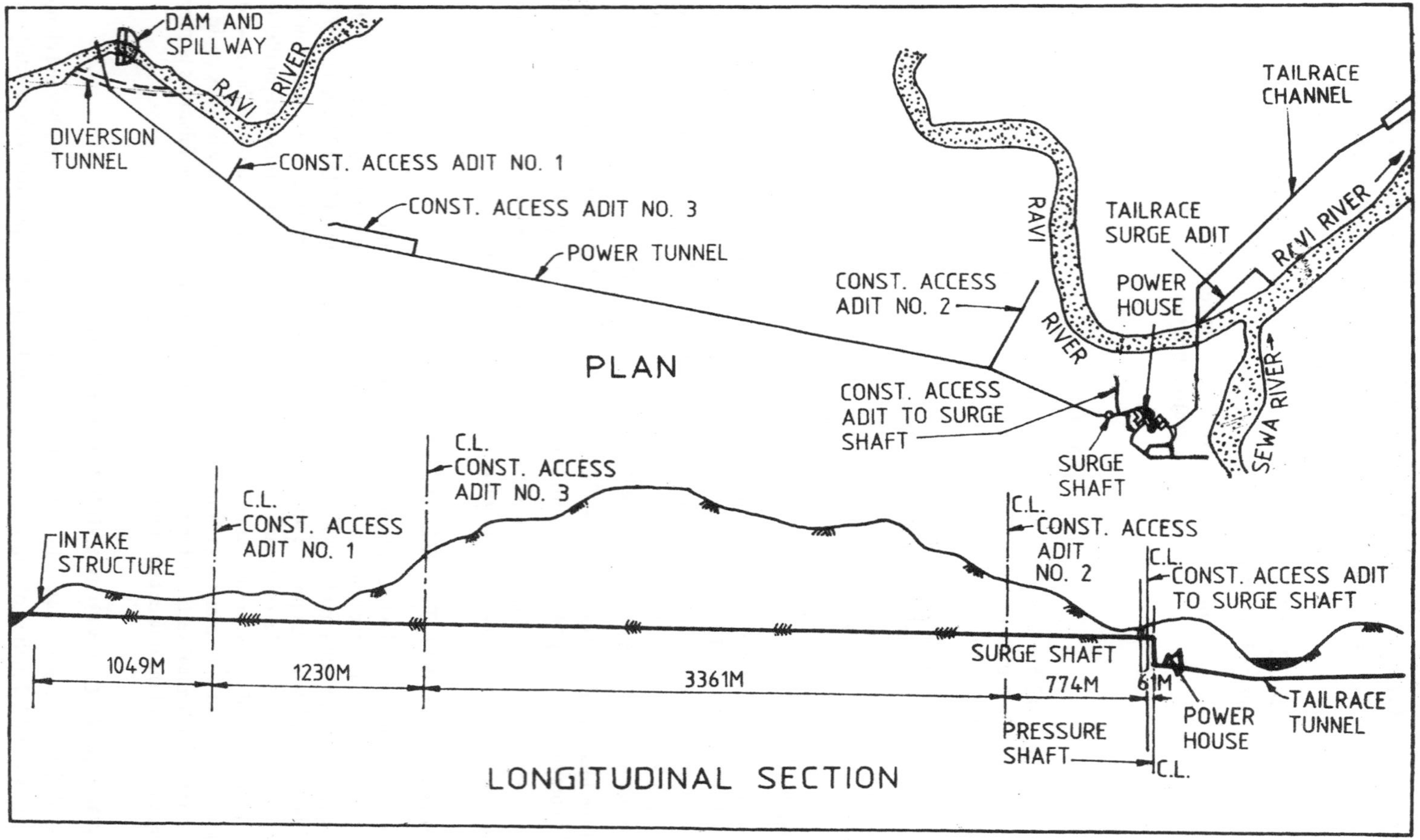

Fig. 1 : Chamera hydroelectric project layout

and shotcreting and extra time required to support the junction with conventional steel rib supports could be averted with great difficulty. However, after establishing the junction, there was hardly any patch could be supported with shotcrete and rockbolts. Steel rib supports, made of RS Joist 200x200 or 200x165, which was circular upto 22.5 deg from horizontal and further extended upto heading excavation level by a 3.1 m long tangent piece, had to be provided. Heading excavation in 600 m reach of the tunnel was so difficult that it became a major cause of concern for the management of NHPC and its collaborators. Monthwise progress of face-3 during heading excavation is shown in Annexure 1.

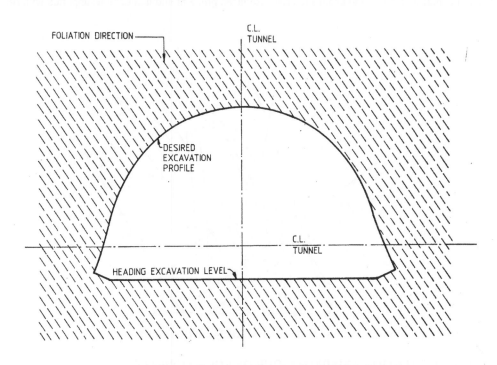

Fig. 2 : Power Tunnel-3: Cross section showing foliation direction

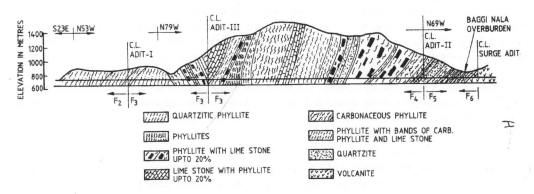

Fig. 3 : Geological section along power tunnel

Excavation in 450m reach of the tunnel face-3 was carried out in full face, 4 stages and 6 stages as per rock conditions. Though the rock as per CSIR classification was class – IV/V and support requirement consisted of systematic strengthening of roof arch and wall with rock bolts, wiremesh and shotcrete (100 to

425

200 mm) alongwith steel ribs, the support provided actually consisted of steel ribs at spacing of 0.5 to 1.0 m backfilled with concrete, boulders and precast lagging. Rockbolting and shotcreting could not be carried out in most of the reaches as relaxation of stresses in the rock along foliations sub-parallel to axis of tunnel and intermittent shear zones alongwith moderate folding of carbonaceous phyllitess often resulted in wedge formation and frequent collapses of the rockmass. A typical geological log at RD 248 is shown in Fig. 4. No stretch in the entire 450 m length could be supported with rockbolt, wiremesh and shotcrete except for 21 m from RD 425 to 464 where tunnel was negotiating a curve of 100 m radius. Later on this reach supported with shotcrete/rockbolt, wiremesh and again shotcrete also faced problem and had to be resupported with ribs.

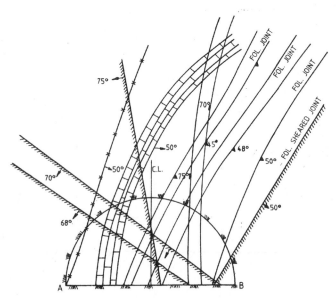

Fig. 4 : Face log at RD 248 m (Face – 3)

4.0 INSTRUMENTATION AND BEHAVIOUR OF ROCK SUPPORT

For studying behaviour of the rock support, J-Hooks were installed for measurement of convergence by tape extensometers. Further, load cells were also installed on the side walls to observe the relaxation of rock stresses. Over a period of time it was observed that heading section is undergoing deformations. The ribs were getting deformed badly and convergency of the steel ribs was to the extent of 0.5 to 1.0 m on sides as well as crown. The deformation in ribs was found to be more prominent at tangent points of the ribs. A typical section showing deformation of rib is shown in Fig. 5. The load cells installed on the side wall got overloaded by rock pressure and at times exhausted its capacity. The face plates of the rockbolt installed were taking the shape of a cusp and very frequently the plates alongwith beveled washers and nuts came out due to shearing of the thread on rockbolts because of excessive loading. Deformation in ribs with lapse of time at various locations in tunnel is shown in Fig. 6.

After studying behaviour of the rockmass and slow progress of excavation, decision was taken in May 1988 to construct a new adit with a cost of 1.5 crores to keep the project abreast with schedule. The adit was so located as to by-pass the geologically poor reach and give two additional faces to make up for the time lost in excavation of face - 3.

Following remedial measures were considered to tackle the distressed reach :-

1. As the deformation in ribs was prominent at tangent point of the rib, the design of rib was changed to semicircular as shown in Fig. 12.

426

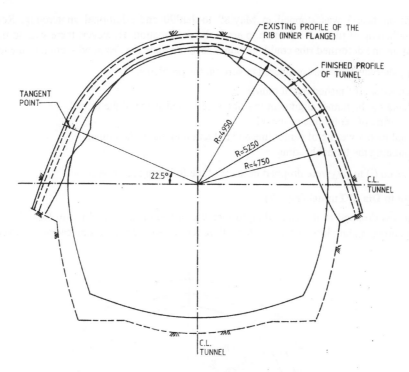

Fig. 5 : Typical deformed shape of RIB

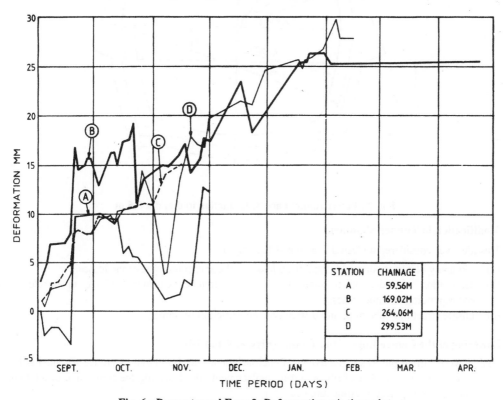

STATION	CHAINAGE
A	59.56M
B	169.02M
C	264.06M
D	299.53M

Fig. 6 : Power tunnel Face-3: Deformation v/s time plot

2. Work on face-3 was stopped in May'89 to Jan'90 and additional shotcreting, Rockbolting and strengthening of ribs was taken up to arrest the deformation. However, there was no improvement in situation and deformed ribs could no longer accommodate the designed section of the tunnel.

Following alternatives were considered to sort out the problems.

(i)   Reduction of finished dia of tunnel.
(ii)   Changing the tunnel alignment to suit the deformed shape of the tunnel.
(iii)   Lowering of center line of tunnel.
(iv)   Combined lowering of center line and reduction in dia of the tunnel.
(v)   Remining the tunnel heading.

Alternative no (i) to (iv) were dropped because of the following reasons :-

4.1 Reduction in Dia of Tunnel (Fig. 7)

This idea was dropped as it was leading to increase in head losses resulting in loss of revenue. Loss of head and corresponding loss of revenue for different reduced diameters is placed at Annexure – 2.

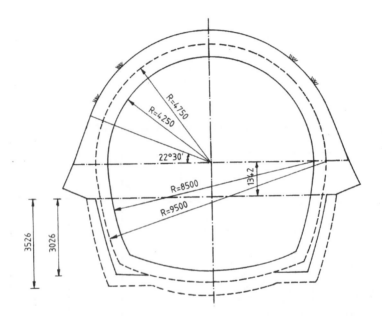

Fig. 7 : Power tunnel Face-3: Reduction in dia, dia 8.5 m

4.2 Modification in Tunnel Alignment

This idea was considered undesirable as it was leading to :-

(a)   too many bends causing additional head losses and corresponding revenue losses.
(b)   association of reduction in diameter in some stretches causing increase in headlosses and corresponding revenue losses.
(c)   unsafe structural behaviour and difficulty in construction of tunnel.

4.3 Lowering of the Centre Line of the Tunnel (Figs. 8, 9 and 10)

Studies were carried out to see the extent of encroachment into the tunnel clearance if tunnel is lowered by 1 m, 1.5 m and 1.7 m. Number of ribs requiring replacement after lowering the center line of the tunnel by 1 m, 1.5 m, and 1.7 m is placed at Annexure - 3.

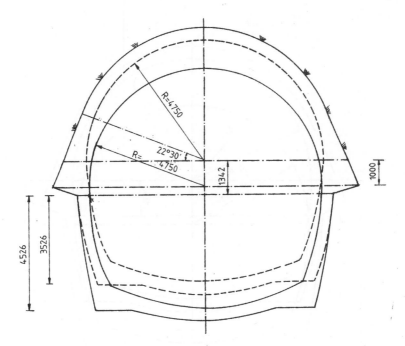

Fig. 8 : Power tunnel, Face-3 C lowering by1.0 m

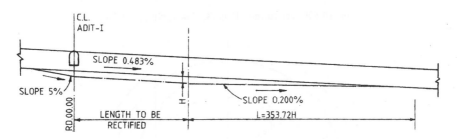

Fig. 9 : Power tunnel, Face-3, invert lowering

Apart from the rib rectification required as per Annexure – '3' lowering of center line of tunnel was leading to :-

(a) increase in depth of benching causing difficulty in support of the side walls and risk of collapse during benching excavation.

(b) requirement of suitable adjustment in the gradient of the tunnel up stream and down-stream of the reach in which center line was to be lowered which involved additional excavation leading to extra expenditure.

(c) requirement of additional concrete upstream and down stream of the troubled reach as it was already excavated as per designed profile and it involved additional expenditure.

4.4 Combined Lowering and Reduction in Diameter of Tunnel (Fig. 11)

This alternative was leading to all the problems mentioned for alternatives (3) and (4) and ultimately causing loss of revenue and hence it was considered undesirable.

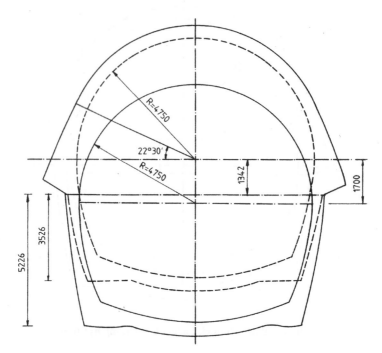

Fig. 10 : Power tunnel, Face-3, C lowering by 1.7 m

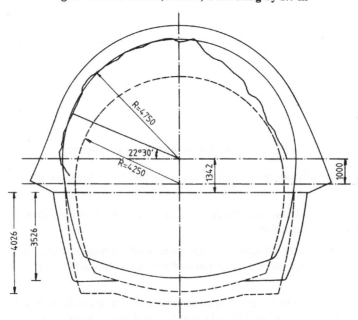

Fig. 11 : Power tunnel, face-3, combined lowering C and reduction in dia (by 1.0 m)

4.5 Remining the Tunnel Heading

Requirement of this alternative was :-

(i)   to remove old and distorted ribs alongwith lagging, backfill concrete, loose rock mass.

(ii)   to re-excavate the tunnel heading to required dimensions.

(iii)   to resupport the finally excavated section through rockbolts, wiremesh, shotcrete, steel ribs and backfill concrete.

Though this proposal was considered very tedious, time consuming and costly, it was approved for implementation as it was leading to :-

(a)   no loss of revenue as there was no head loss.

(b)   safer benching operation which can be carried out after observing rock mass movement in heading.

(c)   easier concreting operation in the tunnel.

(d)   safe structural behaviour of the tunnel.

4.6 Remining

4.6.1 *Proposed Methodology for Rock Support :*

After the decision of remining in the disturbed reaches of power tunnel face-3 was conveyed, designers suggested following sequence of rock support (Fig. 12) :

(i)   Drilling of prob holes 20 m ahead of the portion where remining is to be carried out for advance assessment of type of rock mass, loosefill if any, presence of cavity etc.

(ii)   Grouting through the prob holes to consolidate the loose material and fill the cavity if any, at least for a distance of 10m ahead of the area to be remined.

(iii)   Installation of vertical props u/s and d/s of the rib being remined at least for a distance of 10 m ahead of area to be remined.

(iv)   Removal of precast laggings, ribs, loose fill etc., in the portion marked 'A' keeping the props intact. Preparation of space for required rock profile through chipping with the help of pavement breaker, removal of loose fills etc.

(v)   Apply 50 mm thick shotcrete and install wiremesh. Instal 6 m long fully grouted rockbolts. Apply final layer of shotcrete to make it 100 mm thick. Instal rib in central portion.

(vi)   Remove vertical prop in LHS repeat all the steps from (i) to (v). Install rib segment in LHS.

(vii)   Remove vertical prop in LHS repeat all the steps from (i) to (v). Install rib segments in RHS rib segments on the both sides be secured properly with crown rib.

(viii)   After providing final layer of shotcrete, provide tape extensometer every 10 m / 25 m, take daily convergence measurements. In case measurement show convergency install more no. of rockbolts / longer bolts and provide thicker shotcretre.

4.6.2 *Difficulties in Proposed Methodology and Methodology Adopted at Site*

(i)   As per support sequence proposed by designers, prob holes were drilled to assess the type of rockmass, extent of loose fill, presence of cavities etc. However grouting could not be carried out successfully due to excessive leakage from the gap in between precast blocks used as shuttering for backfill concrete behind ribs. Sealing of the gaps etc. with the help of shotcrete was tried to stop the leakage of grout but to no avail.

(ii)   Because of the excessive overbreaks and irregular rock profile in the tunnel heading excavated earlier wiremesh fixing and shotcreting could not be carried out everywhere. This activity was performed only at places where rock profile was matching with the rib profile or wherever overbreaks were limited. In most of the reaches overbreaks were excessive and therefore backfill concrete with the help of concrete placer was carried out by using steel shutter plates fabricated as per rib profile. Use of precast concrete blocks was avoided so that there is no leakage of grout at later stage. Further, vulnerable locations were provided with wiremesh to reinforce backfill concrete.

431

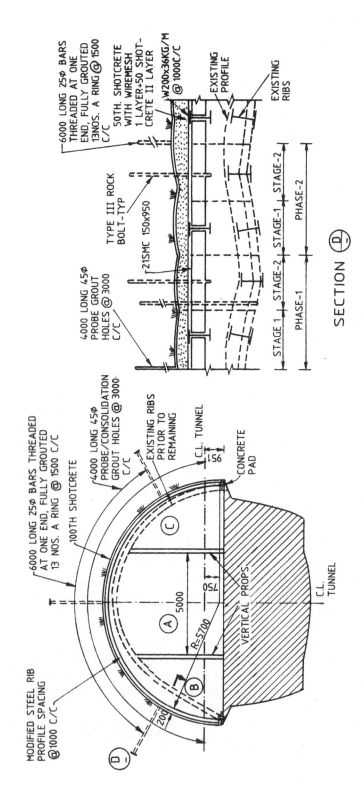

Fig. 12 : Power tunnel, face-3, remaining sequence

(iii)   Installation of props upstream and downstream of the reach taken up for remining was creating a hindrance in movement of machineries. Further installation of part rib, its proper positioning in line and level and its jointing with other rib segments was a difficult job. Also it was experienced that removal of one half of the rib was not giving any trouble and as such no props were provided upstream and downstream of reach under remining. Removal of half portion of the rib alongwith its replacement was taken up thereby simplifying the operation. Chipping of concrete / rock / shotcrete with the help of pavement breaker was too difficult and time consuming and therefore depending on requirement two- three holes were drilled by jack hammer and blasting was carried out using a very low charge just to produce cracks and make the chipping easier.

(iv)   6 m long, 25 mm dia fully grouted TS rock bolts were installed 1.5 m c/c. At times holes were getting clogged due to loose rockmass. In that case holes were grouted and redrilling was carried out to install the bolts.

(v)   5 m long holes were drilled and grouting was carried out to consolidate the loose rockmass and fill the voids, cavities if any.

(vi)   Measurement for convergence was being taken regularly. Though there was movement in the rockmass, it was observed that it subsided with lapse of time. However, some settlement in the crown was observed. On analysis of the problem, it was found that this is mainly due to improper backfilling which could not be carried out in crown due to foliations sub-parallel to tunnel axis. To avoid any encroachment in crown at later stages, a provision for settlement was kept by raising level of ribs by 15 to 20 cms higher than the required level as per design.

(vii)   In the crown portion, where backfilling was not possible and settlement was observed, wiremesh was fixed and proper shotcreting was done to provide proper packing behind steel ribs erected.

(viii)   Wherever there was dripping of water, relief holes were drilled to release water pressure on the rockmass.

4.6.3 *Following Equipments were Utilised on the Work of Remining*

(i)   35 cum cap. front end caterpiller / Michigan loaders

(ii)   Tamrock rockbolt jumbo

(iii)   500 cfm electric compressor 3 nos.

(iv)   80 cum/hr cap Johnson make concrete batching and missing plant.

(v)   Robins / Williams grout pumps for grouting of rock bolts.

(vi)   Foratek grout pumps for consolidation grouting.

(vii)   Aliva-280 shotcrete machine with teledyne robot arm.

(viii)   Scoop tram 3T capacity.

(ix)   Pneumatic concrete placer and concrete pump.

(x)   Jack hammer and pavement breaker.

5.0 CONCLUSIONS

Timely decision of NHPC management to construct additional adit-III removed tunnel construction from critical path and remining could be carried out properly. The above methodology proved to be successful and did not pose any problem. The work of remining in 400 m reach was completed with an average progress of 22 m / month and maximum progress of 30 m / month. Benching excavation was carried out in 4 stages and as a parallel activity to remining. No major cavity was encountered during remining operation or in the course of benching excavation in remined reach.

MONTHLY PROGRESS OF POWER TUNNEL
UPSTREAM PACKAGE FACE –3

Month	Monthly progress (in m)	Cumulative progress (m)	Remarks
February 1987	17		
March 1987	13	30	Full heading excavation
April 1987	9	39	Major cavity at RD +/- 30m
May 1987	24	63	Full face heading excavation
June 1987	23	86	Full face heading excavation
July 1987	24	110	Full face heading excavation
August 1987	30	140	Full face heading excavation
September 1987		175	Upto RD +/- 172 m full face heading
October 1987	Nil	175	Beyond RD +/- 172 m 6 stage excavation. Major cavity at RD +/- 170 m.
November 1987	6	181	
December 1987	14	195	Upto RD +/- 187 m 6 stage excavation. Beyond RD +/- 187 m 4 stage excavation.
January 1988	18	213	4 stage excavation
February 1988	32.5	245.5	4 stage excavation
March 1988	37	282.5	4 stage excavation
April 1988	25.5	308	4 stage excavation
May 1988	3	311	4 stage excavation
June 1988	13	324	4 stage excavation
July 1988	Nil	324	4 stage excavation
August 1988	Nil	324	4 stage excavation
September 1988	7	331	Upto RD +/- 324m four stage excavation and from RD +/-324 to 327 m full heading excavation. Beyond RD +/- 327 m four stage excavation.
October 1988	14	235	4 stage excavation
November 1988	8	353	4 stage excavation
December 1988	18.5	371.5	4 stage excavation.
January 1989	20	391.5	4 stage excavation
February 1989	18.5	410	4 stage excavation

(Contd...)

March 1989	16	426	up to RD +/- 422m 4-stage excavation. From RD+/- 422 m to 434 m full heading excavation.
April 1989	16	442	From RD +/- 434m to 437m 4 stage excavation. Beyond RD +/- 437m full face heading excavation.
May 1989 to January 1990	Nil	442	Full face heading excavation
February 1990	1	43	Full face heading excavation
March 1990	29	472	Full face heading excavation
April 1990	6	478	Full face heading excavation
May 1990	9	487	Upto RD +/- 481m full face heading excavation.
June 1990	15	502	4 stage excavation.
July 1990	13	515	Only two stage completed during the period.
August 1990	20	535	
September 1990	9	544	

Notes :-

1. With effect from September 90 to December 90 balance excavation in two stages was completed.

2. Excavation from RD 544 to 600 (from face-3 side) was carried out from face-3 A, out of which 26 m of excavation was carried out in 4-stages 6 m in six stages and balance 14 m was done full face.

3. A number of loose falls/ smaller cavities were encountered in the course of driving the tunnel heading which have not been mentioned.

4. Remining was not required beyond RD 400 as provision was kept for arresting/ containing rock movement for changing the shape of the rib/ increasing the size of the rib and keeping the rib 15/20 cm. Above the required level.

POWER TUNNEL FACE –3
EFFECT OF REDUCTION IN TUNNEL DIAMETER

Sl. No.	Dia. of Tunnel m	No. of Ribs to be re-moved/rec-tified	Total head loss m	Additional loss of energy (GWH)	Annual loss in revenue (Rs. in lacs)
1.	9.05	533	0.53		
2.	9.0	512	0.77	2.199	21.99
3.	8.5	452	1.12	5.271	52.71
4.	8.0	276	1.61	9.635	96.35
5.	7.5	97	2.33	15.984	159.84

Notes :-
1. Length of tunnel considered is 400 m
2. Energy loss considered assuming annual generation of 1647.36 GWH
3. Cost of energy at bus bar assumed at Re. 1/- per unit.

RIB RECTIFICATION WITH LOWERING OF INVERT AND REDUCTION IN DIAMETER

	Alternatives	No of Ribs to be rectified
1.	Reduction in diameter to 8.5 m	452
2.	Lowering of invert by	
	1.0m	452
	1.5m	183
	1.7m	70
3.	Combined lowering of invert by 1m and reduction of dia to 8.5 m	64
4.	Remining of tunnel	533

CONSTRUCTION METHODOLOGY FOR DESILTING CAVERNS OF 1020 MW TALA HYDROELECTRIC PROJECT, WESTERN BHUTAN

SANJAY K. TURKI
Manager (Contracts)

VINOD K. RAJORA
Works Manager

Hindustan Construction Co. Ltd.
Tala Hydroelectric Project
Bhutan

SYNOPSIS

The paper envisages the construction methodology to be adopted for excavation of three desilting caverns of size L=250 m, B=13.927 m and Headmistress=18.5 m of 1020 MW Tala Hydroelectric Project located in the Western Bhutan Himalayas. The first stage excavation (Headmistress=7 m, B=13.927 m) for all the three caverns is presently being carried out. The methodology adopted for first stage excavation, and the methodology to be adopted for subsequent stages is discussed here in light of the conditions encountered during the adit and the first stage excavation. It is assumed that by adopting the methodology discussed below, rockmass will be least disturbed and smooth profile of the caverns will be achieved.

1.0 INTRODUCTION

The Tala Hydroelectric Project (Fig. 1) is the largest run - of- the- river scheme of the Royal Kingdom of Bhutan. It is located in Chukha Dzongkhag in the Western Bhutan Himalayas. It envisages construction of 92 m high concrete diversion dam across the river Wangchu intercepting a drainage area of about 4,028 sq kms and carrying intake water through three egg shaped desilting chambers (Fig. 2), each of size L = 250m, B = 13.925m and Headmistress = 18.5 m, to remove silt particles larger than 0.20 mm size, 22.884 kms long 6.8 m diameter modified horse shoe shaped concrete lined Head Race Tunnel on the right bank of the river Wangchu with design discharge of 142.5 m³/sec. The Head Race Tunnel is provided with four intermediate D- shaped adits of 7 m x 7 m diameter for facilitating the construction of the tunnel. A 180 m high, 15m diameter under ground restricted orifice type surge shaft is provided at the down stream end of the Head Race Tunnel to alleviate any overpressure from 863 m of water head. It will be connected by two inclined pressure shafts, each 992 m long and 4 m diameter trifurcating near the power house (L = 190 m, B = 18 m and H = 43 m) with an installed capacity of 1020 MW (6 x 170 MW). The water will be discharged back into the river Wangchu through a tail race tunnel 2.2 km long, 7.5 m diameter.

2.0 GEOLOGICAL SET-UP

The Tala Hydroelectric Project area falls within the central crystalline belt of Thimphu Formation and meta-sediments of Paro Formation. The Thimphu Formation comprises a variety of granitoid rocks, such as migmatite, augen-gneiss, banded-gneiss, granite-gneiss, schistose rocks with subordinate quartzite and marble bands. The Paro Formation consists of high grade calcareous rock and meta-sedimentaries such as marble, calc-silicate rock, quartzite, quartz-garnet-staurolite-kyanite-silimanite schist, graphite schist etc. with subordinate felspathic schist and gneiss bands.

The desilting complex mostly lies in biotite-gneiss with intermediate schist bands. The rock is thinly to moderately foliated with several sets of intersecting joints. The orientation of the foliation joints is almost parallel to the cavern alignment.

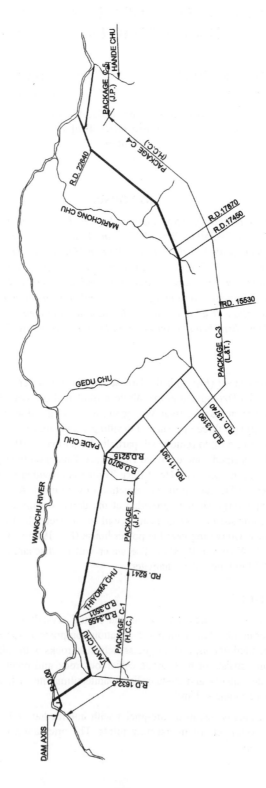

Fig. 1 : Key plan of the project area

438

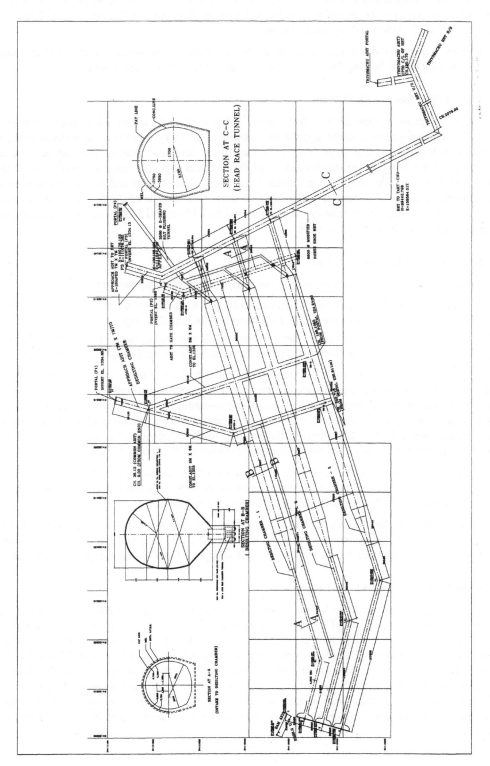

Fig. 2 : Desilting complex arrangement

3.0 EXCAVATION METHODOLOGY FOR DESILTING CAVERNS

A common construction adit (Inv. El. = 1334.85 m) of size 7 m x 7 m and length 36.130 m is provided which bifurcates into two construction adits (named here-in-after Top and Bottom) of size 6 m x 6 m, one reaching Inv. El. of 1338 m and the other 1328 m.

In order to avoid the disturbance of rockmass during cavern excavation and to have the better control for installing the supports, following five stages (Fig. 3) of excavation has been proposed:

3.1 Stage-I (El. 1345 m - El. 1338 m)

The first stage excavation (H=7 m, B=13.90 m …..Fig. 2) of all the three caverns from crown El. 1345 m to El. 1338 m is being carried out through the top adit with invert elevation of this adit perfectly matching with bottom El.1338 m of the first stage. Two faces each were developed one on upstream and the other on downstream of the Top Adit for excavation of all the three caverns. The upstream section of all the three caverns will reach up to portal of 4.9 m diameter, horse shoe inlet tunnels and the downstream section will reach through transition to intakes of 4.9 m diameter horse shoe tunnels of the head race tunnel. Except for Cavern No. III, first stage excavation has almost reached final phase and installation of additional supports is in progress at the time of writing this paper.

3.2 Stage-II (El. 1338 m - El. 1135 m)

The second stage excavation is proposed to be carried out in following two steps :

   (a) Formation of 8 m wide Gullet by vertical drilling in the centre of the cavern from El. 1338 m to El. 1135 m, leaving 3 m on either side for subsequent slashing (2 in Fig. 3).
   (b) Side slashing of the remaining mass by horizontal drilling and subsequent installation of rock bolts in required direction throughout the length of the caverns restricting the rockmass to yield. (2a in Fig. 3).

The excavation of the Gullet will be carried out in a length of 10-15 m per cycle by vertical drilling from the downstream end of the caverns and progressing towards the alignment of top adit. A ramp of slope 1:12 for movement of equipment will be maintained by adjusting the drilling depth, for approaching the top adit. The same procedure for excavation of second stage has to be adopted in the upstream in all the three caverns except at the locations where bottom adit intersects the caverns. At these locations, the width of the Gullet will be reduced to 4 m and will be excavated only on one side for movement of machinery.

The second step will involve the slashing of the left over rockmass on either side of the Gullet by making the vertical face and horizontal drilling with proper trimmers and controlled blasting in order to have better control over the profile and to minimize the disturbance to the rockmass.

After carrying out the second stage excavation from the top adit, only one-meter slab will be left in between the bottom of cavern attained during second stage (El. 1335 m) and the crown of the bottom adit (El. 1334 m). There are chances that the left over slab will collapse by itself during the second stage excavation.

3.3 Stage-III (El. 1335 m- El. 1331.829 m)

The third stage of excavation will be carried out from the bottom adit (Inv. El. 1328 m). Ramp of gradient 1:12 for free movement of tunnelling equipment will be constructed from invert of bottom adit to El. 1331.829 m. It will cover a total length of 45 m on either side of the bottom adit one towards upstream direction and other towards downstream with a width varying from about 5.25 m at El. 1328 m to about 12.5 m at El. 1331.829 m, commensurate with the final profile of the desilting caverns. The third stage of excavation El. 1331.829 m to El. 1335 m; H= 3.171 m, will be carried out by making vertical faces and horizontal drilling with drilling depth of three meters. By this method, maximum control on the shape of the profile will be there, as proper trimmers will be drilled in either wall. The designed supports will be installed in each cycle.

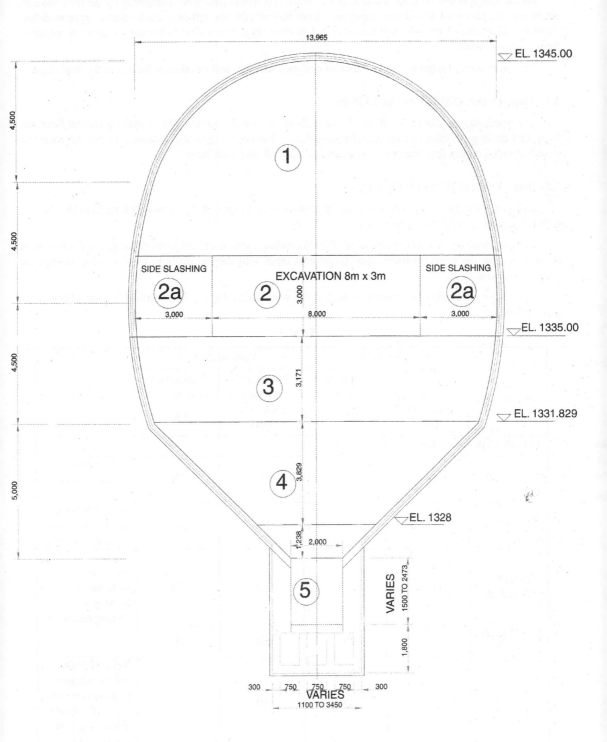

Fig. 3 : Complex section showing stages of desilting cavern excavation

As the first and second stage excavation of cavern –III would have been completed by the time stage-III excavation for caverns-I and II will commence from bottom adit; the left over excavation of caverns falling along the alignment of top adit will be excavated in third stage excavation following the concrete plugging from cavern-III itself.

The excavation of ramp with installation of designed supports will be taken at the end of the third stage.

3.4   Stage-IV (El. 1331.829 m - El. 1328 m)

The fourth stage excavation will also be carried out through the bottom adit by making vertical faces and horizontal drilling in order to minimize the overbreaks. The rock supporting measures as per the rockclass already decided during first stage excavation will be adopted after each blast.

3.5   Stage-V (El. 1328 m - El. 1326.5 m)

Excavation for stage-V (lower trapezoidal portion) will be carried out by vertical drilling from El. 1328 m. with subsequent installation of supports.

The methodology of caverns discussed (Fig. 5) (vertical faces with horizontal drilling) has the following advantages over the conventional method (horizontal faces with vertical drilling) usually being adopted for such excavations.

Stage-wise excavation of caverns in each cycle with horizontal *vis-à-vis* vertical drilling (Table 1) is given below :

TABLE 1

Stages/ elevation (m)		Horizontal drilling		Vertical drilling		Remarks
		Volume to be excavated in each cycle (m³)	Explosive to be consumed (kg)	Volume to be excavated in each cycle (m³)	Explosive to be consumed (kg)	
I / 1345-1338		176	280	-	-	
II/ 1338-1335	II	-	-	360	220	Only central portion will be blasted.
	IIa	31.5	20	135	90	Side slashing in order to maintain the smooth profile.
III/1335-1331.829		145	90	620	375	To avoid rockmass disturbance, horizontal drilling preferred
IV/1331.829-1328		135	80	575	350	
V/1328-1326.50		10	8	90	60	Vertical drilling will be adopted, as there is restriction of the available area and to save the time.

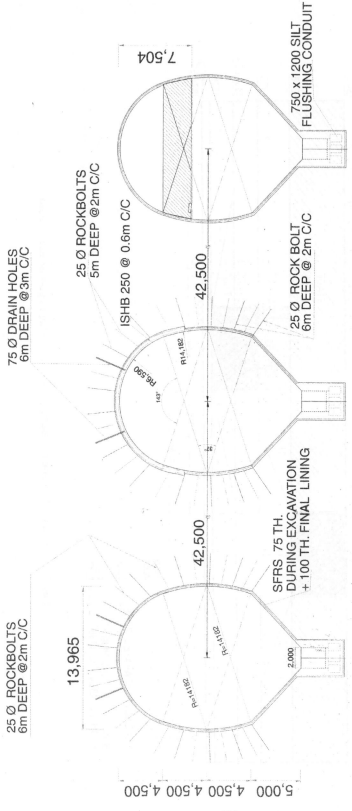

FAIR, GOOD & VERY GOOD ROCK   POOR ROCK   HEADING EXCAVATION

Fig. 4 : Supporting system of desilting caverns

443

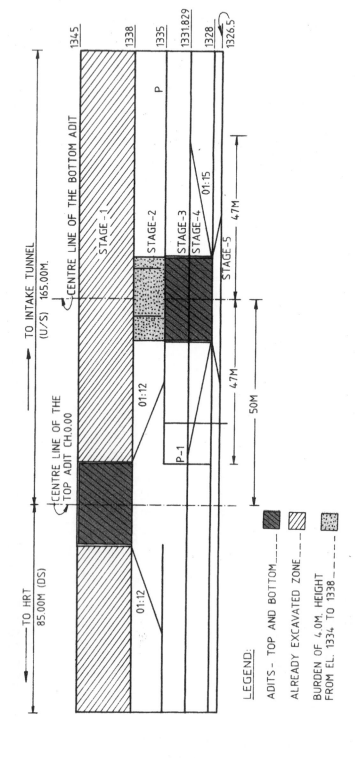

Fig. 5 : Longitudinal section of desilting cavern (showing stage-wise excavation)

LEGEND:

ADITS - TOP AND BOTTOM _____

ALREADY EXCAVATED ZONE _____

BURDEN OF 4.0M. HEIGHT
FROM EL. 1334 TO 1338 _____

4.0 SUPPORT SYSTEM FOR CAVERNS (Fig. 4)

The following supportive measures (Table 2) are being adopted on the basis of rockclass computed during each blast of first stage excavation :

TABLE 2

Rockclass	Support system					
	Rockbolts		Shotcrete (SFRS)		Steel structural support	
	Up to haunch level	Below haunch level	During excavation (mm)	Final lining (mm)	Up to haunch level	Below haunch level
I	25 mm dia grouted, 5000 mm long @ 2000 C/C bothways (staggered)	25 mm dia grouted, 6000 mm long @ 2000 C/C bothways (staggered)	75	100	-	-
II	25 mm dia grouted, 5000 mm long @ 2000 C/C bothways (staggered)	25 mm dia grouted, 6000 mm long @ 2000 C/C bothways (staggered)	75	100	-	-
III	25 mm dia grouted, 5000 mm long @ 2000 C/C bothways (staggered)	25 mm dia grouted, 6000 mm long @ 2000 C/C bothways (staggered)	75	100	-	-
IV	25 mm dia grouted, 5000 mm long @ 2000 C/C bothways (staggered)	25 mm dia grouted, 6000 mm long @ 2000 C/C bothways (staggered)	50 up to haunch level and 75 below haunch level.	100	ISHB 200 @ 600 C/C	-

5.0 CONCLUSIONS

The methodology proposed above has been formulated keeping in view conditions experienced during adit excavation and stage-I excavation of the caverns. It is assumed that by adopting the above methodology, maximum optimization will be achieved as :

(a) Desilting caverns are egg-shaped. Hence vertical drilling in the sides is not possible. But in case of horizontal drilling, it is convenient to drill along the smooth curve of cavern and by reducing the

445

spacing of periphery holes to 300 mm C/C and keeping the alternate hole dummy (i.e., uncharged) we can have better control over profile.

(b) Comparison of quantity of explosive consumption in one cycle, for vertical drilling vs horizontal drilling is shown in Table-1. It is observed that in each cycle, explosive consumption in horizontal drilling is less than vertical drilling. As a result, the rockmass disturbance will be less in case of horizontal drilling.

(c) Required production rate will be achieved for timely completion of excavation work.

(d) There will be better control in maintaining the smooth profile, as it will be always tried to drill the horizontal holes for advancing and there will be better manoeuvreability of machinery for installing the design supports.

(e) Clean house keeping for safety of manpower and machinery will be ensured.

(f) Breakdowns of machinery and thus downtime can be minimized as the activities will be staggered at six fronts and each machinery will be deployed keeping in mind that it gets enough time for maintenance and no machinery at any time remains idle.

ACKNOWLEDGEMENT

Authors are greatly thankful to Mr. S.K. Fotedar, General Manager (O), Hindustan Construction Co. Ltd., Hincon House, LBS Marg, Vikhroli (West), Mumbai for his valuable guidance and inspiration for writing this paper.

CONSTRUCTION OF BOGADA TUNNEL OF
SOUTH CENTRAL RAILWAY (INDIA)

K.J. SINGH
Advisor (Constn.)

Indian Railway Construction Company, New Delhi, India

1.0 INTRODUCTION

The Ministry of Railways undertook the gigantic task for converting metre gauge network of track to broad gauge under Project Unit-Gauge. The South Central Railway (SCR) has taken up the gauge conversion of the section between Giddalur and Nandyal (69 km) which was originally programmed for opening to traffic by end of June, 1996. This target was preponed to February, 1996.

The amount of work in conversion of this section was stupendous and breath-taking and involved construction of two tunnels viz., Bogada 1560 m length, Chelama 280 m length. The job was awarded to Indian Railway Construction Company (IRCON) as this work was of national interest. This work was on 23-km new diversion, which was in thick forest and ghats of Nallamalai hill range.

IRCON have executed various works involved in this project. Construction of Bogada tunnel was perceived as most critical for the timely completion of this work. The actual work on Bogada tunnel was started   on 14th October 1994 after taking environmental clearance and after cutting of trees in approaches to the tunnels. All the works required for opening of the line to traffic were completed by 29th February 1996, and the section was opened to traffic on 9th March 1996 by the then Hon'ble Prime Minister of India. Some residual works of tunnels were continued after opening the section to traffic.

2.0 CONSTRUCTION METHODOLOGY

The major part of the work involved in the construction of B.G. single line Bogada tunnel were as under.

- Drilling trial bores and taking core samples.
- Open execution in cuttings in all types of soil/rock for forming the approaches to the tunnel portal.
- Excavation of the tunnel proper in all types of soil/rock including " trolley and man refuges, niches etc."

The support system of this tunnel consisted of the following :

- Providing temporary and permanent tunnel supports including forepolling wherever necessary which involves fixing of precast RCC laggings, rubble packing, lean concrete filling in the rear of the lagging etc. so as to facilitate the progress of tunnel excavation.
- Rock bolting and grouting in the tunnel strata.
- Lining concrete encasing the permanent steel supports wherever required.
- Shotcreting the tunnel surface as required where permanent steel supports are not provided. Shotcrete was provided with welded wiremesh in the middle where thickness exceeded 100 mm.

The Bogada tunnel was planned to be executed with the following support system -

Cross-sections of tunnel for lined section and reduced section for shotcreting are shown in Fig. 1a and 1b.

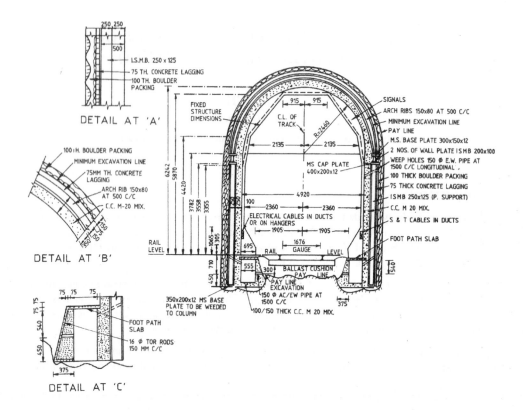

DETAIL AT 'A'

DETAIL AT 'B'

DETAIL AT 'C'

Fig. 1(a) : Cross-section of tunnel (with concrete lining)

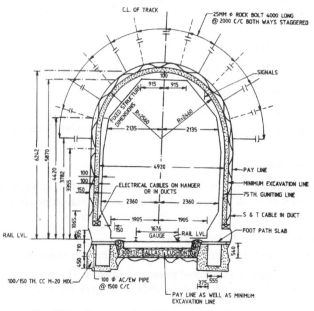

Fig. 1(b) : Cross-section of tunnel with shotcreting

Out of the total length of 1560 m 200 m on either end was to have permanent steel support encased with concrete lining. 1160 m was to have shotcrete of 150-mm thickness in two layers with welded wiremesh reinforcement in between.

And the strata/material through which the tunnel was to be excavated was expected to be hard shale at both the ends, *i.e.*, Portal sites and quartzite bands inter bedded with shales inside the tunnel.

As in any tunnel project, here also the earlier presumptions varied with actual site condition and required modification were done during the excavation of the project as per the requirement of the site conditions. Eventhough trial bores were being taken along the alignment during the formulation time of the project, the details from the bore holes were of a general and overall guide to understand the strata etc. The actual site conditions differed from the general pattern assumed based on the borehole particulars. In this tunnel, only in 2 reaches of 100 m each, the strata was different from the one that was expected. However, the variation between the two reaches was quite vast. In one reach the strata was interbedded more with shales resulting in less progress mainly due to drilled holes getting collapsed/filledup; widespread cracks in the strata leading ineffective blasting etc. Whereas in another reach, in the central portion, the strata met with was very hard quartz variety and consequently the drilling time was more and the wear and tear of the drilling rods and other accessories were also substantially more.

The details of the method of execution of this tunnel and the various problems encountered are enumerated hereunder.

3.0 METHOD OF EXECUTION

The shape of tunnel is 'D' shaped. As the length of tunnel is 1560 m the work of excavation was started from both ends.

Before starting the actual excavation works, the important part of the work was to do the survey and the alignment of tunnel was finalised with respect to other survey marks and establishing the required details, such as horizontal and vertical elevations, distance etc., for executing the work. The survey played a major role from the starting to the end when the excavated tunnel from both ends meet and made through at a designated point in the alignment of the tunnel. Constant monitoring and checking by all concerned resulted in achieving a perfect meeting at the designated point without any error and the tunnel was made through.

As the full face of the tunnel could not be formed from the approach initially, the excavation of tunnel was done by heading and benching method at first. The top half portion of the tunnel is called heading and the bottom half portion is known as benching. The full face of tunnel was established in a month or so.

The process of excavation consists of the following operations for each cycle :

- Surveying and profile marking   ...   1.5 hrs.
- Drilling operation   ...   4.0 hrs. to 6.0 hrs.
- Charging and blasting   ...   2.0 hrs.
- Defuming the gases etc.   ...   1.5 hrs. to 2.0 hrs.
- Scaling operation   ...   1.0 hr.
- Mucking operation   ...   7.0 hrs. to 8.0 hrs.
- Rock bolting/shotcreting   ...   3.5 hrs. to 5.0 hrs.

18.0 hrs. to 24.0 hrs.

As the above operations were repetitive in nature and one set of operations were known as one cycle of operations for the purpose of record and monitoring etc. These cycles of operation were numbered serially. The details of various operations are mentioned below.

3.1 Surveying and Profile Marking

During this operation, the survey team transfers, extend and mark the central line as well as central point of the tunnel on the excavated surface. Then the profile of the tunnel dimension was marked in the surface. These profiles facilitated and guided the drilling operations for proper shape of tunnel and avoided excess or less excavation. The distance/length excavated in the earlier cycle of operation is measured and this length is called as 'Pull of the Cycle'. With details from this survey operations, required corrective actions, if any, were taken then and there to control the working and achieve the desired result.

3.2 Drilling Operation

Depending upon the size of the tunnel, nature of the strata, availability of the resources etc., the drilling pattern of the holes in the rock surface is normally decided. In this project, the 'cut hole' method was used. The typical drilling pattern adopted is as shown in the Fig. 2 a and 2 b.

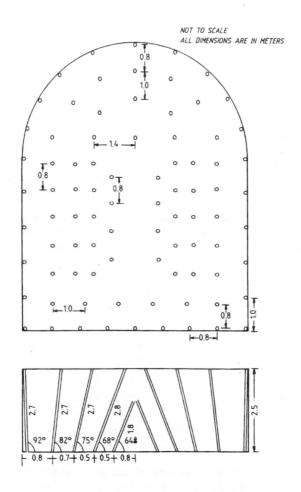

Fig. 2(a) : Drilling pattern for Bogada tunnel

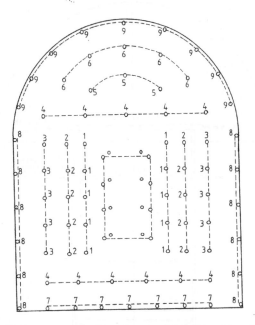

Fig. 2(b) : Delay pattern for Bogada tunnel (use long delay detonators)

The Central holes are drilled at an angle normally from both sides and the other holes are drilled as shown in the Figure. The depth of holes is decided based on the section of the tunnel. Here the drilling depth was about 3 m on the sides and the near by holes except central cut holes which vary depending upon the location. About 90 to 100 holes was drilling for this tunnel.

As the height of tunnel was about 6.25 m the drilling of the top portion of tunnel was done by using a heightened platform called a Drilling Jumbo platform (Fig. 3). This platform was erected on a tipper so that the same could be moved back and kept at a safe distance during the blasting operation and used again. The same platform is used for loose checking, scaling and also for survey, profile marking, rock bolting etc.

For this work only conventional drilling method was adopted. The drilling was done with Jackhammer attached with pusher legs and about 6 to 7 machines were deployed, 2 nos. at the top 2 nos. at the middle and balance in the bottom. If the rock strata is normal, about 3 to 4 hours were taken to complete the drilling operations.

However, when the strata was intermixed with shales or loose materials, it takes long time since the holes gets blocked, drill rods gets jammed, holes have to be shifted / changed etc. Similarly if the strata is very hard, then also drilling time is more as penetration will be less and the wear and tear on the life of drill rods will be more and frequent changes of drill rods were necessitated.

In case of Konkan Railway project where number of tunnels were more and the project was to be completed early, imported tunnel machinery called Boomer was used. In this case, drilling jumbo is not required; cut hole method not adopted instead central hole method adopted.

3.3 Charging and Blasting

As the drilling was over, the drilled holes were cleaned, and then the holes were filled with the explosives. The explosive used here were Gelatin sticks inter mixed with Ammonium Nitrate and electric delay detonators.

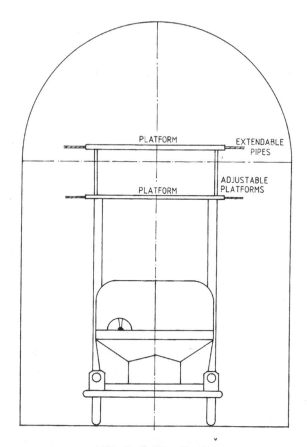

Fig. 3 : Drilling Jumbo

The detonators were connected one by one in series and then the same was connected to electric wires or to a exploder as required and then the same was detonated resulting in a blast. Normally this operation takes about 2 hours.

3.4 Defuming

The fumes, gases etc., of the blasting operation was defumed by means of ventilation fans, ducts which were provided in the tunnel. Ventilation fans were provided for every 200 m to 250 m. This defuming operation took about 1 to 1½ hours.

3.5 Scaling Operation

After the defuming operation was over scaling operation was done. It is a process of removing the loose rock pieces in the top and sides and also checking for misfires etc. to ensure safety for further operation. About an hour will be required for this operation. This scaling operation was done once again during or after the mucking also as a safety measure.

3.6 Mucking Operation

The removal of the blasted rock is called as mucking operation. For this, excavator with front-end bucket and tippers were used. One excavator for each end was used and due to importance of the project, one more excavator was kept as a stand by for the project. 3 to 5 tippers were deployed depending upon the lead. About

7 to 8 hours required for this operation and some time it gets extended when problem like breakdown etc., occur.

- **Rock Bolting** - 3 m long 25 mm dia. rod with plate butting the rock was driven and grouted with cement mortar capsules. It was provided at 1.5 m intervals both length-wise and radially and some places only upto the top half portion.

- **Shotcrete** - 150 mm thick in 2 layers with welded wire mesh in between and 100 mm thick in one layer in certain reaches.

It would be seen that cycle of 18 to 24 hours was adopted for tunnel excavation and support erection.

3.7 Technical Details and Plant and Machinery Used

Size of the tunnel	–	4.92 m width × 6.242 m height
Length of the tunnel	–	1560 m
Shape of the tunnel	–	'D' shaped
Type of rock	–	Shale and Quartzite
Lined portion with permanent support	–	400 m
Shotcreting with wire mesh	–	1160 m

The various plant and equipments, which were deployed for this tunnel work, were as under.

3.8 Survey and Profiling

Theodolite

Distomet

Levelling instrument

3.9 Drilling - Rock Bolting

Compressor - 600 cfm - 3 nos. 1 no. on each side

Compressor - 300 cfm - 1 no. as stand by

Pusher leg with Jack hammer

Drilling Jumbo

Grinder for drill rods

Pumps for dewatering.

3.10 Defuming

Ventilation fans at 250 m intervals and GI ducts between the fans.

3.11 Mucking

Excavator - Ex 100 / CK72 - 3 nos. one no. each end and one no. as stand by.

Tippers - 8 to 10 nos. - 4 to 5 on each side.

3.12 Lining Concrete

Concrete Mixer

Concrete Pneumatic Placer with its accessories.

Moving Shuttering Gantry

Form Vibrators

3.13 Shotcrete

Concrete Mixer

Pneumatic Shotcrete Pump (dry type)

Compressor - 600 cfm.

Mobile platform on rails for shotcreting the roof portion etc.

The original period of completion was upto 30th June 1996 for opening the section of traffic. This was preponed by Ministry of Railways to 29th February 1996. Therefore, IRCON had supplemented the tunnelling machinery of the agency to speed up the work of Bogada tunnel, which was perceived as the most critical activity of the project.

4.0 MODIFICATIONS FOR WORKING UNDER TRAFFIC CONDITIONS

The work could be started only in September 1994 as the environmental clearance was accorded in August 1994 only after a period of about 5 months. After the open excavation was completed the tunnel excavation was started in October 1994. The tunnel was opened to Railway traffic in March 1996. An average progress of 100 m per month was achieved and in certain months the progress was from 120 m to 150 m per month.

The entire work was done through conventional method only. The other methods such as use of tunnel Boomer etc. are costly. The using of Tunnel Boomer or Tunnel Boring Machine is cost effective only when the quantum of work involved is huge and time available is insufficient etc.

Due to the requirements of the Railway, the section was opened to traffic on 9th March 1996 after the tunnel excavation was completed and Railway track was linked and ballasting done. The ancillary works such as permanent supports, concrete lining, shotcrete, drains cable ducts etc., could not be completed during the tunnel excavation because of the movement of the tippers, insufficient time available during drilling when the tippers are not moving and operation like blasting cannot be postponed in the interest of safety in favour of these ancillary works. Hence these works were to be done under traffic conditions by taking permission for block timings between the movement of trains etc. This working under block condition necessitated modification of the working method, additional resources including involving changes in the deployment of the resources etc.

4.1   Permanent Supports and Concrete Lining

For leading the various materials such as structural steel, item including fabricated one, concrete etc. to inside the tunnel the normal Dip trolley frames were modified to suit the requirements of the work and the same were used (Fig. 4).

Similarly the shuttering Gantry for the lining of the roof arch portion of the tunnel was modified taking into consideration the moving dimension of the rail coaches/ wagons. This was done to ensure that the shuttering gantry and its movement did not hinder the traffic. This gantry was made to move at an elevated level i.e. at the springing level of the arch on specially made arrangements instead at bottom level and on rail which is the normal procedure. This shuttering gantry was a hydraulic collapsible shutter with required jacks, Turnbuckles etc.

The concrete was placed by means of concrete placer and its pipeline, operated pneumatically with compressed air. The gantry was fitted with form vibrators fixed in the shutter as per requirement.

The concrete was brought from the batching plant, which is situated outside the tunnel portal, through steel troughs fitted with trolleys and fed to the concrete placer.

Further, as the movement of the concrete troughs on trolley was to be in the trackline only, to have a continuous feeding of concrete to the placer, a 'A' frame type lifting arrangements were erected and used to lift/change the empty steel troughs etc.

454

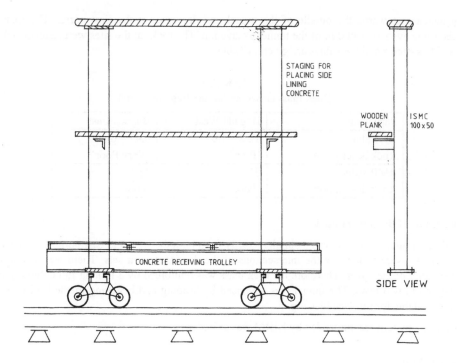

Fig. 4 : Drilling frame

4.2 Rockbolting and Shotcreting

For the rock bolting, a special pipe staging and platform was fabricated by combining two trolleys which can be dismantled and assembled again as per the block condition timings and its requirements.

In case of shotcreting this staging was not suitable as the area of shotcreting to be covered is wider and also the fact that spraying of shotcrete mortar shall be done from a distance of 1.5 m or so that too in a perpendicular fashion to the surface. Hence this needed different type of staging. So on either side of the track near the tunnel surface pipe scaffolding were erected and at the top again a platform was formed by keeping the pipes at closer intervals and timber planks fixed on these pipes. As this staging cannot be moved, assembling/forming of this staging was a continuous process.

5.0 DRILLING PATTERNS – CONSULTANCY FROM NIRM

As the project was of national interest timely completion of the works were very important. Initial blasting patterns for the excavation of the tunnels were designed by the agency M/s Sri Shankarananarayana Construction Company for Bogada tunnel and pull was low and overbreak was high. It was then felt that the tunnel excavation might not be completed in time and overbreak could increase the cost of lining and affect the rock mass condition. Keeping this in view, IRCON approached National Institute of Rock Mechanics (NIRM), Karnataka to design the blasts for overbreak control and high rate of advance in varying rock conditions at minimum costs. The salient features of the field investigation and recommendations of NIRM were as under.

5.1 Site Geology

From the west portal the Bogada tunnel is excavated through a strata of alternating bands of shale and quartzite. The rock formation strikes roughly perpendicular to the tunnel alignment and dips to the North East at 15 degree to 40 degree. The portal portion was highly weathered and has a cover of about 20 m., which gradually increases towards the middle of the tunnel and decreases at the East portal. The rock condition remains the same even after 400 m of the tunnel excavation. The rocks at the sites were categorised according to RMR and Q systems and the values are given in Table 1.

TABLE 1
Rock mass classification for Bogada tunnel

Particulars	Bogada West	Bogada East
'Q' value	1.22 – 1.67	0.47 – 0.56
Rock description	'Poor'	'Very Poor'
RMR value	31	23
Rock description	'Poor'	'Poor'

5.2 Excavation of Bogada Tunnel

The Bogada tunnel was excavated simultaneously from both the west and east portals. At the west portal area, huge blocks were formed by the intersections of the dominant joint sets. There was a serious concern about the stability of the cover. Therefore the rock mass was reinforced by 3 m long full column grouted rock bolts, staggered in two rows. The tunnel was advanced by heading and benching method till good rock was encountered, suitable for full face blasting.

The drilling pattern recommended by NIRM was as under :

5.3 Bogada West

The drilling pattern based upon systematic field investigations in February, 1995 by NIRM and on the observation of the trial blast as per recommended is shown in Fig. 5 (a), (b) and (c) when the face had already advanced to 165 m.

5.4 Design for Reduced Section

Although the rocks fall under very poor to poor, the tunnel faces were generally stable during the excavation. The project authorities felt that the conventional steel and concrete lining might not be necessary for entire length. Though the initial 160 m at Bogada west and 110 m at Bogada east was advanced without support, NIRM suggested to put rock bolts immediately to increase the stability and safety.

The dispensing of the steel supports and concrete lining for central portion of the tunnel led to the reduction in the size of the tunnel. This change in the cross-sectional area necessitated a modification in the drilling and blasting pattern. Fig. 6 (a), (b) and (c) shows the drilling, charging and hook up patterns for reduced section. However, the charging pattern remained the same as that of large cross-section since the depth of the holes remained the same.

5.5 Bogada East

Blasting work commenced in the east portal after the west has advanced to about 200 m. Since the rock mass in Bogada east was highly weathered, heavily jointed and filled with clay, blast designs were more challenging in the Bogada east than in the west, in order to secure the integrity of the rock mass and to improve the stand up time.

The experience in blasting at Bogada west helped to arrive at a near accurate blast design for Bogada east. Since the rock type at Bogada east was highly jointed, multiple wedge cut with staggered pattern of drilling was adopted and the number holes was reduced as compared to that of Bogada west. The drilling depth was restricted to 2.8 m for the cut holes and 2.6 m for the rest of the holes. The explosive consumption was reduced to 85 - 90 kg.

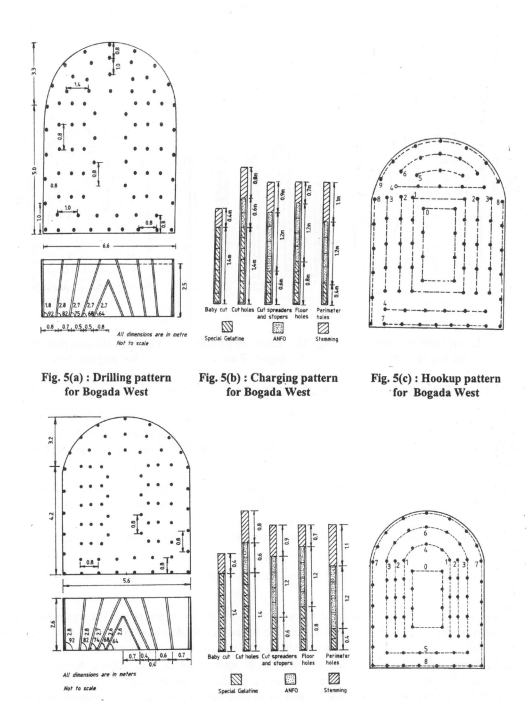

Fig. 5(a) : Drilling pattern for Bogada West

Fig. 5(b) : Charging pattern for Bogada West

Fig. 5(c) : Hookup pattern for Bogada West

Fig. 6(a) : Drilling pattern for Bogada West

Fig. 6(b) : Charging pattern for Bogada West

Fig. 6(c) : Hookup pattern for Bogada West

The drilling, charging and hookup patterns are given in Fig. 7 (a), (b) and (c) respectively. The results were good with respect to fragmentation, throw and muck pile profile. The pull was improved to 2.5 m on an average and overbreak was controlled and hole impressions seen throughout the profile.

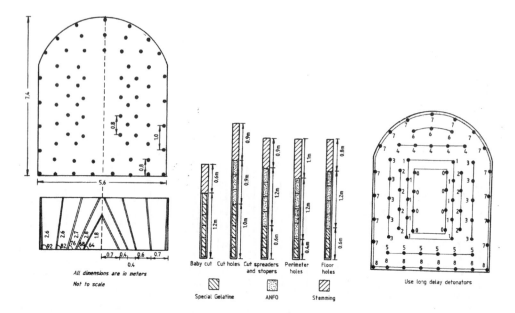

| Fig. 7(a) : Drilling pattern for Bogada East | Fig. 7(b) : Charging pattern for Bogada East | Fig. 7(c) : Hookup pattern for Bogada East |

6.0 SPECIAL PROBLEMS DURING THE EXECUTION OF WORK

The progress of tunnel excavation on 15 th November 1995 was 1086 m out of 1560 m and the support erection was 220 m out of 400 m and tunnel side lining was 146 m out of 400 m and rock bolts 4450 nos out of 11,400 nos. At this rate it was not possible to complete all the works before opening this section to traffic by end of February, 1996.

It was, therefore, decided to concentrate on the tunnels' excavation and complete balance work of excavation and 50 per cent of rock bolting by 31st January 1996 so that the section is safe and can be opened to traffic by 29th February 1996. The balance work of rock bolting, support erection, concrete lining, and shotcreting were therefore planned to be completed after opening the section to traffic by taking traffic blocks.

For improving the progress of tunnel excavation, the drilling pattern recommended by NIRM was followed:

It was also decided that the scope of support system should be reviewed and the minimum work of lining/shuttering consistent with the safety requirement should be completed before opening to traffic on 29th February 1996. It was also considered advisable to take an expert advice about the minimum work of supporting system. For obtaining expert advice, IRCON approached Dr. J.L. Jethwa, Scientist Incharge Central Mining Research Centre, Nagpur, who had vast experience in the design of support system visited the site from 15th to 17th November, 1995 and studied the geology of the rock mass of the tunnels minutely and gave his recommendations which considerably reduced the volume of shotcreting work.

The work was in the reserved forest area. Approval of Ministry of Environment under Govt. of India and taking clearance from Forest Department of State Government was required for cutting trees in the approaches to the tunnels and for access to the tunnels. This clearance took time from 17th March 1994 to 15th October 1994.

458

The construction of tunnels was in the reserve forest area with the law and order problems created by miscreants who could take shelter in forest area and who threatened from time to time to get benefits. Special care was therefore taken for safe custody and handling of explosives and detonators.

ACKNOWLEDGEMENT

The author was Project Head and Coordinator of IRCON for this work at the time of the execution of the above work and expresses thanks to the authorities of SC Railway for drawing material for this paper from various documents. The author is also thankful to Mr. K.R. Mehra, General Manager/IRCON for encouragement to write the paper. The author also acknowledges the assistance of National Institute of Rock Mechanics for their recommendations for speeding up the work of tunnel excavation and the assistance of Dr. J.L. Jethwa, Scientist, Incharge of Central Mining Research Centre, Nagpur for the supporting system.

EXPERIENCE ON THE USE OF RAISE BORER AT DULHASTI HYDROELECTRIC PROJECT

D. PAUL. VERMA
Executive Director (Projects)

M.K. GOEL
Senior Manager (Civil)

National Hydroelectric Power Corporation, Faridabad, Haryana, India

SYNOPSIS

The authors have been associated with the setting up and drilling of pilot hole of surge shaft at Dulhasti HE project in J&K with the help of raise borer. The experience on the use of raise borer has been brought out in this paper. A brief description of various components and specifications of the machine have been given and the initial preparatory works required for setting up the machine have been listed out. The daily drilling and reaming data has been brought out and the average rates of drilling/reaming have been worked out. The authors have also brought out the various precautions required to be taken for effective pilot hole construction with the help of raise borer and its advantage over drill and blast method.

1.0 INTRODUCTION

Dulhasti Hydroelectric Project is located in Kishtwar Tehsil of Distt. Doda in Jammu and Kashmir State. It is a run-of-river scheme on river Chenab and envisages the construction of a 65 m high, 180.5 m long concrete dam, a 10.6 km long head race tunnel and an underground powerhouse with an installed capacity of 3x130 MW (390 MW). The tail water will be discharged through a 275 m long tail race tunnel. The project will generate 1928 MUs of energy annually in a 90% dependable year. The layout of the project is shown in Fig. 1.

The construction of project was awarded on turnkey basis to a French Consortium in October, 1989, with scheduled completion by July, 1994. The civil contractor from the Consortium, M/s DSB stopped work w.e.f. 24 August 1992 due to disturbed law and order conditions at the project site due to militancy in Jammu and Kashmir. The cumulative progress achieved till then was 27% only.

The civil works were then rewarded to a joint venture comprising M/s Jai Prakash Industries Ltd. and M/s Statkraft - Anlegg (Norway) and the works at site were resumed since May, 1997.

The raise borer used at Dulhasti Project was procured by M/s DSB and handed over to NHPC. The raise borer was used by the French Contractor at various sites. However, the present study limits itself to the use of raise borer for boring pilot hole of surge shaft during August/September, 1997.

2.0 DEFINITION AND DESCRIPTION OF THE MACHINE

The raise borer is an electro-hydraulic machine for boring pilot hole of certain diameter in a shaft. The machine available at Dulhasti Project is of robbins make and of 61 R series. At the time of import, the machine was not new but it was an overhauled version supplied by M/s Boretec, U.S.A.

Robbins are the pioneers in the field of raise boring, with the first raise drill having been introduced by the company in 1962. Over the years, improvisation has been done by the company on its line of raise drills to help achieve faster advance rates, with less down time and lower costs.

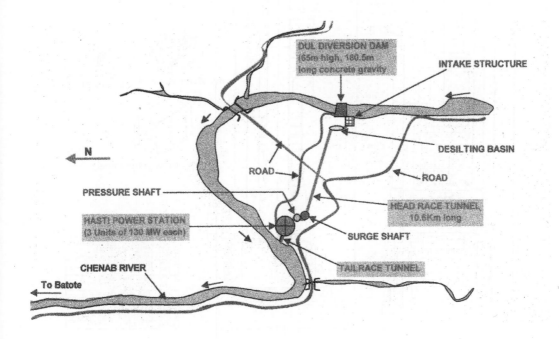

Fig. 1 : Layout plan of Dulhasti Hydroelectric project

The raise borer is quite a robust machine and can be overhauled and reused time and again. The machine in use at Dulhasti Project was manufactured in 1972 and is in constant use since then at various construction sites. Before its use in surge shaft, the machine was used at 5 other places in Dulhasti Project, itself. It was used in excavating pilot holes in three numbers DT gate shafts of 40 m each, TRT gate shaft of 33 m and pressure shaft of 138 m depth. It will be used further for the construction of 416 m deep air exhaust shaft of HRT with modifications.

3.0 VARIOUS COMPONENTS OF RAISE BORER

The general set up and various components of raise borer are shown in Fig. 2 and a photograph of the same is shown in Fig. 3. The components include :

1. Derrick assembly.
2. Hydraulic pack.
3. Electric pack.
4. Operator control or pendant station.
5. Pneumatic crawler erector and hydraulic lift cylinders.
6. Sets of hoses and cables.
7. Basic drill string comprising pipes of 254 mm outer diameter and 1.52 m length.
8. Tri cone bits for drilling.
9. Reamer, 1.82 m diameter with disc or carbide cutters.
10. Other accessories and tools.

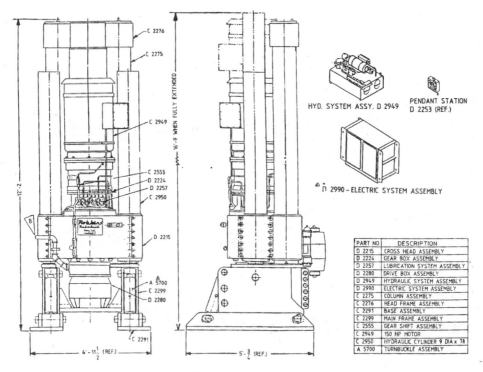

PART NO.	DESCRIPTION
D 2215	CROSS HEAD ASSEMBLY
D 2224	GEAR BOX ASSEMBLY
D 2257	LUBRICATION SYSTEM ASSEMBLY
D 2280	DRIVE BOX ASSEMBLY
D 2949	HYDRAULIC SYSTEM ASSEMBLY
D 2990	ELECTRIC SYSTEM ASSEMBLY
C 2275	COLUMN ASSEMBLY
C 2276	HEAD FRAME ASSEMBLY
C 2291	BASE ASSEMBLY
C 2299	MAIN FRAME ASSEMBLY
C 2555	GEAR SHIFT ASSEMBLY
C 2949	150 HP MOTOR
C 2950	HYDRAULIC CYLINDER 9 DIA x 78
A 5700	TURNBUCKLE ASSEMBLY

Fig. 2 : General set up of raise borer

Fig. 3a : View of raise borer machine

Fig. 3b : View of reamer attached to dril string

462

4.0 SPECIFICATIONS OF ROBBINS RAISE DRILL (MODEL 61R)

1.	Raise diameter	1.82 m (72 in.)
2.	Pilot hole diameter	28 cm (11 in.)
3.	Electric drive	Two speed, constant horsepower, induction motor 50/1500 rpm, 150 hp, 50 cycles.
4.	Drive torque	10,787 kg/m (operating) 78,000 ft. lbs.
5.	Pilot drilling Thrust	95256 kgs (210,000 lbs.) (maximum)
6.	Reaming pull	206388 kgs (455,000 lbs.) (maximum)
7.	Boring rates	0 to 61 cm/min., (0 to 24 in./min.) Pilot hole 0 to 36 cm/min., (0 to 14 in./min.) reaming
8.	Traverse rate	1.68 m/min., (66 in./min.)
9.	Dip angle limit	Vertical to 60 degree from horizontal
10.	Transporter	Air powered crawler.
11.	Electric power	170 kVA, 380 to 550 volts, 3-phase, 50 cycles.
12.	Air consumption	800 cfm free air @ 100 psi or 7 kg/sq cm.
13.	Water consumption	7 kg/cm^2 at (600 lt/min.)
14.	Driving speed	60 rpm.
15.	Reaming speed	8-11 rpm.

5.0 USE OF RAISE BORER FOR PILOT HOLE OF SURGE SHAFT

The surge shaft of Dulhasti Project is completely an underground structure and is not open to the sky. Sections of surge shaft showing various stages of excavation with raise borer have been depicted in Fig. 4. It opens out into the atmosphere through an upper expansion gallery at El. 1286.5 m.

6.0 INITIAL PREPARATORY WORKS DONE FOR SETTING UP OF RAISE BORER

1. Access to the top of shaft, *i.e.*, point from where boring of hole will start and access to bottom of shaft, where the pilot shaft will terminate or where the boring bit will be replaced with the reaming bit were essentially required. In our case, both these accesses were available, these having been constructed by the French Civil Contractor, before abandoning the work.

2. Since the excavation had to be carried out underground, a cavern of particular dimensions was required to be excavated for setting up the machine and other components. The height is governed by the full extension of machine cylinders. The minimum height required is 5.1 m, but considering some headroom at top and depth of foundation at bottom, the height required was about 7 m. The dimensions required at the invert level were 10 m width and 6 m depth into rock.

3. Proper rock support in the form of shotcrete and rockbolts was also provided in this cavern in order to avoid any rock fall and damage to the machine. Overhead rockbolts are also essential for holding reamer bit after completion of reaming.

4. A machine foundation of high quality PCC, *i.e.*, M300, alongwith bolts arrangement as per drawing was constructed. This foundation must be perfectly level.

5. A transformer was also placed about 30 m from the drilling site.

7.0 MOVEMENT OF THE MACHINE INTO THE CAVERN

The machine mounted on crawler transporter was shifted inside by moving an operational diesel compressor 21.25 cum (750 cfm) along with the machine. The compressed air provided power for the crawler movement. The crawler arrangement is depicted in Fig. 5.

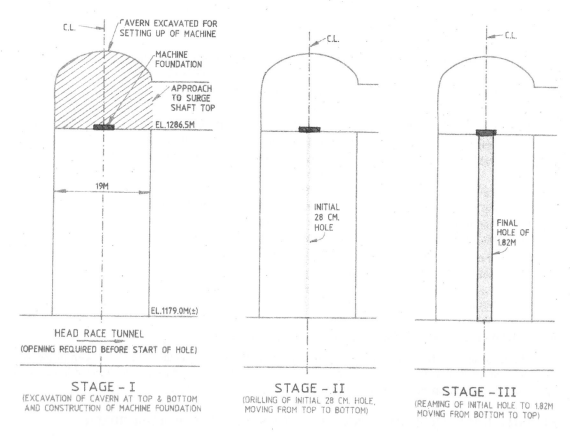

Fig. 4 : Section of surge shaft depicting various stages of excavation of pilot hole with raise borer

8.0 SETTING UP OF THE MACHINE

A base plate assembly was prepositioned on the smooth concrete pad and securely bolted to the rock. The derrick assembly was then transferred from the transporter to the base plate. After transfer, the derrick was adjusted to the dip angle of the hole to be bored. In our case this was vertical. The transport vehicle was then removed from the site.

The hydraulic and electric cabinets were located conveniently within a 9.15 m (30 ft.) radius in the cavern.

The pendant station containing all the controls was also placed in front of the derrick assembly.

9.0 IMPORTANT ACCESSORIES OF THE RAISE BORER

Pipe loader : The pipe loader is in the form of a movable arm on the machine. The pipe loader has a claw to hold the pipe and then place it under the drive box or remove the pipe from the drive box. An electric hoist is mounted on the machine, which lifts the pipe from the ground and mounts it on the pipe loader. The operation of both the pipe loader and electric hoist is remote controlled and can be kept on the pendant station itself.

Wrenches : Separate wrenches are there to hold the pipes on the work table at the bottom of the machine and in the drive box. The verticality of the hole can also be judged from the ease of fitting of the wrench on the pipe in the annular space on the work table. The wrenches are required to be fitted whenever a new pipe is to be added or removed.

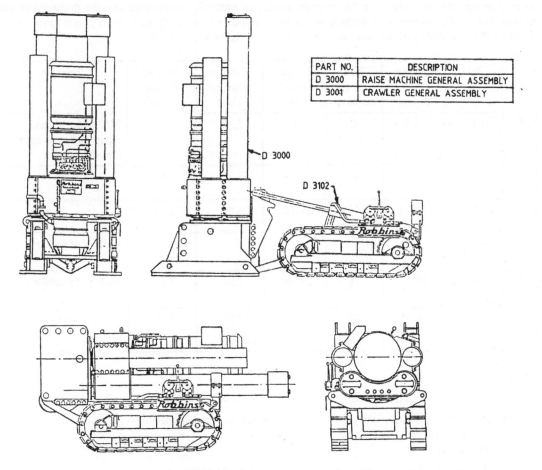

PART NO.	DESCRIPTION
D 3000	RAISE MACHINE GENERAL ASSEMBLY
D 3001	CRAWLER GENERAL ASSEMBLY

Fig. 5 : Crawler arrangement of raise borer

Drill string : The drill string comprises the starter pieces, bottom hole assembly and drill pipes. The starter piece is made to move through a removable bushing on the work table. This fully guided system ensures accurate aiming of the pilot hole. The bottom hole assembly comprises an 27.94 cm (11 in.) pilot bit with jets, attached to the bit sub. Within the bit sub is a float valve preventing reverse circulation immediately above the bit. Rolling cutters and stabilisers are also provided on the bit sub. The bottom hole assembly stabilises the bit and ensures a straight accurate pilot hole. The drill pipe is made from 25.4 cm (10 in.) outer diameter alloy steel tubing with bore replaceable pin ends.

Reamer : The reamer has a central shaft of forged steel equipped with a saver sub. This sub presents a DI-22, 8-1/4 pin and may be replaced as required. In our case the diameter of reamer head was 1.82 m. A photograph of the reamer is shown in Fig. 3.

10.0 DRILLING OF PILOT HOLE

After checking the horizontal and vertical alignment of the machine, drilling of hole was started on 21 August 1997 with special starting arrangement of the hole. This arrangement comprised a special bushing with 24.13 cm (9.5 in.) internal diameter and a corresponding starting pipe with the same diameter. This starting arrangement ensures aiming of vertical hole from the top. After drilling with the first 24.13 cm (9.5 in.) pipe, this pipe was removed and normal drilling with 25.4 cm (10 in.) drill pipes was restored. An air jet pipe was also provided near the machine so that no sand/silt accumulated near the drilling pipe and no backflow of this material into the hole was there. The hole was made through in the power tunnel on 26 August 1997. The daily drilling data and inferences have been tabulated below :

Sl. No.	Date	Pipe Nos.	Total pipes drilled	Length drilled (m)	Time on machine (hrs)	Rate of drilling (m/hr)	Remarks
1.	21.8.97	1-8	8	11.6	5.30	2.11	M/C under b/d and repair during balance period of shift
2.	22.8.97	9-16	8	11.6	3.10	3.66	
3.	23.8.97	17-31	15	21.75	6.00	3.63	
4.	24.8.97	-	-	-	-	-	Holiday of Sunday
5.	25.8.97	32-47	16	23.20	8.00	2.9	No b/d of machine.
6.	26.8.97	48-75	28 Total:	40.6 108.75 m	12.00 34.40 hrs	3.38	- do -

11.0 INFERENCES

    (i)   Average rate of drilling/hour   =   108.75/34.4 = 3.16 m/hr.

    (ii)   Average rate of drilling/day   =   108.75/5 = 21.75m/day
(Work carried out in one shift only).

    (iii) Maximum progress in one   =   40.6 m (28 pipes)
day.

    (iv) Maximum hourly progress   =   4.35 m (3 pipes)

(In this hour time taken to drill each pipe was 10 minutes and time taken to fix each pipe was also 10 minutes).

    (v)   At times drilling had to be stopped due to shortage of water. Hence better progress could have been achieved if this problem was not there.

12.0 VERTICALITY OF THE HOLE

The exact coordinates of centre line of hole drilled from top and centre line of penetration at the bottom were calculated with the help of distomat. It was seen that an overall deviation of 4.63 m occurred in the drilling. This deviation occurred in a length of 108.75 m, the percentage deviation being 4.26 per cent. This deviation is very nominal and will not defeat the purpose of raise boring in any way.

13.0 REAMING

After breaking through in the tunnel on 26 August 1997, 6 more pipes were connected by the operator on 27 August 1997, so that drill bit touched the invert of the tunnel below surge shaft. This was done to facilitate easy fixation of arrangement comprising striking tool and jacks to dismantle 27.94 cm (11 in.) bit

466

from the pipes. Two types of reamers were available at site - òne of robbins make and the other one of sandwik make. It was decided to use the sandwik reamer.

The next 3 days were utilized to change worn out cutters on the reamer and its greasing, etc., and in making arrangement for taking out the tricone bit without disturbing the balance pipes. The bit alongwith one pipe was finally taken out on 30 August 1997. The reamer was mounted on the pipes or drill string on 2 September 1997 and actual reaming of the hole was started on the same day after the removal of protruding pipes.

The following precautions were to be taken while connecting reamers and while reaming :

(i)   Telephone connection was provided between the raise drill operator and the place for connecting the reamer head.The drill operator cannot see the connection process since both the operations are being done underground at different levels.

(ii)   While reaming the drill hole, care was taken to prevent reverse rotation of machine and subsequent loosing of drill string. However, there is always a risk of loosing the drill string and the reamer head. If it so happens, the reamer head and drill string will fall down and will hit the tunnel floor with great violence. To avoid any serious accidents, men and machinery were not allowed in the area during raise boring.

(iii) Mucking out from bottom was be done at the time when the raise drill was not working.

(iv)   The raise drill operator always secured the drill string when leaving the machine.

14.0 PROCESS OF REAMING

The process of reaming involves activities exactly opposite to those involved in boring. An upward thrust is put on the reamer and the reamer, while chipping the rock and widening the hole, moves up by one pipe length at a time (1.45 m). When the full pipe reaches the work table of the machine, the lower reamer and drill strings assembly are held on the machine with the wrench fitted on the work table and the upper pipe is removed with the help of pipe loader and electric hoist and stacked as required.

As mentioned before, reaming of the hole was started on 2 September 1997 and the reaming was completed on 9 September 1997. From 3 September 1997 onwards, the operator worked on the machine for almost two shifts.

The daily reaming data and inferences have been tabulated below :

Sl. No.	Date	Pipe Nos.	Total pipes reamed	Length reamed (m)	Time on machine (hrs)	Rate of reaming (m/hr)	Remarks
1.	2.9.97	1-4	3	4.5	5.20	0.84	Full length of 4th pipe was not reamed.
2.	3.9.97	4-15	12	16.45	14	1.18	
3.	4.9.97	16-24	9	13.05	15.40	0.83	
4.	5.9.97	25-36	12	17.40	13	1.34	
5.	6.9.97	37-47	11	15.95	13	1.23	
6.	7.9.97	48-53	6	8.7	14	0.62	
7.	8.9.97	54-63	10	14.5	15	0.97	
8.	9.9.97	64-75	12	17.4	15	1.16	
			Total:	107.95	105		

15.0 INFERENCES

(i)	Average rate of reaming/hour	=	107.95/ 105 = 1.03 m/hrs.
(ii)	Average rate of reaming/day	=	107.95/8 = 13.5 m/day.
(iii)	Maximum progress in one day	=	17.4 m (12 pipes)
(iv)	Time taken to raise one pipe	=	40 to 75 min.
(v)	Time taken for dismantling one pipe and restart.	=	5 to 10 min.

(vi) The effectiveness of reaming was checked by lowering the reamer after reaming half metre only. Clear impressions of carbide drill bits were observed in the arch created by the reamer.

(vii) There was difficulty in fitting wrench on the worktable, while removing the pipes. Mechanical jacks had to be applied for the same. This was due to the marginal inclination of the hole. Had the 27.93 cm (11 in.) hole been completely vertical, the wrench would have easily fitted in the annular space.

(viii) Though water was not required for actual reaming, some water was used to prevent dust formation in the mucking area of tunnel, below the shaft.

(ix) It can be seen from the table that the progress of reaming on 4 September 1997 and 7 September 1997 was less. This was due to intermittent tripping of the motor due to overheating. To prevent this, the speed and reaming pressure on the machine had to be reduced.

(x) The reaming was continued till reamer contacted the base plate rock bolts. At this time, the bolts started to loosen and there were heavy vibrations in the ground near the machine. Some movement of machine was also noticed. The reaming was finally stopped. In fact though the machine foundation remained intact, small adjoining areas caved in and light could be seen at the bottom of the shaft.

16.0 SHIFTING OUT OF MACHINE AND RETRACTING OF REAMER FROM THE HOLE

The following steps were involved :

(i) After taking out all but one pipe, a pail arrangement provided with the system as an accessory, was connected to the reamer. This pail in turn was tied with the help of slings and D-shackles to rock bolts drilled and installed previously for the purpose, on the front side of the cavern.

(ii) The machine was then released from the foundation plates, mounted back on the crawler erector and shifted out of the tunnel.

(iii) All other accessories like hydraulic pack, electrical pack, pendant station and drill strings were also moved out, thereby clearing the complete area.

(iv) After this, 30.48 cm to 60.96 cm (1 to 2 ft.) holes were drilled with the help of jack hammers in the machine foundation. These holes were charged lightly and blasted. A clear hole around the reamer was obtained. However there was no damage to the reamer.

(v) The reamer was detached from rockbolts and removed from the area with the help of a loader.

17.0 ADVANTAGE OVER DRILL AND BLAST METHOD

As seen from abvoe, the time taken for construction of 109 m pilot hole with raise borer was as under :

Time taken for drilling 28 cm. hole = 6 days

Time taken for reaming = 8 days

Time taken for preparatory and other works of foundation, change over bit, etc. = 15 days

Total time taken for pilot hole of 109 m = 29 days, say one month.

Most of the pilot holes for shafts in our country have been excavated by conventional drill-blast method or with the help of raise climber. For ease of maneuverability in the shaft, a minimum diameter of 2.5 to 3 m is required to be excavated in these cases. Had the pilot hole under consideration been excavated by drill-blast method or with the help of raise climber, the time taken would have been 4 to 5 months. This is based on our experience in construction of pilot holes for shafts by these techniques in other projects.

As compared to drill and blast methods, the raise boring offers a host of time and cost saving benefits, which are faster advance rates, less disturbance to rock structure, cleaner bores, reduced labour costs and above all, greater safety for mining personnel. These benefits have made the raise drill, today's conventional tool for boring pilot holes, air shafts, etc.

Due to drilling of pilot hole with raise borer, the initial installation of winch and cumbersome mucking from top, which would have been required in drill and blast method, are avoided.

Once the construction of pilot hole is completed, slashing of the shaft to required diameter can be started immediately after the installation of overhead hoist arrangement.

18.0 CONCLUSIONS

1. The raise borer was effectively used for excavation of pilot hole of surge shaft at Dulhasti Hydroelectric Project. Maximum drilling progress of 40.6 m/day and maximum reaming progress of 17.4 m/day was achieved.

2. Though there was some inclination in the hole, the hole was within the final diameter of the shaft and the head race tunnel, hence this inclination was immaterial. There have been instances in other projects where deviation of 12 to 15 m has been recorded and additional excavation was required to be done in the lower tunnel to locate the hole.

3. On the basis of experience gained in the use of raise borer, the following precautions are advised for future use of raise borer at other locations :

    (a) The machine foundation should comprise high quality PCC, preferably M300 and shall be perfectly level. This will prevent any movement of the foundation under the weight of machine and application of heavy thrust while drilling.

    (b) Verticality of the raise borer machine is required to be maintained after mounting the machine on the foundation.

    (c) It is essential that only experienced personnel be allowed to operate the machine. Experienced electrician and mechanic are also required, who have proper knowledge of electric and hydraulic circuits of the machine.

4. Ample storage of clean water is required at site for drilling operation. Minimum 600l/min. at 7 bar is required. The boring shall be done only when sufficient water is available.

5. Voltage fluctuations on the machine must be avoided. A suitable transformer must be stationed near the drilling site. Power voltage of 400 volts is required.

6. Sleeve/guides and starter pipe availability is required to be checked before starting the hole in order to aim the hole vertically.

7. The verticality of the hole can be checked to some extent by removing pipes after drilling about 10 m and putting plumb along with light in the hole.

8.   If any deviation is located as mentioned above, the hole may be refilled with concrete and redrilled.

9.   When the machine is required to be reused, complete cleaning and greasing of machine motor and other components must be ensured before reuse, to prevent overheating of motor. Other preventive maintenance must be carried on side by side.

10.   The base of machine foundation must be higher than the adjoining drain into which the water flows in order to prevent backflow of water and sediments into the hole. The drain is also required to be cleaned of the sediment deposit side by side for the same reason. An air jet may be provided near the foundation to prevent accumulation of any silt near the opening of hole.

11.   Precautions to be taken while reaming have already been discussed.

Provided that the above precautions are strictly adhered to an almost vertical pilot hole will be ensured.

SPECIAL ASPECTS

NEW AUSTRIAN TUNNELLING METHOD (NATM) TUNNELLING IN BASOCHHU HYDROELECTRIC PROJECT, BHUTAN

S.K. DESAI   **RAJIV BADAL**

Patel Engineering Ltd., Mumbai, India

SYNOPSIS

Tunnels of water conducting system have a special category within the field of tunnelling technology. The tunnel should be free from water loss and provide stability to the structure from internal and external water pressure. The head race tunnel of Basochhu Hydroelectric project is one of the water conducting tunnel in which the several tunnelling difficulties are being encountered in construction and if a proper tunnel design decisions are not taken such as a concrete lining of the tunnel then it may be a nightmare during operation and maintenance. This text describes some of the problems associated with the tunnel construction and the future problems may be encountered during operation and maintenance of the structure.

1.0 INTRODUCTION

The Basochhu Hydroelectric project is a run of river type scheme located on Basochhu river a tributary of Sankosh river around 22 km south of the Wangdue Phodrang, Bhutan. The project is comprised construction of concrete weir, intake structure, desilting basin, transition chamber, a modified horse shoe shape head race tunnel with inner diameter 2.65 m and a length of around 2600 m, vertical surge shaft, inclined penstock pipeline with 1.4 m inner diameter and length of about 1350 m, a power house structure to accommodate 24 MW power generating units, tail water channel from power house to tail water tunnel portal and tail water tunnel to outfall structure into Rurichhu river.

The New Austrian Tunnelling Method (NATM) constitutes a method by which the surrounding rock or soil formation of a tunnel are integrated into an overall ring like support structure. Thus these formations will themselves be a part of the primary support. The head race tunnel of Basochhu project is being constructed by employing the method of NATM. During the course of construction several tunnelling problems have been encountered - out of which some of them are being described in the text.

2.0 GEOLOGICAL FORMATION

The Basochhu Hydroelectric project is located on the Indian shield of Himalayan ranges. The crystalline rock formations of the area forms the base of the metamorphic Tethyan formation of the Tibetan Himalaya in the north. The rock formations of these crystalline basement are not uniform. They change very frequently and shows a higher tectonic disturbances in practically most of the formations in and around the project. The crystalline complex are mainly comprised quartz biolite gneiss, quartz micacesus gneiss, biolite lschist with quartz.

Because of the higher tectonic disturbances in the area the change in orientation of the schistosity is quite common but most commonly it is oriented towards NW to SW in general. Similarly, varied fault trends can be observed in the region but most commonly they are trending from NW-SE to NNW-SSE.

The head race tunnel alignment mainly comprises the shear zones, transitional zones and metamorphic formations. Shear zones mainly consist of micaceous schist with intercalations of fault gauge and catalistics.

The formations are highly fractured and weathered. A mixed ground condition is most commonly encountered along the tunnel alignment from the south end tunnel portal upto the chainage 1540 m. Some banded and stratified layers of quartz biolite gneisses, quartz micaceous gneiss and quartz garneteferous gneiss etc. have also been encountered in the different stretches of the tunnel section. The variable geological formations and geo-environmental conditions have been experienced upto the excavated tunnel chainage of 1540 m.

3.0 GEOTECHNICAL CLASSIFICATIONS

Based on the geological formations, geo-environmental conditions encountered along the tunnel alignment and structural control on the rock formations have been observed, is classified in two rock classes from the anticipated number of three rock classes *i.e.*, rock class-II and rock class -III. These classifications are based on the geological features perceived by the rock strata and the necessary support treatment required for the stability of the tunnel under the conditions of the strata along the tunnel alignment.

In general the rock strata classification is based on the geological features of the rock formation wore by the tunnel face, not the kind of support required by the rock strata.

4.0 TUNNEL CONSTRUCTION

The head race tunnel with inner diameter 2.65 m and a length of about 2600 m is being constructed by employing NATM. This method is based on the principle to take utmost advantage of the rocks capacity to support itself by careful measures and deliberate guidance of the force during the readjustment process from primary to the secondary state of stress around the excavations. The local progressive loosening can be limited by employing careful excavation methods and timely installation of support elements. The support elements are shotcrete, welded wiremesh, rock bolts, steel ribs, forepoling, steel lagging. The adequacy of the support elements can be assessed by deformation monitoring by installation of extensometer or the tape extensometer.

4.1 Excavation Methods

In the tunnel, the rock excavation is being carried out by commonly employed the conventional methods of drilling and blasting. Wedge and burn cut patterns, both is being employed on the tunnel face according to the conditions of the rock strata encountered. The blast holes drilling are being carried out by high speed jack hammers with an average rate of penetration of about 15 m/hr. But the variability and heterogeneous nature of the rock strata have been encountered along the tunnel alignment for which all the assumptions proved wrong. For excavation of the complex rock strata about 45 to 58 nos. of holes are being drilled on the face to blast according to the rock formation encountered. The number of bore holes drilled on the face depend on the geological conditions of the rock face.

The variable nature of the rock strata along the tunnel alignment is compelled to the blasters to adjust their blast design pattern to cope up with the geological conditions encountered. In fact, these additional holes drilling and charging have consumed more time than the time anticipated to complete a cycle. For the burn cut, the cut holes or reamer holes are drilled with a diameter of 65 mm or 89 mm holes with the help of the Jack Hammers. A systematic blast design pattern which is commonly employed on the tunnel face is shown in the Fig. 1.

Emulsion and special gelatin explosives alongwith Ammonium Nitrate and Fuel Oil (ANFO), are being used in according to the rock strata and environmental conditions persist in the tunnel which is being fired in combinations of the short delay and long delay detonators. The blast holes close to the excavation profile is being charged with less quantity of explosives than production holes, this reduces the rock damage to a minimum on the remaining rock profile. The unfavourable/adverse rock strata conditions which are being encountered alongwith alignment are produced over breaks the pull from each cycle of operation is achieved satisfactory to some extent, however, in some situation when the vertical joint cross the tunnel alignment, through which explosion gasses are escaped resulted in the poor pull, blast misfire or face blowout in the tunnel. Further, under cuts, toe, uneven floor conditions etc is required to be handled separately, which is very costly, time consuming and dangerous to the operators of the blasting operations. The inflow of water is further aggravated to the problems of face drilling, blast hole charging and firring. This has an additional economic burden to the contractor.

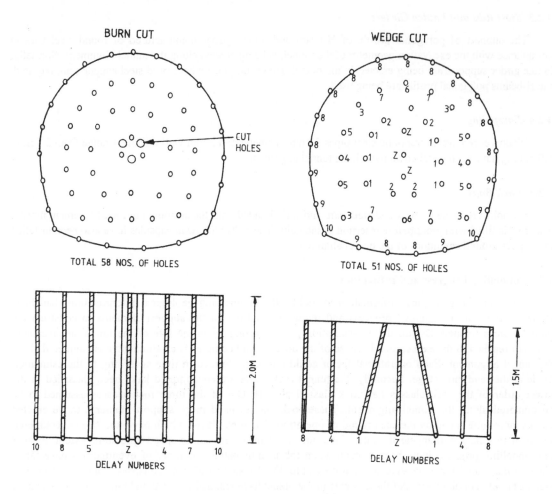

Fig. 1 : A systematic blast design pattern

4.2 Support System

All excavation faces in tunnel have been supported by the minimum recommended support elements of the rock class. These support elements are made up of forepoling, wiremesh, shotcrete, steel ribs and rock bolts. These support elements constitute a layer of support system, which helps the rock strata to achieve a state of equilibrium to establish the excavation profile. At the same time, the primary support elements should be flexible enough to minimize bending movement of the excavation profile during the process of the stress rearrangement.

4.2.1 Shotcrete

Dry to semi-wet shotcrete is forming a good primary support element. The shotcrete lining is in full contact with the exposed rock profile.

4.2.2 Welded Wiremesh

Welded wiremesh of 100 mm x 100 mm, 5 mm dia are being installed close to the excavation profile or previous layer of shotcrete.

4.2.3 Steel Ribs and Lattice Girders

The support of polygonal profile of H-beam and lattice girder manufactured of round steel bars in compliance with the excavation geometry of the tunnel, is being erected along the tunnel alignment. Normally, lattice girder supports are being erected in the tunnel. However, place where rigid steel supports are required, the H-beams polygonal profile are being used.

4.2.4 Forepoling

Wherever required, forepoling steel pipes 32 mm diameter and 3 mm wall thickness range from 2.5 m to 5.0 m long have been installed alongwith the tunnel alignment.

4.2.5 Rock Bolts

SN-bolts are made up of rolled steel bars and fully bonded with the surrounding rock by cement mortar. The role is filled with grout before insertion of the bolts. The bolts with resin capsules have also been installed in the poor ground conditions of the tunnel alignment.

4.3 Tunnelling Progress and Difficulties

Based on the geological information available, the engineering geological interpretations had been made for the head race tunnel progress and required quantities of the support elements to construct the tunnel were estimated. But all these assumption proved wrong when the rate of the tunnel advancement have not reached to the expectation because of the difficult and complex rock strata have encountered along the tunnel alignment. Each operational factor could not be performed in time according to the estimation made at the tendering stage. Secondly, the tunnel rock strata are encountered have been classified in the lower order of the rock classes i.e., rock class-II and III. Out of the three rock classes assumed to be encountered along the tunnel alignment. These rock classes need more support elements to be erected means more time spend on the each round of operation. The rock class-I which have not been encountered so far, if this class will appear the rate of the tunnel advancement can be speed upto some extent. Further the tunnelling progress have been worsen when the unanticipated a number of eventualities have taken place in the tunnel viz. excessive water inflow upto 432 1/s along with face collapses, upheaval of inverts, face instability condition etc. All these events in the tunnel have resulted in a decimal rate of advancement. A statement of the tunnel advancement rate is shown in the Fig. 2.

The design estimated quantities of rock support elements for full length of the tunnel have been installed in nearly fifty percent length of the tunnel construction. This is because of the rock strata along the tunnel installed (maximum upto 35 nos.) with ranging length upto 5 m. Holes drilled for forepoling pipes installation used to collapse, hence new holes are required to be drilled. The water inflow in the tunnel along the alignment have encountered on a number of occasions in much more quantity than predicted in the contract documents. When water inflow quantity was above 400 1/s in the size of the tunnel, the face excavation was stopped and the construction of a by-pass tunnel was necessitated. The number of times water inflow in the tunnel have resulted in an increase of cost by decrease in efficiency of workmen, maintenance of equipments and machines, rail tracks, delay in drilling, charging, blasting, mucking, shotcreting, installation of support elements etc. Similarly at a number of occasion the tunnelling work had been stopped due to the dangerous tunnelling conditions had encountered at the tunnel faces. To overcome these exceptionally dangerous situation had taken several hours extra to overcome.

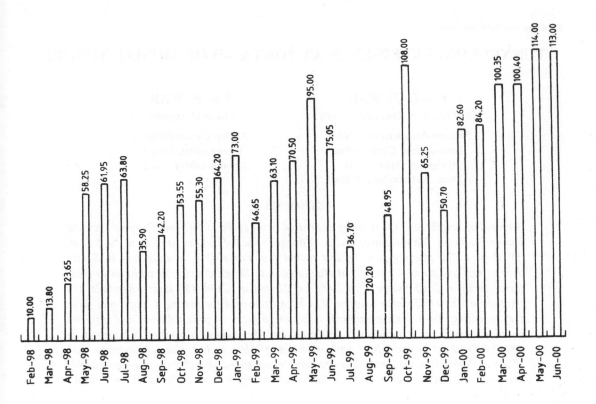

Fig. 2 : Tunnel advancement at Basochhu works

4.4 Engineering Considerations of Tunnel Structures

In Himalayan rock formations a water conducting system is designed and constructed for a life not less than 50 to 70 years to carry water to the appurtment structures of a hydro power project. Initially the water conducting system is supported with the preliminary design support to stabilise the structure. For longer and safe productive safe life of a structure is concrete lined with a thickness ranges from 250 to 400 mm. This concrete lining forms a part of the permanent support of the structure. However, in head race tunnel of Basochhu project, the engineers have removed the concrete lining from the tunnel. The engineers consider that a 50 mm thick semi-wet shotcrete layer is good enough for the structure even in the longer life span but in considered opinion the structure should not be left for experimentation. This design of the structure may be based on the economic reasons, otherwise the structure of this magnitude should not be left without concrete lining.

5.0 CONCLUSION

The head race tunnel of the project have experienced several tunnelling difficulties. These are some of the indications to the designers that the tunnel needs due care not only in construction but also during operation and maintenance. This needs design decisions so that a due care to the structure is taken to avoid the anticipated problems in near future.

TUNNELLING EXPERIENCE AT PORTAL-II OF ARPHAL TUNNEL

V.V. GAIKWAD

Chief Engineer (Specified Project)

**Maharashtra Krishna Valley
Development Corporation
Irrigation Department,
Pune, Maharashtra, India**

P.K. PAWAR

Executive Engineer

**Arphal Canals Division
Karad, Distt. Satara,
Maharashtra, India**

SYNOPSIS

A tunnel of length 16630 m is being excavated between 86 to 102 km of Arphal canal of Kanher dam in Maharashtra, India. Finished size of tunnel is 5.0 m × 5.5 m, 'D' shaped. The starting point of this tunnel (i.e. Portal-I) is at Rajmachi in Satara district and end point (Portal-II) at Hingangaon in Sangli district. In addition to these faces, inclined Shaft (Slope 1 in 5.46) i.e. Shaft-I at Ch.90/830 near Khambale and Shaft-II at Ch. 97/630 near Kadepur has been excavated for getting 4 additional faces for tunnelling. Tunnelling experience at all the 6 faces is quite different and different problems are being experienced. At Portal I of the tunnel, tunnelling is being done smoothly without any problems with minimum overbreaks but with less average pull. While at Portal-II, several rockfalls are being experienced affecting the average monthly progress. Description below discusses observation, experiments at Portal-II and Portal-I of Arphal Tunnel.

1.0 GENERAL

Kanher dam of gross storage 10.10 TMC (286 MCM) has been constructed across the river Venna and is a part of Krishna Irrigation Project in Maharashtra State, India. The left bank canal of this dam when crosses Krihsna river at 20 km, the canal has been named as Arphal canal. Total planned length of this canal is 205 km. The alignment of express Arphal canal from 86 to 131 km was passing through forest land as well as through steep slopes. Also in this length there was no irrigation by Arphal canal. Considering above difficulties, alternatives alignment for canal with a tunnel of length 16.63 km. between 86 to 102 km of Arphal canal was planned to carry 15 cumecs of discharge and is called Arphal tunnel (Fig. 1). It will help to irrigate about 16758 ha. of land in Sangli district of Maharashtra State. The Arphal tunnel is one of the biggest tunnel on canal in Maharashtra.

2.0 GEOLOGY

Arphal tunnel in general passes through jointed compact basalt and through volcanic breccia. At Portal-II jointed basalt is met. Presently tunnel has been excavated between Ch. 102/009 to 101/616. A closely jointed compact basalt is met between Ch. 102/008 to 100/900, while beyond this chainage the rock is block jointed/broadly jointed compact basalt upto Ch. 101/650. Rock met is hydro thermally altered beyond Ch. 101/650. About 33 fractures have been encountered with group of fractures exposed at crown at Ch. 101/962, 101/797, 101/708. Due to fractured closely jointed rock large overbreaks are observed.

3.0 DRILLING AND BLASTING

3.1 Drilling and blasting pattern adopted initially for Portal-II as shown in Fig. 2. The observations for this blasting pattern are shown in Table 1, for tunnel excavation of 685 m through Portal-I and 569 m through Portal-II.

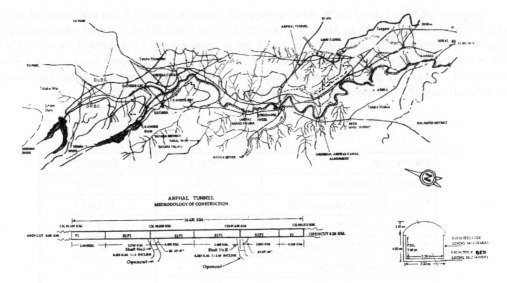

Fig. 1 : Krishna irrigation project index plan

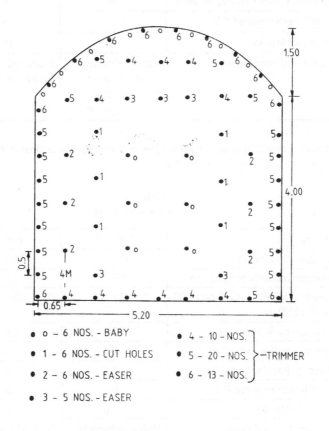

Fig. 2 : Wedge cut pattern

As observed from Table 1 for Portal-II, average cylce time is about 16 hours with achieved average pull of 2.5 m with a drilling depth of 3.0 m. The powder factor at Portal-II is 0.77 kg/m³ while average per cent overbreaks are about 15 per cent. Thus even though pull is same at Portal-I and Portal-II, overbreaks at Portal-I are on very much lower side (2.88%) giving a cycle time of 12 hours and powder factor of 0.74 with monthly progress of 115 m thus giving efficient tunnelling at Portal-I. Table 2 gives the actual observations while tunnelling through Portal-I and II for further chainages.

TABLE 1
Actual observations

Description	Portal-I	Portal-II
Starting chinage, km	85/582.5	102/009
Work executed chainage, km	86/205	101/440
Length executed, m	685	569
Average monthly progress, m	115	80
Average cycle time, hrs.	12	16
Average pull achieved, m	2.5	2.5
Average powder factor, kg./cum	0.74	0.77
% of length required P.S.	1	5
% of overbreaks	2.88	14.96
Geological conditions	Amygdaloidal basalt, massive rock conditions with less jointed	Fractured closely jointed basalt

3.2 As observed from Table 2, it is seen that, for Portal II average cycle time is about 12 hrs with achieved average pull of 2.7 m with a drilling depth of 3.0 m.

TABLE 2
Actual observations

Description	Portal-I	Portal-II
Length executed, m	437	360
Average monthly progress, m	97	72
Progress for the period 24/10/99 to 23/11/99, m	115	129
Average cycle time, hrs.	12	12
Average pull achieved, m	2.2	2.7
Average powder factor, kg./cum	1.2	0.98
% of length required P.S.	0	0
% of overbreaks	3.00	4.79
Geological conditions	Volcanic breccia	Hydrothermally altered basalt

The powder factor at Portal II is 0.98 kg/m^3, while average per cent overbreaks are at 4.79. If these factors are compared for Portal I, where rock is of breccia type, values for pull and overbreaks are on lower side, while powder factor is on higher side. Thus these observations clearly indicate the importance of geological conditions met while tunnelling, which govern the cost of the tunnelling. Due to jointed nature of rock at Portal-II, pull obtained is 2.70 m, giving a good progress of 129 m during the month of November 1999, when the work at Portal II continued without any rockfalls or any other obstruction. But as indicated above, since the rock met at Portal-II is of fractured type, jointed basalt, occasional rock falls, were experienced affecting the average monthly progress of work. If we compare the average progress achieved at Portal I and Portal II it shows that even though pull at Portal I is 2.2 m, due to breccia type rock requiring more explosives, has better average monthly progress of 97 m as compared to the average monthly progress of 72 m at Portal II, even though the pull at Portal II is more. Also per cent overbreaks are minimum at Portal I as compared to Portal-II.

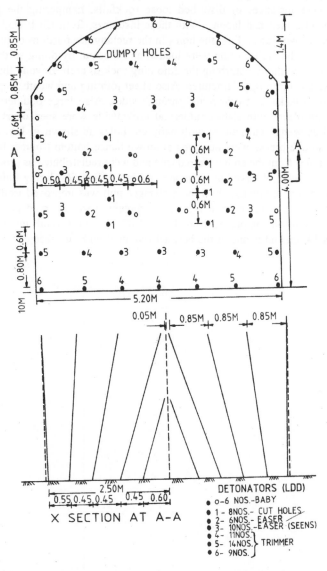

Fig. 3 : Wedge-cut pattern with baby-cut

481

4.0 BLASTING PATTERN TRIALS AND ERRORS

Due to fractured jointed nature of rock, large overbreaks were observed at certain localised spots, causing rock fall and damage to machinery. With a view to improve the tunnelling progress through Portal II, also to reduce the percentage of natural overbreaks, some experiments were tried with original blasting pattern as discussed below.

Initially charge for the periphery holes was reduced gradually to 50 per cent of the initial charge between chainages 101/405 to 101/370. By reducing the charge per hole in the periphery holes the widths were controlled minimising the overbreaks to 1 to 2% (average). With further reduction in the charge, material was not thrown out and hence heavy scaling was required and also some under cuts at springing level were observed.

When tunnelling was in progress beyond chainage 101/355, the rock met was weak, closely fractured and there were instances of disturbed loose rockfalls, hampering the progress of work by damaging the equipment's men and labour. With these instances, from Ch.101/325 to 101/300 loading and blasting was done in two stages. First the holes in the rectangular portions of tunnel were loaded and blasted, then the holes in the arch portion were loaded and blasted, by doing so the material from arch portion was thrown down without disturbing the adjoining rock in arch. Beyond these chainage the rock met was compact basalt with group of fractures. Also sheet jointing and weathering was observed along the fracture surface with leakage at crown and through sides. Also large boulders and undercuts were observed. Also at some places thin slabs/boulders at crown/side were seen overhanging. The blasting pattern was modified as wedge cut pattern with baby cut holes as shown in Fig. 3. Large widths were observed at bottom also overbreaks were observed at crown hence bottom trimmer holes were reduced by two also two trimmer holes from the arch were reduced and were redistributed as production hole as shown in Fig. 3. Also undercuts were observed at springing causing difficulties during lining hence two dummy holes were introduced below springing. With these changes pull was improved from 2.5 m to 2.8 m. per blast. Also widths at bottom were controlled. Also overbreaks were observed to the extent of 2 to 3% (average) and powder factor was improved from 1.2 to 1.07 kg/cum. Overall monthly progress achieved during the month was 144 m and is one of the best achievement with full face excavation by conventional methods.

RISK ANALYSIS AND DECISION MAKING TECHNIQUES FOR LARGE UNDERGROUND PROJECTS

A.K. MISHRA
Chief Engineer (Civil)

B.C.K. MISHRA
Sr. Manager (Civil)

AJAY MATHUR
Manager (Civil)

National Hydro-electric Power Corporation Ltd.
Kurichu Project, Bhutan, India

SYNOPSIS

Risk analysis and decision making is a systematic process of managing an organizations risk exposure to achieve its objectives in a manner consistent with public interest, human safety, environmental factors and the law. It consists of planning, organizing, leading, co-ordinating and controlling activites undertaken with the interest of providing an efficient pre-loss plan that minimizes the adverse impact of risk, risk on organizations resources, earnings and cash flows. In this paper authors have tried to compile all relevant experiences pertaining to risk analysis and its use in decision making. In India huge investments are likely to pour in those projects where underground structures are involved such as hydroelectric projects having underground works worth 50% of total cost of the project, metro rail communication, big size underground structures for storage of LPG etc. This will requir changes in some traditional forms of management, design, procurement and construction. This in turn will require changes in the responsibilities and working relationship between owner's engineers, contractors and consultants, Owner's practice to select lowest bid price has revealed over the years that in this concept contractors generally do not use latest technology and methods.Contractors will tend to use the cheapest machine that they believe will work and then they look for chances to use the contract clauses if problems arise. To overcome this contractual situation, a shift is required from performance to prescription based specifications which can be evolved after having a risk analysis.

1.0 INTRODUCTION

In spite of having a realization in the country that hydro-electric projects are the source of most economic and clean power, the investment from private sector is not coming up. The share of hydro power in the total installed capacity of the country is declining steadily. The hydro share has declined from 44 per cent in 1970 to 25 per cent in 1998. Now the scenario is that to maintain hydro power share at 25 per cent, the capacity addition during 9th and 10th Plan would work out to be 23000 MW. If the share is to be enhanced to 30 per cent, it will require a further addition of 10,000 MW of hydro capacity.

As per present hydro policy capacity addition of hydro power has been planned for 9818 MW in 9th plan (1997-2002), 21060 MW in 10th plan (2002-2007) and 23000 MW in 11th plan (2007-2012). When so much of hydro power potential is planned for exploitation, large underground caverns to contain power houses and long tunnels for conveying water to power houses and further discharging water back to rivers shall have to be excavated. And for successful execution of the project, risk analysis and decision making techniques in underground structures of the project are very relevant.

It is seen that assessment of risks and their adequate provision in the project estimate is missing. Contract documents are framed largely leaving risk handling to the contractor. In turn, the outcome is serious, leading to disputes between owner/client and contractor. Loss of time and increase in cost of the project has become a fate accompli in present environment. Now-a-days, foreign investors are approaching to invest in hydro power but investment is not taking place. In this the major issue is, quantification of risk and its sharing. Until and unless a mechanism is developed for assessment of risks in a particular project and also its cost and time impact, along with the risk sharing formula, the development of hydro projects will not get a boost.

Any investment in today's competitive world requires guaranteed returns and also full safety against risks involved in execution. Investor can come to decision making stage if risks are rationally analysed, quantified and clarity exists on the sharing between owner/client and contractor.

2.0 RELEVANCE OF RISK ANALYSIS AND DECISION MAKING TECHNIQUES IN LARGE UNDERGROUND PROJECTS

Risk analysis and decision making techniques are basically tools for protection of agencies involved in execution of a contract such as contractors, owners, joint venture participants etc. against losses not considered while entering into agreement and not covered in the price bid.

Risk analysis is extremely helpful in minimising risks arising out of negative or adverse events where as decision making techniques will provide ways for responding to risks to the project and maximizing results of positive events.

Important characteristica of underground works which have been observed over the years are as below :

- Complex projects with many variables
- Expensive - many stakeholders involved
- High media and political attention
- Uncertain and variable ground conditions
- Construction sensitive to means, methods and ground conditions
- Laws and regulations may be restrictive
- Similar projects show great variation in different locations

Based on the above characteristics, the major risks often encountered in execution of a project are following:

- Injury or failure with loss of life, extensive property and economic damage
- Significant increase of cost
- Delay in completion
- Inability to meet design, operational,maintainability and quality standards

In the present scenario, where the huge investments are required in underground projects, the risk analysis has become very useful tool to forecast the viability of a project. Now a days it is experienced that most of the projects may be technically feasible but may not be attractive from investment point of view. Therefore, in such situations, risk analysis helps in concluding the decision.

2.1 Risk Analysis

Process of risk analysis involved following steps :

- Identification of risk
- Quantification of risk
- Developing process for responding to risks
- Controlling and managing risks.

A flow chart of project risk management is appended at Annexure-I.

In underground works basically risk identification involves determination of risks, which are likely to affect performance of contractor or joint venture partners. The identification of risks have to be properly documented with their characteristics. Risk identification is a process and not an event. This can be done throughout the project on regular basis.

Risk identification can be accomplished by identifying causes and possible effect *i.e.* what can happen and what will be the result. Also it can be done by analysing effects and probable causes *i.e.* what outcome is to be avoided or encouraged and how it can occur.

Identified risks can be internal or external. Internal risks such as estimation of project cost, manpower planning, equipment planning etc. can be controlled or influenced by project management where as external risks are those which are beyond control of project management or contractor or joint venture partners.

2.2. Inputs required for Risk Identification

2.2.1 Project Description

Description of the project should be properly given to find out potential risks. Description should include :

- location of the project
- altitude
- distance from nearest railhead/airport
- condition of approach roads
- available transport facilities
- condition of bridges en route and their carrying capacity
- climatic condition
- flood conditions
- period of rainy season
- rain fall data
- snow fall data.

The information will provide tool for advance assessment of risks involved in taking up the project.

2.2.2 History of other Projects of Similar Nature in the Area

If some projects of similar nature have been built in the area, historical data pertaining to events causing hurdles to the project and subsequently resulting into losses can be well assessed.

Project records pertaining to risks encountered during execution and available in files can be looked into and probable risks can be identified for the new project. Also individual members of the project team can provide information about various occurrences during executions.

2.2.3 Planning Inputs

Once a project has been planned, input of resources planned for deployment, its analysis based on project description and historical data will help in identification of potential risks. If proven technology is being utilized for execution of project with trained work force, chances of risk hazards are less. Cost estimates if done with conservative approach may lead to estimation of quantities on lesser side and increase in quantum of work will result is time and cost overrun and may ruin project planning.

3.0 RESULTS OF RISKS IDENTIFICATION

Output of the analysis based on above inputs should be utilised to prepare a comprehensive list of sources of risk and potential risk events. Sources of risk should include all identified items regardless of frequency, probability of occurrence and magnitude of gain or loss. Common sources of risk include :

- modification in scope of work as per site requirement
- errors in design

- omissions in work requirement
- improper estimates
- poorly defined roles and responsibilities
- availability of skilled workmen in the area
- ambiguous contract clauses etc.

Potential risk events should include :

- force majeur condition such as excessive rains, snowfall, storm, floods, land slides, road blockade, law and order situation etc.
- geological failure
- probability of occurrence
- expected timing
- anticipated frequency
- results of occurrence.

Potential risk events in underground works of a hydroelectric projects are :

- Risk on account of investigation carried out at feasibility stage

In general, investigations at feasibility stage are carried out by the owner and at the time of execution of work, it is observed that variation takes place between the prediction and the ground realities. This leads to deviation in the planned methods and leads to cost and time over-run. Now the question is who should bear the cost?.

- Ambiguous contract clauses

The contract clauses are written in general and they are not situated specific and result in ambiguity and affect the performance of works. Example is over-break in underground excavation which should range between 5 to 10 per cent but always exceeds. Similarly chimney formation or abnormal cavity formation is never separately defined for measurement and payment purposes.

- Construction planning

Construction schedule is usually chalked out by the owner and contractor is required to implement. By and large it is seen that there is a wide gap between the targeted progress and the achievements by the contractor. Construction schedule with various options should be prepared by the contractor and the best option available among them should be approved by the owner because in both these alternatives of risk remains undefined, but in second alternative the level of risk can be brought down if the contractor is made clear through contract regarding the quantification of risk and its sharing responsibility.

- Design changes and omission in work requirement

It is seen that final designs go on concomitant with the progress of work but at times it is seen that designed alignment of tunnels required looping due to geological reasons. Similarly underground caverns excavation methodology changed due to change in final designs. Such design changes turns cost and time and both the parties of contract gets affected but final decision of sharing time and cost remains with owner and till end of contract remains unclear on account of risk sharing due to above changes.

4.0 RISK QUANTIFICATION

Sources of risk and potential risk events with interactions should be evaluated and risks should be quantified to assess the possible outcomes. Assessment of risk quantification is very complicated because of various reasons. For example a simple risk event of cavity formation can cause multiple effects such as idling of equipments, idling of manpower, delay in schedule, penalty an account of delays etc.

Also opportunities and threats resulting from any potential event can interact. For example delay in construction schedule of a underground excavation project may force the owner to change in strategy which may result into early commissioning of project but impact of such changes on cost remains a dispute between owner and contractor.

After risk quantification financial implication likely to arise out of its occurrence should be calculated. Provision should be kept in cost estimate of the project under a special head or under contingency to meet expenses in the event of occurrence of the risk.

Risk workshops can be organized by the owner in order to allow the project team to quickly identify, quantify and evaluate potential threats, develop possible mitigation or risk reduction strategies, determine cost/benefit of these strategies and decide a prudent course of action. The risk workshops work to :

- Identify potential threats to project
- Examine related linkages and casual risk drivers
- Quantify the impact of such threats if they should occur
- Evaluate the probability of the threats occurring
- Determine risk - defined as the product of impact and probability
- Prioritize risks for action
- Determine action plans for top ranked risks
- Determine cost and cost/benefit for each action plan

5.0 RISK RESPONSE DEVELOPMENT

Risk response development involves defning enhancememt steps for opportunities and response to threats. Responses to threats generally fall into one of the three categories:

- **Avoidance** - eliminating a specific threat, usually by eliminating the cause. The project management can never eliminate all risk, but specific risk events can often be eliminated.
- **Mitigation** - reducing the expected montary value of a risk event by reducing the probability of occurrence by using proven technology, reducing the risk event value by taking insurance coverage
- **Acceptance** - accepting the consequences. This acceptance can be active by developing a contingency plan to execute should the risk event occur or this can ne passive by accepting a lower profit it some activities overrun.

The following tools and techniques can be used in risk response development:

(i) **Procurement**

Acquiring goods or services from outside the immediate project organization, is often an appropriate response to some types of risk. For example, risks associated with using a particular technology may be mitigated by contracting with an organization that has experience with that technology.

(ii) **Contingency Planning**

Contingency planning involves defining action steps to be taken if an unidentified risk event shoukd occur.

(iii) **Alternative Strategies**

Risk events can often be prevented or avoided by changing the planned apporach. For example in an underground project additional design work may decrease the number of changes which must be handled during the implemenation or construction phase.

(iv) **Insurance**

Insurance or an insurance-like-arrangement such as bonding is often available to deal with some categories of risk. The type of coverage available and the cost of coverage varies by application area.

The risk response development process results in a risk management plan. The risk management plan should document the procedures that will be used to manage risk throughout the project. In addition to documenting the results of risk identification and risk quantification processes, it should also cover who is responsible for managing various area of risk, how the initial identification and quantification outputs will be maintained, how contingency plans will be implemented and how the reserves kept for such eventualities will be allocated.

6.0 RISK SHARING

Risk sharing between client / owner and contractor has always been a subject of debate and most of the contracts being framed and under execution are having non-transparency. For example excavation of a chimney or cavity formation in a tunnel or underground cavern is paid at 50 per cent of rate whereas concerting is paid at 75 per cent of rate. This is a risk sharing by an owner but on lower side, especially when geology is investigated by the owner. Crux of the dispute is transfer of the risk from one to other. In such situations both parties suffer loss due to absence of quantification in the beginning and non-existence of the provision in Bill of Quantities invites claim and counter claims. Therefore it is advisable to follow two principles in allocating the risks while framing contract documents :

- Risk should be allocated and managed by the party that is in the best position to control the risk.
- Risk should not be allocated to a party that could not survive the financial consequences if the risk materialized (unless insurance cover is made mandatory).

When the risks are shared then it should be clearly specified. Ideally each party should have a stake in minimizing the consequences. An example of shared risk is the extension of time clause found in contracts under which the owner shares the risk due to delay in time but the contractor bears the cost of such delays.

In Indian scenario most of the large infra-structural projects including big hydro power projects are being undertaken mostly by Government agencies and for successful and timely completion of these projeccts the following risks ought to be taken by the owner :

(i)   Acts of god/ force majeure.
(ii)   Adequacy of project budget/ funding.
(iii)   Adequacy of design.
(iv)   Ambiguous contract documents.
(v)   Changes in the law after the tender date.
(vi)   Variations.
(vii)   Delay in decision making.
(viii)   Delays in transmitting information.
(ix)   Site condition more onerous than predicted.
(x)   Ground characterization including sub-soil conditions.
(xi)   Unforeseen existing utilities and underground facilities.
(xii)   Quality of construction management/ administrtion.
(xiii)   Timeliness of inspections/ decision making.
(xiv)   Site availabilty

The attitude that imposing the risk on the contractor leads to price certainty and cost saving is incorrect. It is inevitable that where more risks are imposed contractors tender prices will be higher. The foundation is also laid for a dispute-riddled project and at worst the owner faces a defaulting contractor. All the above mentioned risks should be borne by the owner because not only are they risks which are not reasonably foreseeable by the contractor at the time he enters into the tender, but more importantly, the Government ot its consultant are best able to deal with those risks if they arise.

No contractor likes losing money. The more money the contractor stands to lose on a project the more likely he is to attempt to challenge the owner. In every contract, there are possible loopholes or ways out. A dispute prone project will often lead to significant extra costs and long delays.

The risk of claims and disputes is always present and taking a construction dispute to arbitration or litigation is commonly taking several years and costs several millions of rupees. While claims and disputes will never be entirely eliminated but principals or owners can reduce the risk by ensuring that risks are fully identified and appropriately allocated, that an appropriate contract delivery approach is adopted and that contract documents are carefully drafted by experienced and qualified persons.

7.0 DECISION-MAKING TECHNIQUES

Developing a process for responding to risks and controlling and managing risks come under decision-making techniques. These techniques are required to be chosen properly for long term survival. Techniques can be simple or very sophisticated depending on importance and complexity of the problem. As process of decision-making at the last moment is not liked by many and no one likes to take decision until and unless there is crisis it is always better to do advance planning for decision making. Advance planning makes you to take decisions in a very comfortable and intelligent way by deciding guide lines in advance and goals for decisions. As such advance planning can be termed as decision simplification technique. Following are the four benefits of advance planning for decision-making :

7.1 Replacement for Crisis Management

Replacement for crisis management or management by fire fighting is replaced by planned proactive (taking control of situation) and conscious series of options, for example in a project a major cavity was encountered and tunnel driving came to halt. There was no cash flow of contractor as progress of tackling of cavity was very slow process. Tunnel excavation was being carried out in multiple drifts and there was no provision in the contract for this method of excavation. Establishment expenses of the contractor were very high and the contractor was not able to make two ends meet. He had no option but to stop the work. The crisis could be averted only after rates were decided. Had there been provision in the contract for multi drift excavation this situation would not have arisen. This decision could have very well been taken in advance either while preparing the estimate or at tender stage itself.

7.2 Provides A Guide Line for Measurement

Once a risk is identified and quantified in advance and one decides mode of measurement if risk is encountered, it can be used for drawing a guide line for measurement when same risk event or risk event of similar nature is met with in the course of project execution.

6.3 Allows Conversion of Values to Action

When one encounters many decisions to a problem, for proper decision one can refer his plans and decide which decision is best suited to the objective of the project.

7.4 Planning of Decision will Lead to Systematic and Economic Utilization of Resources

In every project, contractor wants to keep optimum resources and make best use of it. Apart from main work, many items of work arise to which contractor is not able to attend due to limited manpower, time constraints and finances. If planned decisions exist they can be implemented to avoid idling of resources when any risk event is encountered.

7.5 Levels of Decision-making Techniques

There are three levels of decision-making techniques :

7.5.1 *Strategic Decision*

These decisions are of most important nature. In such decisions one decides directions, and targets to be achieved. They are very risky and out come is very uncertain.

7.5.2   *Tactical Decision*

These decisions are to be taken to support and implement strategic decisions.

7.5.3   *Operational Decision*

These decisions are required to be taken on day to day basis to support tactical decisions and in turn strategic decisions.

Any risk event should be examined at all the above three levels of decisions and should be sequentially followed from strategic to operational decision.

7.6   Techniques of Decision-making

A few practical methods which can be used to take decision of simple or complex risk events are listed below which will provide a systematic and structured set of factors involved in the decision. Decision-maker can consider them in thoughtful and coherent way to arrive at a decision.

7.6.1   *T-Chart*

. A T-chart is an orderly representation of alternative points involved in a decision. Drawing up such chart will ensure that positive and negative aspects of decisions can be considered.

For example the decision to buy a Tunnel Boring Machine (TBM) or a Drilling Jumbo, for a large underground work, so as to minimise risk can be analysed through T-chart as below :

Drilling Jumbo		Tunnel Boring Machine
Cost	-	Less / risk less	High / risk high
Performance	-	Certain (risk less)	Uncertain (risk high)
O&M charges	-	Less	High
Know how	-	Can be easily made available	Can be had from specific company
Progress	-	Will be less compared to TBM	Will be high compared to Drill Jumbo

If there are more than two choices available, the same can be added in the table and can be analysed and as per actual quantification of risk, decision can be taken.

7.6.2   *PMI*

T-chart idea was refined and a three part structure called PMI (Plus, Minus and Interesting) was formed to list all plus points, minus points and interesting points of an event so that areas of uncertainty or consequences are determined and decision taken.

7.6.3   *Buriden's Ass Analysis*

The name has been picked up from an old fable of an ass placed between two equally nice bales of hay. The ass could not decide which bale to turn to because they were both so attractive and it starved to death from in-decision. This method of analysis of decision is used when two or more alternatives of equal attraction are encountered. In this method all minus or bad points of the decisions are listed and one with minimum drawback is finalised.

7.6.4 *Measured Criteria Technique*

In this method each of the criteria, one's decision is required to meet, is assigned points based on relative importance and then each alternative is given a certain number of points according to how fully it meets the criteria. Decision is taken with alternative having maximum points.

7.6.5 *Weighted Decision Table*

It is a sophisticated version of the measured criteria technique. In this method a table is set up with each criteria with given weightage depending on its importance in the decision. Now each alternative is given a ranking for that criteria according to how well it meets the named criteria in each case. After ranking all the alternatives, total points are added and one with maximum points is chosen.

8.0 CONCLUSIONS

Risk analysis and decision making based on the analysed results will lead to a new era of technological age where nothing would be hidden in terms of time and cost. Instead of sharing risks as owner on award of an arbitration or litigation case, let us underline risk component in the beginning and eliminate contractual disputes.

Serious thinking is required on the lines of creating risk funding agencies, different from insurance companies and a tripartite agreement among owner, contractor and risk funding agency will entirely cover all possible risks and hydro projects could be executed at a fast pace.

PROJECT RISK MANAGEMENT

Risk Identification

1. Inputs
    1. Product description
    2. Other planning outputs
    3. Historical information

2. Tools and Techniques
    1. Checklists
    2. Flowcharting
    3. Interviewing

3. Outputs
    1. Sources of risk
    2. Potential risk events
    3. Risk symptoms
    4. Inputs to other processes

Risk Quantification

1. Inputs
    1. Stakeholder risk tolerances
    2. Sources of risk
    3. Potential risk events
    4. Cost estimates
    5. Activity duration estimates

2. Tools and Techniques
    1. Expected monetary value
    2. Statistical sums
    3. Simulation
    4. Decision trees
    5. Expert judgement

3. Outputs
    1. Opportunities to pursue, threats to respond to
    2. Opportunities to ignore, threats to accept

Risk Response Development

1. Inputs
    1. Opportunities to pursue, threats to respond to
    2. Opportunities to ignore, threats to accept

2. Tools and Techniques
    1. Procurement
    2. Contingency planning
    3. Alternative strategies
    4. Insurance

3. Outputs
    1. Risk management plan
    2. Inputs to other processes
    3. Contingency plans
    4. Reserves
    5. Contractual agreements

Risk Response Control

1. Inputs
    1. Risk management plan
    2. Actual risk events
    3. Additional risk identification

2. Tools and Techniques
    1. Workarounds
    2. Additional risk response development

3. Outputs
    1. Corrective action
    2. Updates to risk management plan

ADMINISTERING COST IN SOIL TUNNELLING

G. NARAYANAN

Chief Project Manager/Construction

Southern Railway, India

SYNOPSIS

Cost and time overruns are inherent in major tunnelling projects in India. This subject has assumed more seriousness and significance in projects funded by non-governmental resources like raising loans through bonds. Repayment of loan and interest liability will have serious set back in case of abnormal cost and time overruns. In soft ground tunnelling, uncertainties and risks are more compared to rock tunnelling and probable contingencies to be kept in view while costing the tunnelling works in soft ground are discussed in this paper.

1.0 INTRODUCTION

Tunnelling has been undertaken since pre-historic times. Industrialisation and commercialisation have invariably given an impetus to the development of road and rail networks, hydropower generation, irrigation, mining etc. which involve construction of tunnels. Poor tunnelling rate leads to long gestation periods and high financial costs. As tunnelling in soil has lot of uncertainties, cost conscious approach is necessary right from planning to completion of tunnelling with maintenance period. Rock tunnelling can be executed without much overruns as tunnelling technology is well established whereas in soil tunnelling, geological and geotechnical risks are more and there is no well established time tested method of excavation and supporting in all weather conditions. Controlling of cost overruns and associated cost consciousness right from planning to completion are discussed in this paper.

2.0 ALIGNMENT

2.1 The choice of the alignment is the first and foremost important aspect having a bearing on cost. A right decision on the alignment at the initial stage itself by an experienced engineer in tunnelling will pay rich dividends in reduction in overall cost of construction. It is better to increase earthwork and bridges rather than going in for long tunnels especially in soft soils.

2.2 If tunnelling in inescapable, it shall be located in a harder strata even with a slight detour or level change which is strategically better than facing tunnel hazards typical in soft grounds.

2.3 With geological mapping of area and remote sensing satellite imagery, alignment in stable ground can be chosen out of various alternative alignments. Alignment giving shortest length of soil tunnel need not be the best if hard stratum is available in longer length. Lined tunnel in soft ground costs three times that of lined tunnel in hard ground and five times that of unlined tunnel in hard rock. Conventional tunnelling in hard rock will give a progress of about 80 m per month and with electrohydraulic drilling jumbos and loaders progress can be as high as 160 m per month whereas progress in soil tunnelling can vary widely from 10 m per month to 45 m per month depending on face and side soils stability. With this in background, soil tunnel shall be proposed only if it is unavoidable and alignment shall be so chosen to have minimum hassles. Tunnel length alone is not the criteria for choice of alignment but associated cost of tunnelling in various strata shall also be considered. Tunnelling costs in various types of soils and at various moisture contents are unfortunately not available as experience is limited in India and therefore client as well contractor have to make guess only on cost estimate.

3.0 INVESTIGATION

3.1 When land is acquired, associated geological uncertainties and geotechnical complexities are also acquired. It is difficult for the client to fairly estimate the cost as well as for the tenderer to quote the rates realistically with these uncertainties and possible geological surprises while tunnelling through soft terrain. There are many contractors who willingly come forward for rock tunnelling but seldom competitive offers are received for soil tunnelling.

3.2 Method of excavation, type of supporting system and thereby the rate for tunelling cannot just be decided with conventional soil borings at 100 m intervals. For fair costing, soil explorations shall be maximum at 30 m intervals and shall cover along the cross section also. Assessment of stand uptime depends on behaviour of soil in summer as well as in monsoon seasons. Ground radar equipments supplemented with electrical resistivity test equipments can be used for soil property mapping in three dimensions. To decide on shield tunnelling (open / closed shield) or conventional heading and benching or multidrift system or ground improvements prior to excavation, soil properties at various moisture contents are necessary. Hence soil investigation plays a major role in deciding the method of tunnelling and thereby the cost of tunnelling.

3.3 Geological mapping of terrain is necessary for assessing instability during tunnelling and in service. Geologically unstable overburden may cause landslides and location of portals shall duly take care of this aspect. To cater for possible instability of overburden, tunnel length is advanced artificaly by about 30 m and tunnel entry will have only minimum damages in the event of major landslides on top of tunnel. This can be an optional item in bill of quantities.

3.4 In one of the soil tunnelling projects, due to inadequate assessment of soil stability at face, full face excavation with open shield was not found possible, resulting in change of method of tunnelling during the course of tunnelling and original cost and time scheduling with shield were very much affected. From standard soil test results, only an experienced tunnel engineer can fairly predict the stand up time which is a function of overburden, heterogenity of overburden and depth of excavation.

3.5 Fair assessment of stand up time leads to fairly reliable method of supporting immediately after excavation and thereby costing of works. Amount spent in 'in-depth' investigation will never be infructuous and on the contrary, will reduce the element of uncertainty while costing the works.

4.0 PROGRESS DEPENDENT DECISIONS

During the course of tunnelling, decisions of minor or major nature might be taken depending on soil stability. Major decisions of converting open shield to closed shield due to heavy face collapse and then resorting to conventional heading and benching due to poor progress with closed shield and then opening more fronts to accelerate the progress were taken in one of the soil tunnelling projects. Cost re-scheduling of the project arising out of redundant infrastructure of shield and segment erector has to be carefully considered. Similarly, a major decision of cut and cover for part length may be taken depending on safety and progress of works which also affects the cost scheduling seriously. Such eventualities are quite possible in soil tunnelling but the same cannot be quantified and loaded into offered rates. This uncertainty has to be inbuilt in risk clause of the contract and not to be included in estimated cost / rate offered.

5.0 TUNNEL HAZARDS

The following hazards are possible in soil tunnelling and tenderer and client have to be aware of such occurances:

(i) Sinking of wall plates.
(ii) Squatting of ribs.
(iii) Buckling of ribs / posts.
(iv) Inward movement of ribs / posts.
(v) Cracking / bending of laggings.

(vi)   Face / side soil collapse (heavy).

(vii)   Heavy mud flow.

(viii)   Gushing out of water / gas / oil.

(ix)   Bulging / cracking of lining.

(x)   Ground subsidences / land slides on top of tunnel.

(xi)   Nose diving of shield.

(xii)   Craters formation.

(xiii)   Chimney formation/ day lighting.

(xiv)   Cavities in strata arising out of tunnelling.

(xv)   Jumping of well shafts / tilting.

Prevention is better than cure. Most of the foregoing failures can be prevented by thorough geological and geotechnical mapping of tunnel zone. Arch action, whether possible or not, has to be analysed from soil investigation of overburden strata before designing the temporary and permanent support systems. However, some geological surprises are inevitable as every foot of the strata cannot be explored practically. For certain uncertainties, cost of remedial measures cannot be padded up in rate offered and such an eventuality has to be tackled on cost plus basis only with the same tunnel agency or through single / limited offers from other agencies if tunnel agency is not in a position to execute the remedial work. Items (x), (xii) to (xv) can happen unexpectedly whereas for other items, failures can be forecast based on geological and geotechnical investigations.

6.0 RISK PLANNING

6.1 Soil tunnelling has lot of risks compared to rock tunnelling as soil strata and behaviour can vary drastically with respect to space and time.

6.2 Loss of life, shield becoming unserviceable or redundant, damages to costly machineries, full work or part of work executed becoming redundant or to be redone, unpredictable geological and geotechnical failures, public litigations, environmental objections, change in government policy drastically affecting the rates offered, delayed progress due to uncontrollable parameters etc. are some of the risk elements contractor has to face during soil tunnelling. One fatal accident may result in loss of few weeks work, loss of morale of other workers in addition to loss of precious human life. Further, in case of loss of life, workforce may not turn up for several weeks due to psychological fear. Same is true with possible infectious diseases in underground working. Sometimes opposition may be there in use of local labour force due to such eventualities and labour has to be imported from other states.

6.3 Insurance cover is therefore necessary against such risks. Contract clauses may explicitly cover shared risks or risk of individual parties to the contract. System of insuring performance is necessary and premium payable against such insurances has to be forecast in costing by tenderer as well as client as the case may be.

7.0 TENDER CONDITIONS

7.1 Tenderer will offer a competitive rate if his financial hardship is taken care of by proposing following contract conditions :

(i)   Mobilisation advance against Bank Guarantee.

(ii)   Machinery advance against Bank Guarantee.

(iii)   Acceleration advance for speeding up progress more than that originally contemplated.

(iv)   Advances for executing works arising out of tunnel hazards.

(v)   Stage payments with respect to major items of works.

(vi)   Supply of materials – cement, structural steel, diesel for running machineries – for enabling works.

(vii)   Concessional recovery of advances as per progress of works.

(viii)   Payment of 75 per cent of bill amount within 24 hours of preferring the bill and balance 25 per cent payment within three days.

(ix)   Price variation clause with components of works as applicable to soil tunnelling and not general tunnelling.

(x)   Reimbursement of taxes leviable on works contract.

(xi)   Licensing / leasing of land from government / private parties for contractors' work area.

(xii)   Licensing / leasing of quarries from government departments for quarry products.

(xiii)   Land availability for immediate commencement of schedule works and for dumping cut spoils.

(xiv)   Incentive / bonus on accelerated completion and penalty on delayed completion.

7.2   Some contractors are demanding concessional rate of interest against advances which is not possible as the client himself is borrowing capital at market rates of interest.

7.3   In many cases, clients are not settling the rates for new items of works in time and satisfactorily. Time bound settlement of rates and associated penal interest shall be specified in the special conditions of contract.

7.4   The unit of measurement for billing may be either volume in cum. of earth excavated or per metre length of tunnel. Generally in soil tunnelling, maintaining fixed excavation profile is very difficult and such extra quantities are not easily amenable for measurements. Hence unit of measurement for excavation and lining can be per metre length of tunnel.

7.5   Quantity variations due to denser spacing of ribs / posts as per site conditions and associated costs have to be settled without delay. Quantity of blocking concrete / spalls may be varied depending on cavities behind extrados. Such contingency expenditure may be upto 15 per cent of cost of lining.

7.6   The components – labour, fuel, explosives, detonators, other materials and fixed amount of standard price variation clause may be different for rock and soil tunnelling and even in soil tunnelling, these may vary depending on methodology adopted. While floating tenders, the client may not be knowing in greater detail the methodology likely to be adopted as it is dependent on tenderer's expertise and resources available with him and hence price variation clause as proposed in the tender may not be exactly comparable to components of actual execution of work. This likely variation has to be inbuilt in tender rates by tenderers.

7.7   Forming bank out of tunnel cut spoils has certain uncertainties as percentage of usable cut spoils may vary along the length of tunnel. Cut spoils may have to be temporarily dumped and rehandled in which case large dumping ground is required. Long leads may not be favourable to a tunnel contractor and leading through narrow roads or busy roads may invite opposition from the public. These aspects have to be kept in view while stipulating contract clauses / schedule for banking.

8.0   PRE-TENDER MEETING

Pre-tender meeting with prospective tenderers will help in reducing element of uncertainties. Opinions can be taken on :

(i)   adequacy of data proposed to be supplied to the tenderers or need for more investigation

(ii)   grouping of tunnel works spread over different sites for competitive bidding

(iii)   achievable rates of soil tunnelling, appropriate resource requirements as well as methodology and realistic completion time

(iv)   extent of financial assistance proposed by client, sufficiency of it, comments on risk sharing, relevance of components of price variation clause proposed by client, incentive / bonus clause

(v)   ground improvement as part of tunnelling work and in advance of excavation or only as post hazard management

(vi)   clarifications on government policies, statutory obligations, environmental issues, local public and labour issues, law and order problems etc.

Consultations with prospective tenderers will enable changes in proposed contract clauses to be made as applicable to the tunnel under reference through special conditions of contract. On broad terms, consensus can be evolved on tackling some uncertainties and to this extent tenderer will be clear while quoting the rates.

9.0 CONCLUSIONS

Costing is mainly evaluating and monitoring the cost of the project during various stages, right from conceptual stage of the project till its completion and its importance is emphasised in this paper in respect of soil tunnelling projects in which cost and time overruns have not been unusual. Exhaustive geological and geotechnical studies are necessary to reduce the uncertainties, typical in soil tunnelling and thereby reduce cost of tunnelling. Common failures associated with soil tunnelling are listed out in this paper to enable the client and tenderer to plan the methodology of tunnelling to suit site conditions and soil behaviour and to control the cost overruns. Works arising out of unpredictable ground conditions shall be executed on cost plus basis duly incorporating in contract clauses. Tenderer will have more confidence while quoting the rates, if shared risk is specified in contract clauses.

MANAGING HAZARDS IN URBAN TUNNELLING

R. PRASAD

Rail India Technical and Economic
Services (RITES), New Delhi, India

K. NAYAN

Delhi Metro Rail Corporation
New Delhi, India

SYNOPSIS

Tunnelling or construction of underground structures in urban areas is vastly different from tunnelling in non-urban or "greed field" sites. In addition to safeguarding the construction workers in the tunnel, urban area tunnelling must also safeguard the overlying utilities, roads and structures and ensure appropriate construction methods to control ground and ground water movements. This paper identifies the hazards involved with urban tunnelling and the measures necessary during the planning, design and construction of the project that will reduce the risk of occurrence of hazards. The concept of a Project Safety Review is also discussed whereby hazards are identified and risk mitigation measures reviewed at each stage of the project and at regular intervals during construction.

1.0 INTRODUCTION

Urban areas are characterized by a highly complex and complicated network of public utility services such as water mains, sewerage, electrical and communication cables, road network and concentration of buildings and structures in the vicinity. Tunnelling or construction of underground structures in urban areas is vastly different from tunnelling in non-urban areas because urban tunnelling brings with it many hazards in the form of damage due to subsidence to buildings, roads, other structures and utilities due to construction activities which may lead to loss of property and some times loss of life. Minimizing disruption to public life and ensuring full safety of the public is an important factor in urban tunnelling and plays a vital role in planning, design and selection of construction methodology. These can be ensured by suitable studies such as sub-surface investigations, traffic studies, utility layout investigations, building condition surveys, selection of suitable construction methodology, providing suitable measures/regulations in bid documents and detailed instrumentation during construction.

2.0 HAZARDS

BS 4778 defines a hazard as a situation that could occur during the lifetime of a product, system (includes construction phase of a project) or plant that have the potential of human injury, damage to property, damage to the environment or economic loss.

Appreciation of potential hazards is a necessary preliminary to the control of hazards. Following are some of the major hazards in the urban tunnelling which may cause damage to property and loss of human life.

2.1 Ground Movement/Ground Loss

The urban tunnels are either cut and cover or bored tunnels but the construction of both types of tunnels inevitably causes some movement. The construction of cut and cover tunnels involves settlement of adjacent ground due to movement of temporary support walls. Different wall construction methods will lead to different degrees of wall stiffness and associated movements. Diaphragms walls and secant pile walls are rigid, watertight and will minimise deflections in soft ground. Sheet piles are reasonably watertight but will lead to large deflections and ground movement in soft ground. Soldier piles and lagging are not watertight and can only be used in reasonably firm ground. The bored tunnels can be constructed either by New Austrian Tunnelling Machine (NATM) or by the use of Tunnel Boring Machines (TBMs). Ground movement occurs with both methods due to the movement of surrounding soil into working space or annular space left by the tail clearance.

Ground movement also occurs because of ground loss due to seepage of ground water. Seepage in cut and cover tunnel or bored tunnels may lead to ground loss due to carriage of soil particles with the ground water. This ground loss also has its implications during the operation phase of the tunnel and should be avoided. However, the impact during construction could be quite serious leading to localised excessive ground settlement and the development of voids below the ground surface.

In urban areas ground movements will effect the building and utilities along the alignment thereby endangering the structures and the occupants.

There are many instances where water ingress in to a tunnel through porous strata has cause extensive ground loss and subsidence. Fig. 1 illustrates the subsidence occurred during extension of under ground Munich Metro 1994, in which a crowded bus drove into a hole that suddenly opened in the middle of the road due to ingress of water, resulting in death of two persons and thirty were seriously injured. Fig. 2 shows the ground conditions and section through the resulting tunnel collapse. The ground water weakened the marl rock and allowed the water bearing gravels above the tunnel crown to flood into the tunnel excavation.

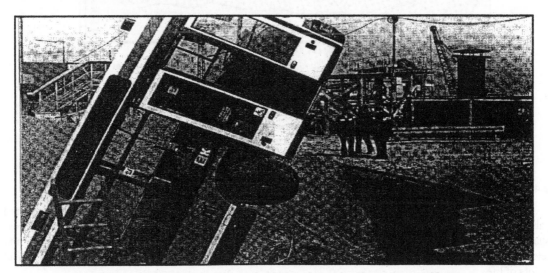

Fig. 1 : Photograph showing subsidence of road leading to bus accident

2.2 Ground Water Lowering

The threat from ground water is always a factor of paramount importance. In cut and cover tunnels the dewatering may lead to ground water lowering which may cause settlement of building and other structures due to consolidation. Any ground water based water supply system may also be effected. There is not much effect on ground water table in tunnelling by TBM.

2.3 Vibrations

Vibrations due to construction activity may effect the structures and utilities specially the water supply, sewage utilities. These can be induced from blasting in rock excavation, the use of heavy construction equipment, breakers etc. or from tunnelling machine. Vibrations become critical for old buildings with masonry foundation and for buildings, which already show signs of distress and can lead to increased damage. Old utilities are also sensitive and can fail if subjected to vibrations from construction works.

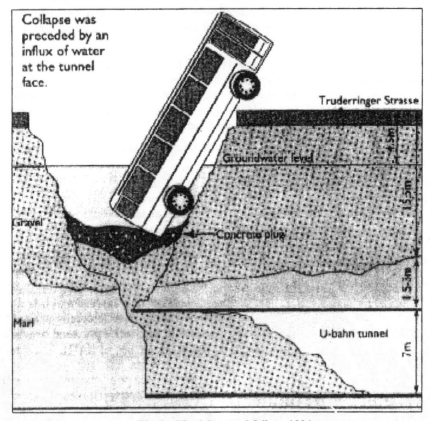

Fig. 2 : Munich tunnel failure 1994

2.4 Uncharted Utilities

One nearly ubiquitous hazard of urban projects concerns the presence of pre existing service pipes and cables either unrecorded by relevant authority or misplaced sufficiently to cause delay in construction of access shafts in bored tunnels and temporary wall in cut and cover construction. Misplaced underground power and telecom cables may lead to large financial loss from damage to cables, misplaced power cables may even lead to loss of life.

2.5 Heave due to Grouting

Grouting in various forms such as compensation grouting, jet-grouting, permeation grouting etc. is employed in almost all urban tunnelling projects to protect structures at the surface as well as to strengthen the ground to ensure safe excavation. However sometimes grouting may cause ground heave leading to damage to structures and utilities. Jet grouting in weak clays has caused ground heave to a considerable extent during construction stage of Singapore Mass Rapid Transit System in 1980's.

2.6 Compressed Air Blows

Compressed air was commonly used till recent time for stability of the excavation face in tunnelling by NATM or open face Tunnel Boring Machines. Although use of compressed air has declined due to medical concerns, examples of blows continue to occur where air finds a path to escape to the surface and can create a large cavity. A compressed air blow occurred during the construction of a railway tunnel in London in 1995 (Fig. 3).

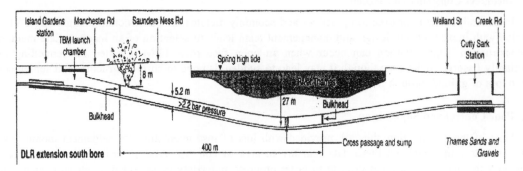

Fig. 3 : Compressed air blow in docklands light railways

2.7 Face Collapse

Face collapse is a risk more associated with NATM and open face bored tunnelling although it can also occur with the more modern closed face TBM's. This generally occurs due to low stand-up time of the excavation face soil and large unsupported excavation. The face collapse may cause large subsidence on the surface in addition to causing dangerous condition for the workers inside tunnel.

2.8 Contaminated Soil / Inflammable Gas

A particular hazard is the presence of methane gas associated either with landfill areas or old coal / oil workings, which can lead to an explosion. As methane is odourless and colourless gas circumstances leading to explosion may develop without any warning unless necessary steps have been taken against such incidences.

3.0 WHAT CAUSES THE HAZARDS ?

The principal causes of hazards in tunnelling may be expressed as :

Inadequate design

Unforeseen ground condition

Incorrect construction methods

Poor control /supervision

3.1 Inadequate Design

The designer must have a clear understanding of the ground conditions, the ground behaviour, water conditions and, most important in tunnelling works, he must understand the method of construction and its limitations.

3.2 Unforeseen Ground Conditions

Any amount of soil investigation cannot predict fully the sub surface condition to be encountered during urban tunnelling as many areas are built up and access is difficult, also the ground geology does not follow a set pattern and can change abruptly. For example, narrow buried valley containing alluvial deposit may be struck in a completely rocky terrain or a thin layer of weathered mica schist may exist in folded quartzite formation which may provide a passage for water and result in sudden large inflow of soil and water inside the tunnel during tunnelling.

3.3 Incorrect Construction Method

In general, current construction practices and economy dictates the construction methodology. This coupled with inexperienced design and management team leads to selection of an inappropriate method of construction. This situation can occur when an open face rock TBM is opted for tunnelling in marginally fissured rocks even though it may lead to sudden damage to surface structure and settlement due to loss of ground through fissures.

3.4 Poor Control and Supervision

Most of the construction sites are provided with lots of instrumentation for settlement monitoring, however follow up action on these data remains an area of concern. Also in order to achieve targets sometimes the management gives priority to faster progress and safety is neglected. A recent example of such lapse occurred during construction of Heathrow tunnel in London (1994) where indications from monitoring instrumentation of consistent circumferential movement of tunnel lining coupled with depression of crown were overlooked during NATM construction. Prompt reaction from the supervisors would have averted the collapse of tunnel.

4.0 HOW TO MANAGE HAZARDS ?

To control the hazards a holistic approach is required at each stage- planning, design and construction.

4.1 Planning Stage

During planning an investigation of site in the form of geo-technical investigation, building condition survey, utility layout survey and other topographical features is necessary to reduce the effect of unforeseen factors. At this stage in the project, it may not be possible to incur large costs as the project may not be fully sanctioned, however the following should be done if possible.

4.1.1 Geotechnical Investigation

To reduce the risks due to unforeseen ground conditions it is imperative to do geotechnical investigation along the alignment to ascertain the geological succession, geotechnical properties in terms of strength, deformability, permeability and spatial distribution of such materials.

The expected ground condition needs to be ascertained primarily by the use of site investigation and with additional information from historical records of other project in the area and other geological reports relevant to the area. Such reports may include details of watercourses, previous ground movements and previous construction problems encountered in the area.

However the geotechnical properties cannot be ascertained completely over the entire alignment as these investigations give only a fair amount of idea about the ground conditions. More intensive investigation is required where variation in the geological strata is noticed.

4.1.2 Building Condition Survey

In urban areas ground settlement will affect the buildings along the alignment. To ascertain the sensitivity of the buildings to the settlement condition, a survey of buildings is necessary. Buildings are then to be categorized as per the severity of the damage from "negligible" to "severe" as per Clarification of Damage to Buildings by Burland *et al.* (1974). A three-stage approach to building condition may be adopted to demarcate the high-risk buildings. Special protection measures may be required for these high-risk buildings. Normally the construction technology should be such as not to cause more than "Slight" damage to buildings. However for buildings of historical or other significance the degree of severity may be further limited. This approach will help in identifying the buildings, which require to be monitored intensively and require special measures to be adopted to minimize the risk during construction.

4.1.3 Utility Survey

It is necessary to chart all utilities *e.g.* water mains, sewer lines, power cables, telecom lines both above ground and underground at planning stage so that measures required to protect them are incorporated in construction. Physical verification of utility based on information received from utility agencies should be done by digging trenches. In old urban area the risk of striking an uncharted utility cannot be completely avoided because of non-availability of old records, non recording of modifications and changes made from time to time in layout / size of utility by utility agencies, incorrectness of location etc., but can be reduced to a large extent by doing physical verification.

4.1.4 Other Investigations

In urban areas wells, abandoned or in use, open bore wells along the alignment, landfills pose a potential hazard as such investigation of wells and landfills should be undertaken. The wells may cause the slurry to come out or provide a easy passage for the ground water to flood the tunnel. Where landfills are encountered the possibility of methane leakage during construction needs to be taken care of. Local archives and records should also be reviewed to gather as much information as possible.

All these factors should be taken in consideration while finalising the alignment of the tunnel and the alignment should be such that it leads to minimum hazards.

4.2 Design Stage

At design stage the consideration of risk must be at the heart of designer. The design should take into account.

- The geology along the length of the route
- The hydrology and ground permeability
- Maximum depth of construction
- Control over heave and instability of the base of excavation
- Measures against floatation

Design in case of cut and cover tunnels should define the type of temporary walls, methods of securing against floatation, measures against heave and instability, water tightness of structures and measures to control deflection of temporary walls.

Design for bored tunnel should define scheme of excavation in case of NATM method and in case of TBM design features for TBM apart from location of tunnel, distance between two tunnels, cover to tunnels, measures where cover is less, effect on heritage buildings and special measures for protection of heritage structures.

For tunnelling in urban residential areas the analysis of ground settlement and the assessment of related risks is a critical design issue. Combination of numerical and analytical methods are used for settlement analysis and estimation of the ground surface settlement in the transverse section of the bored tunnels is generally carried out following normal Gaussian distribution curve (Fig. 4).

The amount of ground loss in case of bored tunnelling is governed by nature of surrounding soil including the ground water condition and is strongly influenced by methods and details of construction. The ground loss may vary. Typical values of ground loss for different tunnelling methods based on tunnelling history on similar ground conditions are very helpful in reducing the risk. Modern TBM with face support using slurry or earth pressure should be able to limit face loss to around 1% (O'Reilly and New, 1982).

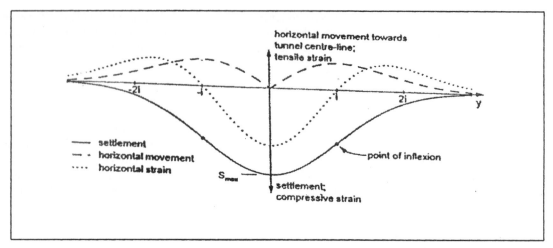

Fig. 4 : Transverse settlement trough

At the design stage, the designer should review the earlier risk assessment and update it in the line of his intended design to see how the hazards will now be affected with the design measures he proposes.

Stemming from the risk analysis the designer should review his understanding of the ground and carry out further bore holes in the areas where risk is unacceptable.

4.3 Provisions in Contract Documents

To control the risks the contract documents need to specify the following -

(a) Permissible ground settlement and damage to general buildings and historical and other significant buildings

(b) Draw down of water table

(c) Vibration limits for the buildings in the form of different peak particle velocity depending upon the importance and condition of the structure. Methods of limiting vibrations such as limiting explosive charges, operating pile driven equipment at lower energy levels, use of expensive agents or mechanical excavation methods may be specified at certain critical locations.

(d) Ground treatment at break-in and break-out location and ground treatment in advance of the tunnel excavation in certain vulnerable locations.

(e) Various requirements of monitoring and instrumentation during and after construction including provision of review levels such as trigger, design and allowable levels and specific action to be taken at various levels.

4.4 Construction Stage

During construction stage Instrumentation and Monitoring is extremely significant to ensure safety in urban areas. Early warning of any excessive movement, settlement, deflection or deformation during construction may help in avoiding an unforeseen hazard. Buildings along the alignment need to be monitored for settlement. Heritage and other buildings, which are very sensitive to settlement and vibration etc., need to be rigorously monitored.

Geological observations make direct contribution to the specific information required for successful advances of tunnel. During construction stage verification of initial design parameters need to be done and design should be optimized suitably. Following measures help in containing hazards in urban tunnelling during construction.

(1) Measures required for control of ground movement

    (a) Preloading of struts to control deflection of temporary walls.

    (b) Forward probing in case of bored tunnelling

    (c) Monitoring and controlling the earth / slurry pressure in EPBM / slurry machine

    (d) Monitoring the forward progress of TBM vis-a-vis spoil removal to ensure that soil beyond tunnel circumference is not removed

    (e) Continuous grouting between annular space behind the segments

    (f) Supporting the tunnel before removing segments for cross passage construction

    (g) Convergence monitoring in NATM to avoid collapse due to release of hoop stress

    (h) Large excavations to be split into small headings in NATM

    (i) Monitoring the shape of the headings in NATM

    (j) Preventing the compressed air to accumulate at any level at a pressure equal to or greater than overburden.

(2) Measures to control the ground water such as :

    (a) Jet grouting

    (b) Tube-a-manchette grouting

    (c) Ground freezing

    (d) Use of compressed air.

5.0 MEASURES PROPOSED FOR MANAGEMENT OF HAZARDS IN DELHI METRO

(1) All structures in zone of influence to be investigated to establish allowable settlement criteria. Zone of influence has been defined as 50 m on either side from tunnel center line or line of excavation plus a spread of 45^0, which ever is more.

(2) Design of all ground support wall shall limit

    (i) Settlement in the adjacent structures or ground to 25mm maximum

    (ii) Angular distortion in adjacent structures to 1:2000 maximum

(3) Construction - induced vibration to existing adjacent structures is limited as follows :

	Maximum allowable peak particle velocity (mm/sec)
Most structures in "good" condition	25
Most structures in "fair" condition	12
Most structures in "poor" condition	5
Water-supply structures	5
Heritage structures/bridge structures	5

(4) Limit on drawdown of water outside the excavation walls.

(5) Three-stage assessment mandatory to study the effect of settlement on buildings.

(6) The construction methodology should be such that no structure should undergo damage more than "slight" as per damage classification based on Burland building classification system. Damage classification is "negligible" for heritage buildings/structures.

(7) Early warning system must for recording any undue ground movement.

(8) Review levels for instrumentation defined. Monitoring of structures to start before the commencement of works. Monitoring equipments to be provided are inclinometers, extensometers, load cells, tilt meters, crack and wall settlement monitoring, piezometers, precise levelling points etc.

6.0   PROJECT SAFETY REVIEW (PSR)

Project Safety Review is a continuous process of carrying out risk assessments and then reviewing and modifying the project at each stage of planning, design and construction.

The work of tunnelling entails a degree of uncertainty. Risks may be foreseeable in nature but not in extent or precise occasion and may not be practically capable of total elimination. Mitigation is best achieved by :

1. At the planning stage describing the nature of the hazards, their probable or possible incidence and extent in relation to different forms of tunnelling, indicating the circumstances in which the hazards are likely to occur. At this stage hazards will be general in nature, such as buried valleys, boulders, soft ground etc. but the measures to reduce the uncertainty can be specified - such as more site investigation and testing.

2. At design stage the same review should take place of hazards but this time the additional measures identified at the planning stage should have been completed and more information is available. The hazard of a buried valley may now have been confirmed or it may still be a residual risk. The designer should therefore list out the "revised" hazards and also identify the measures needed to minimise these in tender documents. For example if the buried valley is now known, it may be necessary to specify ground treatment.

3. At the construction stage the same process is repeated in a more detailed manner as the actual construction methods proposed are developed. The risk of a buried valley may remain, but the construction methods can now be chosen to accommodate this risk, *e.g.* selection of a full face close TBM, forward probing or further site investigation. The monitoring data collected during initial drive can be used to verify the initial parameters, the predicted values to be compared with initial monitoring results and the design may be reviewed.

Using this structured approach to manage the hazards of tunnelling, the risks should be better understood and the means to tackle the risks should be identified.

7.0   CONCLUSION

Uncertainty is pervasive. Unforeseen things always happen and the risk cannot be avoided completely but this approach will minimize the risk and give greater certainty of project success.

REFERENCES

Marchini H. (1990) Ground Treatment in "Civil Engineering for under ground Rail Transport" pp. 57-86.

Alan Muir Wood Tunnelling Management by Design pp. 243 - 248.

Burland J.B. and Wroth, C.P (1974). Settlement of buildings and associated damage. SOA Review, Conf. Settlement of Structures, Cambridge, Pentech Press. London, pp. 611-654.

O'Reily, M.P. and New, B.M., (1982). Settlements above tunnels in the United Kingdom - their magnitude and prediction. Tunnelling '82, London, IMM 173 - 181.

UNDERGROUND TUNNELLING WORKS WITH TALA HYDROELECTRIC PROJECT AUTHORITY

P.E. SAGAR LAL
Directorate General Border Road (DSS Cell)
Delhi Cantt. New Delhi, India

1.0 INTRODUCTION

A 1020 MW hydel power project is under construction by Tala Hydroelectric Project Authority (THPA) which involves a total length of 22.64 km of tunnelling work. BRO is executing the road works for this project to provide road network for letting the private agencies like L & T, HCC and Jaiprakash Industries to execute their tunnelling work under contract packages.

Basic machinery and equipment concerning the tunnelling works had been procured by THPA from well-known foreign companies mastering tunnelling works and supplied to its contractors for speeding up the works, cost of which is debited later from the contractors.

Type of works, machinery and equipment, techniques involved and stores for tunnelling works are given in brief in the following paragraphs.

2.0 DRILLING OPERATIONS FOR TUNNELLING

Cross-section of approaches to the main head race tunnel (HRT) known as Adit is of semi-circular roof with vertical walls on both sides, where as actual cross-section of HRT is of horse shoe shape, Fig. 1 showing the cross sectional view of HRT in detail. Drilling through the rock face is basically dependent on the rock condition which has been divided into six types from Class I to VI as given below :

(a)	Class I	-	Very good rock
(b)	Class II	-	Good rock
(c)	Class III	-	Fair rock
(d)	Class IV	-	Poor rock
(e)	Class V	-	Poor rock/very poor rock
(f)	Class VI	-	Very poor rock

Rock condition is assessed by the Geologist/Engineer-in-Charge, while tunnelling is in progress by drilling probe holes of variable lengths. If the rock is of extremely poor condition then multi-drifting Process is used to excavate the loose material to obtain the desired shape of tunnel. Details of various internal support system for above classification of rock as shown in Fig. 2.

Drilling operation include the following :

(a) Drilling
(b) Loading
(c) Blasting
(d) Mucking

2.1 Drilling

(a) Drilling pattern to get the desired cross-section of tunnel is done in two types known as standard cut and parallel cut. Details of these patterns are shown in Figs. 3 and 4 respectively.

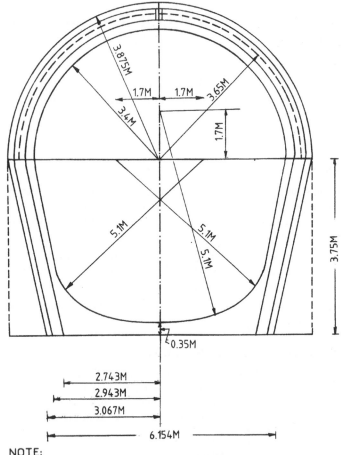

Fig. 1 : Cross-section of HRT

(b)   By drilling holes of 4.0 m long, pull achieved in 3.5 m Drill holes of 45 mm dia, 4.0 m long are drilled on the face of the tunnel with centre holes tapering inwards and diverging outwards gradually to become parallel at the periphery of the tunnel basically designed so as to pull the rock from the face of the tunnel. This has also been shown in Fig. 3.

(c)   Approximately 76 to 80 nos. holes are drilled at the face of the rock.

(d)   Drilling is being carried out by 2 Boom Hydraulic drill Jumbo having following specifications:-

(i)   Make Atlas Copco

(ii)   Type boomer 352

(iii) Feed length 4.3 m

(iv)   Rate of drilling per boom is 1 m/min

(v)   Above boomer has centre basket which is utilised to accommodate persons for carrying out loading of explosive into the drill holes.

(e)   Drilling time is approximately 2 ½ hours.

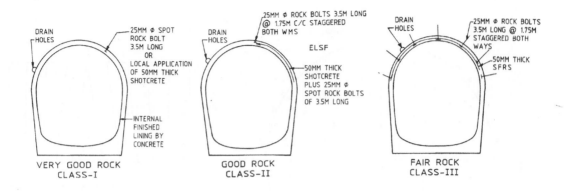

Fig. 2 : Classification of rock

2.2 Loading and Blasting

(a)   Loading of holes with explosive, detonators and blasting is carried out manually and consumes approximately 2 hours.

(b)   Powergel of diameter 25 mm, 32 mm and 40 mm are used as explosive. In place of ordinary detonators electric and non-electric detonators are used with a delay ranging from 0 to 11 secs. Use of electric detonators is restricted as unwanted explosion takes place accidentally due to charging of explosive by thundering occurring in this area, which is a common phenomenon during monsoon.

(c)   Placement of delay detonators in the holes is also clearly shown in Figs. 3 and 4 as pull achieved during blasting is directly ruled by the placement of these delay detonators.

2.3 Mucking

(a)   After the blasting is carried out 15 to 30 minutes is given for de-fuming from the face of the tunnel. Then mucking is carried by loader L & T make Pocklain type CK-90-3 (Elect) with 100 HP electric motor of 415 volts.

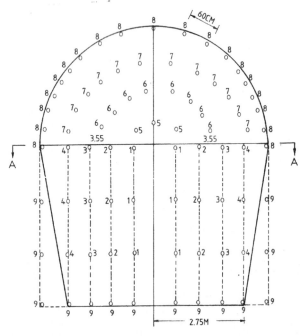

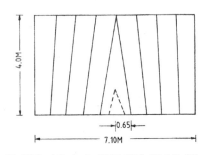

X-SEC. AT A-A SHOWING PLAN OF
DRILLING HOLES

NOTE:-

DOTTED LINE HOLES ARE KNOWN AS BABY CUT
USED ONLY WHEN ROCK IS QUITE HARD FOR
PULLING THE ROCK FROM THE CENTRE.

PATTERN OF DRILLING HOLES ON THE FACE OF TUNNEL

NOTE:-

1. NOS. BESIDE THE HOLES INDICATE THE NO OF DELAY DETONATORS TO BE
   USED FOR OBTAINING THE PULL OF ROCK FROM FACE OF THE TUNNEL.
2. DIA OF HOLES IS 45MM ϕ.
3. EXCAVATION LINE CATERS FOR 250MM RIB, 75MM RCC LAGGING AND 50-175MM
   SFRS I.E, APPROX 3.95M AT SPRINGING LINE AND 3.15M AT BASE.

4. ROOF HOLE (NO. 8)	- 18 NOS.
WALL HOLE (NO. 9)	- 06 NOS.
FLOOR HOLE (NO. 9)	- 08 NOS.
CUT HOLE (NO. 1 & 2)	- 12 NOS.
STOPPING UPWARD HZ. (NO. 384)	- 12 NOS.
STOPPING DOWNWARD HZ. (NO. 687)	- 20 NOS.
	76 NOS.

Fig. 3 : Standard pattern of drilling

(b) The muck is loaded by this Pocklain in following vehicles and taken outside the tunnel :-

    (i) WAGON MT 420

        (aa) Make Wagnor Corporation USA

        (ab) Four wheel drive, diesel powered with 20MT capacity

        (ac) With use of 4 Nos for a lead of 500 m, time taken is approximately 2 ½ ours to clear the muck from the end of tunnel.

    (ii) DEICEDUMPERS

        (aa) Italian make 12 ton capacity

        (ab) Approximate time taken to clear same muck as given above is 5 to 6 hours.

    (iii) Required equipment is used accordingly for gaining time and progress to be achieved.

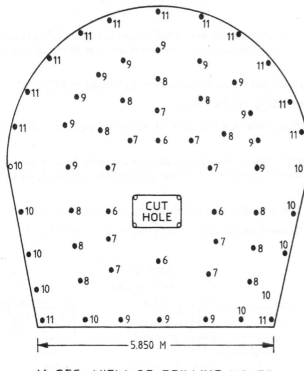

X-SEC. VIEW OF DRILLING HOLES

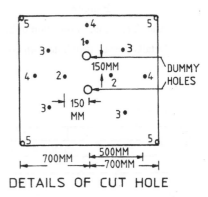

DETAILS OF CUT HOLE

Fig. 4 : Parallel cut

3.0 ROCKBOLTING

The procedure to execute rock bolting on the interior face of the tunnel is as follows:-

(a) Drilling of holes

(b) Shooting of resin and cement capsules

(c) Bolt insertion

(d) Plate tightening and rotation of rock bolt

(e) Torquing

1. Entire procedure is carried out mechanically by a single equipment known as BOLTEC having following specifications:-

(a) Atlas Copco BOLTEC 435 H

(b) Rate of drilling and bolting is approximately 6 bolts/hour for fixing a bolt of 3.5 m length in a hole of 38mm dia.

3.1 Rockbolts

(a) Bolting is successful by mechanical means upto class III rock. For latter classes of rock it is done manually.

(b) Rock bolts shall be tensioned to a load of 8 tons

(c) Bolts are of 25 mm dia, high yield strength tossed steel confirming to IS: 1786-1985. One end threaded with a coarse thread over a length of 20 cm.

3.2 Capsules

(a) Resin capsules are inserted into the drilling holes as end anchorages to contain 1/3 volume of the entire empty space after insertion of the rock bolt. Different types of resin capsules in use are :-

    (i) SIKA resin capsules of 28 x 250mm make SIKA Qualcerte Ltd..-Rokkon RFS

    (ii) Rock seal by Shree Vinayaka Engg. Pvt. Ltd., Delhi

(b) Approximately 4.6 capsules (resin) come in a rock hole of 3.5 m Setting time is 40 to 50 seconds.

(c) For balance $2/3^{rd}$ volume of empty space, cement capsules are inserted. Type of capsules being used are Combextra Capsule Std Set (25 x 250 mm) of FORSOC Chemicals (I) Ltd., Banglore.

After 2 to 3 min of insertion of bolts in the hole, and tightening of bolt, torque is applied to the bolt.

Details of rock drill hole with bolt in position and showing resin/cement capsules are given in Fig. 5.

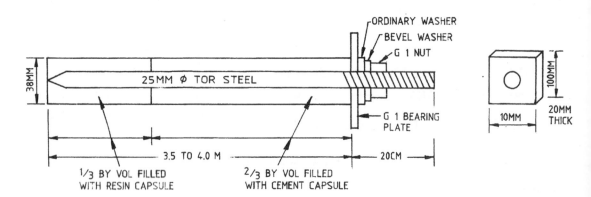

Fig. 5 : Placement of rock bolt and capsules in rock drill hole

Rock bolting is carried out only above the springing line of the tunnel staggering both ways @1.25 m c/c to 1.750 m c/c depending upon the rock condition. Details are also shown in Fig. 2.

4.0 REINFORCEMENT

4.1 Shotcrete

As soon as the rock face is cut and muck cleared from the end of the tunnel, the roof and sides of tunnel are given an application of shotcrete. It is of two varieties. One is plain and second is steel fibre reinforced shotcrete (SFRS):

(a) Plain Shotcrete : Specification of M-25, A-10 grade concrete. Accelerators of sodium silicate like SUPA set L-10, by Structural Water Proofing Co. Pvt. Ltd., Calcutta are used for quick setting.

(b) SFRS : Steel fibers are added in the plain shotcrete which acts as grouting carried out on link chain placed on the face of the rock. This avoids usage of link chain as these steel fibres are included within the concrete due their miniature size structure and inter locking: -

    (i) Steel Fiber : Supplied by Dramix Company looks like stappler pin big size of length 40 mm approximately.

    (ii) Details of steel fiber used in SFRS is given in Fig. 6

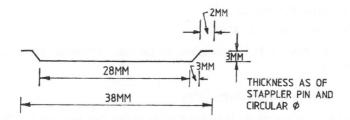

Fig. 6 : Steel fibre in SFRS

(c) Equipment: Shotcreting machine is of Italian make CIFA SPRITZ System Type C S S 2 electric powered with concrete pumping pressure of 70bar. Output is 30 cum/ hr.

(d) Thickness of shotcrete varies from 50 mm to 175 mm as per the classification of rock. This is given in details in Fig. 2.

4.1 Ribbing

(a) For rock classifications Class-IV and above, ribs are to be provided for reinforcing the entire structure of the tunnel internally. ISMB 250 x 125 and ISHB 200 x 125 ribs confirming IS: 226-1975 are placed on the interior face of the tunnel and anchored to the rock surface by 20 mm dia rock botls upto 1.0 m depth. They are placed at a distance of 600 mm c/c to 750 mm c/c, as per the classifications of rock and interlocked by ISMC 100 and tie rods of 25 mm dia plain steel rods with threading on both ends to set the c/c distance between the ribs.

(b) Pre-cast R C C lagging slab of cross-section 150 mm x 75 mm of variable length are placed between the ribs one above the other. Length of lagging is 40 mm lesser than the gap between the ribs.

(c) The gap between the face of the rock and supporting portal frame obtained by inter linking the ribs and RCC lagging is filled with M-15, A-10 grade concrete which is pumped by concrete pump having following specifications:-

    (i) Concrete pump greaves make BP-350E.

    (ii) Capacity 30 cum/hr

    (iii) Run by 45 kV/60 HP electric motor with rpm 1475

(d) The ribs are moulded to the desired shape of the tunnel required such as horse shoe.

(e) Transit concrete mixers like FIORI-SPA Modena Italian and DEICI;Italian make convey concrete from batching plants to the end of the tunnel of concreting work.

5.0 VENTILATION SYSTEM

Due to requirement of fresh air at the end face of the tunnel all the time and especially for de-fuming poisonous gases due to explosion, ventilation duct is placed continuously at the crown having following specifications:-

(a) 1600 mm dia flexible ventilation duct VENTIFLEX-PROTON NORWAY 62 FR 1600.

(b) Hung from crown with hook supports placed at 0.5 m interval.

Air is generally pumped in from outer end of tunnel through this ventilation duct by high capacity fans having following specifications:-

(a) ZITRON made from Spain

(b) Capacity 25 cum/sec with working pressure of 2100 pascals.

(c)   75 kV/1440 rpm motor

(d)   Input power supply of 415V

Air velocity of 0.3 m/sec is to be maintained all the time. If methane gas is detected then it is to be increased upto 0.5 m/sec.

6.0 OTHER MAJOR EQUIPMENTS/PLANTS

Batching Plant for concrete: Shirke Batching Plant by Shirke Structural Pvt. Ltd. Pune with capacity of 30 cum/hr.

Scissor Lifter : USHA Atlas Company make Scissor Lifter fitted on TATA chassis. Has a platform on the top which can be hydraulically raised to accommodate men for doing operations at the crown of the tunnel especially for fitting the ribs oat the crown point above the springing line of tunnel.

Air Compresseor : Atlas Copco GA 1107 Standard Elect with capacity of 600 cfm, working pressure 7 bar and 110 kV elect motor.

Electric Generator : 625 kVA Kirloskar Cummins 415 V 3 phase utilised as an alternative in case of main electric supply of 11kV fails.

7.0 OTHER ACCESSORIES

For supplying compressed air from air compressor to the end of the tunnel where the work is in progress 6 inch dia MS pipe line is used.

For supplying water to the working site where drilling is in progress, 4 inch dia MS pipe-line is used.

For stepping down power from main electric supply of 11kV, XLPE cables of 3 core 50 mm sq are used.

For conveyance of power from transformer to inside of tunnel following cables are used:

(a)   400 mm sq LT 3 ½ core PVC insulated cable 3 phase, 415 V are used from transformer outside to the tunnel to control panel inside the tunnel.

(b)   From the control panel to actual working site for equipments at the last stretches where work is in progress 240/185mm sq 3 ½ core LT PVC insulated cables are used.

8.0 MULTI-DRIFTING SYSTEM

This type of drilling system is applied when the rock condition is extremely poor. Cross-section of tunnel is divided into various segments and then excavation is carried out in each segment starting from the roof of the tunnel. Excavation of rock above the springing line is called heading and that below the springing line is called Benching. Immediate reinforcement is done in each segment to avoid collapse of tunnel at any stage.

9.0 CONCLUSION

Cross-section of tunnel for adits i.e., semi circular roof with vertical sides can be used in BRO for conveyance of traffic. There is no requirement of internal lining of concrete which otherwise is a must of conveyance of water in hydel project tunnels.

Well advanced Hi-Fi survey instruments of Leika company and Laser lights are required to monitor the alignment of tunnel and gradient for greater accuracy.

SELECTION OF OPTIMUM DIAMETER OF TUNNELS
FOR HYDRO GENERATING STATIONS

M.G. SHARMA
Research Scholar
Water Resources Development Training Centre

J.D SHARMA
Professor
Electric Engg. Deptt.

University of Roorkee, Roorkee, India

SYNOPSIS

A methodology for finding optimum diameter of tunnels on the basis of economic analysis is presented. The benefits of hydroelectric power plants can be maximized by minimizing the cost of civil works, water conductor system, electro-mechanical equipments and power evacuation system, which is directly related with the selection of optimum size/capacity of equipments. The selection of optimum diameter of tunnel has been achieved on the basis of friction losses, seepage losses and cost of construction of tunnel.

1.0 INTRODUCTION

Electric energy is a basic need for economic development of any country. In India installed capacity reached a level of about 90,000 MW and will require additional generation of 70,000 MW by 2010. Now a days it is becoming difficult to go for thermal power plants as fossil fuels are becoming scarce and causing irrepairable damage to environment, hence emphasis is being made to tap the water potential by installing hydroelectric power plants [HEP] for meeting the growing energy demand. The benefits from [HEP] can be maximized by selection of suitable size of equipments.

The study conducted by (Egil and Skog 1997) and the conventional methods (by Nigam, by CBIP and by Therianos) for calculation of optimum diameter of tunnels are not associated with the seepage loss, whereas seepage loss is one of the decisive factor for optimization of diameter of tunnels especially in case of high head power plants and authors have taken the infeasible range of time span for calculation of frictional losses.

The main objectives of this study is :

- To develop the methodology for calculation of optimum diameter of tunnel based on the economic analysis.

- The objective function consists of the cost of construction and loss of revenue through the friction losses and seepage losses in tunnels.

1.1 Optimization of Diameter of Tunnel

The optimization of diameter of tunnel is complicated process, because large number of variable are involved. It is essential to go for most economic diameter of tunnels, because by changing the dimensions or slope of the tunnel, the engineering cost and benefits of whole project will change.

To formulate the mathematical model for this problem, the objective function is to minimize the cost Z.

$$\text{Min } Z = \sum_{j=1}^{n} C_j (D) \qquad \qquad \dots (1)$$

Subject to

$$D \geq 2.5 \qquad \qquad \dots (2)$$

Where

$$D = f(Q, S^1, A) \qquad \qquad \dots (3)$$

S^1 = tunnel gradient
Q = discharge (m^3/s)
A = area of the tunnel
C_j = cost function for jth items

It can be seen that the problem is about the optimization non-linear function which can be solved for least cost per kWh.

Economic design / optimum cross-section area of hydropower tunnel is influenced by several factors. The requirement of accurate / better data for cost estimation and economic design combined with new methods for calculating the roughness factors, friction losses and seepage losses for which zeological conditions of tunnel should be known. The economic cross-sectional area of hydropower tunnel basically depends upon cost involving head loss, generation loss due to seepage, and construction cost.

(i) Calculation of head loss due to friction per m length of tunnel :

The head loss due to friction per m length of tunnel can be calculated by using the Manning's formula, i.e.

$$H = \frac{Q^2 N^2}{\left(\frac{\pi D^2}{4}\right)^2 \left(\frac{D}{4}\right)^{4/3}} = 10.246 \ D^{-16/3} \ N^2 Q^2 \qquad \qquad \dots (4)$$

Where

H = Head loss due to friction
Q = Discharge m^3/s
N = Rugosity coefficient
D = Diameter of tunnel

Power loss due to friction in the tunnel = 9.8 Qi Hη.

Let the operation hours during the year be T. The friction losses will occur only for the period when the powerhouse is in operation, which can be ascertained from the operation schedule of powerhouse (for calculation purpose it can be taken 70%).

Annual energy loss due to friction in the tunnel = 9.8 Q Hη T

Substituting the value to T and H from eqn. (4) we have

$$= 6.157 \times 10^5 \ Q^3 \ N^2 \ \eta D^{-16/3} \ Et \qquad \qquad \dots (5)$$

η is the efficiency of Electro-Mechanical equipment's of power house;
Et is the cost of electrical energy per kWh.

(ii) Seepage Loss

As the seepage is directly effected by the increase in diameter / area of tunnel, hence it is to be considered for evaluation of economic diameter / area of the tunnel. According to Darcy's Eqn. (6) :

$$q_i = \frac{k\Delta h A}{d} \qquad \qquad \dots (6)$$

Where

k is the coefficient of permeability;

516

q_i is the leakage across concrete lining(m^3/s);

Δh is the drop in hydraulic head through the concrete lining (mtr)

i.e. Δh = (hi-he), hi is the internal and he is the external hydraulic head;

A is the area of cross-section(m^2);

d is the thickness of concrete lining(mtr);

Δh will depend upon the zeological conditions which can be ascertained after fixing the alignment of tunnel by initial investigations. The seepage will be maximum when external hydraulic pressure will be equal to zero (he=0). In that case the equation will be:

$$q_i = \frac{khi\,\pi D}{d}$$

hi will depend upon the reservoir level, friction head loss in the tunnel and slope of the tunnel.

Power loss due to seepage in the tunnel = $9.8q_i$ hη

Substituting the value of q_i we have:

$$= 9.8h\eta\frac{khi\,\pi D}{d}$$

Annual energy loss due to seepage $= 8.5848 \times 10^4 \; h\eta \; \dfrac{Khi\pi D}{d} \; Et$ (6)

3.   Evaluation of cost of construction of tunnel per mtr length:

(a)   Cost of Excavation :

$$= \frac{E\pi \; (D + 2d)^2}{4} \; Rs/m$$

Where:

        E is the mean price of tunnel excavation.

(b)   Cost of Lining :

$$= L\pi\left[\frac{(D + 2d)^2 - D^2}{4}\right]Rs/m$$

Where:

        L is the unit price of tunnel lining.

(c)   Cost of Grouting :

$$= G\pi \; (D + d)$$

Where :

        G is the unit price of grouting.

Total cost per linear mtr :

$$Cc = \frac{E\pi \; (D + 2d)^2}{4} + L\pi\left[\frac{(D + 2d)^2 - D^2}{4}\right] + G\pi(D + d)$$

$$= \frac{E\pi D^2}{4} + \pi D(Ed + Ld + G) + \pi(Ed^2 + Ld^2 + 2Gd)$$ ... (7)

The above-evaluated cost be increased by C for contingencies and by S for supervision charges.

Overall cost = Cc (1+C) (1+S) per meter.

Annual charges on tunnel due to depreciation and interest:

$$= \text{Cc} (1+C)(1+S)\,\text{CRF}$$

$$= \text{Cc}(1+C)(1+S)\left[\frac{i(1+i)^n}{(1+i)^n - 1}\right] \text{ per meter}$$

O&M Cost :

This can be taken as percentage of the gross annual cost and expressed as = Cc (1+C) (1+S) O

Where :

O is the operation and maintenance cost.

Total annual cost :

$$T_1 = \text{Cc}(1+C)(1+S)\left[\frac{i(1+i)^n}{(1+i)^n - 1}\right] + \text{Cc}(1+C)(1+S)O + 6.157 \times 10^5\,\eta N^2 Q i^3 D^{-16/3}\,Et$$

$$+ 8.5848 \times 10^4\,h\eta\,\frac{Kh i \pi D}{d}\,Et$$

Where :

$$\text{Cc} = \frac{E\pi D^2}{4} + \pi D(Ed + Ld + G) + \pi(Ed^2 + Ld^2 + 2Gd)$$

For minimization of cost: $\dfrac{dT1}{dD} = 0$

i.e.

$$= D + \frac{2d(E+L) + 2G}{E} + \frac{1.716 \times 10^5\,h\eta EtKhi}{dE(1+C)(1+S)(CRF+O)} - \frac{20.88 \times 10^5\,\eta N^2 Q^3 D^{-19/3}\,Et}{E(1+C)(1+S)(CRF+O)} = 0 \qquad \ldots (8)$$

The equation (8) can be simplified and can be written as :

$$mD^{-19/3} - D - K' = 0 \qquad \ldots (9)$$

Where

$$m = \frac{20.88 \times 10^5\,\eta N^2 Q i^3\,Et}{E\,(1+C)(1+S)(CRF+O)}$$

$$K' = \frac{2d(E+L) + 2G}{E} + \frac{1.716 \times 10^5\,h\eta EtKhi}{dE(1+C)(1+S)(CRF+O)}$$

This is non linear equation and can be solved for finding optimum diameter of tunnel.

2.0 METHOD

The way of optimization of suggested method is presented by numerical example.

For calculations following is presupposed :

E=Rs.900.00/-cum.	C=5%	Et=Rs. 1.50.
L=Rs.1600.00/-cum.	S=15%	$\eta = 0.85$
G=Rs. 800.00/-cum.	O=8%	K = 20*10[-9] [7]
	n=50 years	h=200mtrs
	i=18%	hi=80mtrs.

518

From the results shown in the Table (1.1) it can be observed that :

• Seeapage has direct impact on selection of optimum diameter/ area of the tunnel and cannot be ignored.

The methodology explained in the paper is a simplified and systematic approach in dealing with calculation of optimum diameter of tunnels and can be easily implemented in practice.

TABLE 1

S.No./Power Plant	Discharge Q_i m^3/sec	Diameter of tunnel (from CBIP manual), m	Diameter tunnel (from Nigam) m	Diameter tunnel with proposed method, m
P1	20.00	3.66	4.13	3.32

3.0 CONCLUSION

The methodology developed in this paper will give the optimum diameter of tunnels used for transfer of water / water conductor system for hydroelectric power plants. The method has been derived considering friction losses, seepage losses and cost of construction of tunnel. Seepage is more likely to occur as head / internal water pressure increases[8] hence act as a decisive factor for selection of optimum diameter of tunnels. The method is simple in computation and can be practically implemented.

REFERENCES

P.S Nigam, "Hand Book of Hydro electric Engineering", Nem Chand and Brothers, Roorkee (India).

CBIP Manual, "Manual on Planning and Design of Hydraulic Tunnels", Publication No. 178.

A.D Therianos, "Economic Tunnel Diameter for a Hydroelectric Project", Water Power, October 1967, (p. 418).

Pal Egil and Magne Skog, Economic Design of Hydro Power Tunnels", Hydro Power 1997 (p.-687).

L. Douglas James Robert R. lee, "Economics of Water Resources Planning", McGraw Hill Publishing Co. Ltd., New Delhi.

A.M. Neville and J.J. Brooks, "Concrete Technology", Longman, Scientific and Technical Co., published in the United States with John Wiley and Sons, New York.

George W.Washa , "Concrete Construction Handbook", Technical services Engineers, river division, American Cement Corporation , California. McGraw Hill Book Company New York.

Emil Mosonyi, "Water power development high head power plants," Akademiai Kiado Publishing House of Hangarian Academy of Science Budapeast.

DELHI METRO CUT AND COVER TUNNELS PLANNING FOR MAJOR UTILITY X-ING LOCATIONS

MANGU SINGH
Chief Project Manager (Metro)

RAJESH AGARWAL
Deputy Chief Engineer (Metro)

Delhi Metro Rail Corporation Ltd., New Delhi, India

SYNOPSIS

Major cross utilities normally pose great difficulties in construction of temporary support wall (diaphragm wall) etc. for underground cut and cover tunnels. In Calcutta, during execution of the first metro of the country, at such locations discontinuities in the diaphragm wall were left. These discontinuities in the diaphragm wall were later on plugged by steel plates also called as lagging plate. These discontinuities (lagging zones) had led to serious problems during excavation. Present paper deals with planning in Delhi Metro to tackle such locations to ensure construction of continuous diaphragm wall and avoid related problems during execution.

1.0 BACKGROUND

Major utilities like water mains of more than 300 mm dia, power cable of 11 kV and more, clusters of telecommunication cables wherever crossing the cut and cover tunnel alignment cannot be diverted in advance as it is required to be diverted over the completed portion of diaphragm wall. Therefore, diversions have to be programmed alongwith the main work of construction of diaphragm wall. The diversions of such utilities during execution of diaphragm wall are difficult on following account.

- The diversion needs shut down and hence suspension of services to the public and, therefore, at short notice it is difficult to make the utility agencies agree for the same.

- The diaphragm wall construction on both sides of the cut is normally not taken up simultaneously on account of non-availability of full road width for construction at a time. Therefore, even if one side panel is ready to support the new diverted cross alignment, the other side panel is not ready and, therefore, diversion cannot be affected unless both panels are completed.

- Utility owning agencies are normally reluctant to allow for diversion since alternative of supporting the utility and ground treatment by the contractor in the discontinuity zone is available.

- Since this involves very close interfacing, any delay in diversion affect the contractor's overall programme and therefore, he prefers alternative like supporting of utility and ground treatment in the lagging zone.

- Work of supporting of utility and ground treatment as an alternative to diversion remains fully within the control of contractor. On the other hand, for diversion he has to depend upon the utility agency. In-fact in Calcutta Metro the construction of diaphragm wall work was held up many times on account of non- diversion of utilities by the concerned agencies.

In view of the above, supporting major utilities with discontinuities in diaphragm wall is preferred as compared to diversion by the contractors.

2.0 METHODOLOGY ADOPTED AND ITS REPERCUSSIONS

However, during execution of work in Calcutta Metro it was found that these discontinuities in diaphragm wall had caused serious difficulties. This discontinuity, also called lagging zone, was later on plugged by special arrangement made on adjacent panels. The RSJ are kept embedded in the adjoining diaphragm wall panels as shown in Fig. 1. As the excavation proceeds, the RSJ are exposed by chipping the concrete which is quite time consuming. Thereafter, the exposed surface of RSJ is brushed and cleaned to make it weldable. Thereafter the steel plate between the RSJs is welded as shown in Fig. 1. Due to above, there is always a time lag between excavation and fixing of the steel plate. This zone, however, remains structurally weakest point in the diaphragm wall length and cases of failure of lagging plate are quite common. The above arrangement has following disadvantages :

1. The exposed soil face remains unsupported till the lagging plate is welded which results in soil loss into the cut.

2. Even after welding of lagging plate there is a gap between soil face and that of steel plate resulting into nil support to the soil mass initially.

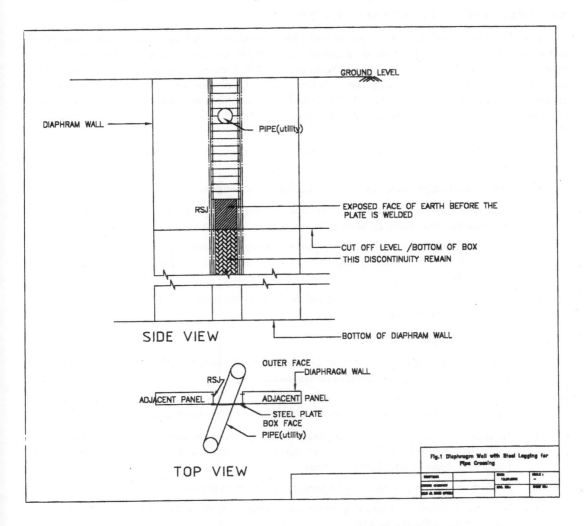

Fig. 1 : Diaphragm wall and steel lagging for pipe crossing

3. Even after the soil settles down to butt the steel plate, the soil loss continue to take place from the bottom due to upheaval.

4. The situation is further aggravated with the surface run off percolating between steel plate laggings and soil face resulting in soil loss from outside of the box into the cut.

5. All above results into excessive ground settlement with following repercussions :

    (i)   Collapse of the adjacent utility and resultant dislocation of the service.

    (ii) Severe damages to the adjacent structure.

    (iii) Unevenness in the road (sometime collapses).

    (iv) Damage to the surface drainage system.

Every such problem forms a vicious circle and resultant dislocation to the services is normally more than what would have been for the originally planned diversion.

3.0 CURRENT SCENARIO AND NEW PLANNING

A total of 11.0 km of North South line of Delhi Metro involves 5.6 km length of cut and cover tunnels and stations. Utility mapping has been done and all major utilities are already charted. These involve water mains upto 1200 mm dia. The typical layout of water mains at VW station is shown in the Fig. 2. Utility of 300 dia and 750 dia are crossing the alignment, whereas utility of 600 mm dia is partly coinciding with the alignment crosses the box required for VW Station and utility of 1000 mm dia is partly coinciding with the alignment as well as crossing the alignment. The diversion for these utilities has been planned as explained in Fig. 3 and below :

*   Two 'Y' Junctions are inserted on either side of the cut in the pipeline with valves as shown in the Fig. 3.

*   Diversion of length coinciding with the diaphragm wall alignment.

*   Shut down for the above works has been planned well in advance and availed as per the convenience of the utility agencies in the off-peak demand period.

*   Once the diaphragm wall panel at proposed diversion locations are completed, pipeline is connected over completed diaphragm wall location and diversion is commissioned with the help of valves without affecting the service even for a second.

*   The existing cross line is dismantled and diaphragm wall panels are constructed continuously *i.e.* without any lagging zone.

With above following will be achieved :

1. Utility diversions during execution of the work are not dependent on the shut down and, therefore, will not affect overall programme of the contractor.

2. With the provision of the valves on both sides of the cut, the work area can be isolated in case of any emergency like failure of pipe etc. to avoid flooding of the work area.

3. In absence of discontinuity in diaphragm wall, problems related with the discontinuity are completely avoided and hence chances of excessive settlement and associated repercussions outside the diaphragm wall are minimized.

4. Uninterrupted service is maintained during the construction period.

5. There is a psychological advantage to the construction team due to proper support of utility and no discontinuity in diaphragm wall.

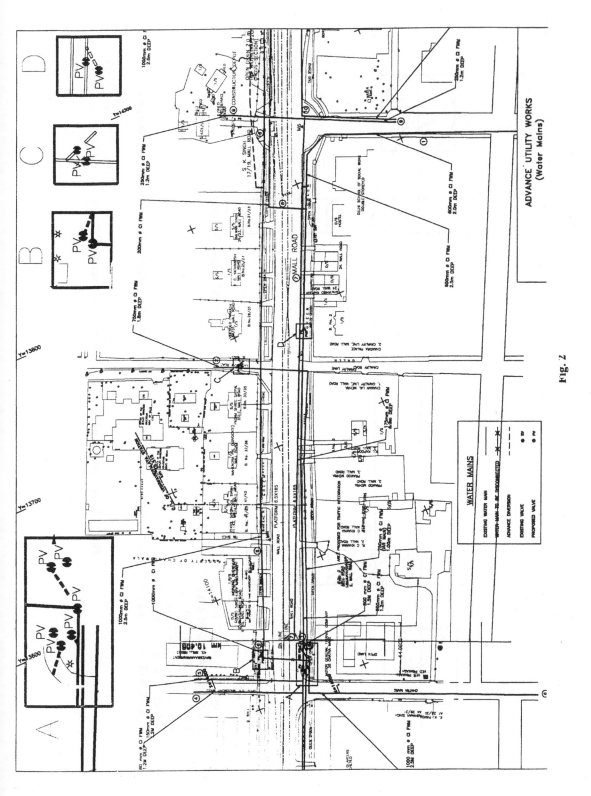

ADVANCE UTILITY WORKS
(Water Mains)

Fig. 2

523

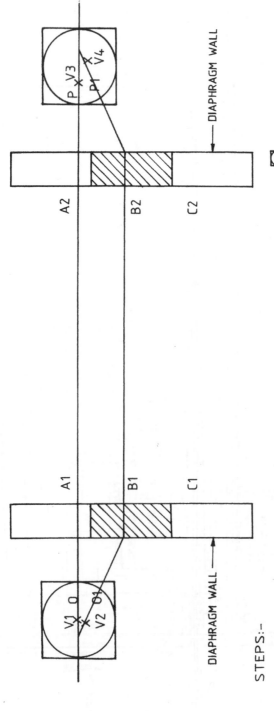

STEPS:-

1. WORK OF FIXING Y-JUNCTION & FIXING VALVES IN EXISTING LINE IS DONE IN ADVANVE SHOWN IN

2. D/WALL PANELS B1 & B2 COMPLETED.

3. LINE O1P1 CONNECTED.

4. LINE OP DISMANTELLED.

5. PANELS A1 & A2 COMPLETED.

6. RECONNECT OP (IF REQUIRED) AND RESTORE ORIGINAL PIPE LINE OR ALT. O1P1 CAN REMAIN AS PERMANENT RESTORATION.

Fig. 3 : Typical plan of advance utility diversion arrangement

4.0 EXECUTION OF LATEST PLANNING

The implementation of the above concept as an advance work has been taken up by DMRC and water mains of 300 to 1000 mm dia has been tackled successfully. The execution of the same requires meticulous planning and mobilization of resources. The pipeline under consideration is traced from origin to destination alongwith the positioning of sluice valves, scour valves, pumping installation and area under its influence which forms important consideration for the management of shut down. A trial pit at the proposed location is made to find out the exact material and dimension of the pipe to be replaced by 'Y' along with adjacent utilities, over-head obstacles, drainage facilities, levels of road and proposed work so as to finalize the exact requirement. Thereafter, the drawings are developed and approval is obtained from the utility owning authority.

Now, as per approved plan, the necessary preparatory work, mobilization of required tools and plants, equipments, consumables, labour and specials are done. Simultaneously permissions from PWD, traffic authority and any other local bodies are obtained. Finally the shut down is taken to execute the work in the least possible time. After the new assembly is commissioned, the necessary chambers, thrust block are constructed and restoration of road, footpath etc. is done. The above work has been included as information in the main tender under the heading 'Advance Utility Diversion Works' for facilitating the "would be contractor" to plan his tender more effectively. The tender specification also provides, the construction of diaphragm wall without any discontinuity, which shall be possible at no extra cost due to above advance works planned by DMRC.

5.0 CONCLUSION

Construction of a continuous diaphragm wall, even at locations of major utilities without affecting the service at the time of construction and without the interruption to the contractors' overall programme of execution, is possible. This, in turn, will avoid troubles of subsidence and associated repercussions.

REFERENCES

The Calcutta Metro, Dum Dum – Tollygunge – Design and Construction Phase –I.

Technical Paper on Utility Diversion Works (Water Mains), DMRC – May 2000.

TRENCHLESS TECHNOLOGY IN INDIA — AN OVERVIEW

A.K. SARKAR

Chairman and Managing Director

National Building Construction Corporation Ltd., New Delhi, India

1.0 INTRODUCTION

Liberalization, globalization and privatization policy of the Government of India has generated positive response for the foreign participation in development of Indian economy and industry and has created unprecedented opportunities for industrial and economic growth. It has attracted MNCs and NRIs to invest in the infrastructure projects with new technologies. India is the most populous country in the world next only to China, and has been the emerging market, in the world. The country is also now emerging as a major information technology power. With interest shown in Indian telecommunication, electronic, automobile and transport sectors, it is clear that India is the destination for many advanced countries to grab a share of new emerging market.

All these changes lead to unprecedented emphasis on basic Infrastructural base. Government has also laid stress on the infrastructure development, particularly water supply, drainage, road communication, telecommunication services, roads, power and environment pollution control etc. Common to all these infrastructure projects is the network of underground services that needs immediate attention of the government, planners and the engineers. The same is to be developed and maintained giving due importance to environment friendly methods which are of least inconvenience to public. Digging of open trenches till recently, had been a common method for laying the water supply, sewage and drainage pipe lines, power and communication cables, oil and gas pipe lines. Due to the inherent disadvantages of the open trench method, many of the developed countries like Japan, Germany, UK, USA, Singapore and Australia had resorted to new method of NO Dig or Trenchless Technology way back since 1970's. In India, this technology although laid its feet only around 1990's, it has in a span of just few years progressed very well is continuously gaining popularity.

2.0 TRENCHLESS TECHNOLOGY

It is an innovative method of laying underground utilities without digging the ground or with minimum excavation. It is also called the "No-Dig method". This can be used for new installation and also for rehabilitation or renovation.

3.0 STEPS TO LAY A TRENCHLESS UTILITY

1. Underground survey is done using ground probing radar and other tracking instruments.
2. The pits are dug at two ends and the equipment goes underground to lay the utility service pipeline.
3. Now a days there are equipments by which we can lay services without even digging the pits.

Speedier implementation of construction projects without cost over-runs and disruption of utility services, traffic and environment is the major concern faced today. This increasing concern is more evident in construction activities involving extensive excavation through open cut method that result in disruption in the normal life particularly in congested areas.

4.0 CATEGORIES OF TRENCHLESS TECHNOLOGY

Trenchless Technology methods/systems have been categorized into two broad groups according to their utility :

- **New Installation.**
- **Rehabilitation and renovation of the existing installations.**

New Installation Methods

IA	IB	IC
Impact Moling	Guided Boring	Micro Tunnelling
and	and	and
Pipe Ramming	Directional Drilling	Pipe Jacking

Rehabilitation Methods

A. Slip lining
B. Inline replacement
C. Cured in place pipe (CIPP)
D. Close fit pipe
E. Point source repair

4.1 New Installation Methods

4.1.1 *Impact Moling and Pipe Ramming*

Impact Moling: Impact moling or earth piercing is defined as the creation of bore by use of a tool which comprises a percussive hammer within a suitable cylindrical casing, generally torpedo shaped. It is a non-steered or a limited streering device without rigid attachment to the launch pit. It relies for forward movement upon internal hammer action to overcome the frictional resistance of the ground. During the operation, the soil is displaced (not removed). (Two pits are dug: 1. Launch Pit and 2. Reception Pit. The mole starts at the launch pit and goes upto the reception pit), Fig. 1.

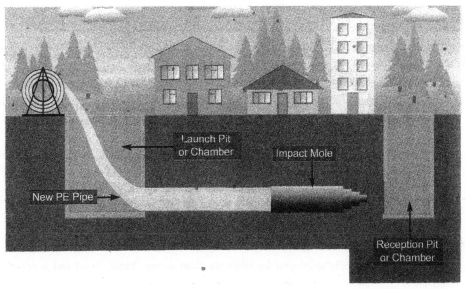

Fig. 1 : Pipe installation by impact moling

The basic mechanism of impact moling is the reciprocating action of the pneumatically or hydraulically powered hammer within the cylindrical steel body. The piston is driven forward and on striking forward, the unit imparts its kinetic energy to the body which is driven forward. The energy of the piston for the return stroke is regulated so as to reposition it for the next forward stroke.

Repeated impact of the hammer piston advances the whole unit through the ground. As forward movement takes place, the soil in front of the mole is forced aside and compacted by the conical or stepped nose to form the walls of the bore. The power of the unit is also often used to pull the product pipe, cable or cable duct through the above at the same time as the impact mole advances.

Either an unsupported bore is to be formed in a suitable ground or a pipe may also be drawn or pushed-in immediately behind the impact moling tool. Cables may also be pulled in.

Another category of moles which operate by expansion rather than hammer action are used for pipe bursting applications. (Renovation) rather than new installations.

Ground Conditions

- Moles can operate only in soils that can be compressed or displaced.
- Thorough ground investigation is essential prior to work commencement in order to establish a clear route.

Applications

- Because the impact-moling is generally un-steered, the technique is most suitable for shorter bores.
- Diameter range: About 45 to 200 mm depending upon the pipe or cable being installed.

Cost Comparision

Impact moling can be a very cost-effective method of installing small to medium size pipes, ducts and cables for a broad range of utilities including gas, electricity, water, and telecommunications. The technique is in common use for simple road crossings and installation of service connections between main lines and individual properties. Moles are relatively easy to use, monitor and maintain.

4.1.2 *Pipe Ramming Method*

Pipe ramming is a non-steerable system of forming a bore driving steel casing usually open ended, using a percussive hammer from a driver pit. The soil may be removed from an open-ended casing by auguring, jetting (with water) or compressed air. In appropriate ground conditions a closed casing may be used, Fig. 2.

- A typical ramming operation requires a solid base normally a concrete mat on the launch side of the installation either against the side of a slope or in a start pit. After this guide rails are set to the line of the bore.
- The first length of steel pipe is positioned, then the guide rails and cutting edge is fitted to end of the pipe. The ramming hammer is attached to the rear end of the pipe and the hammer is started. This forces the steel pipe into the ground along the line dictated by the guide rails. When one pipe has been driven, the hammer is stopped and removed and the next length of steel pipe is welded in place. The cycle is repeated until the leading edge of the first pipe arrives at the reception end of the shaft.

Applications

- Pipe ramming is generally employed for relatively short drives (about 50 m) and is not a steerable technique.
- It is common method of installing steel casings (as no other material is strong enough to withstand the impact forces generated by the hammer) in a straight line and is frequently used under railways and road embankments and also for underwater pipe laying applications.
- Once the steel pipe is installed it can be used as a pipe line in its own right or as a ducting for most types of pipes or cables.
- Bores upto 2000 mm diameters have been installed in suitable conditions.

Fig. 2 : Installation of a steel casing by a pneumatic ramming hammer

4.1.3 *Guided Boring and Directional Drilling*

Guided boring and directional drilling techniques are used for installation of new pipelines, ducts and cables. The drill path may be straight or gradually curved, and the direction of drilling lead can be adjusted at any stage during the bore to steer around obstacles or under highways, rivers or railways. Drilling can be carried out either between pre-excavated launch pit and reception pit, or from the surface by setting the machine to drill into the ground at a shallow angle. The scale and capability of guided boring and directional drilling is some where between the techniques of 'impact moling' (lowerend) and micro tunnelling (higher end).

Although the term Directional Drilling is frequently used to describe bigger projects such as major river, canal and high way crossings often covering long distances, it is almost inter changeable with Guided Boring.

Stages of Installation

    (i)   The first stage consists of drilling a pilot hole along the required path.

    (ii)   The second stage involves enlarging the pilot hole. The bore is back - reamed to a larger diameter to accommodate the product pipe.

During the pull back stage, the pipe is attached to the reamer by means of swivel connector, and is pulled into the enlarged bore. In the cases where the bore enlargement is considerable, there may be one or more intermediate reaming stages during which the bore diameter is increased progressively, Fig. 3.

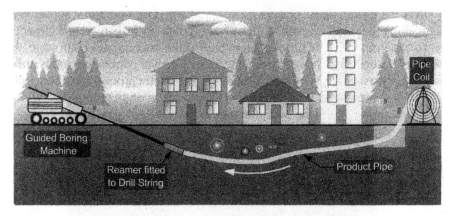

Fig. 3 : Backreaming and pulling in the product pipe

Fig. 4 : Microtunnelling machines with various cutting head designs

Highlights of Operation

Both surface launched and pit launched drilling machines are available, the choice depending on the nature of the project.

Machine range from compact rigs suitable for small bores and operations in restricted space, to extremely large units designed for large diameter and long distance crossings.

Drilling fluid lubricates and stabilizes the bore and also conveys the excavated material in suspension. Some rigs are also designed for dry operation.

Drill pipes should be chosen to provide the right combination of strength and flexibility. The maximum length of individual drill pipes depends on the types of machine and the operating space.

Applications

- Until recently guided boring was used mainly for the installation of pressure pipes and cable ducts, where precise gradients are not usually critical. However, some of the latest drilling machines offer improved accuracy for gravity pipelines as well.

- Apart from the obvious environmental benefits of trenchless installation, the relative cost of guided boring has fallen in many parts of the world below that of trenching for many applications, even ignoring the social costs of traffic and delay.

- It can be used for installation of pipe ducts and cables in most diameters and over distances of 1 km.

4.1.4 *Microtunnelling and Pipe Jacking*

The term microtunnelling is used to describe methods of Horizontal earth boring, which are highly sophisticated laser guided, and remote controlled. These methods permit accurate monitoring and adjusting of the alignment and grades as the work proceeds so that pipe can be installed on precise line and grade. Microtunnelling involves the works for less than 900 mm in diameter however, identical tunnelling equipment is used on larger pipes exceeding 2250 mm (89 inch) in dia. They are called automated tunnel boring machine (TBM). The pipes which may be used are reinforced concrete pipe, vitrified clay pipe, glass reinforcement plastic pipe, fiber glass pipe (Hobbs), ductile iron pipe, steel pipe and asbestos cement pipes. Most microtunnelling equipments are designed to operate in soft-ground soil conditions. It can be used in very difficult ground conditions without expensive de-watering systems or compressed air. Since workers are not required to enter the trenches, it enhances the job safety. The basic process involves hydraulically jacking the pipe into the place, following the microtunnel-boring machine from a bore pit. The bore pit must be structurally stable. All the systems like jacking, steering, mechanical earth pressure counter balance, slurry are controlled at the operation board, which is at the surface of the bore pit. The cutting head is designed to mix the cuttings with the slurry so that the spoil can be removed by pumping through the discharge lines into the slurry separation tank. The pumping pit bye pass unit guides flow and pressure control of the slurry system.

A pipe laying laser is used for alignment and grade control. The indicator panel at the rear end of the machine houses numerous pressure gauges as well as the laser target. The operators control the process by reading the indicating panel. The target at the rear end of the machine is monitored by a close circuit TV camera. There are two methods of spoil transportation (i) Slurry spoil removal (ii) Auger spoil removal.

Most microtunnelling drives are straight between shafts, although specialized systems are available for curved drives. In such cases where because of curvature of tunnel, line of slight is not possible between the drive shaft and microtunnelling machine, alignment systems based on Gyroscopic devices may be used as an alternative to the more usual laser equipment.

The major disadvantage of microtunnelling method is that the machine capital cost is high.

Highlights

- Pipe Jacking the microtunnelling can be cost effective methods of installing new pressure or gravity pipe lines through most soil types and at virtually any depth (diameter 150 mm upwards).

- Precise control of gradient and alignment is possible, and the techniques are particularly suitable for medium to large diameter gravity sewers.

- Field experience has often proved to be the biggest potential cost saver on any particular project.

4.1.5 *Pipe Jack*

A Pipe Jack is defined as a system of directly installing pipes behind a shield machine by hydraulic jacking from a drive shaft. It requires the workers inside the pipe. The pipes are prefabricated and jacked into position from a jacking pit. The excavation can be either mechanical or manual. The pipes form a continuous string in the ground. These pipes joint mostly by flush-fitting collars. These pipes are specially designed to withstand the jacking force likely to be encountered during the installation. Primarily, this method has been used to install reinforced concrete pipe. However, it is appropriate for steel pipe. Hobbs (fiberglass) pipe, ductile iron pipe, and corrugated metal pipe, depending on the project conditions.

Excavation is normally accomplished inside the articulated shield. This shield can be guided with individually controlled hydraulic steering jacks. The shield is designed to provide a safe working environment for the workers and allow both to stay open for the pipe to be jacked in place. It is a cyclic procedure of using thrust power of hydraulic jacks to force the pipe forward. In unstable conditions the face is excavated simultaneously with jacking operation to minimize over excavation and risk of face collapse. In stable ground conditions excavation may precede the jacking process. The spoil is removed through the inside of the pipe to the jacking pit. After a section of pipe has been installed the rams are retracted and another joint is placed into the position so that the thrust operation can be started again.

Guide rails are propositioned so that the jacking pipe will be properly aligned when it is placed on the guide rails. Much success in a pipe jacking operation is based on ensuring proper alignment. The application of Bentonite slurry to the outside skin of the pipe may reduce frictional forces.

5.0 POTENTIAL MARKET

India offers a potential market for trenchless technology in all the capital cities like Delhi, Mumbai, Calcutta and Chennai. These cities have already overgrown their master plans limits. The water supply and sewage system of these cities are over burdened and need augmentation, repair and renovation without causing much hindrance to the society, commerce, trade and environment. The trenchless technology offers a better and a suitable alternative to open trench digging method.

5.1 Present Scenario

Initially adoption of trenchless technology was a chicken and egg syndrome in the sense that decision makers could not go for the trenchless technique, if companies did not have presence in India and companies would not cherish the risk of trying unknown market. Whatever extent of intelligence a nation may have, without the availability of appropriate equipment or materials of construction, the commitment to growth of infrastructure development can not be fulfilled. Bringing together the user departments (clients), contractors as well as equipment manufacturers has been a Herculean task indeed. User department, today, are showing their absolute preference for trenchless technology. Foreign manufacturers of trenchless equipment have stated to appreciate market potential of their equipments in India. Major international manufacturers of the equipment now have their base in India. Many works are being carried out using trenchless technology across the country and this technique has received tremendous boost. Not only a number of contractors are now working in the field, some companies are also planning to manufacturing the equipments, locally. The trend started initially using simple methods *i.e.*, Moling and Pipeline, where equipment cost were relatively lower. The telecom department (MTNL) and Electricity Department (DVB) started floating their tenders for laying their cables using these technologies. Afterwards, the Development Authority (DDA) used for laying sewer line across National Highways using Pipe Ramming and Auger Boring Machine. Municipal Corporation (NDCM) followed this for laying water lines. The Oil Companies have also awarded projects for laying Gas and Oil Pipelines using trenchless techniques across various river crossing. The projects have been carried out successfully. The thrust and trust on the technology has laid to use directional drilling across the busy roads (100mm wide) and various rivers crossing. The Municipal Corporation of Greater Mumbai has awarded a US $ 8 Million worth Micro-tunnelling contract for installing approximately 3700 long (dia ranging from 350 mm to 1400 mm) sewage pipelines. Even the missing links between various sewer lines that could not be laid by using open trenches are planned to be installed using pipe jacking and micro tunnelling methods.

6.0 INVOLVEMENT OF NBCC IN TRENCHLESS METHODS

NBCC, a premier construction agency in India, has focused on Hi-tech projects using innovative technologies. In its efforts towards the nation building using new technologies. NBCC has pioneered in adoption and propagation of Trenchless Technology. NBCC was appointed as a Nodal agency for adoption of trenchless technology in India by Ministry of Urban Affairs and Employment and a Technical Committee was formed under the Chairmanship of Shri A.K. Sarkar, CMD, NBCC Limited to formulate the draft guidelines on adoption of trenchless technology in India. The same were submitted during August, 1996. NBCC is also hosting Indian Society for Trenchless Technology (INDSTT), which is a non-profit body affiliated to the International Society for Trenchless Technology (ISTT) U.K.

NBCC has been executing number of projects by using various methods of trenchless technology such as Impact Moling, Pipe Ramming, Directional Drilling, Auger Boring for departments like, DVB, MTNL, DDA, and NDMC etc.

7.0 COST COMPARISON

For common utility services like electricity and telephone cables, small dia pipeline, the cost difference between open cut method and no dig method is comparable if not favourable where mass execution is involved. The trenchless technology scores better if the social cost and time saving are also accounted for.

8.0 CONCLUSION

There is a large market yet to tapped in the country. The use of technology is still confined to urban areas or areas where open trench methods are not feasible. The gigantic task of spreading the technique all over India is our "mission". We now need to develop related industries like, geo-technical investigators and material manufactures. I am sure with the support of all our valuable clients we shall make "NO-DIG" a Millennium mission and re-write the future - "The No-Dig Way".

PREDICTION OF ENGINEERING CLASSIFICATION OF WEDGE TERRAINS OF EASTERN GHATS

K. RAMA SARMA

Associate Professor

Civil Engineering Department, K.I.S.T., B.P. 3900, Kigali, Rwanda (C. Africa)

SYNOPSIS

The classification system may be considered as a tool to facilitate diagnostic and design work. An experienced engineer who really understands geological materials can apply principles of geotechnics without a formal classification. Unfortunately, experienced tunnel engineers are few and inexperienced engineers are often given large responsibilities.In addition, the designers of tunnel support systems are usually structural engineers with little knowledge of the important aspects of geological material and the data on which they base their design are collected by geotechnical engineers who often have little appreciation of structural concepts. The problem of communication between these groups of engineers is major and real. The classifications serve to point out important parameters and problems and constitute effective means of communication. In this study an approach is made to adopt a simple classification method like rock quality designation, rock structure rating, rock mass index, rock mass rating, rock quality (Q-system) etc, from the tests conducted on several samples in the wedge terrains of eastern ghats, particularly in tunnel reaches where distress was observed. From this study the RMR classification system seems more suitable classification method to be adopted while planning under ground structures/tunnels in this area.

1.0 INTRODUCTION

Until a few decades back, rock was considered to be an unquestionably safe foundation material. However, the in-depth study of rocks revealed the weakness of rock as a result of joint planes and brought to light the concept of the rock mass. The analysis of the failure of Vaijont reservoir in Italy and the catastrophe of Malpasset dam in France reiterated the significance of joints in the analysis.

During the feasibility and preliminary design stages of a project, when very little detailed information on the rock mass and its stress and hydrologic characteristics is available, the use of a rock mass classification scheme can be of considerable benefit. At its simplest, this may involve using the classification scheme as a checklist to ensure that all-relevant information has been considered. At the other end of the spectrum, one or more rock mass classification schemes can be used to build up a picture of the composition and characteristics of a rock mass to provide initial estimates of support requirements and to provide estimates of the strength and deformation properties of the rock mass.

The need for classification of rocks according to their characteristics has long been recognized, as such a classification provides an "effective communication between the engineer and the geologist or between the designer and the contractor". In general, the rock classification should fulfill the following requirements :

(i)   Provision of a common basis for communication,

(ii)   Categorization of rock masses into groups and

(iii)   Provision of a basis for understanding the characteristics of each group.

More over, it may yield quantitative data and provide an empirical design approach for geotechnical projects. In fact, for several projects, the classification approach serves as the only practical basis for the design of complex underground systems. Most of the tunnels constructed at present make use of some classification system. An advantage in rock mass classification is in the selection of more significant factors controlling the engineering behaviour of a rock mass. There appears to be no single factors or index which can fully and quantitatively describe a rock mass. It is important to understand that the use of a rock mass classification scheme does not replace some of the more elaborate design procedures. However, the use of these design procedures requires access to relatively detailed information on in situ stresses, rock mass properties and planned excavation sequence, none of which may be available, the use of the rock mass classification schemes should be updated and used in conjunction with site specific analyses.

The factors, which are commonly employed in the current classification system, are :

(i)   Spacing of discontinuity,
(ii)   Orientation of discontinuity,
(iii)   Condition of discontinuity,
(iv)   Strength of rock material and
(v)   Ground water conditions.

2.0 STATEMENT OF PROBLEM

Rapid industrial development in the coastal states of Andhra Pradesh and Orissa is generating heavy traffic. Besides this there is an abnormal increase in the land cost and land requirement. As the existing roads and rail tracks are inadequate to cater to the present and future traffic requirements, it is imperative to construct new railway lines, roads, tunnels, etc. Particularly in Visakhapatnam, Andhra Pradesh, India, where liquid petroleum gas storage is being planned in the depths of the rock mass of Dolphin's nose hill, in addition to vast industrial development, the feasibility of road tunnels and metro railways are already under investigation. Due to lack of proper understanding of rocks and their behaviour, the recently completed railway project (Koraput-Rayagada line) was delayed by about 6 years, there by escalating the estimated cost by 3 times. It is aimed to suggest the most probable classification system for these terrains, there by simplifying the analysis of tunnels in this area.

3.0 GEOLOGY

The Eastern Ghat granulate terrain occurs as a linear band along the east coast of India and exposes varieties of ortho and para-gneisses intruded by anorthosities and alkaline rocks. A detailed structural and geo chronological study on the eastern ghat granulates and associated magmatic rocks are meager and are summarized. In Visakhapatnam- Koraput section principal rock types include Khondalite (garnet – sillimanite-perthite-quartz), Orthopyroxene granulate (garnet- Orthopyroxene- plagioclase-perthite-rutile-illuminate), cate-granulates (calcite-scapolite-plagioclase-diopside), mafic granulate (Orthopyroxene-clinopyroxene-garnet-plagioclase), spinel granulate (spinel-cordierite-Orthopyroxene-sillimanite-quartz; spinel-garnet-sillimanite quartz), Leptynites, quartz silliminate schists, granetiferous quartzites with minor insertions of charnokite and granite gneiss. The rock had undergone polyphase deformation as seen from the presence of intricate folding. The rock had suffered weathering and erosion, the weathered zones some times extending to great depth. The majority of the tunnels are offshore type and as the hill slopes are covered with slope wash debris and as the rocks suffer from geological defects like joints shears and faults presenting zones of weakness.

4.0 METHODOLOGY

The author during the execution of the project studied the rock masses insitu and 45 bore holes were driven at various locations in the region from Visakhapatnam to Koraput. Rock samples were collected from each bore hole at various depths depending on the rock strata. The total depth of bore hole varied from 30 to

46 m. About 25 to 30 rock core samples were obtained from each bore hole. On this rock core samples various experiments were conducted to determine engineering properties like field density, specific gravity, porosity, natural moisture content, uniaxial compressive strength, and point load index, standard penetration value, quantity of water discharge.

In addition to the above experiments various geological and engineering parameters like rock quality designation (RQD), rock number of joints (JN), joint spacing (Js), joint orientation (Jo), joint roughness number (Jr), general area geology (A), joint pattern and direction of drive (B), ground water condition (C), joint water reduction number (Jw), condition of discontinuities, ground water inflow per 10 m length of tunnel (l/m), spacing of discontinuities, strike and dip direction of discontinuities, joint set number (J_n), stress reduction factor (SRF), joint alteration number (J_a), joint length (J_1), block volume (V_b), joint condition factor (J_c), joint parameter (J_p) and degree of weathering were determined on the core samples. After knowing all these parameters, various engineering classification methods like rock quality designation, rock structure rating, rock mass index, rock mass rating, rock quality (Q-system) etc, were used to classify the rock masses of theses terrains.

5.0 OBSERVATIONS AND DISCUSSIONS

The previous history (1963), during construction of three tunnels in K-K-Railway line between Tummanavalasa and Sivalingapuram of wedge terrians of eastern ghats, huge chimney formation and fall of muck occurred. Similar chimney formations were also observed in these tunnels during construction stage, as told by engineers, who worked at that time in those tunnels. Rock mechanics subject was not developed during that period. They adopted open cut methods in those tunnels (oral discussion with engineers). During construction of tunnels in K-R Railway line, roof falls, under ground caverns of water, squeezing ground conditions have been encountered and the geological report as extracted below turned out to be true.

A detailed investigation and study of the geology of this region is revealed as follows.

The study area consists of Khondalite suit of rock comprising quartz-sillimanite schist and granetiferous quartzite with minor intrusive body of charnokite from the Korapute end. The balance major portion of the tunnel alignment passes through charnokite suit of rock with minor amount of gneiss and quartz-sillimanite schist.

Boring at various locations revealed different rock strata such as charnokite, basic charnokite, granitic ferrous gneiss, biotite gnesis, quartz silliminate schist, quartz-garnet sillimanitic schist, quartz etc. Quartz silliminate schists and grantifers quartzites with minor intrusion of charnokites and granite gneiss were present in some bore hole locations. The rocks had undergone polyphase deformations seen from the presence of interacting folding. The rock had suffered weathering and erosion, the weathered zones some times extending to great depths and the rocks also suffered from geological defects like joints, shears, and faults presenting zones of weakness. Massive, hard and jointed quartz feldspar-garnetiferous gneiss were exposed. The gneiss exposed was highly weathered.

At some locations of this region revealed that the zone is well foliated, moderately weathered khondalite at some places. The khondalite type of rocks chiefly comprised of quartz-garnet-biotite-silliminate-feldsar gneiss and quartz-feldspar-biotite-garnet-gneiss. Weathered gneiss exposed is of quatz-felaspar-garnet –gneiss. Khondalite rock was considerably decomposed, but physically intact faces were exposed. Also, a shear zone has been encountered. It was uncertain as to how long the shear zone would extend. It consists of clay and rock fragments which were seen from the muck that accumulated in front of the face in the form of a heap from the crown of the Tunnel-23 (K-R Railway line) to the bottom after seeping out of the shear zone.

Form the case studies of this region, it was noticed that funnel formation was more common due to presence of weak shear zones and in this region it was also observed in a few bore holes, unweathered charnokites existed in between completely and highly weathered rock masses. Completely disintegrated rock and hard rock coexist at very close spacing. This condition prohibits the use of explosives and thus makes whole of tunneling process very slow and costly. The weathering of rock through the tunnel was heavy and thus the tunnel was excavated under flowing ground conditions, which is highly hazardous. It is likely that the sheet rock exposed near the exit portal of T-25 (K-R-Railway line), may form the roof of the tunnel. As most of the joints were open at the surface, certain amount of grout was required at the portal site.

From the bore holes 2, 13, 15, 17, 26, 34 and 36 completely weathered and highly weathered rock were observed. From all classifications rock quality was found to be poor to fair. From the bore holes 1, 9 and 10 it was observed that mostly the strata was completely weathered and highly weathered rock. From all classifications rock quality was found to be poor to fair. From the bore hole 3, it was observed weak strata was found 33 m to 36 m and all around these strata the rock quality was good. From the bore hole 12, it was observed below 38.75 m the rock mass was very poor and above that the rock mass was good. From the bore hole 14, it was observed in between 21 to 24 m, the rock mass was very poor. From the bore hole 18, it was observed in between 34.5 to 36.5 m, the rock mass was poor. From the bore hole 20, it was observed in between 23.5 to 27.4 m that the rock mass was good and rest of rock mass was poor.

From the bore hole 21, it was observed in between 28.5 to 36.8 m, that the rock mass was poor. From the bore hole 22, 24 and 39, it was observed that the rock mass in this bore hole was alternatively poor, good fair and very poor. From the bore hole 23, it was observed the rock mass in between 42.8 to 45 m varied from poor to fair from most of classifications. From the bore hole 27, it was observed that below 33.5 m the poor from all the classification.

From the bore hole 29 and 30, it was observed that the rock mass in the top layers was very poor to poor and the rock mass thereafter varied fair to good from all the classifications. From the bore hole 31, it was observed that the rock mass in between 23 to 26.5 m was good from all the classification. From the bore hole 32, it was observed the rock mass in between 21.8 to 32 m was good from all the classification. From the bore hole 38, it was observed that the rock mass in between 20 to 22m was good from all the classification.

From the bore hole data and case histories of this region, the rock mass in this region is of poor quality and the rock masses in which the tight interlocking has been partially destroyed by shearing or weathering, the rock mass has no tensile strength or cohesion and specimens will fall apart without confinement.

It was also observed from bore hole data of this region, intact rock mass of this region contains single/two joint sets. Hence RMR classification may be used to estimate the strength of rock masses. Knowing the strength parameters Hoek and Brown (1980) and Hoek (1994) failure criterion can be applied. More over the RMR system is very simple to use and the classification parameters are easily obtained from either bore hole data or underground mapping. It can also apply to slope stability, foundation stability and tunnelling.

6.0 CONCLUSIONS

The empirical approach used most frequently is to base the design on present practice by means of a general assessment of the characteristics of the rock mass. Where the rock can be shown to have the same overall "rating" or "quality" as a similar rock mass at the site of a previously completed project, then the adoption of a similar or the same design approach is justified. The most well known examples of this approach are Bieniawski's rock mass rating (RMR) system, and Barton, Lien and Lunde's tunnelling quality (Q) system. A second approach is to use the classification system to estimate equivalent rock mass strength parameters for the rock mass, which may then be used in a deterministic model or a numerical analysis.

In all circumstances, the use of rock mass classification systems should be preceded or accompanied by a careful appraisal of the geological conditions at the site in question, as it is very easy to lose sight of these fundamental data, once observations at the site are treated quantitatively.

The samples obtained from various locations at various depths from bore holes indicated that the degree of weathering was moderate to complete. The weathering front in all these rock masses are extended from few meters below the present ground level to deeper layers. Thus this heterogeneity of rock masses and the railway line has resulted in caving in at some of the tunnel sites. Any classification based on stochastic methods needs more samples. The earlier investigations, due to shortage of time, were restricted to less number of bore holes. Probably the analysis made at the time of sanctioning the project was not sufficient in revealing the rock mass discontinuities.

From the study of rock masses in the wedge terrains of Eastern ghats, the RMR classification system seems more suitable classification method to be adopted while planning under ground structures/tunnels. The rock mass rating also happens to be relatively simple classification than Q-system, rock mass index etc. Whenever it is not possible to adopt RMR rock mass classification system, RSR rock mass classification system may be adopted, and rock mass rating can be predicted using correlation between 'RSR and RMR'. This will enable the construction engineers to adopt a more appropriate and rational method for safe execution of construction projects in this terrain viz. Tunnelling, under ground storage facilities etc.

8.0 REFERENCES

Hoek E. (1994) Strength of rock and rockmasses, vol.2, no. 2, News Journal of ISRM, pp. 4-15.

Hoek, E. and Brown, E.T. (1980) Empirical strength criterion for rockmasses, J. Geotech. Engg., ASCE, 106, GT 9, pp 1013-1035.

BENCHMARKING MEASURES FOR TBM AND DBM TUNNELLING

A. SRIVIDYA   B.A. METRI

**Reliability Engineering Group, Indian Institute of Technology Bombay
Powai, Mumbai, India**

SYNOPSIS

In this paper, present tunnelling practice has been highlighted and rationale of benchmarking has been addressed. Hard and soft benchmarking measures have been identified for both TBM (Tunnel Boring Machine) and DBM (Drill –Blast - Muck) tunnelling and performance benchmarks have been presented for typical case using the data collected from tunnelling projects and the manufacturers of tunnelling equipment. Directions for future work has been suggested and finally, it is concluded that benchmarking measure play a key role in identifying the superior performance in TBM and DBM tunnelling.

1.0 INTRODUCTION

The construction of tunnels for the infrastructures is the key to the economic development of the country. In India, as on today, bulk of tunnelling work has been done using conventional Drill-Blast–Muck (DBM) method Metri B.A. and Patil B C.(1979). In most of the projects, like hydropower, Railway and Highway, tunnelling emerged as the critical activity due to slow progress rate of conventional cyclic method. Ultimately, it became the cause for delay in completion of these projects and consequent delay in commercial utilization of projects. In India, still bulk of tunnelling is to be done for hydropower, highway, railway and metropolitan transportation and utility systems. With improvement in tunnel technology and rapid increase in labour rate, there is an increasing trend of mechanization in tunnel construction. Now-a-days, developing countries like India TBM (Tunnel Boring Machine) tunnelling has become inevitable due to new challenges of metropolitan transportation and utility systems. Tunnelling contractors in developing countries are not inclined to adopt TBM tunnelling due to its high initial costs. However, TBM tunnelling is most common and successful in European countries and America. In this context, benchmarking measures helps the tunnelling community in realizing the superior performance of TBM and DBM tunnelling. In this paper, hard and soft performance benchmarking measures have been identified and general benchmarks for the measures have been presented along with the performance graphs for both the tunnelling.

2.0 RATIONALE OF BENCHMARKING

Benchmarking originated in Japan and was first pioneered at Xerox Corporation, USA in 1979. Benchmarking is the search for the best practices that will lead to superior performance Camp (1989). It is the process that establishes landmarks of performance and standards of excellence for the services or operations or process or activities used in the projects. These benchmarks are reference points, based on the proven industry practices, that the company or one of its divisions, sections or departments, must endeavour to attain or even surpass. Benchmarking is a very powerful tool in helping organizations to optimize their capability to deliver by developing all internal processes to be superior, consistent and very effective. By optimizing delivery to the owner there is a clear move from merely fulfilling basic requirements to ensuring total satisfaction and even customer delight.

Benchmarking motivates organizations to be constantly externally focused and working on identifying gaps in the performance and developing the right strategies for closing them. It certainly opens windows of opportunities and makes strategy formulation and deployment a more systemized process. This is a tool for survival in a highly competitive environment. It is more applicable to any organization, which is committed to continuous improvement. The potential of benchmarking is still not recognized in construction industry Metri (1999). It is high time to adopt it as a standard tool to improve the overall performance of the industry including tunnelling industry. Initial step towards this is performance benchmarking *i.e.* to identify measures and compare the performance with others.

3.0 BENCHMARKING MEASURES

The key to success in benchmarking is to collect the right information. This should be collected both quantitatively and qualitatively. It is important to establish measures for performance evaluation, which has become more critical in recent years with the increased emphasis on change implementation in the search for continuous improvement. The performance of any operation usually includes the quantity of its output and its quality. The quality of performance evaluation is presently restricted by the nature of the information available and that too mainly on the hard or count data. Although, a review of hard data or measure provides an easy means of measurable comparisons, what it does not identify is 'how the performance measures are achieved.' If on the other hand, the soft measures or qualitative data are considered and analysed, the reasons for the performance gap would be clear from the output. The issue of soft or subjective data itself is important and demands attention because it is more useful for evaluation and decision making process (Cook, 1995). Therefore, the information found in any project for performance evaluation can be divided into two types: Hard Metrics – Hard information based on complete counts of quantitative facts and Soft Metrics – soft information based on Qualitative Facts. These two-metrics combine form an information system useful in evaluating the performance. These metrics help to assess the current status and identify strength and weaknesses. Keeping general tunnel project objectives in mind, a list of some sample of hard and soft metrics have been identified and presented in Tables 1 and 2.

TABLE 1
Benchmarking measures - hard metrics

Parameters	Metrics	Units
Costs	Excavation cost	Rs. per cum
	Labour cost	Rs. per cum
	Concrete lining cost	Rs. per cum
	Supporting cost	Rs. per cum
	Ventilation cost	Rs. per cum
	Maintenance cost	Rs. per cum
Economy	Net saving	Percentage
	Break –even length	Meters
Speed	Cycle time	Hours
	Advance rate	m per day
Waste	Over-break	Percentage

4.0 PERFORMANCE BENCHMARKS

In the context of tunnelling industry, competitive performance benchmarking is comparing one's own performance measure or metrics with that of competitors *i.e.* companies in the tunnelling industry. These metrics are financial or relative to a function or process of tunnel projects. In case of tunnelling techniques, use of these metrics for performance benchmarking, show how the TBM tunnelling is performing in relation to DBM tunnelling. The term "Benchmark" designates a "Best Value" for performance comparisons between self and others. For present case it designates best value between TBM and DBM Tunnelling.

TABLE 2
Benchmarking measures – soft metrics

Parameters	Metrics	Linguistic rating scale (Likert scale) : 1 - 5				
		1	2	3	4	5
Quality	Alignment Accuracy	Very low	Low	Medium	High	Very high
	Excavation preciseness	Very low	Low	Medium	High	Very high
	Tunnel surface	Very rough	Rough	Medium	Smooth	Very smooth
Safety	Ground disturbance	Much more	More	Little	Very little	Nil
	Accidental risks	Very high	High	Low	Very low	Nil
	Lost time incidents	Much more	More	Little	Very little	Nil
	Vibration shocks	Very high	High	Low	Very low	Nil
Environmental	Noise pollution	Very high	High	Low	Very low	Nil
	Air pollution	Very high	High	Low	Very low	Nil
	Housekeeping level	Very low	Low	Medium	High	Very high
Productivity	Productive time	Very low	Low	Medium	High	Very high
	Capacity utilization	Very low	Low	Medium	High	Very high
	Muck size uniformity	Very low	Low	Medium	High	Very high
Stability	Section stability	Very low	Low	Medium	High	Very high

For the purpose of performance benchmark study, 9 hard metrics and 20 soft metrics have been identified and presented in Tables 3 and 4 respectively. Data has been collected by author, Metri (1995) for the purpose of his dissertation work from tunnelling equipment manufacturers and then on going hard rock tunnelling projects in Maharashtra. Performance measures for hard metrics have been developed for a case of 15000 m long, 6.6 m dia. with 300mm concrete lined tunnel from the data analysis and presented in Table 3. Also, using the same data, cost performance of both TBM and DBM tunnelling has been analyzed for 5000 m to 40,000 m long tunnels and presented in Fig. 1. (For detail discussions on analysis refer Metri, (1995). Due to qualitative nature, soft metrics have been evaluated using linguistic rating by the project in-charge of both TBM and DBM tunnelling projects and actual site observations. The results have been presented in Table 4 and Fig. 2. In Tables 3 and 4, performance benchmarks have been shown in bold letters.

TABLE 3
Performance benchmarks for hard metrics

Sr. No.	Metrics	TBM	DBM	Best performer
1.	Labour cost	Rs. 4 per cum	Rs. 10 per cum	TBM
2.	Concrete lining cost	**Rs. 10500 per cum**	Rs. 16500 per cum	TBM
3.	Supporting cost	**Average 75 % of DBM**	Depending on rock strata	TBM
4.	Ventilation cost	**Rs. 2.00 per cum**	Rs. 10.00 per cum	TBM
5.	Maintenance cost	**Rs. 19.50 per cum**	Rs. 49.00 per cum	TBM
6.	Cycle time	**0.33 hours**	16.00 hours	TBM
7.	Advance rate	**Min. 15 m /day**	1.5 m /day	TBM
8.	Overbreak	**2.5 0 %**	20.00 %	TBM
9.	Normal maximum gradient	10^0	30^0	DBM

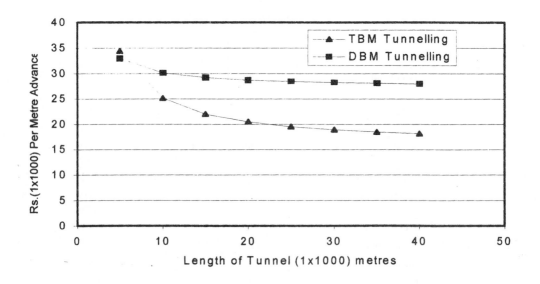

Fig. 1 : Cost performance of TBM comparison to DBM

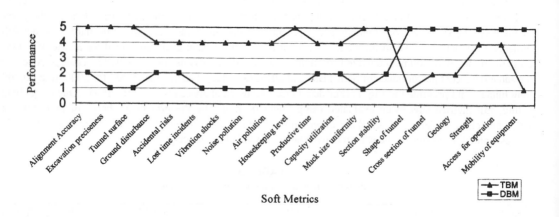

Fig. 2 : Performance of TBM and DBM tunnelling

From above tables and figures, it is clear that in general, TBM has definitely an edge over DBM tunnelling. However, DBM is superior for various shape, cross section, geology and mobility. The focus of this study is to present how this metric tool can be used to change implementation and improve performance in tunnelling. Tunnelling community generally face a problem in making decision regarding TBM and DBM tunnelling for medium to long tunnel projects. Developing performance benchmarks will therefore provide the necessary information to take a realistic decision on the tunnelling techniques.

TABLE 4
Performance benchmarks for soft metrics

Sr. No.	Metrics	Performance on Likert Scale : 1 - 5		Best performer
		TBM	DBM	
1.	Alignment accuracy	5	2	TBM
2.	Excavation preciseness	5	1	TBM
3.	Tunnel surface	5	1	TBM
4.	Ground disturbance	4	2	TBM
5.	Accidental risks	4	2	TBM
6.	Lost time incidents	4	1	TBM
7.	Vibration shocks	4	1	TBM
8.	Noise pollution	4	1	TBM
9.	Air pollution	4	1	TBM
10.	Housekeeping level	5	1	TBM
11.	Productive time	4	2	TBM
12.	Capacity utilization	4	2	TBM
13.	Muck size uniformity	5	1	TBM
14.	Section stability	5	2	TBM
15.	Shape of tunnel	1	5	DBM
16.	Cross section of tunnel	2	5	DBM
17.	Geology	2	5	DBM
18.	Strength	4	5	DBM
19.	Access for operation	4	5	DBM
20.	Mobility of equipment	1	5	DBM

5.0 LIMITATIONS AND DIRECTIONS FOR FUTURE WORK

In the present study some sample of metrics have been presented for hard and soft measures. The identified performance benchmarks are the only indicative. Information on more number of tunnelling projects for particular rock strata will give accurate performance benchmarks.

Metrics identification can be extended in details in the process measure categories for establishing performance benchmarks. Process benchmarking study for both TBM and DBM tunnelling against world leader or best performer will identifies the root causes for the performance gaps and subsequently helps in incorporating best practices in tunnelling processes.

The standardization of benchmarks across the industry is the need of the hour. The performance benchmarks sharing database is required to promote the sharing of practices, knowledge, know-how and experience, which will prove to be valuable to the tunnelling industry. The benchmark database provides insight into what has already been done and what may be possible rather than trying to provide " the right answer" at the touch of a button. It gives all users a wide range of information than that has ever been available.

6.0 CONCLUSIONS

Establishment of benchmarking measure is an important aspect to realize the performance of tunnelling. If measures are created properly, they will provide valuable information about the performance. They can help tunnelling community to determine the effectiveness of the typical technique so that appropriate actions may be taken in the beginning to ensure the realization of established goals. The selection of appropriate tunnelling technique based on performance benchmarking measure is of paramount importance, particularly in developing countries like India to get tangible and intangible benefits. While identifying metrics, project objectives should be kept in mind. Benchmarking measures, which include hard and soft metrics that, encompass quantitative and qualitative data are presented. These metrics enable a systematic appraisal of performance that is not possible at present. Although performance benchmarks presented here are based on specific case, the methodology is applicable for all variety tunnelling projects.

In India, currently benchmarking standards are not available for tunnelling. This necessitates to create dynamic benchmarks *i.e.* benchmarking standards for the industry on continuous basis to improve productivity and quality of tunnelling. It is, therefore, high time to set up and adopt dynamic benchmarks to achieve superior quality performance in tunnelling projects and enable our firms to compete with the multinational companies and foreign firms of developed countries.

REFERENCES

Metri. B.A. and Patil, B.K. (1997). "Economics and benefits associated with TBM for hydro power projects in India." 2nd International CBIP's R & D Conference, Vadodara, India. Vol. 2, pp. 532-539.

Camp, Robert. C. (1989). "Benchmarking: The search for best practices that lead superior performance." Part 1 Quality progress. 22(1), pp. 61-68.

Metri, B.A. (1999). "Quality benchmarking studies in construction industry." Ph.D. Seminar Report, Reliability Engineering Group, IIT Bombay, Mumbai, India.

Cook Sarah. (1995). Practical benchmarking. Kogan page, London. UK.

Metri, B.A. (1995). "Mechanization in Tunnel Construction" M.E. Dissertation. Shivaji Univ., Kolhapur, India.

AUTHOR INDEX

T - #0307 - 101024 - C0 - 254/178/30 [32] - CB - 9789058092281 - Gloss Lamination